Lecture Notes in Mathematics

Edited by J.-M. Morel, F. Takens and B. Teissier

Editorial Policy for Multi-Author Publications: Summer Schools / Intensive Courses

1. Lecture Notes aim to report new developments in all areas of mathematics and their applications – quickly, informally and at a high level. Mathematical texts analysing new developments in modelling and numerical simulation are welcome. Manuscripts should be reasonably self-contained and rounded off. Thus they may, and often will, present not only results of the author but also related work by other people. They should provide sufficient motivation, examples and applications. There should also be an introduction making the text comprehensible to a wider audience. This clearly distinguishes Lecture Notes from journal articles or technical reports which normally are very concise. Articles intended for a journal but too long to be accepted by most journals, usually do not have this "lecture notes" character.

2. In general SUMMER SCHOOLS and other similar INTENSIVE COURSES are held to present mathematical topics that are close to the frontiers of recent research to an audience at the beginning or intermediate graduate level, who may want to continue with this area of work, for a thesis or later. This makes demands on the didactic aspects of the presentation. Because the subjects of such schools are advanced, there often exists no textbook, and so ideally, the publication resulting from such a school could be a first approximation to such a textbook.
Usually several authors are involved in the writing, so it is not always simple to obtain a unified approach to the presentation.
For prospective publication in LNM, the resulting manuscript should not be just a collection of course notes, each of which has been developed by an individual author with little or no co-ordination with the others, and with little or no common concept. The subject matter should dictate the structure of the book, and the authorship of each part or chapter should take secondary importance. Of course the choice of authors is crucial to the quality of the material at the school and in the book, and the intention here is not to belittle their impact, but simply to say that the book should be planned to be written by these authors jointly, and not just assembled as a result of what these authors happen to submit.
This represents considerable preparatory work (as it is imperative to ensure that the authors know these criteria before they invest work on a manuscript), and also considerable editing work afterwards, to get the book into final shape. Still it is the form that holds the most promise of a successful book that will be used by its intended audience, rather than yet another volume of proceedings for the library shelf.

3. Manuscripts should be submitted (preferably in duplicate) either to Springer's mathematics editorial in Heidelberg, or to one of the series editors (with a copy to Springer). Volume editors are expected to arrange for the refereeing, to the usual scientific standards, of the individual contributions. If the resulting reports can be forwarded to us (series editors or Springer) this is very helpful. If no reports are forwarded or if other questions remain unclear in respect of homogeneity etc, the series editors may wish to consult external referees for an overall evaluation of the volume. A final decision to publish can be made only on the basis of the complete manuscript, however a preliminary decision can be based on a pre-final or incomplete manuscript. The strict minimum amount of material that will be considered should include a detailed outline describing the planned contents of each chapter.
Volume editors and authors should be aware that incomplete or insufficiently close to final manuscripts almost always result in longer evaluation times. They should also be aware that parallel submission of their manuscript to another publisher while under consideration for LNM will in general lead to immediate rejection.

Continued on inside back-cover

Lecture Notes in Mathematics 1831

Editors:
J.-M. Morel, Cachan
F. Takens, Groningen
B. Teissier, Paris

Subseries:
Fondazione C.I.M.E., Firenze
Adviser: Pietro Zecca

Springer
Berlin
Heidelberg
New York
Hong Kong
London
Milan
Paris
Tokyo

A. Connes   J. Cuntz   E. Guentner
N. Higson   J. Kaminker   J. E. Roberts

Noncommutative Geometry

Lectures given at the
C.I.M.E. Summer School
held in Martina Franca, Italy,
September 3-9, 2000

Editors: S. Doplicher
         R. Longo

Fondazione
C.I.M.E.

Springer

Editors and Authors

Alain Connes
Collège de France
11, place Marcelin Berthelot
75231 Paris Cedex 05, France

e-mail: connes@ihes.fr

Joachim Cuntz
Mathematisches Institut
Universität Münster
Einsteinstr. 62
48149 Münster, Germany

e-mail: cuntz@math.uni-muenster.de

Sergio Doplicher
Dipartimento di Matematica
Università di Roma "La Sapienza"
Piazzale A. Moro 5
00185 Roma, Italy

e-mail: doplicher@mat.uniroma1.it

Erik Guentner
Department of Mathematical Sciences
University of Hawaii, Manoa
2565 McCarthy Mall
Keller 401A, Honolulu
HI 96822, USA

e-mail: erik@math.hawaii.edu

Nigel Higson
Department of Mathematics
Pennsylvania State University
University Park, PA 16802
USA

e-mail: higson@psu.edu

Jerome Kaminker
Department of Mathematical Sciences
IUPUI, Indianapolis
IN 46202-3216, USA

e-mail: kaminker@math.iupui.edu

Roberto Longo
John E. Roberts
Dipartimento di Matematica
Università di Roma "Tor Vergata"
Via della Ricerca Scientifica 1
00133 Roma, Italy

*e-mail: longo@mat.uniroma2.it
roberts@mat.uniroma2.it*

Cataloging-in-Publication Data applied for
Bibliographic information published by Die Deutsche Bibliothek

Die Deutsche Bibliothek lists this publication in the Deutsche Nationalbibliografie;
detailed bibliographic data is available in the Internet at http://dnb.ddb.de

Mathematics Subject Classification (2000): 58B34, 46L87, 81R60, 83C65

ISSN 0075-8434
ISBN 3-540-20357-5 Springer-Verlag Berlin Heidelberg New York

Springer-Verlag is a part of Springer Science + Business Media

springeronline.com

© Springer-Verlag Berlin Heidelberg 2004
Printed in Germany

Typesetting: Camera-ready TeX output by the authors

SPIN: 10967928   41/3142/du - 543210 - Printed on acid-free paper

Preface

If one had to synthesize the novelty of Physics of the XX century with a single magic word, one possibility would be "Noncommutativity".

Indeed the core assertion of Quantum Mechanics is the fact that observables ought to be described "by noncommuting operators"; if you wished more precision and said "by the selfadjoint elements in a C*-algebra $\mathcal{A}$, while states are expectation functionals on that algebra, i.e. positive linear forms of norm one on $\mathcal{A}$", you would have put down the full axioms for a theory which includes Classical Mechanics if $\mathcal{A}$ is commutative, Quantum Mechanics otherwise.

More precisely, Quantum Mechanics of systems with finitely many degrees of freedom would fit in the picture when the algebra is the collection of all compact operators on the separable, infinite-dimensional Hilbert space (so that all, possibly unbounded, selfadjoint operators on that Hilbert space appear as "generalized observables" affiliated with the enveloping von Neumann algebra); the distinction between different values of the number of degrees of freedom requires more details, as the assignment of a dense Banach *-algebra (the quotient, obtained by specifying the value of the Planck constant, of the L^1-algebra of the Heisenberg group).

Quantum Field Theory, as explained in Roberts' lectures in this volume, fits in that picture too: the key additional structure needed is the local structure of $\mathcal{A}$. This means that $\mathcal{A}$ has to be the inductive limit of subalgebras of local observables $\mathcal{A}(\mathcal{O})$, where $\mathcal{O} \mapsto \mathcal{A}(\mathcal{O})$ maps coherently regions in the spacetime manifold to subalgebras of $\mathcal{A}$.

As a consequence of the axioms, as more carefully expounded in this book, $\mathcal{A}$ is much more dramatically noncommutative than in Quantum Mechanics with finitely many degrees of freedom: $\mathcal{A}$ cannot be any longer essentially commutative (in other words, it cannot be an extension of the compacts by a commutative C*-algebra), and actually turns out to be a simple non type I C*-algebra.

In order to deal conveniently with the natural restriction to locally normal states, it is also most often natural to let each $\mathcal{A}(\mathcal{O})$ be a von Neumann algebra, so that

$\mathcal{A}$ is not norm separable: for the sake of both Quantum Statistical Mechanics with infinitely many degrees of freedom and of physically relevant classes of Quantum Field Theories - fulfilling the split property, cf. Roberts' lectures - $\mathcal{A}$ can actually be identified with a universal C*-algebra: the inductive limit of the algebras of all bounded operators on the tensor powers of a fixed infinite dimensional separable Hilbert space; different theories are distinguished by the time evolution and/or by the local structure, of which the inductive sequence of type I factors gives only a fuzzy picture. The actual local algebras of Quantum Field Theory, on the other side, can be proved in great generality to be isomorphic to the unique, approximately finite dimensional III_1 factor (except for the possible nontriviality of the centre).

Despite this highly noncommutative ambient, the key axiom of Quantum Field Theory of forces other than gravity, is a demand of commutativity: local subalgebras associated to causally separated regions should commute elementwise. This is the basic Locality Principle, expressing Einstein Causality.

This principle alone is "unreasonably effective" to determine a substantial part of the conceptual structure of Quantum Field Theory. This applies to Quantum Field Theory on Minkowski space but also on large classes of curved spacetimes, where the pseudo-Riemann structure describes a classical external gravitational field on which the influence of the quantum fields is neglected (cf Roberts' lectures). But the Locality Principle is bound to fail in a quantum theory of gravity.

Mentioning gravity brings in the other magic word one could have mentioned at the beginning: "Relativity".

Classical General Relativity is a miracle of human thought and a masterpiece of Nature; the accuracy of its predictions grows more and more spectacularly with years (binary pulsars are a famous example). But the formulation of a coherent and satisfactory Quantum Theory of all forces including Gravity still appears to many as one of the few most formidable problems for science of the XXI century.

In such a theory Einstein Causality is lost, and we do not yet know what really replaces it: for the relation "causally disjoint" is bound to lose meaning; more dramatically, spacetime itself has to look radically different at small scales. Here "small" means at scales governed by the Planck length, which is tremendously small but is there.

Indeed Classical General Relativity and Quantum Mechanics imply Spacetime Uncertainty Relations which are most naturally taken into account if spacetime itself is pictured as a Quantum Manifold: the commutative C*-algebra of continuous functions vanishing at infinity on Minkowski space has to be replaced by a noncommutative C*-algebra, in such a way that the spacetime uncertainty relations are implemented [DFR]. It might well turn out to be impossible to disentangle Quantum Fields and Spacetime from a common noncommutative texture.

Quantum Field Theory on Quantum Spacetime ought to be formulated as a Gauge theory on a noncommutative manifold; one might hope that the Gauge principle, at the basis of the point nature of interactions between fields on Minkowski space and hence of the Principle of Locality, might be rigid enough to replace locality in the world of quantum spaces.

Gauge Theories on noncommutative manifolds ought to appear as a chapter of Noncommutative Geometry [CR,C].

Thus Noncommutative Geometry may be seen as a main avenue from Physics of the XX century to Physics of the XXI century; but since it has been created by Alain Connes in the late 70s, as expounded in his lectures in this Volume, it grew to a central theme in Mathematics with a tremendous power of unifying disparate problems and of progressing in depth.

One could with good reasons argue that Noncommutative Topology started with the famous Gel'fand-Naimark Theorems: every commutative C*-algebra is the algebra of continuous functions vanishing at infinity on a locally compact space, every C*-algebra can be represented as an algebra of bounded operators on a Hilbert space; thus a noncommutative C*-algebra can be viewed as "the algebra of continuous functions vanishing at infinity" on a "quantum space".

But it was with the Theory of Brown, Douglas and Fillmore of Ext, with the development of the K-theory of C*-algebras, and their merging into Kasparov bivariant functor KK that Noncommutative Topology became a rich subject. Now this subject could hardly be separated from Noncommutative Geometry.

It suffices to mention a few fundamental landmarks: the discovery by Alain Connes of Cyclic Cohomology, crucial for the lift of De Rham Theory to the noncommutative domain, the Connes-Chern Character; the concept of spectral triple proved to be central and the natural road to the theory of noncommutative Riemannian manifolds.

Since he started to break this new ground, Connes discovered a paradigm which could not have been anticipated just on the basis of Gel'fand-Naimark theory: Noncommutative Geometry not only extends geometrical concepts beyond point spaces to "noncommutative manifolds", but also permits their application to singular spaces: such spaces are best viewed as noncommutative spaces, described by a noncommutative algebra, rather than as mere point spaces.

A famous class of examples of singular spaces are the spaces of leafs of foliations; such a space is best described by a noncommutative C*-algebra, which, when the foliation is defined as orbits in the manifold $\mathcal{M}$ by the action of a Lie group G and has graph $\mathcal{M} \times G$, coincides which the (reduced) cross product of the algebra of continuous functions on the manifold by that action. The Atiyah - Singer Index Theorem has powerful generalizations, which culminated in the extension of its local form to transversally elliptic pseudodifferential operators on the foliation, in terms of the cyclic cohomology of a Hopf algebra which describes the transverse geometry [CM].

There is a maze of examples of singular spaces which acquire this way nice and tractable structures [C]. But also discrete spaces often do: Bost and Connes associated to the distribution of prime numbers an intrinsic noncommutative dynamical system with phase transitions [BC]. Connes formulated a trace formula whose extension to singular spaces would prove Riemann hypothesis [Co]. The geometry of the two point set, viewed as "extradimensions" of Minkowski space, is the basis for the Connes and Lott theory of the standard model, providing an elegant motivation for the form of the action including the Higgs potential [C]; this line has been further developed by Connes into a deep spectral action principle, formulated on Euclidean, compactified spacetime, which unifies the Standard model and the Einstein Hilbert action [C1].

Thus Noncommutative Geometry is surprisingly effective in providing the form of the expression for the action. But if one turns to the Quantum Theory it has a lot to say also on Renormalization. Connes and Moscovici discovered a Hopf algebra associated with the differentiable structure of a manifold, which provides a powerful organizing principle which was crucial to the Transverse Index Theorem; in the case of Minkowski space, it proved to be intimately related with Kreimer's Hopf algebra associated to Feynman graphs. Developing this connection, Connes and Kreimer could cast Renormalization Theory in a mathematically sound and elegant frame, as a Riemann - Hilbert problem [CK].

The relations of Noncommutative Geometry to the Algebraic Approach to Quantum Field Theory are still to be explored in depth. The first links appeared in supersymmetric Quantum Field Theory: the non polynomial character of the index map on some K groups associated to the local algebras in a free supersymmetric massive theory [C], and the relation to the Chern Character of the Jaffe Lesniewski Osterwalder cyclic cocycle associated to a super Gibbs functional [JLO,C].

More generally in the theory of superselection sectors it has long been conjectured that localized endomorphisms with finite statistics ought to be viewed as a highly noncommutative analog of Fredholm operators; the discovery of the relation between statistics and Jones index gave solid grounds to this view. While Jones index defines the analytical index of the endomorphism, a geometric dimension can also be introduced, where, in the case of a curved background, the spacetime geometry enters too, and an analog of the Index Theorem holds [Lo]. One can expect this is a fertile ground to be further explored.

Noncommutative spaces appeared also as the underlying manifold of a quantum group in the sense of Woronowicz; noncommutative geometry can be applied to those manifolds too. Most recent developments and discoveries can be found in the Lectures by Connes.

Noncommutative Geometry and Noncommutative Topology merge in the celebrated Baum - Connes conjecture on the K-Theory of the reduced C* algebra of any discrete group. While it has been realized in recent years that one cannot extend this conjecture to crossed products ("Baum - Connes with coefficients"), the original conjecture is still standing, a powerful propulsion of research in Index Theory, Discrete Groups, Noncommutative Topology. The lectures of Higson and Guentner expound that subject, with a general introduction to K-Theory of C*-algebras, E-theory, and Bott periodicity. Aspects of the Baum - Connes conjecture related to exactness are dealt with by Guentner and Kaminker.

K-Theory, KK-Theory and Connes - Higson E-Theory are unified in a general approach due to Cuntz and Cuntz - Quillen; a comprehensive introduction to these theories and to cyclic cohomology can be found in Cuntz's lectures.

Besides the fundamental reference [C] we point out to the reader other references related to this subject [GVF,L,M]. Since the theory of Operator Algebras is so intimately related to the subject of these Lecture Notes, we feel it appropriate to bring to the reader's attention the newly completed spectacular treatise on von Neumann Algebras ("noncommutative measure theory") by Takesaki [T].

Of course this volume could not by itself cover the whole subject, but we believe it is a catching invitation to Noncommutative Geometry, in all of its aspects from Prime Numbers to Quantum Gravity, that we hope many readers, mathematicians and physicists, will find stimulating.

Sergio Doplicher and Roberto Longo

References

[BC]   J.B. Bost & A. Connes, *Hecke Algebras, type III factors and phase transitions with spontaneous symmetry breaking in number theory*, Selecta Math. **3**, 411-457 (1995).

[C]   A. Connes, "Noncommutative Geometry", Acad. Press (1994).

[C1]   A. Connes, *Gravity coupled with matter and the foundation of noncommutative geometry*, Commun. Math. Phys. **182**, 155-176 (1996), and refs.

[Co]   A. Connes, *Trace formula in noncommutative geometry and the zeros of the Riemann zeta function*, Selecta Math. **5** (1999), no. 1, 29-106.

[CK]   A. Connes, *Symetries Galoisiennes et renormalisation*, Seminaire Poincaré Octobre 2002, math.QA/0211199 and refs.

[CM]   A. Connes & H. Moscovici, *Hopf algebras, cyclic cohomology and the transverse index theorem*, Commun. Math. Phys. **198**, 199-246 (1998), and refs.

[CR]   A. Connes & M. Rieffel, *Yang Mills for noncommutative two tori*, in: "Operator Algebras and Mathematical Physics", Contemp. Math. **62**, 237-266 (1987).

[DFR] S. Doplicher, K. Fredenhagen & J.E. Roberts, *The quantum structure of spacetime at the Planck scale and quantum fields*, Commun. Math. Phys. **172**, 187-220 (1995).

[GVF] J.M. Gracia-Bondia, J.C. Varilly & H. Figueroa: "Elements of Noncommutative Geometry", Birkhaeuser (2000).

[JLO] A. Jaffe, A. Lesniewski & K. Osterwalder, *Quantum K-theory I. The Chern character*, Commun. Math. Phys. **118**, 1-14 (1988).

[L]   G. Landi, "An Introduction to Noncommutative Spaces and their Geometries", Springer, LNP monographs 51 (1997).

[Lo]   R. Longo, *Notes for a quantum index theorem*, Commun. Math. Phys. **222**, 45-96 (2001).

[M]   J. Madore: "An Introduction to Noncommutative Differential Geometry and its Physical Applications", LMS Lecture Notes (1996).

[T]   M. Takesaki, "Theory of Operator Algebras" vol. I, II, III, Springer Encyclopaedia of Mathematical Sciences 124 (2002), 125, 127 (2003).

CIME's activity is supported by:

Ministero dell'Università Ricerca Scientifica e Tecnologica, COFIN '99;
Ministero degli Affari Esteri - Direzione Generale per la Promozione e la Cooper-
azione - Ufficio V;
Consiglio Nazionale delle Ricerche;
E.U. under the Training and Mobility of Researchers Programme;
UNESCO-ROSTE, Venice Office.

Contents

Group C*-Algebras and K-Theory

Cyclic Cohomology, Noncommutative Geometry and Quantum Group Symmetries

Alain Connes

Collège de France
11, place Marcelin Berthelot
75231 Paris Cedex 05
connes@ihes.fr

Abstract We give an introduction to the basic notions of noncommutative geometry including the calculus of infinitesimals with operators, cyclic cohomology and the local index formula. We also explain in details how the infinitesimal calculus based on operators gives a natural home for the infinitesimal line element ds of geometry and leads one to the basic notion of spectral triple, which is the basic paradigm of noncommutative geometry. In order to illustrate these general concepts we then analyse the noncommutative space underlying the quantum group $SU_q(2)$ from this spectral point of view, and show how the general theory developped in our joint work with H. Moscovici applies to the specific spectral triple defined by Chakraborty and Pal. This provides the pseudo-differential calculus, the Wodzciki-type residue, and the local cyclic cocycle giving the index formula. This specific example allows to illustrate the general notion of locality in noncommutative geometry. The formulas computing the residue are "local". Locality by stripping all the expressions from irrelevant details makes them easily computable. The original Chern character is non-local and the cochain whose coboundary is the difference between the original Chern character and the local one is much harder to compute than the local cochains. It is given by the remainders in the rational approximation of the logarithmic derivative of the Dedekind eta function. The key feature of this spectral triple is its equivariance, i.e. the $SU_q(2)$-symmetry. We explain how this leads naturally to the general concept of invariant cyclic cohomology in the framework of quantum group symmetries and relate this notion to previous work.

Contents

1 Introduction

Our purpose in these notes is to give a detailed introduction to basic notions of non-commutative geometry and to illustrate them in a concrete manner in a very specific example.

In noncommutative geometry a geometric space is described from a spectral point of view, as a triple $(\mathcal{A}, \mathcal{H}, D)$ consisting of a $*$-algebra $\mathcal{A}$ represented in a Hilbert space $\mathcal{H}$ together with an unbounded selfadjoint operator D, with compact resolvent, which interacts with the algebra in a bounded fashion. This spectral data embodies both the metric and the differential structure of the geometric space.

An essential ingredient of the general theory is the Chern character in K-homology which together with cyclic cohomology and the spectral sequence relating it to Hochschild cohomology, were defined in 1981 (cf. [9],[10],[11]). The essence of the theory is to allow for computations of differential geometric nature in the non-commutative framework.

There is a wealth of examples of noncommutative spaces, the basic ones are coming from arbitrary foliated manifolds. Their transverse geometry is described by a spectral triple ([16]) whose analysis has been completed in full generality in ([17]). While easier basic examples such as the non-commutative tori were analysed as early as 1980 (cf. [8]), and have far reaching generalisations ([23],[24]) the case of the underlying noncommutative-spaces to quantum groups has been left aside till recently, mainly because of the "drop of dimension" which occurs when the deformation parameter q affects non-classical values $q \neq 1$. Thus for instance the Hochschild dimension of $SU_q(2)$ drops from the classical value $d = 3$ to $d = 1$ so that these noncommutative-spaces seem at first rather esoteric.

A very interesting spectral triple for $SU_q(2)$, $q \neq 1$, has recently been proposed in [7]. The algebra $\mathcal{A}$ is the algebra of functions on $SU_q(2)$ and the representation in $\mathcal{H}$ is the coregular representation of $SU_q(2)$. The operator D is very simple, and is invariant under the action of the quantum group $SU_q(2)$. (The Anzats proposed in a remark at the end of [23] provides the right formula for $|D|$ but not for the sign of D as pointed out in [30]).

Our purpose in this paper is to show that the general theory developped by Henri Moscovici and the author (cf.[16]) applies perfectly to the above spectral triple.

The power of the general theory comes from general theorems such as the local computation of the analogue of Pontrjagin classes: *i.e.* of the components of the cyclic cocycle which is the Chern character of the K-homology class of D and which make sense in general. This result allows, using the infinitesimal calculus, to go from local to global in the general framework of spectral triples $(\mathcal{A}, \mathcal{H}, D)$. The notion of locality which is straightforward for classical spaces is more elaborate in the non-commutative situation and relies essentially on the non-commutative integral which is the Dixmier trace in the simplest case and the analogue of the Wodzicki residue in general. Its validity requires the discreteness of the dimension spectrum, a subset of $\mathbb{C}$ which is an elaboration of the classical notion of dimension. At an intuitive level this subset is the set of "dimensions", possibly complex, in which the noncommutative-space underlying the spectral triple manifests itself non-trivially. At the technical level it is the set of singularities of functions,

$$\zeta_b(z) = \text{Trace}\,(b|D|^{-z}) \qquad \text{Re}\,z > p\,,\ b \in \mathcal{B}. \tag{1}$$

where $b \in \mathcal{B}$ varies in a suitable algebra canonically associated to the triple and allowing to develop the pseudo-differential calculus.

Our first result is that in the above case of $SU_q(2)$, the dimension spectrum is simple and equal to $\{1, 2, 3\} \subset \mathbb{C}$. Simplicity of the dimension spectrum means that the singularities of the functions (1) are at most simple poles. It then follows from the general results of [16] that the equality,

$$\int P = \text{Res}_{z=0}\,\text{Trace}(P|D|^{-z}) \tag{2}$$

defines a trace on the algebra generated by $\mathcal{A}$, $[D, \mathcal{A}]$ and $|D|^z$, where $z \in \mathbb{C}$.

Our second result is the explicit computation of this functional in the above case of $SU_q(2)$. In doing so we shall also determine the analogue of the cosphere bundle in that example and find an interesting space S_q^*. This space is endowed with a one parameter group γ_t of automorphisms playing the role of the geodesic flow, and is intimately related to the product $D_{q+}^2 \times D_{q-}^2$, of two NC-two-disks, while the coproduct gives its relation to $SU_q(2)$. The formulas computing the residue will be "local" and very simple, locality by stripping all the expressions from irrelevant details makes them computable.

Our third result which is really the main point of the paper, is the explicit formula for the local index cocycle, which owing to the metric dimension 3 is a priori given by the following cocycle,

$$\varphi_1(a^0, a^1) = \int a^0[D, a^1]|D|^{-1} - \frac{1}{4}\int a^0 \nabla([D, a^1])\,|D|^{-3} \tag{3}$$

$$+ \frac{1}{8}\int a^0 \nabla^2([D, a^1])\,|D|^{-5}$$

and,

$$\varphi_3(a^0, a^1, a^2, a^3) = \frac{1}{12}\int a^0[D, a^1][D, a^2][D, a^3]\,|D|^{-3}, \tag{4}$$

where $\nabla(T) = [D^2, T] \quad \forall\,T$ operator in $\mathcal{H}$.

In contrast with the local index cocycle, the non-local original formula for the Chern character in cyclic cohomology is much harder to compute. Its computation will be achieved by an explicit formula for a cochain realising the homology between the two cocycles. The formula for that cochain will involve the logarithmic derivative of the eta function of Dedekind and its approximation by rational functions.

We shall begin in section 2 by giving a thorough introduction to cyclic cohomology and its original description in terms of cycles and their cobordism. In section 3 we shall recall the basic calculus of infinitesimals based on compact operators, the Dixmier trace and commutators. In section 4 we recall the basic framework for the metric aspect of noncommutative geometry, based on spectral triples. Section 5 is the general local index formula of Henri Moscovici and the author. In the next sections 6 and 7 we begin the analysis of the example of $SU_q(2)$ in the degenerate case $q = 0$ with a luxury of details, mainly to show that the numerical coefficients involved in the above formula are in fact unique in order to get a (non-trivial) cocycle. The coboundary involved in the formula (theorem 8) will then be conceptually explained (in section 8) and the specific values $\zeta(0) = -\frac{1}{2}$ and $\zeta(-1) = -\frac{1}{12}$ of the Riemann Zeta function will account for the numerical coefficients encountered in the coboundary.

We shall then move on in section 9 to the general case $q \in]0, 1[$ and construct the pseudo-differential calculus on $SU_q(2)$ following the general theory of [16]. We shall determine the algebra of complete symbols by computing the quotient by smoothing operators. This will give the cosphere bundle S_q^* of $SU_q(2)$ already mentionned above. The analogue of the geodesic flow will give a one-parameter group of automorphisms γ_t of $C^\infty(S_q^*)$. We shall also construct the restriction morphism r to the product of two non-commutative 2-disks,

$$r : C^\infty(S_q^*) \to C^\infty(D_{q+}^2 \times D_{q-}^2) \tag{5}$$

We shall then show (section 10) that the dimension spectrum of $SU_q(2)$ in the above spectral sense is $\{1, 2, 3\}$ and compute the residues in terms of the symbol $\rho(b) \in C^\infty(S_q^*)$ of the operator b of order 0. If one lets $\rho(b)^0$ be the component of degree 0 for the geodesic flow γ_t, the formulas for the residues are,

$$\int b \, |D|^{-3} = (\tau_1 \otimes \tau_1)(r\rho(b)^0)$$

$$\int b \, |D|^{-2} = (\tau_1 \otimes \tau_0 + \tau_0 \otimes \tau_1)(r\rho(b)^0)$$

$$\int b \, |D|^{-1} = (\tau_0 \otimes \tau_0)(r\rho(b)^0) \, ,$$

where r is the above restriction map to $D_{q+}^2 \times D_{q-}^2$. The algebras $C^\infty(D_{q\pm}^2)$ are Toeplitz algebras and as such are extensions of the form,

$$0 \longrightarrow \mathcal{S} \longrightarrow C^\infty(D_{q\pm}^2) \overset{\mathfrak{s}}{\longrightarrow} C^\infty(S^1) \longrightarrow 0 \tag{6}$$

where the ideal $\mathcal{S}$ is isomorphic to the algebra of matrices of rapid decay. The functional τ_1 is the trace obtained by integrating $\mathfrak{s}(x)$ on S^1, while τ_0 is a regularized form of the trace on the ideal $\mathcal{S}$. Due to the need of regularization, τ_0 is not a trace but its Hochschild coboundary (which measures the failure of the trace property) is easily computed in terms of the canonical morphism $\mathfrak{s}$.

A similar long exact sequence, and pair of functionals τ_j make sense for $\mathcal{A} = C^\infty(SU_q(2))$. They are invariant under the one parameter group of automorphisms generated by the derivation ∂, which rotates the canonical generators in opposite ways.

In section 10 we shall use this derivation together with the second derivative of $\mathfrak{s}(x)$ to define the differential. We then show how to construct a one dimensional cycle (in the sense of ([10])) whose character is extremely simple to compute. This shows how to bypass the shortage of traces on $\mathcal{A} = C^\infty(SU_q(2))$ to obtain a significant calculus.

Our main result (theorem 10) is that the local formula for the Chern character of the above spectral triple gives exactly the above cycle, thus completing the original computation. Another quite remarkable point is that the cochain whose coboundary is the difference between the original Chern character and the local one is given by the remainders in the rational approximation of the logarithmic derivative of the Dedekind eta function. The computation of this non-local cochain is very involved (theorem 11).

One fundamental property of the above spectral triple is its equivariance ([7]) under the action of the quantum group $SU_q(2)$. In the last section we shall use this example to obtain and explain in general a new concept of quantum group invariance in cyclic cohomology.

Given an algebra $\mathcal{A}$ on which a quantum group G is acting, the new theory is obtained by considering the crossed product $\mathcal{A} \rtimes G$ of $\mathcal{A}$ by the quantum group action and restricting to the cochains of the cyclic complex for the crossed product which vanish if any of the arguments belongs to the quantum group algebra.

There are several qualitatively different available completions ranging from the von-Neumann algebra of G to the universal envelopping algebra $\mathcal{U}$ and we shall show how the above notion depends upon this qualitative choice, by exhibiting concrete examples of cocycles.

The new theory obviously maps by restriction to the ordinary cyclic theory $HC^*(\mathcal{A})$,

$$HC^*_{\mathcal{U}}(\mathcal{A}) \xrightarrow{\rho} HC^*(\mathcal{A}), \tag{7}$$

In the case of the envelopping algebra $\mathcal{U} = U_q(SL(2))$, we shall show that $HC^*_{\mathcal{U}}(\mathcal{A})$ also maps in fact to the "twisted" cyclic cohomology $HC^*_{\text{inv}}(\mathcal{A}, \theta)$ proposed in [37],

$$HC^*_{\mathcal{U}}(\mathcal{A}) \xrightarrow{\rho} HC^*_{\text{inv}}(\mathcal{A}, \theta), \tag{8}$$

where θ is the inner automorphism implemented by the diagonal element k^2.

This will allow to put the proposal of [37] in the correct perspective. The main drawback of this easy variation on ([11]) is that it lacks the relation to K-theory which is the back-bone of cyclic cohomology.

This was a good reason to refrain from developing such a "twisted" form of the general theory (Note its previous appearance in ([18]) equation 2.28 p.14).

However, the twisted theory has the merit of connecting with the various "differential calculi" on quantum groups ([48],[49]) which certainly helps in the understanding of these developments.

The theory which we propose is intimately connected with K-theory thanks to the restriction map (7) and thus does not suffer from the drawback of the twisted theory. The above map (8) shows that it would be very interesting to use the twisted theory as a "detector" of classes in $HC_{\mathcal{U}}^*(\mathcal{A})$. Once lifted to $HC_{\mathcal{U}}^*(\mathcal{A})$ such classes would have K-theoretical meaning and could then be used for some relevant purpose.

2 Cyclic Cohomology

In the commutative case, for a compact space X, we have at our disposal in K-theory a tool of great relevance, the Chern character

$$\text{ch} : K^*(X) \to H^*(X, \mathbb{Q}) \tag{1}$$

which relates the K-theory of X to the cohomology of X. When X is a smooth manifold the Chern character may be calculated explicitly by the differential calculus of forms, currents, connections and curvature. More precisely, given a smooth vector bundle E over X, or equivalently the finite projective module, $\mathcal{E} = C^\infty(X, E)$ over $\mathcal{A} = C^\infty(X)$ of smooth sections of E, the Chern character of E

$$\text{ch}(E) \in H^*(X, \mathbb{R}) \tag{2}$$

is represented by the closed differential form:

$$\text{ch}(E) = \text{trace}\left(\exp(\nabla^2/2\pi i)\right) \tag{3}$$

for any connection ∇ on the vector bundle E. Any closed de Rham current C on the manifold X determines a map φ_C from $K^*(X)$ to $\mathbb{C}$ by the equality

$$\varphi_C(E) = \langle C, \text{ch}(E) \rangle \tag{4}$$

where the pairing between currents and differential forms is the usual one.

One obtains in this way numerical invariants of K-theory classes whose knowledge for arbitrary closed currents C is equivalent to that of $\text{ch}(E)$. The noncommutative torus gave a striking example where it was obviously worthwhile to adapt the above construction of differential geometry to the noncommutative framework ([8]). As an easy preliminary step towards cyclic cohomology one can reformulate the essential ingredient of the construction without direct reference to derivations in the following way ([10],[11]).

By a cycle of dimension n we mean a triple $(\Omega, d, \int)$ where (Ω, d) is a graded differential algebra, and $\int : \Omega^n \to \mathbb{C}$ is a closed graded trace on Ω.

Let $\mathcal{A}$ be an algebra over $\mathbb{C}$. Then a cycle over $\mathcal{A}$ is given by a cycle $(\Omega, d, \int)$ and a homomorphism $\rho : \mathcal{A} \to \Omega^0$.

Thus a *cycle* over an algebra $\mathcal{A}$ is a way to embed $\mathcal{A}$ as a subalgebra of a differential graded algebra (DGA). We shall see in f) below the role of the graded trace.

The usual notions of connection and curvature extend in a straightforward manner to this context ([10],[11]).

Let $\mathcal{A} \xrightarrow{\rho} \Omega$ be a cycle over $\mathcal{A}$, and $\mathcal{E}$ a finite projective module over $\mathcal{A}$. Then a connection ∇ on $\mathcal{E}$ is a linear map $\nabla : \mathcal{E} \to \mathcal{E} \otimes_\mathcal{A} \Omega^1$ such that

$$\nabla(\xi x) = (\nabla \xi)x + \xi \otimes d\rho(x), \quad \forall \xi \in \mathcal{E}, \quad x \in \mathcal{A}. \tag{5}$$

Here $\mathcal{E}$ is a *right* module over $\mathcal{A}$ and Ω^1 is considered as a bimodule over $\mathcal{A}$ using the homomorphism $\rho : \mathcal{A} \to \Omega^0$ and the ring structure of Ω^*. Let us list a number of easy properties ([10],[11]):

a) Let $e \in \mathrm{End}_\mathcal{A}(\mathcal{E})$ be an idempotent and ∇ a connection on $\mathcal{E}$; then $\xi \mapsto (e \otimes 1)\nabla \xi$ is a connection on $e\mathcal{E}$.

b) Any finite projective module $\mathcal{E}$ admits a connection.

c) The space of connections is an affine space over the vector space

$$\mathrm{Hom}_\mathcal{A}(\mathcal{E}, \mathcal{E} \otimes_\mathcal{A} \Omega^1). \tag{6}$$

d) Any connection ∇ extends uniquely to a linear map of $\widetilde{\mathcal{E}} = \mathcal{E} \otimes_\mathcal{A} \Omega$ into itself such that

$$\nabla(\xi \otimes \omega) = (\nabla \xi)\omega + \xi \otimes d\omega, \quad \forall \xi \in \mathcal{E}, \quad \omega \in \Omega. \tag{7}$$

e) The map $\theta = \nabla^2$ of $\widetilde{\mathcal{E}}$ to $\widetilde{\mathcal{E}}$ is an endomorphism: $\theta \in \mathrm{End}_\Omega(\widetilde{\mathcal{E}})$ and with $\delta(T) = \nabla T - (-1)^{deg T} T\nabla$, one has $\delta^2(T) = \theta T - T\theta$ for all $T \in \mathrm{End}_\Omega(\widetilde{\mathcal{E}})$.

f) For n even, $n = 2m$, the equality

$$\langle [\mathcal{E}], [\tau] \rangle = \frac{1}{m!} \int \theta^m, \tag{8}$$

defines an additive map from the K-group $K_0(\mathcal{A})$ to the scalars.

Of course one can reformulate f) by dualizing the closed graded trace $\int$, i.e. by considering the homology of the quotient $\Omega/[\Omega, \Omega]$ ([34]) and one might be tempted at first sight to assert that a noncommutative algebra often comes naturally equipped with a natural embedding in a DGA which should suffice for the Chern character. This however would be rather naive and would overlook for instance the role of *integral* cycles for which the above additive map only affects *integer* values.

The starting point of cyclic cohomology is the ability to compare different cycles on the same algebra. In fact the invariant of K-theory defined in f) by a given cycle only depends on the multilinear form

$$\varphi(a^0, \ldots, a^n) = \int \rho(a^0) \, d(\rho(a^1)) \, d(\rho(a^2)) \ldots d(\rho(a^n)) \quad \forall a^j \in \mathcal{A} \tag{9}$$

(called the character of the cycle) and the functionals thus obtained are exactly those multilinear forms on $\mathcal{A}$ such that

φ is *cyclic* i.e.

$$\varphi(a^0, a^1, \ldots, a^n) = (-1)^n \, \varphi(a^1, a^2, \ldots, a^0) \qquad \forall \, a_j \in \mathcal{A}, \tag{10}$$

$b\varphi = 0$ where

$$(b\varphi)(a^0, \ldots, a^{n+1}) = \sum_0^n (-1)^j \, \varphi(a^0, \ldots, a^j a^{j+1}, \ldots, a^{n+1})$$
$$+ (-1)^{n+1} \, \varphi(a^{n+1} a^0, a^1, \ldots, a^n).$$

This second condition means that φ is a Hochschild cocycle. In particular such a φ admits a Hochschild class

$$I(\varphi) \in H^n(\mathcal{A}, \mathcal{A}^*) \tag{11}$$

for the Hochschild cohomology of $\mathcal{A}$ with coefficients in the bimodule $\mathcal{A}^*$ of linear forms on $\mathcal{A}$.

The n-dimensional *cyclic cohomology* of $\mathcal{A}$ is simply the cohomology $HC^n(\mathcal{A})$ of the *subcomplex* of the Hochschild complex given by cochains which are *cyclic* i.e. fulfill 10. One has an obvious "forgetful" map

$$HC^n(\mathcal{A}) \xrightarrow{\ I\ } H^n(\mathcal{A}, \mathcal{A}^*) \tag{12}$$

but the real story starts with the following long exact sequence which allows in many cases to compute cyclic cohomology from the B operator acting on Hochschild cohomology ([10],[11]),

Theorem 1. *The following triangle is exact:*

$$
\begin{array}{ccc}
 & H^*(\mathcal{A}, \mathcal{A}^*) & \\
{\scriptstyle B}\swarrow & & \nwarrow{\scriptstyle I} \\
HC^*(\mathcal{A}) & \xrightarrow{\ S\ } & HC^*(\mathcal{A})
\end{array}
$$

The operator S is obtained by tensoring cycles by the canonical 2-dimensional generator of the cyclic cohomology of $\mathbb{C}$.

The operator B is explicitly defined at the cochain level by the equality

$$B = AB_0, \quad B_0 \, \varphi(a^0, \ldots, a^{n-1}) = \varphi(1, a^0, \ldots, a^{n-1}) - (-1)^n \, \varphi(a^0, \ldots, a^{n-1}, 1)$$

$$(A\psi)(a^0, \ldots, a^{n-1}) = \sum_0^{n-1} (-1)^{(n-1)j} \psi(a^j, a^{j+1}, \ldots, a^{j-1}).$$

Its conceptual origin lies in the notion of cobordism of cycles which allows to compare different inclusion of $\mathcal{A}$ in DGA as follows. By a *chain* of dimension $n+1$ we shall mean a quadruple $(\Omega, \partial\Omega, d, \int)$ where Ω and $\partial\Omega$ are differential graded algebras of dimensions $n+1$ and n with a given surjective morphism $r : \Omega \to \partial\Omega$ of degree 0, and where $\int : \Omega^{n+1} \to \mathbb{C}$ is a graded trace such that

$$\int d\omega = 0, \quad \forall \omega \in \Omega^n \text{ such that } r(\omega) = 0. \tag{13}$$

By the *boundary* of such a chain we mean the cycle $(\partial\Omega, d, \int')$ where for $\omega' \in (\partial\Omega)^n$ one takes $\int' \omega' = \int d\omega$ for any $\omega \in \Omega^n$ with $r(\omega) = \omega'$. One easily checks, using the surjectivity of r, that $\int'$ is a graded trace on $\partial\Omega$ and is closed by construction.

We shall say that two cycles $\mathcal{A} \xrightarrow{\rho} \Omega$ and $\mathcal{A} \xrightarrow{\rho'} \Omega'$ over $\mathcal{A}$ are *cobordant* if there exists a chain Ω'' with boundary $\Omega \oplus \widetilde{\Omega'}$ (where $\widetilde{\Omega'}$ is obtained from Ω' by changing the sign of $\int$) and a homomorphism $\rho'' : \mathcal{A} \to \Omega''$ such that $r \circ \rho'' = (\rho, \rho')$.

The conceptual role of the operator B is clarified by the following result, ([10],[11])

Theorem 2. *Two cycles over $\mathcal{A}$ are cobordant if and only if their characters* $\tau_1, \tau_2 \in HC^n(\mathcal{A})$ *differ by an element of the image of B, where*

$$B : H^{n+1}(\mathcal{A}, \mathcal{A}^*) \to HC^n(\mathcal{A}).$$

The operators b, B given as above by

$$(b\varphi)(a^0, \dots, a^{n+1}) =$$

$$\sum_0^n (-1)^j \varphi(a^0, \dots, a^j a^{j+1}, \dots, a^{n+1}) + (-1)^{n+1} \varphi(a^{n+1} a^0, a^1, \dots, a^n)$$

$$B = AB_0, \quad B_0\, \varphi(a^0, \dots, a^{n-1}) = \varphi(1, a^0, \dots, a^{n-1}) - (-1)^n \varphi(a^0, \dots, a^{n-1}, 1)$$

$$(A\psi)(a^0, \dots, a^{n-1}) = \sum_0^{n-1} (-1)^{(n-1)j} \psi(a^j, a^{j+1}, \dots, a^{j-1})$$

satisfy $b^2 = B^2 = 0$ and $bB = -Bb$ and periodic cyclic cohomology which is the inductive limit of the $HC^n(\mathcal{A})$ under the periodicity map S admits an equivalent description as the cohomology of the (b, B) bicomplex.

With these notations one has the following formula for the Chern character of the class of an idempotent e, up to normalization one has

$$Ch_n(e) = (e - 1/2) \otimes e \otimes e \otimes \dots \otimes e, \tag{14}$$

where $\otimes$ appears 2n times in the right hand side of the equation.

Both the Hochschild and Cyclic cohomologies of the algebra $\mathcal{A} = C^\infty(V)$ of smooth functions on a manifold V were computed in ([10],[11]).

Let V be a smooth compact manifold and $\mathcal{A}$ the locally convex topological algebra $C^\infty(V)$. Then the following map $\varphi \to C_\varphi$ is a canonical isomorphism of the continuous Hochschild cohomology group $H^k(\mathcal{A}, \mathcal{A}^*)$ with the space of k-dimensional de Rham currents on V:

$$\langle C_\varphi, f^0 \, df^1 \wedge \dots \wedge df^k \rangle = \frac{1}{k!} \sum_{\sigma \in S_k} \varepsilon(\mathfrak{s}) \, \varphi(f^0, f^{\sigma(1)}, \dots, f^{\sigma(k)})$$

$\forall f^0, \ldots, f^k \in C^\infty(V)$.

Under the isomorphism C the operator $I \circ B : H^k(\mathcal{A}, \mathcal{A}^*) \rightarrow H^{k-1}(\mathcal{A}, \mathcal{A}^*)$ is (k times) the de Rham boundary b for currents ([10],[11]).

Theorem 3. *Let $\mathcal{A}$ be the locally convex topological algebra $C^\infty(V)$. Then*

1) *For each k, $HC^k(\mathcal{A})$ is canonically isomorphic to the direct sum*

$$\mathrm{Ker}\, b \oplus H_{k-2}(V, \mathbb{C}) \oplus H_{k-4}(V, \mathbb{C}) \oplus \cdots$$

where $H_q(V, \mathbb{C})$ is the usual de Rham homology of V and b the de Rham boundary.

2) *The periodic cyclic cohomology of $C^\infty(V)$ is canonically isomorphic to the de Rham homology $H_*(V, \mathbb{C})$, with filtration by dimension.*

As soon as we pass to the noncommutative case, more subtle phenomena arise. Thus for instance the filtration of the periodic cyclic homology (dual to periodic cyclic cohomology) together with the lattice $K_0(\mathcal{A}) \subset HC_{\mathrm{ev}}(\mathcal{A})$, for $\mathcal{A} = C^\infty(\mathbb{T}_\theta^2)$, gives an even analogue of the Jacobian of an elliptic curve. More precisely the filtration of HC_{ev} yields a canonical foliation of the torus HC_{ev}/K_0 and one can show that the foliation algebra associated as above to the canonical transversal segment $[0, 1]$ is isomorphic to $C^\infty(\mathbb{T}_\theta^2)$.

A simple example of cyclic cocycle on a nonabelian group ring is provided by the following formula. Any *group cocycle* $c \in H^*(B\Gamma) = H^*(\Gamma)$ gives rise to a cyclic cocycle φ_c on the algebra $\mathcal{A} = \mathbb{C}\Gamma$

$$\varphi_c(g_0, g_1, \ldots, g_n) = \begin{cases} 0 & \text{if } g_0 \ldots g_n \neq 1 \\ c(g_1, \ldots, g_n) & \text{if } g_0 \ldots g_n = 1 \end{cases}$$

where $c \in Z^n(\Gamma, \mathbb{C})$ is suitably normalized, and the formula is extended by linearity to $\mathbb{C}\Gamma$. The cyclic cohomology of group rings is given by,

Theorem 4. [21] *Let Γ be a discrete group, $\mathcal{A} = \mathbb{C}\Gamma$ its group ring.*

a) *The Hochschild cohomology $H^*(\mathcal{A}, \mathcal{A}^*)$ is canonically isomorphic to the cohomology $H^*((B\Gamma)^{\mathfrak{S}^1}, \mathbb{C})$ of the free loop space of the classifying space of Γ.*

b) *The cyclic cohomology $HC^*(\mathcal{A})$ is canonically isomorphic to the $\mathfrak{S}^1$ - equivariant cohomology $H_{\mathfrak{S}^1}^*((B\Gamma)^{\mathfrak{S}^1}, \mathbb{C})$.*

The role of the free loop space in this theorem is not accidental and is clarified in general by the equality

$$B\Lambda = BS^1$$

of the classifying space $B\Lambda$ of the *cyclic category* with the classifying space of the compact group S^1. We refer to ([12]) for this point.

The integral curvature of vector bundles on $\mathbb{T}_\theta^2$ surprisingly gives an integer, in spite of the irrationality of θ ([8]). The conceptual understanding of this type of integrality result lies in the existence of a natural lattice of *integral cycles* which we now describe.

Definition 1. *Let A be an algebra, a Fredholm module over A is given by:*

1) *a representation of A in a Hilbert space $\mathcal{H}$;*

2) *an operator $F = F^*$, $F^2 = 1$, on $\mathcal{H}$ such that*

$$[F, a] \text{ is a compact operator for any } a \in A.$$

Such a Fredholm module will be called *odd*. An *even* Fredholm module is given by an odd Fredholm module $(\mathcal{H}, F)$ as above together with a $\mathbb{Z}/2$ grading γ, $\gamma = \gamma^*$, $\gamma^2 = 1$ of the Hilbert space $\mathcal{H}$ such that:

a) $\gamma a = a\gamma \ \forall \ a \in A$

b) $\gamma F = -F\gamma$.

The above definition is, up to trivial changes, the same as Atiyah's definition [1] of abstract elliptic operators, and the same as Kasparov's definition [35] for the cycles in K-homology, $KK(A, \mathbb{C})$, when A is a C^*-algebra.

The main point is that a Fredholm module over an algebra A gives rise in a very simple manner to a DGA containing A. One simply defines Ω^k as the linear span of operators of the form,

$$\omega = a^0 [F, a^1] \dots [F, a^k] \qquad a^j \in A$$

and the differential is given by

$$d\omega = F\omega - (-1)^k \omega F \qquad \forall \ \omega \in \Omega^k.$$

One easily checks that the ordinary product of operators gives an algebra structure, $\Omega^k \Omega^\ell \subset \Omega^{k+\ell}$ and that $d^2 = 0$ owing to $F^2 = 1$.

Moreover if one assumes that the size of the differential $da = [F, a]$ is controlled, i.e. that

$$|da|^{n+1} \quad \text{is trace class,}$$

then one obtains a natural closed graded trace of degree n by the formula,

$$\int \omega = \text{Trace}(\omega)$$

(with the supertrace Trace $(\gamma\omega)$ in the even case, see [19] for details).

Hence the original Fredholm module gives rise to a *cycle* over A. Such cycles have the remarkable *integrality* property that when we pair them with the K theory of A we only get *integers* as follows from an elementary index formula ([19]).

We let $Ch_*(\mathcal{H}, F) \in HC^n(A)$ be the character of the cycle associated to a Fredholm module $(\mathcal{H}, F)$ over A. This formula defines the Chern character in K-homology.

Cyclic cohomology got many applications [39], it led for instance to the proof of the Novikov conjecture for hyperbolic groups [15]. Basically, by extending the Chern-Weil characteristic classes to the general framework it allows for many concrete computations of differential geometric nature on noncommutative spaces. It also showed the depth of the relation between the classification of factors and the geometry of foliations.

Von Neumann algebras arise very naturally in geometry from foliated manifolds (V, F). The von Neumann algebra $L^\infty(V, F)$ of a foliated manifold is easy to describe, its elements are random operators $T = (T_f)$, i.e. bounded measurable families of operators T_f parametrized by the leaves f of the foliation. For each leaf f the operator T_f acts in the Hilbert space $L^2(f)$ of square integrable densities on the manifold f. Two random operators are identified if they are equal for almost all leaves f (i.e. a set of leaves whose union in V is negligible). The algebraic operations of sum and product are given by,

$$(T_1 + T_2)_f = (T_1)_f + (T_2)_f, \ (T_1 T_2)_f = (T_1)_f (T_2)_f, \tag{15}$$

i.e. are effected pointwise.

All types of factors occur from this geometric construction and the continuous dimensions of Murray and von-Neumann play an essential role in the longitudinal index theorem.

Using cyclic cohomology together with the following simple fact,

$$\text{"A connected group can only act trivially on a homotopy} \atop \text{invariant cohomology theory",} \tag{16}$$

one proves (cf. [13]) that for any codimension one foliation F of a compact manifold V with non vanishing Godbillon-Vey class one has,

$$\text{Mod}(M) \text{ has finite covolume in } \mathbb{R}_+^*, \tag{17}$$

where $\text{Mod}(M)$ is the flow of weights of $M = L^\infty(V, F)$.

In the recent years J. Cuntz and D. Quillen ([25] [26] [27]) have developed a powerful new approach to cyclic cohomology which allowed them to prove excision in full generality.

3 Calculus and Infinitesimals

The central notion of noncommutative geometry comes from the identification of the noncommutative analogue of the two basic concepts in Riemann's formulation of Geometry [44], namely those of manifold and of infinitesimal line element. Both of these noncommutative analogues are of spectral nature and combine to give rise to the notion of spectral triple and spectral manifold, which will be described below. We shall first describe an operator theoretic framework for the calculus of infinitesimals which will provide a natural home for the line element ds.

Let us first start a little excursion, and go back to infinitesimals in a way which is as naive as possible. We want to ask an extremely naive question about the notion of infinitesimal variable. Let me first explain one answer that was proposed for this intuitive idea of infinitesimal variable and why this answer is not satisfactory. We shall then give another really satisfactory answer and use it as the corner stone of the general theory.

So, I remember quite a long time ago to have seen an answer which was proposed by non standard analysis. The book I was reading [4] was starting from a variant of the following problem:

You play a game of throwing darts at some target called Ω

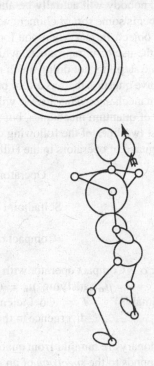

and the question which is asked is: what is the probability $dp(x)$ that actually when you send the dart you land exactly at a given point $x \in \Omega$? Then the following argument was given: certainly this probability $dp(x)$ is smaller than $1/2$ because you can cut the target into two equal halves, only one of which contains x. For the same reason $dp(x)$ is smaller than $1/4$, and so on and so forth. So what you find out is that $dp(x)$ is smaller than any positive real number ε. On the other hand, if you give the answer that $dp(x)$ is 0, this is not really satisfactory, because whenever you send the dart it will land somewhere. So now, if you ask a mathematician about this naive question, he might very well answer: well, $dp(x)$ is a 2-form, or it's a measure, or something like that. But then you can try to ask him more precise questions, for instance "what is the exponential of $-\frac{1}{dp(x)}$". And then it will be hard for him to give a satisfactory answer, because you know that the Taylor expansion of the function $f(y) = e^{-\frac{1}{y}}$ is zero at $y = 0$. Now the book I was reading claimed to give an answer, and it was what is called a non standard number. So I worked on this theory for some time, learning some logics, until eventually I realized there was a very bad obstruction preventing one to get concrete answers. It is the following: it's a little lemma that one can easily prove, that if you are given a non standard number you can canonically produce a subset of the interval which is not Lebesgue measurable. Now we know from logic (from results of Paul Cohen and Solovay) that it will forever be impossible to produce explicitely a subset of the real numbers, of the interval $[0, 1]$, say, that is not Lebesgue measurable. So, what this says is that for instance in this example, nobody will actually be able to name a non standard number. A nonstandard number is some sort of chimera which is impossible to grasp and certainly not a concrete object. In fact when you look at nonstandard analysis you find out that except for the use of ultraproducts, which is very efficient, it just shifts the order in logic by one step; it's not doing much more. Now, what I want to explain is that to the above naive question there is a very beautiful and simple answer which is provided by quantum mechanics. This answer will be obtained just by going through the usual dictionary of quantum mechanics, but looking at it more closely. So, let us thus look at the first two lines of the following dictionary which translates classical notions into the language of operators in the Hilbert space $\mathcal{H}$:

Complex variable	Operator in $\mathcal{H}$
Real variable	Selfadjoint operator
Infinitesimal	Compact operator
Infinitesimal of order α	Compact operator with characteristic values μ_n satisfying $\mu_n = O(n^{-\alpha})$, $n \to \infty$
Integral of an infinitesimal of order 1	$\int T =$ Coefficient of logarithmic divergence in the trace of T.

The first two lines of the dictionary are familiar from quantum mechanics. The range of a complex variable corresponds to the *spectrum* of an operator. The holomorphic functional calculus gives a meaning to $f(T)$ for all holomorphic functions f on the spectrum of T. It is only holomorphic functions which operate in this generality which reflects the difference between complex and real analysis. When $T = T^*$ is selfadjoint then $f(T)$ has a meaning for all Borel functions f.

The size of the infinitesimal $T \in \mathcal{K}$ is governed by the order of decay of the sequence of characteristic values $\mu_n = \mu_n(T)$ as $n \to \infty$. In particular, for all real positive α the following condition defines infinitesimals of order α:

$$\mu_n(T) = O(n^{-\alpha}) \qquad \text{when } n \to \infty \tag{1}$$

(i.e. there exists $C > 0$ such that $\mu_n(T) \le C n^{-\alpha} \ \forall \, n \ge 1$). Infinitesimals of order α also form a two–sided ideal and moreover,

$$T_j \text{ of order } \alpha_j \to T_1 T_2 \text{ of order } \alpha_1 + \alpha_2. \tag{2}$$

Hence, apart from commutativity, intuitive properties of the infinitesimal calculus are fulfilled.

Since the size of an infinitesimal is measured by the sequence $\mu_n \downarrow 0$ it might seem that one does not need the operator formalism at all, and that it would be enough to replace the ideal $\mathcal{K}$ in $\mathcal{L}(\mathcal{H})$ by the ideal $c_0(\mathbb{N})$ of sequences converging to zero in the algebra $\ell^\infty(\mathbb{N})$ of bounded sequences. A variable would just be a bounded sequence, and an infinitesimal a sequence $\mu_n, \mu_n \to 0$. However, this commutative version does not allow for the existence of variables with range a continuum since all elements of $\ell^\infty(\mathbb{N})$ have a point spectrum and a discrete spectral measure. Only *noncommutativity* of $\mathcal{L}(\mathcal{H})$ allows for the coexistence of variables with Lebesgue spectrum together with infinitesimal variables. As we shall see shortly, it is precisely this lack of commutativity between the line element and the coordinates on a space that will provide the measurement of distances.

The integral is obtained by the following analysis, mainly due to Dixmier ([29]), of the logarithmic divergence of the partial traces

$$\text{Trace}_N(T) = \sum_0^{N-1} \mu_n(T) , \ T \geq 0. \tag{3}$$

In fact, it is useful to define $\text{Trace}_\Lambda(T)$ for any positive real $\Lambda > 0$ by piecewise affine interpolation for noninteger Λ.

Define for all order 1 operators $T \geq 0$

$$\tau_\Lambda(T) = \frac{1}{\log \Lambda} \int_e^\Lambda \frac{\text{Trace}_\mu(T)}{\log \mu} \frac{d\mu}{\mu} \tag{4}$$

which is the Cesaro mean of the function $\frac{\text{Trace}_\mu(T)}{\log \mu}$ over the scaling group $\mathbb{R}_+^*$. For $T > 0$, an infinitesimal of order 1, one has

$$\text{Trace}_\Lambda(T) \leq C \log \Lambda \tag{5}$$

so that $\tau_\Lambda(T)$ is bounded. The essential property is the following *asymptotic additivity* of the coefficient $\tau_\Lambda(T)$ of the logarithmic divergence (4):

$$|\tau_\Lambda(T_1 + T_2) - \tau_\Lambda(T_1) - \tau_\Lambda(T_2)| \leq 3C \, \frac{\log(\log \Lambda)}{\log \Lambda} \tag{6}$$

for $T_j \geq 0$.

An easy consequence of (6) is that any limit point $\tau = \text{Tr}_\omega$ of the nonlinear functionals τ_Λ for $\Lambda \to \infty$ defines a positive and linear trace Tr_ω, called a Dixmier trace on the two-sided ideal of infinitesimals of order 1. We shall use the generality in which such a trace is defined in the last section of this paper.

In most concrete examples however the choice of the limit point τ is irrelevant because T is a *measurable* operator, i.e.:

$$\tau_\Lambda(T) \text{ converges when } \Lambda \to \infty. \tag{7}$$

Thus the value $\tau(T)$ is independent of the choice of the limit point τ and is denoted

$$\int\!\!\!\!\!- \, T \, . \tag{8}$$

The first interesting example is provided by pseudodifferential operators T on a differentiable manifold M. When T is of order 1 in the above sense, it is measurable and $\int\!\!\!\!\!- T$ is the non-commutative residue of T ([47]). It has a local expression in terms of the distribution kernel $k(x, y)$, $x, y \in M$. For T of order 1 the kernel $k(x, y)$ diverges logarithmically near the diagonal,

$$k(x, y) = -a(x) \log |x - y| + 0(1) \text{ (for } y \to x) \tag{9}$$

where $a(x)$ is a 1–density independent of the choice of Riemannian distance $|x - y|$. Then one has (up to normalization),

$$\int\!\!\!\!\!- \, T = \int_M a(x). \tag{10}$$

The right hand side of this formula makes sense for all pseudodifferential operators (cf. [47]) since one can see that the kernel of such an operator is asymptotically of the form

$$k(x, y) = \sum a_k(x, x - y) - a(x) \log |x - y| + 0(1) \tag{11}$$

where $a_k(x, \xi)$ is homogeneous of degree $-k$ in ξ, and the 1–density $a(x)$ is defined intrinsically.

The same principle of extension of $\int\!\!\!\!\!-$ to infinitesimals of order < 1 works for hypoelliptic operators and more generally as we shall see below, for spectral triples whose dimension spectrum is simple.

We can now go back to our initial naive question about the target and the darts, we find that quantum mechanics gives us an obvious infinitesimal which answers the question: it is the inverse of the Dirichlet Laplacian for the domain Ω. Thus there is now a clear meaning for the exponential of $\frac{-1}{dp}$, that's the well known heat kernel which is an infinitesimal of arbitrarily large order as we expected from the Taylor expansion.

From the H. Weyl theorem on the asymptotic behavior of eigenvalues of Δ it follows that dp is of order 1, and that given a function f on Ω the product $f \, dp$ is measurable, while

$$\int\!\!\!\!\!- \, f \, dp = \int_\Omega f(x_1, x_2) \, dx_1 \wedge dx_2 \tag{12}$$

gives the ordinary integral of f with respect to the measure given by the area of the target.

4 Spectral Triples

In this section we shall come back to the two basic notions introduced by Riemann in the classical framework, those of *manifold* and of *line element* ([44]). We shall see that both of these notions adapt remarkably well to the noncommutative framework and this will lead us to the notion of spectral triple which noncommutative geometry is based on.

In ordinary geometry of course you can give a manifold by a cooking recipe, by charts and local diffeomorphisms, and one could be tempted to propose an analogous cooking recipe in the noncommutative case. This is pretty much what is achieved by the general construction of the algebras of foliations and it is a good test of any general idea that it should at least cover that large class of examples.

But at a more conceptual level, it was recognized long ago by geometors that the main quality of the homotopy type of an oriented manifold is to satisfy Poincaré duality not only in ordinary homology but also in K-homology. Poincaré duality in ordinary homology is not sufficient to describe homotopy type of manifolds [42] but D. Sullivan [46] showed (in the simply connected PL case of dimension ≥ 5 ignoring 2-torsion) that it is sufficient to replace ordinary homology by KO-homology. Moreover the Chern character of the KO-homology fundamental class contains all the rational information on the Pontrjagin classes.

The characteristic property of *differentiable manifolds* which is carried over to the noncommutative case is *Poincaré duality* in KO-homology [46].

Moreover, as we saw above in the discussion of Fredholm modules, K-homology admits a fairly simple definition in terms of Hilbert space and Fredholm representations of algebras, as gradually emerged from the work of Atiyah ([1]), Singer ([45]), Brown-Douglas-Fillmore ([5]), Miscenko ([43]), and Kasparov ([35]).

For an ordinary manifold the choice of the fundamental cycle in K-homology is a refinement of the choice of orientation of the manifold and in its simplest form is a choice of Spin-structure. Of course the role of a spin structure is to allow for the construction of the corresponding Dirac operator which gives a corresponding Fredholm representation of the algebra of smooth functions. The origin of the construction of the Dirac operator was the extraction of a square root for a second order differential operator like the Laplacian.

What is rewarding is that this will not only guide us towards the notion of noncommutative manifold but also to a formula, of operator theoretic nature, for the line element ds. In the Riemannian case one gives the Taylor expansion of the square ds^2 of the infinitesimal line element, in our framework the extraction of square root effected by the Dirac operator allows us to deal directly with ds itself.

The infinitesimal unit of length"ds" should be an infinitesimal in the sense of section 3 and one way to get an intuitive understanding of the formula for ds is to consider Feynman diagrams which physicists use currently in the computations of quantum field theory. Let us contemplate the diagram:
which is involved in the computation of the self-energy of an electron in QED. The two points x and y of space-time at which the photon (the wiggly line) is emitted and reabsorbed are very close by and our ansatz for ds will be at the intuitive level,

$$ds = \times\!\!-\!\!\!-\!\!\times \, . \tag{1}$$

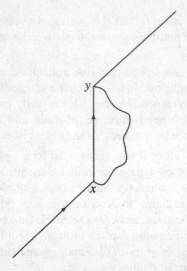

The right hand side has good meaning in physics, it is called the Fermion propagator and is given by

$$\times\!\!-\!\!\!\times = D^{-1} \tag{2}$$

where D is the Dirac operator.

We thus arrive at the following basic ansatz,

$$ds = D^{-1}. \tag{3}$$

In some sense it is simpler than the ansatz giving ds^2 as $g_{\mu\nu}\, dx^\mu\, dx^\nu$, the point being that the spin structure allows really to extract the square root of ds^2 (as is well known Dirac found the corresponding operator as a differential square root of a Laplacian).

The first thing we need to do is to check that we are still able to measure distances with our "unit of length" ds. In fact we saw in the discussion of the quantized calculus that variables with continuous range cant commute with "infinitesimals" such as ds and it is thus not very surprising that this lack of commutativity allows to compute, in the classical Riemannian case, the geodesic distance $d(x, y)$ between two points. The precise formula is

$$d(x, y) = \mathrm{Sup}\left\{ |f(x) - f(y)| \,;\, f \in \mathcal{A},\, \|[D, f]\| \le 1 \right\} \tag{4}$$

where $D = ds^{-1}$ as above and $\mathcal{A}$ is the algebra of smooth functions. Note that if ds has the dimension of a length L, then D has dimension L^{-1} and the above expression for $d(x, y)$ also has the dimension of a length.

Thus we see in the classical geometric case that both the fundamental cycle in K-homology and the metric are encoded in the *spectral triple* $(\mathcal{A}, \mathcal{H}, D)$ where $\mathcal{A}$ is the algebra of functions acting in the Hilbert space $\mathcal{H}$ of spinors, while D is the Dirac operator.

To get familiar with this notion one should check that we recover the volume form of the Riemannian metric by the equality (valid up to a normalization constant [19])

$$\fint f \, |ds|^n = \int_{M_n} f \sqrt{g} \, d^n x \tag{5}$$

but the first interesting point is that besides this coherence with the usual computations there are new simple questions we can ask now such as "what is the two-dimensional measure of a four manifold" in other words "what is its area ?". Thus one should compute

$$\fint ds^2 \tag{6}$$

It is obvious from invariant theory that this should be proportional to the Hilbert–Einstein action but doing the direct computation is a worthwile exercice (cf. [36]), the exact result being

$$\fint ds^2 = \frac{-1}{48\pi^2} \int_{M_4} r \sqrt{g} \, d^4 x \tag{7}$$

where as above $dv = \sqrt{g} \, d^4 x$ is the volume form, $ds = D^{-1}$ the length element, *i.e.* the inverse of the Dirac operator and r is the scalar curvature.

In the general framework of Noncommutative Geometry the confluence of the Hilbert space incarnation of the two notions of metric and fundamental class for a manifold led very naturally to define a geometric space as given by a *spectral triple:*

$$(\mathcal{A}, \mathcal{H}, D) \tag{8}$$

where $\mathcal{A}$ is a concrete algebra of coordinates represented on a Hilbert space $\mathcal{H}$ and the operator D is the inverse of the line element.

$$ds = 1/D. \tag{9}$$

This definition is entirely spectral; the elements of the algebra are operators, the points, if they exist, come from the joint spectrum of operators and the line element is an operator.

The basic properties of such spectral triples are easy to formulate and do not make any reference to the commutativity of the algebra $\mathcal{A}$. They are

$$[D, a] \text{ is bounded for any } a \in \mathcal{A}, \tag{10}$$

$$D = D^* \text{ and } (D + \lambda)^{-1} \text{ is a compact operator } \forall \lambda \notin \mathbb{C}. \tag{11}$$

(Of course D is an *unbounded* operator).

There is no difficulty to adapt the above formula for the distance in the general non-commutative case, one uses the same, the points x and y being replaced by arbitrary states φ and ψ on the algebra $\mathcal{A}$. Recall that a state is a normalized positive linear form on $\mathcal{A}$ such that $\varphi(1) = 1$,

$$\varphi : \bar{\mathcal{A}} \to \mathbb{C}, \ \varphi(a^* a) \geq 0, \quad \forall a \in \bar{\mathcal{A}}, \ \varphi(1) = 1. \tag{12}$$

The distance between two states is given by,

$$d(\varphi, \psi) = \mathrm{Sup}\, \{|\varphi(a) - \psi(a)|\; ;\; a \in \mathcal{A}\,,\; \|[D, a]\| \leq 1\}\,. \tag{13}$$

The significance of D is two-fold. On the one hand it defines the metric by the above equation, on the other hand its homotopy class represents the K-homology fundamental class of the space under consideration.

It is crucial to understand from the start the tension between the conditions (10) and (11). The first condition would be trivially fulfilled if D were bounded but condition (11) shows that it is unbounded. To understand this tension let us work out a very simple case. We let the algebra $\mathcal{A}$ be generated by a single unitary operator U. Let us show that if the index pairing between U and D, i.e. the index of PUP where P is the orthogonal projection on the positive eigenspace of D, *does not vanish* then the number $N(E)$ of eigenvalues of D whose absolute value is less than E grows at least like E when $E \to \infty$. This means that in the above circumstance $ds = D^{-1}$ is of order one or less.

To prove this we choose a smooth function $f \in C_c^\infty(\mathbb{R})$ identically one near 0, even and with Support $(f) \subset [-1, 1]$. We then let $R(\varepsilon) = f(\varepsilon D)$. One first shows ([19]) that the operator norm of the commutator $[R(\varepsilon), U]$ tends to 0 like ε. It then follows that the trace norm satisfies

$$\|[R(\varepsilon), U]\|_1 \leq C \varepsilon N(1/\varepsilon) \tag{14}$$

as one sees using the control of the rank of $R(\varepsilon)$ from $N(1/\varepsilon)$. The index pairing is given by $-\frac{1}{2} \mathrm{Trace}\,(U^*[F, U])$ where F is the sign of D and one has,

$$\mathrm{Trace}\,(U^*[F, U]) = \lim_{\varepsilon \to 0} \mathrm{Trace}\,(U^*[F, U]\, R(\varepsilon)) = \lim_{\varepsilon \to 0} \mathrm{Trace}\,(U^*\, F[U, R(\varepsilon)])\,. \tag{15}$$

Thus the limit being non zero we get a lower bound on the trace norm of $[U, R(\varepsilon)]$ and hence on $\varepsilon\, N\left(\frac{1}{\varepsilon}\right)$ which shows that $N(E)$ grows at least like E when $E \to \infty$. This shows that ds cannot be too small (it cannot be of order $\alpha > 1$). In fact when ds is of order 1 one has the following index formula,

$$\mathrm{Index}\,(PUP) = -\frac{1}{2} \int U^{-1}[D, U]\,|ds|\,. \tag{16}$$

The simplest case in which the index pairing between D and U does not vanish, with ds of order 1, is obtained by requiring the further condition,

$$U^{-1}[D, U] = 1\,. \tag{17}$$

It is a simple exercise to compute the geometry on $S^1 = \mathrm{Spectrum}\,(U)$ given by an irreducible representation of condition (17). One obtains the standard circle with length 2π.

The above index formula is a special case of a general result ([19]) which computes the n-dimensional Hochschild class of the Chern character of a spectral triple of dimension n.

Theorem 5. *Let* $(\mathcal{H}, F)$ *be a Fredholm module over an involutive algebra* $\mathcal{A}$. *Let* D *be an unbounded selfadjoint operator in* $\mathcal{H}$ *such that* D^{-1} *is of order* $1/n$, $\mathrm{Sign}\, D = F$, *and such that for any* $a \in \mathcal{A}$ *the operators* a *and* $[D, a]$ *are in the domain of all powers of the derivations* δ, *given by* $\delta(x) = [|D|, x]$. *Let* $\tau_n \in HC^n(\mathcal{A})$ *be the Chern character of* $(\mathcal{H}, F)$.

For every n-*dimensional Hochschild cycle* $c \in Z_n(\mathcal{A}, \mathcal{A})$, $c = \sum a^0 \otimes a^1 \ldots \otimes a^n$, *one has* $\langle \tau_n, c \rangle = f \sum a^0 [D, a^1] \ldots [D, a^n] |D|^{-n}$.

We refer to [19] for the precise normalization and to [32] for the detailed proof. By construction, this formula is scale invariant, i.e. it remains unchanged if we replace D by λD for $\lambda \in \mathbb{R}_+^*$. The operators T_c of the form

$$T_c = \sum a^0 [D, a^1] \ldots [D, a^n] |D|^{-n} \tag{18}$$

are *measurable* in the sense of section 3.

The long exact sequence of cyclic cohomology (Section 2) shows that the Hochschild class of τ_n is the obstruction to a better summability of $(\mathcal{H}, F)$, indeed τ_n belongs to the image $S(HC^{n-2}(\mathcal{A}))$ (which is the case if the degree of summability can be improved by 2) if and only if the Hochschild cohomology class $I(\tau_n) \in H^n(\mathcal{A}, \mathcal{A}^*)$ is equal to 0.

In particular, the above theorem implies nonvanishing of residues when the cohomological dimension of $\mathrm{ch}_*(\mathcal{H}, F)$ is not lower than n:

Corollary 1. *With the hypothesis of Theorem 5 and if the Hochschild class of* $\mathrm{ch}_*(\mathcal{H}, F)$ *pairs nontrivially with* $H_n(\mathcal{A}, \mathcal{A})$ *one has*

$$f |D|^{-n} \neq 0. \tag{19}$$

In other words the residue of the function $\zeta(s) = \mathrm{Trace}\,(|D|^{-s})$ at $s = n$ cannot vanish.

5 Operator Theoretic Local Index Formula

Given a spectral triple $(\mathcal{A}, \mathcal{H}, D)$, with $D^{-1} \in \mathcal{L}^{(p, \infty)}$, the precise normalization for its Chern character in cyclic cohomology is obtained from the following cyclic cocycle τ_n, $n \geq p$, n odd,

$$\tau_n(a^0, \ldots, a^n) = \lambda_n \, \mathrm{Tr}' \left(a^0 [F, a^1] \ldots [F, a^n] \right), \qquad \forall a^j \in \mathcal{A}, \tag{1}$$

where $F = \mathrm{Sign} D$, $\lambda_n = \sqrt{2i} \, (-1)^{\frac{n(n-1)}{2}} \, \Gamma \left(\frac{n}{2} + 1 \right)$ and

$$\mathrm{Tr}'(T) = \frac{1}{2} \, \mathrm{Trace}(F(FT + TF)) \tag{2}$$

If one wants to regard the cocycle τ_n of (1) as a cochain of the (b, B) bicomplex, one uses instead of λ_n, the normalization constant $\mu_n = (-1)^{[n/2]} (n!)^{-1} \lambda_n = \sqrt{2i} \frac{\Gamma(\frac{n}{2}+1)}{n!}$.

It is difficult to compute the cocycle τ_n in general because the formula (1) involves the ordinary trace instead of the local trace f and it is crucial to obtain a local form of the above cocycle.

The problem of finding a local formula for the *cyclic cohomology* Chern character, i.e. for the class of τ_n is solved by a general formula [16] which is expressed in terms of the (b, B) bicomplex and which we now explain.

Let us make the following regularity hypothesis on $(\mathcal{A}, \mathcal{H}, D)$

$$a \text{ and } [D, a] \in \cap \operatorname{Dom} \delta^k, \ \forall \, a \in \mathcal{A} \tag{3}$$

where δ is the derivation $\delta(T) = [|D|, T]$ for any operator T.

We let $\mathcal{B}$ denote the algebra generated by $\delta^k(a)$, $\delta^k([D, a])$. The usual notion of *dimension* of a space is replaced by the *dimension spectrum* which is a subset of $\mathbb{C}$. The precise definition of the dimension spectrum is the subset $\Sigma \subset \mathbb{C}$ of singularities of the analytic functions

$$\zeta_b(z) = \operatorname{Trace}(b|D|^{-z}) \qquad \operatorname{Re} z > p \, , \, b \in \mathcal{B}. \tag{4}$$

Note that D may have a non-trivial kernel so that $|D|^{-s}$ is ill defined there. However the kernel of D is finite dimensional and the poles and residues of the above function are independent of the arbitrary choice of a non-zero positive value $|D| = \varepsilon$ on this kernel. The dimension spectrum of a manifold M consists of relative integers less than $n = \dim M$; it is simple. Multiplicities appear for singular manifolds. Cantor sets provide examples of complex points $z \notin \mathbb{R}$ in the dimension spectrum.

We assume that Σ is discrete and simple, i.e. that ζ_b can be extended to $\mathbb{C}/\Sigma$ with simple poles in Σ. In fact the hypothesis only matters in a neighborhood of $\{z, \operatorname{Re}(z) \geq 0\}$.

Let $(\mathcal{A}, \mathcal{H}, D)$ be a spectral triple satisfying the hypothesis (3) and (4).

We shall use the following notations:

$$\nabla(a) = [D^2, a] \quad ; \quad a^{(k)} = \nabla^k(a) \, , \quad \forall \, a \text{ operator in } \mathcal{H} \, .$$

The local index theorem is the following, [16]:

Theorem 6.

1. *The equality*

$$\int P = \operatorname{Res}_{z=0} \operatorname{Trace}(P|D|^{-z})$$

defines a trace on the algebra generated by $\mathcal{A}$, $[D, \mathcal{A}]$ and $|D|^z$, where $z \in \mathbb{C}$.

2. *There is only a finite number of non–zero terms in the following formula which defines the odd components $(\varphi_n)_{n=1,3,...}$ of a cocycle in the bicomplex (b, B) of $\mathcal{A}$,*

$$\varphi_n(a^0, \ldots, a^n) = \sum_k c_{n,k} \int a^0 [D, a^1]^{(k_1)} \ldots [D, a^n]^{(k_n)} |D|^{-n-2|k|} \quad \forall a^j \in \mathcal{A}$$

where k is a multi-index, $|k| = k_1 + \ldots + k_n$,

$$c_{n,k} = (-1)^{|k|} \sqrt{2i} (k_1! \ldots k_n!)^{-1} ((k_1+1) \ldots (k_1+k_2+\ldots+k_n+n))^{-1} \Gamma\left(|k| + \frac{n}{2}\right)$$

3. *The pairing of the cyclic cohomology class $(\varphi_n) \in HC^*(\mathcal{A})$ with $K_1(\mathcal{A})$ gives the Fredholm index of D with coefficients in $K_1(\mathcal{A})$.*

For the normalization of the pairing between HC^* and $K(\mathcal{A})$ see [19]. In the even case, i.e. when $\mathcal{H}$ is $\mathbb{Z}/2$ graded by γ,

$$\gamma = \gamma^*, \ \gamma^2 = 1, \ \gamma a = a\gamma \quad \forall a \in \mathcal{A}, \ \gamma D = -D\gamma,$$

there is an analogous formula for a cocycle (φ_n), n even, which gives the Fredholm index of D with coefficients in K_0. However, φ_0 is not expressed in terms of the residue $\int$ because the character can be non-trivial for a finite dimensional $\mathcal{H}$, in which case all residues vanish.

This theorem has been successfully applied to obtain the transverse local index formula for hypoelliptic operators associated to arbitrary foliations ([17]). The corresponding computations are in fact governed by a Hopf algebra symmetry which underlies transverse geometry in codimension n. It also led to a cyclic complex naturally associated to Hopf algebras which allowed to develop the theory of characteristic classes in the context of Hopf-actions and cyclic cohomology ([17])

To show the power and generality of the above theorem, we shall undertake the computation in yet another example, that of the quantum group $SU_q(2)$. Its original interest is that it lies rather far from ordinary manifolds and is thus a good test case for the general theory.

6 Dimension Spectrum of $SU_q(2)$: Case $q = 0$

Let q be a real number $0 \le q < 1$. We start with the presentation of the algebra of coordinates on the quantum group $SU_q(2)$ in the form,

$$\alpha^*\alpha + \beta^*\beta = 1, \ \alpha\alpha^* + q^2 \beta\beta^* = 1, \ \alpha\beta = q\beta\alpha, \ \alpha\beta^* = q\beta^*\alpha, \ \beta\beta^* = \beta^*\beta. \quad (1)$$

Let us recall the notations for the standard representation of that algebra. One lets $\mathcal{H}$ be the Hilbert space with orthonormal basis $e_{ij}^{(n)}$ where $n \in \frac{1}{2}\mathbb{N}$ varies among half-integers while $i, j \in \{-n, -n+1, \ldots, n\}$.

Thus the first elements are,

$$e_{00}^{(0)}, \; e_{ij}^{(1/2)}, \; i,j \in \left\{ -\frac{1}{2}, \frac{1}{2} \right\}, \ldots$$

The following formulas define a unitary representation in $\mathcal{H}$,

$$\alpha e_{ij}^{(n)} = a_+(n,i,j)\, e_{i-\frac{1}{2},j-\frac{1}{2}}^{(n+\frac{1}{2})} + a_-(n,i,j)\, e_{i-\frac{1}{2},j-\frac{1}{2}}^{(n-\frac{1}{2})} \tag{2}$$

$$\beta e_{ij}^{(n)} = b_+(n,i,j)\, e_{i+\frac{1}{2},j-\frac{1}{2}}^{(n+\frac{1}{2})} + b_-(n,i,j)\, e_{i+\frac{1}{2},j-\frac{1}{2}}^{(n-\frac{1}{2})}$$

where the explicit form of $a_\pm$ and $b_\pm$ is,

$$a_+(n,i,j) = q^{2n+i+j+1} \frac{(1-q^{2n-2j+2})^{1/2}(1-q^{2n-2i+2})^{1/2}}{(1-q^{4n+2})^{1/2}(1-q^{4n+4})^{1/2}} \tag{3}$$

$$a_-(n,i,j) = \frac{(1-q^{2n+2j})^{1/2}(1-q^{2n+2i})^{1/2}}{(1-q^{4n})^{1/2}(1-q^{4n+2})^{1/2}}$$

and

$$b_+(n,i,j) = -q^{n+j} \frac{(1-q^{2n-2j+2})^{1/2}(1-q^{2n+2i+2})^{1/2}}{(1-q^{4n+2})^{1/2}(1-q^{4n+4})^{1/2}} \tag{4}$$

$$b_-(n,i,j) = q^{n+i} \frac{(1-q^{2n+2j})^{1/2}(1-q^{2n-2i})^{1/2}}{(1-q^{4n})^{1/2}(1-q^{4n+2})^{1/2}}.$$

Note that a_- does vanish if $i = -n$ or $j = -n$, which gives meaning to $a_-(n,i,j)\, e_{i-\frac{1}{2},j-\frac{1}{2}}^{(n-\frac{1}{2})}$ for these values while $i - \frac{1}{2} \notin \left[-\left(n - \frac{1}{2}\right), n - \frac{1}{2} \right]$ or $j - \frac{1}{2} \notin \left[-\left(n - \frac{1}{2}\right), n - \frac{1}{2} \right]$. Similarly b_- vanishes for $j = -n$ or $i = n$.

Let now as in ([7]), D be the diagonal operator in $\mathcal{H}$ given by,

$$D(e_{ij}^{(n)}) = (2\,\delta_0(n-i) - 1)\, 2n\, e_{ij}^{(n)} \tag{5}$$

where $\delta_0(k) = 0$ if $k \neq 0$ and $\delta_0(0) = 1$. It follows from [7] that the triple

$$(\mathcal{A}, \mathcal{H}, D) \tag{6}$$

is a spectral triple.

In order to simplify we start the discussion with the case $q = 0$. We then have the simpler formulas,

$$a_+(n,i,j) = 0 \tag{7}$$
$$a_-(n,i,j) = 0 \text{ if } i = -n \text{ or } j = -n$$
$$a_-(n,i,j) = 1 \text{ if } i \neq -n \text{ and } j \neq -n$$

$$b_+(n,i,j) = 0 \text{ if } j \neq -n \qquad (8)$$
$$b_+(n,i,j) = -1 \text{ if } j = -n$$
$$b_-(n,i,j) = 0 \text{ if } i \neq -n \text{ or } j = -n$$
$$b_-(n,i,j) = 1 \text{ if } i = -n, j \neq -n.$$

Thus for $q = 0$ the operators α and β in $\mathcal{H}$ are given by,

$$\alpha e_{ij}^{(n)} = e_{i-\frac{1}{2},j-\frac{1}{2}}^{(n-\frac{1}{2})} \qquad \text{if} \quad i > -n, j > -n \qquad (9)$$

and $\alpha e_{ij}^{(n)} = 0$ if $i = -n$ or $j = -n$.

$$\beta e_{ij}^{(n)} = 0 \qquad \text{if} \quad i \neq -n \quad \text{and} \quad j \neq -n \qquad (10)$$

$$\beta e_{-n,j}^{(n)} = e_{-(n-\frac{1}{2}),j-\frac{1}{2}}^{(n-\frac{1}{2})} \qquad \text{if} \quad j \neq -n \qquad (11)$$

and

$$\beta e_{i,-n}^{(n)} = -e_{i+\frac{1}{2},-(n+\frac{1}{2})}^{(n+\frac{1}{2})}. \qquad (12)$$

By construction $\beta\beta^* = \beta^*\beta$ is the projection e on the subset $\{i = -n \text{ or } j = -n\}$ of the basis.

Also α is a partial isometry with initial support $1 - e$ and final support $1 = \alpha\alpha^*$. The basic relations between α and β are,

$$\alpha^*\alpha + \beta^*\beta = 1, \ \alpha\alpha^* = 1, \ \alpha\beta = \alpha\beta^* = 0, \ \beta\beta^* = \beta^*\beta. \qquad (13)$$

For $f \in C^\infty(S^1)$, $f = \sum \widehat{f}_n e^{in\theta}$, we let

$$f(\beta) = \sum_{n>0} \widehat{f}_n \beta^n + \sum_{n<0} \widehat{f}_n \beta^{*(-n)} + \widehat{f}_0 e \qquad (14)$$

and the map $f \to f(\beta)$ gives a (degenerate) representation of $C^\infty(S^1)$ in $\mathcal{H}$.

Now let $\mathcal{A}$ be the linear space of sums,

$$a = \sum_{k,\ell \geq 0} \alpha^{*k} f_{k\ell}(\beta) \alpha^\ell + \sum_{\ell \geq 0} \lambda_\ell \alpha^\ell + \sum_{k>0} \lambda_k' \alpha^{*k} \qquad (15)$$

where λ and λ' are sequences (of complex numbers) of rapid decay and $(f_{k\ell})$ is a sequence of rapid decay with values in $C^\infty(S^1)$.

We let A be the C^* algebra in $\mathcal{H}$ generated by α and β.

Proposition 1. *The subspace $\mathcal{A} \subset A$ is a subalgebra stable under holomorphic functional calculus.*

Proof. Let $\mathfrak{s}$ be the linear map from $\mathcal{A}$ to $C^\infty(S^1)$ given by,

$$\sigma(a) = \sum_{\ell \geq 0} \lambda_\ell \, u^\ell + \sum_{k > 0} \lambda'_k \, u^{-k} \tag{16}$$

where $u = e^{i\theta}$ is the generator of $C^\infty(S^1)$. Let $\mathcal{J} = \mathrm{Ker}\,\mathfrak{s}$. For $a \in \mathcal{J}$ one has $a = \sum \alpha^{*k} f_{k\ell} \, \alpha^\ell$ and the equality,

$$\alpha^{*k} f_{k\ell} \, \alpha^\ell \alpha^{*k'} \, g_{k'\ell'} \, \alpha^{\ell'} = \delta_{\ell,k'} \, \alpha^{*k} f_{k\ell} \, g_{k'\ell'} \, \alpha^{\ell'} \tag{17}$$

shows that $\mathcal{J}$ is an algebra and is isomorphic to the topological tensor product

$$C^\infty(S^1) \otimes \mathcal{S} = C^\infty(S^1, \mathcal{S}) \tag{18}$$

where $\mathcal{S}$ is the algebra of matrices of rapid decay.

Since $\mathcal{S}$ is stable under holomorphic functional calculus (h.f.c.) in its norm closure $\mathcal{K}$ (the C^* algebra of compact operators), it follows from (18) that $\mathcal{J}$ is stable under h.f.c. in its norm closure $\overline{\mathcal{J}} \subset A$.

The equalities $\alpha f(\beta) = 0 \,\forall f \in C^\infty(S^1)$ and $\alpha\alpha^* = 1$ show that $\mathcal{J}$ is stable under left multiplication by α^* and α. It follows using (13) that $\mathcal{A}$ is an algebra, $\mathcal{J}$ a two sided ideal of $\mathcal{A}$ and that one has the exact sequence,

$$0 \longrightarrow \mathcal{J} \longrightarrow \mathcal{A} \overset{\mathfrak{s}}{\longrightarrow} C^\infty(S^1) \longrightarrow 0\,. \tag{19}$$

By construction $\mathcal{A}$ is dense in A. Let us check that it is stable under h.f.c. in A. Let $a \in \mathcal{A}$ be such that $a^{-1} \in A$. Let us show that $a^{-1} \in \mathcal{A}$.

Let ∂_α be the derivation of $\mathcal{A}$ given by,

$$\partial_\alpha \alpha = \alpha\,, \ \partial_\alpha \beta = 0\,. \tag{20}$$

The one parameter group $\exp(it\partial_\alpha)$ of automorphisms of $\mathcal{A}$ is implemented by unitary operators in $\mathcal{H}$ (cf.(49) below) and extends to A. Moreover $\mathcal{A}$ is dense in the domain,

$$\mathrm{Dom}\,\partial_\alpha^j = \{x \in A\,;\; \partial_\alpha^j x \in A\} \tag{21}$$

in the graph norm.

Since $a^{-1} \in \mathrm{Dom}\,\partial_\alpha^j$ we can, given any $\varepsilon > 0$, find $b \in \mathcal{A}$ such that,

$$\|\partial_\alpha^j (b - a^{-1})\| < \varepsilon \qquad j = 0, 1, 2\,. \tag{22}$$

Thus, given $\varepsilon > 0$, we can find $b \in \mathcal{A}$ such that, with $x = ab$,

$$\|\partial_\alpha^j (x - 1)\| < \varepsilon \qquad j = 0, 1, 2\,. \tag{23}$$

For ε small enough it follows that if we let $\sigma(x^{-1})^\wedge_n$ be the Fourier coefficients of $\sigma(x^{-1})$,

$$c = \sum_{n \geq 0} \sigma(x^{-1})_n^{\wedge} \alpha^n + \sum_{n < 0} \sigma(x^{-1})_n^{\wedge} \alpha^{*-n} \tag{24}$$

is an element of $\mathcal{A}$, invertible in A, such that,

$$\sigma(c) = \sigma(x^{-1}). \tag{25}$$

(Since one controls $n^2 \sigma(x^{-1})_n^{\wedge}$ from $\|\partial_\alpha^j (x^{-1} - 1)\|$.)

Thus $\sigma(xc) = 1$ and since xc is invertible in A (by (23), (25)) and $1 - xc \in \mathcal{J}$ the stability of $\mathcal{J}$ under h.f.c. shows that $(xc)^{-1} = y \in \mathcal{A}$. Then $abcy = 1$ and $a^{-1} = bcy \in \mathcal{A}$.

Our next result determines the dimension spectrum of the spectral triple $(\mathcal{A}, \mathcal{H}, D)$ defined above in (6),

Theorem 7. *The dimension spectrum of the spectral triple* $(\mathcal{A}, \mathcal{H}, D)$ *is simple and equal to* $\{1, 2, 3\} \subset \mathbb{C}$.

Thus we let $\mathcal{B}$ be the algebra generated by the

$$\delta^k(a), \ \delta^k([D, a]), \ a \in \mathcal{A}, \ k \in \mathbb{N} \tag{26}$$

where δ is the unbounded derivation of $\mathcal{L}(\mathcal{H})$ given by the commutator with $|D|$,

$$\delta(T) = |D| T - T |D|. \tag{27}$$

(It is part of the statement that the elements in (26) are in the domain of δ^k.)

For $b \in \mathcal{B}$ we consider the function,

$$\zeta_b(s) = \text{Trace}\,(b\,|D|^{-s}) \tag{28}$$

where we take care of the eigenvalue $D = 0$ by replacing $|D|$ by an arbitrary $\varepsilon > 0$ there. The statement of the theorem is that all the functions $\zeta_b(s)$ which are a priori only defined for $\text{Re}\,(s) > 3$, do extend to meromorphic function on $\mathbb{C}$ and only admit *simple* poles at the 3 points $\{1, 2, 3\} \subset \mathbb{C}$.

To prove it we shall first describe the algebra $\mathcal{B}$. We let,

$$F = \text{Sign}\,D \tag{29}$$

so that $F = 2P - 1$ where P is the orthogonal projection on the subset $\{i = n\}$ of the basis. Concerning the generator α one has,

$$\delta(\alpha) = -\alpha, \ \delta(\alpha^*) = \alpha^*, \ [F, \alpha] = 0. \tag{30}$$

It follows that $[D, \alpha^*] = F\delta(\alpha^*) = F\alpha^* = \alpha^* F$. Thus $\alpha\,[D, \alpha^*] = \alpha\alpha^* F = F$,

$$F = \alpha\,[D, \alpha^*]. \tag{31}$$

This shows that $F \in \mathcal{B}$.

Concerning the generator β one has

$$\delta(\beta) = \beta K \,, \ \delta(\beta^*) = -K\beta^* \tag{32}$$

where K is the multiplication operator,

$$K(e_{ij}^{(n)}) = k(n, i, j)\, e_{ij}^{(n)} \tag{33}$$

with

$$k(n, i, j) = 0 \qquad \text{unless } i = -n \ \text{ or } \ j = -n \tag{34}$$

$$k(n, -n, j) = -1 \qquad \text{if } j \neq -n$$

$$k(n, i, -n) = 1\,.$$

Thus the support of K is $e = \beta^*\beta$ and,

$$K^2 = e\,. \tag{35}$$

We let $e_0 = \frac{1}{2}\left(K - \beta K \beta^*\right)$. It is the orthogonal projection on the subset of the basis $\{i = -n$ and $j = -n\}$. For each m one lets,

$$e_m = \beta^m e_0 \beta^{*m} \tag{36}$$

and the e_m are pairwise orthogonal projections such that,

$$\sum_{m \in \mathbb{Z}} e_m = e\,. \tag{37}$$

We let $\mathcal{L}$ be the algebra of double sums with rapid decay,

$$\mathcal{L} = \left\{\sum \lambda_{n,m}\, \beta^n e_m \,;\, \lambda \in \mathcal{S}\right\} \tag{38}$$

(where $\beta^{-\ell} = \beta^{*\ell}$ for $\ell > 0$).
One has $\delta(e_n) = 0$, $[K, \beta] = 2e_0\beta$, $Ke_m = \operatorname{sign}(m)e_m$ and using (32),

$$\delta(\beta^n) = n\beta^n K \qquad \text{modulo } \mathcal{L}\,. \tag{39}$$

Thus $\mathcal{L}$ is invariant (globally) under δ. Also for any $f \in C^\infty(S^1)$ one has

$$[K, f(\beta)] \in \mathcal{L} \tag{40}$$

and the algebra B_0,

$$B_0 = \{f_0(\beta) + f_1(\beta)\, K + h \,;\, f_j \in C^\infty(S^1),\ h \in \mathcal{L}\} \tag{41}$$

is stable by the derivation δ.
A similar result holds if we further adjoin the operator

$$F_1 = eF = Fe\,. \tag{42}$$

Indeed $e_0' = \frac{1}{2}(F_1 - \beta F_1 \beta^*)$ is the projection on the element $\{e_{0,0}^{(0)}\}$ of the basis and the $e_n' = \beta^n e_0' \beta^{*n}$ are pairwise orthogonal projections on the one dimensional subspaces spanned for $n > 0$ by $e_{n/2,-n/2}^{(n/2)}$ and for $n < 0$, $n = -k$, by $e_{-k/2,k/2}^{(k/2)}$.

We let,

$$\mathcal{L}' = \left\{ \sum \lambda_{n,m} \beta^n e_m' \, ; \, \lambda \in \mathcal{S} \right\}. \tag{43}$$

One has,

$$[F, f(\beta)] \in \mathcal{L}' \qquad \forall f \in C^\infty(S^1) \tag{44}$$

and $F_1 \mathcal{L}' = \mathcal{L}' F_1 = \mathcal{L}'$.

Also $e_n' \leq e_n$ for each n so that $\mathcal{L}\mathcal{L}' \subset \mathcal{L}'$ and $\mathcal{L}'\mathcal{L} \subset \mathcal{L}'$ which shows that the sum,

$$\mathcal{L}'' = \mathcal{L} + \mathcal{L}' \tag{45}$$

is an algebra.

Thus the algebra generated in $e\mathcal{H}$ by the $\delta^k(f)$, $f \in C^\infty(S^1)$ and F_1, is contained in the algebra B_1

$$B_1 = \{ f_0 + f_1 K + f_2 F_1 + h \, ; \, f_j \in C^\infty(S^1), \, h \in \mathcal{L}'' \}. \tag{46}$$

Note that $F_1 K = 1 + 2(K - F_1)$ so that we do not need terms in $F_1 K$.

We then let B be the algebra of double sums,

$$B = \left\{ \sum \alpha^{*k} b_{k\ell} \alpha^\ell + A_0 + A_1 F \right\} \tag{47}$$

where $b_{k\ell} \in B_1$ and the sequence $(b_{k\ell})$ is of rapid decay while A_0, A_1 are sums of rapid decay of the form,

$$A = \sum_{\ell \geq 0} a_\ell \alpha^\ell + \sum_{k > 0} a_{-k} \alpha^{*k}.$$

Since F commutes with α and α^* it commutes with A_j. Thus one checks that B is an algebra, that it is stable under δ and contains both F and $\mathcal{A}$, thus it contains $\mathcal{B}$.

Let then $b \in B$ and consider the function,

$$\zeta_b(s) = \mathrm{Trace}\,(b\,|D|^{-s}) \tag{48}$$

which is well defined for $\mathrm{Re}\,(s) > 3$.

There is a natural bigrading corresponding to the degrees in α and β. It is implemented by the following action of $\mathbb{T}^2$ in $\mathcal{H}$,

$$V(u,v)\,e_{k,\ell}^{(n)} = \exp i(-u(k+\ell) + v(k-\ell))\,e_{k,\ell}^{(n)}. \tag{49}$$

Note that both $k + \ell$ and $k - \ell$ are integers so that one gets an action of $\mathbb{T}^2$.

The indices k, ℓ are transformed to $k - \frac{1}{2}, \ell - \frac{1}{2}$ by α, so that

$$V(u,v)\,\alpha(e_{k,\ell}^{(n)}) = \exp i(-u(k+\ell-1) + v(k-\ell))\,\alpha(e_{k,\ell}^{(n)})$$

and we get,

$$V(u,v)\,\alpha\,V(-u,-v) = e^{iu}\alpha. \tag{50}$$

The indices k, ℓ are transformed to $k+\frac{1}{2}, \ell-\frac{1}{2}$ by β and,

$$V(u,v)\,\beta(e_{k,\ell}^{(n)}) = \exp i(-u(k+\ell) + v(k-\ell+1))\,\beta(e_{k,\ell}^{(n)})$$

so that,

$$V(u,v)\,\beta\,V(-u,-v) = e^{iv}\beta. \tag{51}$$

Moreover, since V is a multiplication operator it commutes with $|D|$, D, and $F = 2P - 1$.

Using the restriction of this bigrading to B (which gives bidegree $(0,0)$ for diagonal operators, $(1,0)$ for α and $(0,1)$ for β) one checks that homogeneous elements of bidegree $\neq (0,0)$ satisfy $\zeta_b(s) \equiv 0$, thus one can assume that b is of bidegree $(0,0)$. Any $b \in B^{(0,0)}$ is of the form,

$$b = \sum \alpha^{*k} b_k \alpha^k + a_0 + a_1 F \tag{52}$$

where a_0, a_1 are *scalars* and (b_k) is a sequence of rapid decay with $b_k \in B_1^{(0,0)}$. Elements c of $B_1^{(0,0)}$ are of the form,

$$c = \lambda_0 + \lambda_1 K + \lambda_2 F_1 + h \tag{53}$$

where λ_j are *scalars* and $h \in \mathcal{L}''^{(0,0)}$. Finally elements of $\mathcal{L}''^{(0,0)}$ are of the form,

$$h = \sum h_n e_n + \sum h'_m e'_m \tag{54}$$

where (h_n) and (h'_m) are *scalar* sequences of rapid decay. The equality,

$$|D|^z \alpha^{*k} = \alpha^{*k}(|D|+k)^z \qquad z \in \mathbb{C} \tag{55}$$

is checked directly ($k \geq 0$).

Using $\alpha^k \alpha^{*k} = 1$ it follows that with b as in (52),

$$\text{Trace}\,(b\,|D|^{-s}) = \text{Trace}\,((a_0 + a_1 F)\,|D|^{-s}) + \sum_{k \geq 0} \text{Trace}\,(b_k(|D|+k)^{-s}). \tag{56}$$

Now for h as in (54) one has

$$\text{Trace}\,(h\,|D|^{-s}) = \sum h_n \text{Trace}\,(e_n\,|D|^{-s}) + \sum h'_m \text{Trace}\,(e'_m\,|D|^{-s}).$$

Moreover

$$\text{Trace}\,(e_n\,|D|^{-s}) = \sum_{\ell=0}^{\infty} \frac{1}{(|n|+\ell)^s} = \zeta(s) - \left(\sum_{0}^{|n|-1} \frac{1}{r^s}\right).$$

But $\sum h_n \left(\sum_{0}^{|n|-1} \frac{1}{r^s}\right) = \rho_1(s)$ is a holomorphic function of $s \in \mathbb{C}$ and similarly,

since $\text{Trace}\,(e'_m\,|D|^{-s}) = \frac{1}{|m|^s}$, the function $\rho_2(s) = \sum h'_m \text{Trace}\,(e'_m\,|D|^{-s})$ is holomorphic in $s \in \mathbb{C}$. Thus modulo holomorphic functions one has,

$$\text{Trace}\,(h\,|D|^{-s}) \sim \left(\sum h_n\right)\zeta(s). \tag{57}$$

Next,

$$\text{Trace}\,(e\,|D|^{-s}) = \sum_{0}^{\infty} \frac{2n+1}{n^s} = 2\zeta(s-1) + \zeta(s) + \varepsilon^{-s} \tag{58}$$

$$\text{Trace}\,(K\,|D|^{-s}) = \sum_{n\in\mathbb{Z}}\sum_{\ell\geq 0} \frac{\text{sign}(n)}{(|n|+\ell)^s} = \zeta(s) + \varepsilon^{-s}$$

and with $F = 2P - 1$ we also have,

$$\text{Trace}\,(eP\,|D|^{-s}) = \zeta(s) + \varepsilon^{-s}. \tag{59}$$

Thus, with c as in (53) we get,

$$\zeta_c(s) = \lambda\,\zeta(s-1) + \mu\,\zeta(s) + \rho(s) \tag{60}$$

where λ, μ are scalars and ρ is a holomorphic function of $s \in \mathbb{C}$.
A similar result holds for

$$\sum_{k\geq 0} \text{Trace}\,(b_k(|D|+k)^{-s}).$$

For instance one rewrites the double sum

$$\sum h_{n,k}\,\text{Trace}\,(e_m(|D|+k)^{-s}) = \sum_{n,k,\ell} h_{n,k}\frac{1}{(|n|+k+\ell)^s}$$

as

$$\sum_{m}\left(\sum_{|n|+k\leq m} h_{n,k}\right)\frac{1}{m^s} = a\,\zeta(s) + \rho(s)$$

where $a = \sum h_{n,k}$ and ρ is holomorphic in $s \in \mathbb{C}$.
Finally

$$\text{Trace}\,(P\,|D|^{-s}) = \sum_{0}^{\infty} \frac{(n+1)}{n^s} = \zeta(s-1) + \zeta(s)$$

and

$$\text{Trace}\,(|D|^{-s}) = \sum_0^\infty \frac{(n+1)^2}{n^s} = \zeta(s-2) + 2\,\zeta(s-1) + \zeta(s)\,.$$

Thus we conclude that for any $b \in B$ one has

$$\zeta_b(s) = \lambda_3\,\zeta(s-2) + \lambda_2\,\zeta(s-1) + \lambda_1\,\zeta(s) + \rho(s) \tag{61}$$

where the λ_j are scalars and ρ is a holomorphic function of $s \in \mathbb{C}$, thus proving theorem 7.

7 The Local Index Formula for $SU_q(2), (q = 0)$

In this section we shall compute the local index formula for the above spectral triple. Since the dimension spectrum is simple and equal to $\{1, 2, 3\} \subset \mathbb{C}$ the cyclic cocycle given by the local index formula has two components φ_1 and φ_3 of degree 1 and 3 given, up to an overall multiplication by $(2i\pi)^{1/2}$, by

$$\varphi_1(a^0, a^1) = \int a^0[D, a^1]|D|^{-1} - \frac{1}{4} \int a^0 \nabla([D, a^1])\,|D|^{-3} \tag{1}$$

$$+ \frac{1}{8} \int a^0 \nabla^2([D, a^1])\,|D|^{-5}$$

and,

$$\varphi_3(a^0, a^1, a^2, a^3) = \frac{1}{12} \int a^0[D, a^1][D, a^2][D, a^3]\,|D|^{-3}\,. \tag{2}$$

With these notations the cocycle equation is,

$$b\varphi_1 + B\varphi_3 = 0\,. \tag{3}$$

The following formulas define a cyclic cocycle τ_1 on $\mathcal{A}$,

$$\tau_1(\alpha^{*k}, x) = \tau_1(x, \alpha^{*k}) = \tau_1(\alpha^l, x) = \tau_1(x, \alpha^l) = 0, \tag{4}$$

for all integers k, l and any $x \in \mathcal{A}$,

$$\tau_1(\alpha^{*k} f(\beta)\alpha^\ell, \alpha^{*k'} g(\beta)\alpha^{\ell'}) = 0 \tag{5}$$

unless $\ell' = k, \, k' = \ell$ and

$$\tau_1(\alpha^{*k} f(\beta)\alpha^\ell, \alpha^{*\ell} g(\beta)\alpha^k) = \frac{1}{\pi i} \int_{S^1} f \, dg\,.$$

Let φ_0 be the 0-cochain given by $\varphi_0(\alpha^{*k} f(\beta)\alpha^\ell) = 0$ unless $k = \ell$ and,

$$\varphi_0(\alpha^{*k} f(\beta)\alpha^k) = \rho(k) \frac{1}{2\pi} \int_{S^1} f \, d\theta, \tag{6}$$

where $\rho(j) = \frac{2}{3} - j - j^2$. Finally, let φ_2 be the 2-cochain given by the pull back by $\mathfrak{s}$ of the cochain $\frac{-1}{24} \frac{1}{2\pi i} \int f_0 f_1' f_2'' \, d\theta$ on $C^\infty(S^1)$.

Our next task is to prove the following result,

Theorem 8. *The local index formula of the spectral triple $(\mathcal{A}, \mathcal{H}, D)$ is given by the cyclic cocycle τ_1 up to the coboundary of the cochain (φ_0, φ_2).*

The precise equations are,

$$\varphi_1 = \tau_1 + b\varphi_0 + B\varphi_2, \qquad \varphi_3 = b\varphi_2. \tag{7}$$

The proof is a computation but we shall go through it in details in order to get familiar with various ways of computing residues and manipulating "infinitesimals" in the sense of the quantized calculus. In other words our purpose is not concision but rather a leisurly account of the details.

7.1 Restriction to $C^\infty(\beta)$

Let us first concentrate on the restriction of the cocycle φ to the subalgebra $C^\infty(\beta)$ generated by β and β^*. To see the subspace of $\mathcal{H}$ responsible for the non-triviality of that cocycle we follow the action of β on the vectors,

$$\xi_{-n} = e_{-\frac{n}{2}, \frac{n}{2}}^{\left(\frac{n}{2}\right)} \qquad n \geq 0, \, n \in \mathbb{N}. \tag{8}$$

and,

$$\xi_n = e_{n/2, -n/2}^{(n/2)} \qquad n \geq 0, \, n \in \mathbb{N} \tag{9}$$

For $n > 0$, (11) shows that $\beta(\xi_{-n}) = \xi_{-(n-1)} = \xi_{-n+1}$, with $\beta(\xi_{-1}) = e_{0,0}^{(0)} = \xi_0$. Next, $\beta(\xi_0) = \beta(e_{0,0}^{(0)}) = -e_{(1/2), -1/2}^{(1/2)} = -\xi_1$, and for $n > 0$ (12) shows that $\beta(\xi_n) = -\xi_{n+1}$. Thus,

$$\beta(\xi_n) = -\mathrm{sign}(n) \, \xi_{n+1} \qquad (\mathrm{sign}(0) = 1) \tag{10}$$

We let $\ell^2(\mathbb{Z}) = \mathcal{H}_0 \subset \mathcal{H}$ be the subspace of $\mathcal{H}$ spanned by the ξ_n and rewrite the above equality as,

$$\beta = -UH \text{ on } \ell^2(\mathbb{Z}) \subset \mathcal{H} \tag{11}$$

where H is the sign operator and U the shift,

$$U\xi_n = \xi_{n+1}. \tag{12}$$

The operator D also restricts to the subspace $\ell^2(\mathbb{Z}) = \mathcal{H}_0 \subset \mathcal{H}$ and its restriction D_0 is given by,

$$D_0\xi_n = \mathrm{sign}(n) \, |n| \, \xi_n = n \, \xi_n \qquad \forall n. \tag{13}$$

The unitary $W = e^{i\frac{\pi}{2}(|D_0| - D_0)}$ commutes with D_0 and conjugates U to $-UH$,

$$WUW^* = -UH \tag{14}$$

Thus the triple $(\beta, \mathcal{H}_0, D_0)$ is isomorphic to,

$$\left(e^{i\theta}, \; L^2(S^1), \; (-i)\frac{\partial}{\partial\theta} \right). \tag{15}$$

In particular the index and cyclic cohomology pairings with the restriction to $\mathcal{H}_0$ are non trivial and we control,

$$\text{Res}_{s=1} \, \text{Trace}_{\mathcal{H}_0} \left(\beta^*[D_0, \beta] \, |D_0|^{-s} \right) = 2. \tag{16}$$

This however does not suffice to get the non-triviality of the restriction of φ to $C^\infty(\beta)$ since we need to control the residues on $e\mathcal{H}$ where, as above e is the support of β. To see what happens we shall conjugate the restriction of both β and D to the orthogonal complement of $\mathcal{H}_0$ in $e\mathcal{H}$ with a very simple triple. Let us define for each $k \in \mathbb{N}$ the vectors,

$$\xi_{-n}^{(k)} = e_{-\left(\frac{k+n}{2}\right), \frac{n-k}{2}}^{\left(\frac{k+n}{2}\right)} \qquad n \geq 0 \tag{17}$$

$$\xi_n^{(k)} = e_{\frac{n-k}{2}, -\left(\frac{k+n}{2}\right)}^{\frac{k+n}{2}} \qquad n \geq 0$$

so that $\xi_0^{(k)} = e_{-k/2, -k/2}^{(k/2)}$.

For $n > 0$ one has

$$\beta(\xi_{-n}^{(k)}) = \beta\left(e_{-\left(\frac{k+n}{2}\right), \frac{n-k}{2}}^{\left(\frac{k+n}{2}\right)} \right) = e_{-\left(\frac{k+n-1}{2}\right), \frac{n-1-k}{2}}^{\left(\frac{k+n-1}{2}\right)} = \xi_{-(n-1)}^{(k)} = \xi_{-n+1}^{(k)}.$$

For $n = 0$,

$$\beta(\xi_0^{(k)}) = \beta(e_{-k/2, -k/2}^{(k/2)}) = -e_{-\frac{k}{2}+\frac{1}{2}, -\left(\frac{k+1}{2}\right)}^{\left(\frac{k+1}{2}\right)} = \xi_1^{(k)}$$

and,

$$\beta(\xi_n^{(k)}) = \beta\left(e_{\frac{n-k}{2}, -\left(\frac{n+k}{2}\right)}^{\left(\frac{k+n}{2}\right)} \right) = -e_{\frac{n+1-k}{2}, -\left(\frac{n+1+k}{2}\right)}^{\left(\frac{k+n+1}{2}\right)} = -\xi_{n+1}^{(k)}.$$

Thus, as in (10) we have,

$$\beta(\xi_n^{(k)}) = -\text{sign}\,(n)\,\xi_{n+1}^{(k)}. \tag{18}$$

Now $\beta(e_{ij}^{(m)}) = 0$ unless $i = -m$ or $j = -m$ and for any $m \in \frac{1}{2}\mathbb{N}$ the vectors $e_{-m,j}^{(m)}$ and $e_{i,(-m)}^{(m)}$ are of the form $\xi_n^{(k)}$. Indeed in the first case one takes $n = m+j$, $k = m - j$ which are both in $\mathbb{N}$, and $\xi_{-n}^{(k)} = e_{-m,j}^{(m)}$. In the second case $n = m + i$, $k = m - i$ are both in $\mathbb{N}$ and $\xi_n^{(k)} = e_{i,-m}^{(m)}$.

We then let $\mathcal{H}_k$ be the span of the $\xi_n^{(k)}$, $n \in \mathbb{Z}$ and

$$\mathcal{H}' = \bigoplus_{k \geq 1} \mathcal{H}_k = \ell^2(\mathbb{Z}) \otimes \ell^2(\mathbb{N}^+). \tag{19}$$

The operator D restricts to $\mathcal{H}'$ and is given there by,

$$D' = |D_0| \otimes 1 + 1 \otimes N \tag{20}$$

where N is the number operator $N\varepsilon_k = k\varepsilon_k$.

Also β is $-UH \otimes 1$ and we can conjugate it as in (14) back to $U \otimes 1$.

Thus the triple $(\beta, \mathcal{H}', D')$ is isomorphic to

$$(U \otimes 1, \ \ell^2(\mathbb{Z}) \otimes \ell^2(\mathbb{N}^+), \ |D_0| \otimes 1 + 1 \otimes N) \tag{21}$$

The metric dimension is 2 in this situation, and the contribution φ' of $\mathcal{H}'$ to the restriction of φ to $C^\infty(\beta)$ only has a one dimensional component φ_1' which involves the two terms,

$$\varphi_1'(a^0, a^1) = \int a^0 [D', a^1] |D'|^{-1} - \frac{1}{4} \int a^0 \nabla([D', a^1]) |D'|^{-3}, \ \ a^j \in C^\infty(\beta) \tag{22}$$

Since D' is positive, it is K-homologically trivial and the above cocycle must vanish identically on $C^\infty(\beta)$. As we shall see this vanishing holds because of the precise ratio $-\frac{1}{4}$ of the coefficients in the local index formula.

To see this, we need to compute the poles and residues of functions of the form Trace $((T \otimes 1) |D'|^{-s})$ for operators T in $\ell^2(\mathbb{Z})$. For that purpose it is most efficient to use the well known relation between residues of zeta functions and asymptotic expansions of related theta functions. More specifically, for $\lambda > 0$ and $Re(s) > 0$, the equality,

$$\lambda^{-s} = \frac{1}{\Gamma(s)} \int_0^\infty e^{-t\lambda} t^s \frac{dt}{t} \tag{23}$$

gives

$$\text{Trace}\,((T \otimes 1) |D'|^{-s}) = \frac{1}{\Gamma(s)} \int_0^\infty \text{Trace}\,((T \otimes 1) e^{-tD'}) t^s \frac{dt}{t}$$

$$= \frac{1}{\Gamma(s)} \int_0^\infty \text{Trace}\,(T e^{-t|D_0|}) \left(\frac{1}{e^t - 1}\right) t^s \frac{dt}{t}.$$

Thus if we assume that one has an expansion of the form Trace $(Te^{-t|D_0|}) = \frac{a}{t} + b + ct + 0(t^2)$ one gets, using

$$\frac{1}{e^t - 1} = \frac{1}{t} - \frac{1}{2} + \frac{t}{12} + 0(t^2) \tag{24}$$

the equality modulo holomorphic functions of s, $Re(s) > 0$,

$$\text{Trace}\,((T \otimes 1)\,|D'|^{-s}) \sim \frac{1}{\Gamma(s)} \int_0^1 \varphi(t)\, t^s\, \frac{dt}{t},$$

where,

$$\varphi(t) = \frac{a}{t^2} + \frac{\left(b - \frac{a}{2}\right)}{t} + \frac{a}{12} - \frac{b}{2} + c\,.$$

One has $\int_0^1 t^\alpha \frac{dt}{t} = \frac{1}{\alpha}$, thus one gets 2 poles $s = 2$ and $s = 1$, and the expansion,

$$\text{Trace}\,((T \otimes 1)\,|D'|^{-s}) \sim \frac{a}{\Gamma(s)(s-2)} + \left(b - \frac{a}{2}\right)\frac{1}{\Gamma(s)(s-1)} + \cdots$$

so that,

$$\text{Res}_{s=2}(\text{Trace}\,((T \otimes 1)|D'|^{-s})) = a \tag{25}$$

$$\text{Res}_{s=1}(\text{Trace}\,((T \otimes 1)|D'|^{-s})) = b - \frac{a}{2} \tag{26}$$

Let us compute $\varphi'_1(\beta^*, \beta)$. The first term in (22) is $\int \beta^*[D', \beta]|D'|^{-1}$. Thus we take $T = U^*[|D_0|, U]$. One has $[|D_0|, U] = UH$, $T = H$ and

$$\text{Trace}\,(Te^{-t|D_0|}) = \sum_{\mathbb{Z}} \text{sign}\,(n)\, e^{-t|n|} = -\sum_1^\infty e^{-tk} + 1 + \sum_1^\infty e^{-tk} = 1\,. \tag{27}$$

Thus in that case $a = 0$, $b = 1$ and,

$$\int \beta^*[D', \beta]|D'|^{-1} = 1\,. \tag{28}$$

The second term in (22) is $\int \beta^* \nabla([D', \beta])\,|D'|^{-3}$ where ∇ is the commutator with D'^2. If we let as above δ be the commutator with $|D'|$, one has

$$\nabla T = \delta(T)\,|D'| + |D'|\,\delta(T)\,.$$

Thus, permuting $|D'|$ modulo operators of lower order, we get,

$$\int \beta^* \nabla([D', \beta])\,|D'|^{-3} = 2 \int \beta^* \delta([D', \beta])\,|D'|^{-2} \tag{29}$$

To compute the r. h. s. we take $T = U^* \delta^2(U)$ and look at the residue at $s = 2$. One has $\delta(U) = UH$, $\delta^2(U) = (UH)H = U$ and $T = 1$. Thus

$$\text{Trace}\,(Te^{-t|D_0|}) = \sum_{\mathbb{Z}} e^{-t|n|} = 1 + 2\sum_1^\infty e^{-tn}$$

$$= 1 + \frac{2}{e^t - 1} \sim \frac{2}{t} + 0(t)$$

so that $a = 2$, $b = 0$. Thus we get,

$$\oint \beta^* \delta([D', \beta]) \, |D'|^{-2} = \mathrm{Res}_{s=2}(\mathrm{Trace}\,((U^* \delta^2(U) \otimes 1) \, |D'|^{-s}) = 2 \qquad (30)$$

Thus we get,

$$\oint_{\mathcal{H}'} \beta^* \, [D, \beta] \, |D|^{-1} = 1 \, , \quad \oint_{\mathcal{H}'} \beta^* \, \nabla([D, \beta]) \, |D|^{-3} = 4 \qquad (31)$$

and $\varphi_1'(\beta^*, \beta) = 0$ precisely because of the coefficient $\frac{-1}{4}$ in (22).

One proceeds similarly to compute $\varphi_1'(\beta, \beta^*)$. The first term in (22) comes from $T = U\delta(U^*)$. In the canonical basis ε_n of $\ell^2(\mathbb{Z})$ one has $U\delta(U^*)\,\varepsilon_n = (|n-1| - |n|)\,\varepsilon_n = -\mathrm{sign}\,(n-1)\,\varepsilon_n$ and $\mathrm{Trace}\,(Te^{-t|D_0|}) = \sum(-\mathrm{sign}\,(n-1))\,e^{-t|n|} = 1$. Thus,

$$\oint \beta \, [D', \beta^*] \, |D'|^{-1} = 1 \, . \qquad (32)$$

Since $U\delta^2(U^*) = 1$ the computation of the second term of (22) is the same as above and we get,

$$\oint_{\mathcal{H}'} \beta \, [D, \beta^*] \, |D|^{-1} = 1 \, , \quad \oint_{\mathcal{H}'} \beta \, \nabla([D, \beta^*]) \, |D|^{-3} = 4 \qquad (33)$$

so that $\varphi_1'(\beta, \beta^*) = 0$.

Let us now take $n > 0$ and compute $\varphi_1'(\beta^{*n}, \beta^n)$. The first term of (22) involves $T = U^{*n}\delta(U^n)$. One has,

$$\delta(U^n) = UHU^{n-1} + U^2HU^{n-2} + \cdots + U^jHU^{n-j} + \cdots + U^nH = \sum_{j=1}^{n} U^jHU^{n-j} \, ,$$

$$U^{*n}\delta(U^n) = (U^*)^{n-1}HU^{n-1} + \cdots + H \, .$$

One has $U^{*k}HU^k\varepsilon_\ell = \mathrm{sign}(k+\ell)\,\varepsilon_\ell$ and,

$$\mathrm{Trace}\,(U^{*k}HU^k e^{-t|D_0|}) = \mathrm{Trace}\,(He^{-t|D_0|}) + 2\sum_1^k e^{-t|j|} \sim 1 + 2k + 0(t) \, .$$

Thus,

$$\mathrm{Trace}\,(U^{*n}\delta(U^n)\,e^{-t|D_0|}) \sim \sum_{j=1}^{n} (2(n-j)+1) + 0(t) = \sum_{k=0}^{n-1} (2k+1)! + 0(t) = n^2 + 0(t) \, ,$$

and $(n > 0)$

$$\oint \beta^{*n} \, [D', \beta^n] \, |D'|^{-1} = n^2 \, , \qquad (34)$$

Now modulo finite rank operators one has $\delta(U^n) = nU^nH$ and $\delta^2(U^n) = n^2U^n$, thus as above,

$$\oint \beta^{*n} \, \nabla([D', \beta^n]) \, |D'|^{-3} = 4n^2, \tag{35}$$

so that $\varphi'_1(\beta^{*n}, \beta^n) = 0$.

Finally the computation of $\varphi'_1(\beta^n, \beta^{*n})$ involves $T = U^n \delta(U^{*n})$. One has

$$U^n \delta(U^{*n}) = -\sum_{k=1}^n U^k H U^{*k},$$

and $U^k H U^{*k} \varepsilon_\ell = \mathrm{sign}\,(\ell - k)\, \varepsilon_\ell$ so that,

$$\mathrm{Trace}\,(U^k H U^{*k} e^{-t|D_0|}) - \mathrm{Trace}\,(H e^{-t|D_0|}) = -2 \sum_0^{k-1} e^{-tj}$$

and,

$$\mathrm{Trace}\,(U^k H U^{*k} e^{-t|D_0|}) \sim 1 - 2k + 0(t).$$

Thus,

$$\mathrm{Trace}\,(U^n \delta(U^{*n})\, e^{-t|D_0|}) = -\sum_{k=1}^n \mathrm{Trace}\,(U^k H U^{*k} e^{-t|D_0|})$$

$$\sim -\sum_{k=1}^n (1 - 2k) + 0(t) \sim n^2 + 0(t),$$

and,

$$\oint \beta^n \, [D', \beta^{*n}] \, |D'|^{-1} = n^2. \tag{36}$$

Also as above,

$$\oint \beta^n \, \nabla([D', \beta^{*n}]) \, |D'|^{-3} = 4n^2. \tag{37}$$

so that we get the required vanishing,

$$\varphi'_1 = 0$$

What is instructive in the above computation is that this vanishing which is required by theorem 6, involves because of the factorisation (21) terms such as "Trace(H)" which appear in equation (27) and are similar to eta-invariants.
We have thus shown that $\varphi_1 = \tau_1$ on $C^\infty(\beta)$, or equivalently that,

$$\oint \beta^{-n} \, [D, \beta^n] \, |D|^{-1} - \frac{1}{4} \oint \beta^{-n} \, \nabla([D, \beta^n]) \, |D|^{-3} = 2n. \tag{38}$$

7.2 Restriction to the Ideal $\mathcal{J}$

Let us extend this computation to $\mathcal{J} = \text{Ker } \mathfrak{s}$. The component φ_3 vanishes on $\mathcal{J}$ and we just need to compute $\varphi_1 = \varphi_1^{(0)} - \frac{1}{4} \varphi_1^{(1)} + \frac{1}{8} \varphi_1^{(2)}$. We begin by $\varphi_1^{(0)}(\mu', \mu) = \mathop{\rlap{$\int$}\mkern4mu}\nolimits \mu'[D, \mu]|D|^{-1}$, and need only consider the case where $\mu = \alpha^{*k} \beta^n \alpha^\ell$ and $\mu' = \alpha^{*k'} \beta^{n'} \alpha^{\ell'}$ are monomials. (As above $\beta^{-n} = \beta^{*n}$ and $\beta^0 = e$).

With $F = 2P - 1$ one has $[F, \alpha] = 0$ and

$$[D, \alpha^\ell] = -\ell F \alpha^\ell = -\ell \alpha^\ell F \tag{39}$$

and

$$[D, \alpha^{*k}] = kF\alpha^{*k} = k\alpha^{*k} F. \tag{40}$$

Thus,

$$[D, \mu] = [D, \alpha^{*k}] \beta^n \alpha^\ell + \alpha^{*k}[D, \beta^n] \alpha^\ell + \alpha^{*k} \beta^n [D, \alpha^\ell]$$
$$= k\alpha^{*k} F \beta^n \alpha^\ell + \alpha^{*k}[D, \beta^n] \alpha^\ell - \ell \alpha^{*k} \beta^n F \alpha^\ell,$$

so that

$$[D, \mu] = \alpha^{*k}(kF\beta^n + [D, \beta^n] - \ell \beta^n F) \alpha^\ell \tag{41}$$

The bigrading (49) shows that $\mathop{\rlap{$\int$}\mkern4mu}\nolimits \mu'[D, \mu]|D|^{-1}$ vanishes unless both total degrees are 0, i.e.

$$\ell + \ell' - k - k' = 0, \; n' = -n. \tag{42}$$

The element $X = (kF\beta^n + [D, \beta^n] - \ell \beta^n F)$ satisfies $eXe = X$ so that the product

$$\mu'[D, \mu] = \alpha^{*k'} \beta^{n'} \alpha^{\ell'} \alpha^{*k} X \alpha^\ell,$$

vanishes unless $\ell' = k$. Combining with (42) we get $\ell = k'$, and can assume that $\mu' = \alpha^{*\ell} \beta^{*n} \alpha^k$. Then $\mu'[D, \mu] = \alpha^{*\ell} \beta^{*n} X \alpha^\ell$ so that we just need to compute,

$$\mathop{\rlap{$\int$}\mkern4mu}\nolimits \alpha^{*\ell} \beta^{*n} X \alpha^\ell |D|^{-1}.$$

Now by (55) one has,

$$|D|^{-1} \alpha^{*\ell} = \alpha^{*\ell}|D|^{-1} - \ell \alpha^{*\ell}|D|^{-2} + 0(|D|^{-3}) \tag{43}$$

Thus,

$$\mathop{\rlap{$\int$}\mkern4mu}\nolimits \mu'[D, \mu]|D|^{-1} = \mathop{\rlap{$\int$}\mkern4mu}\nolimits \beta^{*n} X |D|^{-1} - \ell \mathop{\rlap{$\int$}\mkern4mu}\nolimits \beta^{*n} X |D|^{-2} \tag{44}$$

where

$$X = (kF\beta^n + [D, \beta^n] - \ell \beta^n F).$$

Note that α, α^* have now disappeared so that we can compute using the subspaces $\mathcal{H}_0$ and $\mathcal{H}'$ of $e\mathcal{H}$. Note also that on $\mathcal{H}'$ one has $F = -1$ since $P = 0$. Only $\mathcal{H}'$ matters for $\mathop{\rlap{$\int$}\mkern4mu}\nolimits \beta^{*n} X |D|^{-2}$. One has $\mathop{\rlap{$\int$}\mkern4mu}\nolimits \beta^{*n} kF\beta^n |D|^{-2} = -k \mathop{\rlap{$\int$}\mkern4mu}\nolimits_{\mathcal{H}'} |D|^{-2}$ since

$F \sim -1$, and $\fint \beta^{*n}(-\ell\beta^n F)|D|^{-2} = \ell \fint_{\mathcal{H}'}|D|^{-2}$. Also $\fint \beta^{*n}[D, \beta^n]|D|^{-2} = 0$ since $\fint_{L^2(S^1)} U^{-n}[|D_0|, U^n]|D_0|^{-1} = 0$. Thus,

$$\fint \beta^{*n} X |D|^{-2} = (\ell - k) \fint_{\mathcal{H}'} |D|^{-2}. \tag{45}$$

One has

$$\fint \beta^{*n} X |D|^{-1} = \fint \beta^{*n} k F \beta^n |D|^{-1} + \fint \beta^{*n}[D, \beta^n]|D|^{-1} - \ell \fint e F |D|^{-1}$$

where the $\fint$ are on $\mathcal{H}' \oplus \mathcal{H}_0$. One has $\fint_{\mathcal{H}_0} \beta^{*n} F \beta^n |D|^{-1} = 0$ and $\fint_{\mathcal{H}_0} F |D|^{-1} = 0$. On $\mathcal{H}'$ one has $F = -1$, thus,

$$\fint k \beta^{*n} F \beta^n |D|^{-1} = -k \fint_{\mathcal{H}'} |D|^{-1} \tag{46}$$

and

$$-\ell \fint e F |D|^{-1} = \ell \fint_{\mathcal{H}'} |D|^{-1} \tag{47}$$

so that,

$$\fint \beta^{*n} X |D|^{-1} = \fint \beta^{*n}[D, \beta^n]|D|^{-1} + (\ell - k) \fint_{\mathcal{H}'} |D|^{-1}. \tag{48}$$

We now need to compute $-\frac{1}{4} \fint \mu' \nabla([D, \mu])|D|^{-3}$ with μ, μ' monomials in $\mathcal{J}$ as above.

Since $|D|^{-2}$ is order 1 on $\mathcal{J}$ we can replace the above by $-\frac{1}{2} \fint \mu' \delta([D, \mu])|D|^{-2}$ where $\delta(x) = [|D|, x]$. Moreover, using (41),

$$\delta([D, \mu]) = \delta(\alpha^{*k} X \alpha^\ell) = k \alpha^{*k} X \alpha^\ell + \alpha^{*k} \delta(X) \alpha^\ell - \ell \alpha^{*k} X \alpha^\ell,$$

$$\delta([D, \mu]) = (k - \ell) \alpha^{*k} X \alpha^\ell + \alpha^{*k} \delta(X) \alpha^\ell \tag{49}$$

with $X = k F \beta^n + [D, \beta^n] - \ell \beta^n F$.

As above for $\varphi_1^{(0)}$ we get that $\varphi_1^{(1)}$ vanishes unless $\ell' = k$, $k' = \ell$, $n' = -n$ so that $\mu' = \alpha^{*\ell} \beta^{*n} \alpha^k$ and we can replace $\mu' \delta([D, \mu])|D|^{-2}$ by

$$(k - \ell) \beta^{*n} X |D|^{-2} + \beta^{*n} \delta(X)|D|^{-2}. \tag{50}$$

Now by (45),

$$\fint \beta^{*n} X |D|^{-2} = (\ell - k) \fint_{\mathcal{H}'} |D|^{-2}.$$

Moreover one has $\fint \beta^{*n} \delta(F \beta^n)|D|^{-2} = \fint \beta^{*n} \delta(\beta^n F)|D|^{-2} = 0$ since only the $\fint$ on $\mathcal{H}'$ matters and $\fint_{L^2(S^1)} U^{-n}[|D_0|, U^n]|D_0|^{-1} = 0$. Thus,

$$-\frac{1}{4} \fint \mu' \nabla([D, \mu])|D|^{-3} = -\frac{1}{2} (k - \ell)(\ell - k) \fint_{\mathcal{H}'} |D|^{-2}$$

$$-\frac{1}{2} \fint \beta^{*n} \delta([D, \beta^n])|D|^{-2}. \tag{51}$$

Since $\varphi_1^{(2)} = 0$ on $\mathcal{J}$ we get,

$$\varphi_1(\mu',\mu) = \int\!\!\!\!\!\!\!\!-\; \beta^{*n}[\dot{D},\beta^n]\,|D|^{-1} + (\ell-k)\int_{\mathcal{H}'}|D|^{-1} - \ell\,(\ell-k)\int_{\mathcal{H}'}|D|^{-2}$$
$$-\frac{1}{2}\,(k-\ell)(\ell-k)\int_{\mathcal{H}'}|D|^{-2} - \frac{1}{2}\int\!\!\!\!\!\!\!\!-\;\beta^{*n}\delta([D,\beta^n])\,|D|^{-2}\,.$$

Now by (38),

$$\int\!\!\!\!\!\!\!\!-\;\beta^{*n}[D,\beta^n]\,|D|^{-1} - \frac{1}{2}\int\!\!\!\!\!\!\!\!-\;\beta^{*n}\delta([D,\beta^n])\,|D|^{-2} = 2n \tag{52}$$

Thus we get,

$$\varphi_1(\mu',\mu) = 2n + (\ell-k)\int_{\mathcal{H}'}|D|^{-1} - \frac{1}{2}(\ell^2-k^2)\int_{\mathcal{H}'}|D|^{-2}\,. \tag{53}$$

Let us show that φ_1 is cohomologous to τ_1 on $\mathcal{J}$. Indeed, let $\rho(k)$ be an arbitrary sequence of polynomial growth and φ_0 be the 0-cochain given by,

$$\varphi_0(\alpha^{*k}\beta^n\alpha^\ell) = 0 \text{ unless } k = \ell, \qquad \varphi_0(\alpha^{*k}\beta^0\alpha^k) = \rho(k)\,. \tag{54}$$

Then

$$(b\varphi_0)(\mu',\mu) = \varphi_0(\mu'\mu) - \varphi_0(\mu\mu')$$

and both terms vanish unless $k = \ell'$, $k' = \ell$, $n' = -n$. Moreover in that case

$$\mu\mu' = \alpha^{*k}\beta^n\alpha^\ell\alpha^{*k'}\beta^{n'}\alpha^{\ell'} = \alpha^{*k}\beta^0\alpha^k$$

while

$$\mu'\mu = \alpha^{*\ell}\beta^{-n}\alpha^k\alpha^{*k}\beta^n\alpha^\ell = \alpha^{*\ell}\beta^0\alpha^\ell$$

so that

$$(b\varphi_0)(\mu',\mu) = \rho(\ell) - \rho(k)\,.$$

Thus, with,

$$\rho(k) = k\int_{\mathcal{H}'}|D|^{-1} - \frac{1}{2}k^2\int_{\mathcal{H}'}|D|^{-2} \tag{55}$$

we have, on the ideal $\mathcal{J}$,

$$\varphi_1 = \tau_1 + b\varphi_0\,. \tag{56}$$

Let us now extend this equality to the case when only one of the variables μ,μ' belongs to the ideal $\mathcal{J}$.

Assuming first that μ belongs to the ideal $\mathcal{J}$, we just need to compute $\varphi_1(\mu',\mu)$ for $\mu = \alpha^{*k}\beta^0\alpha^\ell$ and $\mu' = \alpha^{*k'}$ if $k' = \ell - k \geq 0$ or $\mu' = \alpha^{\ell'}$ if $\ell' = k - \ell > 0$. One has by (41), $[D,\mu] = \alpha^{*k}X\alpha^\ell$, $X = (k-\ell)\beta^0 F$ since $[D,\beta^0] = 0$ and $[F,\beta^0] = 0$. Thus, for $k' \geq 0$,

$$\int\!\!\!\!\!\!\!\!-\;\mu'[D,\mu]\,|D|^{-1} = \int\!\!\!\!\!\!\!\!-\;\alpha^{*k'}\alpha^{*k}(k-\ell)\,\beta^0 F\alpha^\ell\,|D|^{-1} = (k-\ell)\int\!\!\!\!\!\!\!\!-\;\alpha^{*\ell}\beta^0 F\alpha^\ell\,|D|^{-1}$$

since $k + k' = \ell$.

For $k' < 0$, $\ell' = k - \ell > 0$ one gets

$$\fint \alpha^{\ell'} \alpha^{*k} (k - \ell) \beta^0 F \alpha^\ell \, |D|^{-1} = (k - \ell) \fint \alpha^{*\ell} \beta^0 F \alpha^\ell \, |D|^{-1} \, .$$

Thus in both cases we get, using,

$$\fint \alpha^{*\ell} \beta^0 F \alpha^\ell \, |D|^{-1} = \fint \beta^0 F \, |D|^{-1} - \ell \fint \beta_0 F \, |D|^{-2}$$

$$= - \fint_{\mathcal{H}'} |D|^{-1} + \ell \fint_{\mathcal{H}'} |D|^{-2} \tag{57}$$

the formula,

$$\fint \mu' [D, \mu] \, |D|^{-1} = (\ell - k) \fint_{\mathcal{H}'} |D|^{-1} + \ell(k - \ell) \fint_{\mathcal{H}'} |D|^{-2} \, . \tag{58}$$

One has

$$\delta([D, \mu]) = \delta((k - \ell) \alpha^{*k} \beta^0 F \alpha^\ell) = (k - \ell)^2 \alpha^{*k} \beta^0 F \alpha^\ell$$

so that

$$\fint \mu' \delta([D, \mu]) \, |D|^{-2} = -(k - \ell)^2 \fint_{\mathcal{H}'} |D|^{-2} \, . \tag{59}$$

Thus

$$\varphi_1(\mu', \mu) = (\ell - k) \fint_{\mathcal{H}'} |D|^{-1} + \left(\ell(k - \ell) + \frac{1}{2}(k - \ell)^2 \right) \fint_{\mathcal{H}'} |D|^{-2}$$

$$= (\ell - k) \fint_{\mathcal{H}'} |D|^{-1} + \frac{1}{2}(k^2 - \ell^2) \fint_{\mathcal{H}'} |D|^{-2} \, .$$

Now one has $\tau_1(\mu', \mu) = 0$ and,

$$b\varphi_0(\mu', \mu) = \varphi_0(\mu'\mu) - \varphi_0(\mu\mu') = \varphi_0(\alpha^{*\ell} \beta^0 \alpha^\ell) - \varphi_0(\alpha^{*k} \beta^0 \alpha^k) = \rho(\ell) - \rho(k) \, .$$

Thus we check that,

$$\varphi_1(\mu', \mu) = \tau_1(\mu', \mu) + b\varphi_0(\mu', \mu) \, . \tag{60}$$

Let us now assume that μ' belongs to the ideal $\mathcal{J}$. We take $\mu' = \alpha^{*k'} \beta^0 \alpha^{\ell'}$ and μ to be α^{*k} if $k = \ell' - k' \geq 0$ and α^ℓ if $\ell = k' - \ell' > 0$. Assume first $k \geq 0$. One has $[D, \mu] = k \alpha^{*k} F$ and $\fint \mu' [D, \mu] \, |D|^{-1} = k \fint \alpha^{*k'} \beta^0 \alpha^{k'} F \, |D|^{-1}$. Thus using (57) we get,

$$\fint \mu' [D, \mu] \, |D|^{-1} = k \left(- \fint_{\mathcal{H}'} |D|^{-1} + k' \fint_{\mathcal{H}'} |D|^{-2} \right) \, .$$

But $k = \ell' - k'$ so that,

$$\fint \mu' [D, \mu] \, |D|^{-1} = (k' - \ell') \fint_{\mathcal{H}'} |D|^{-1} + k'(\ell' - k') \fint_{\mathcal{H}'} |D|^{-2} \, . \tag{61}$$

Also $\delta([D, \mu]) = k^2 \alpha^{*k} F$ and

$$\fint \mu' \delta([D, \mu]) |D|^{-2} = k^2 \fint \alpha^{*k'} \beta^0 \alpha^{k'} F |D|^{-2} = -k^2 \fint_{\mathcal{H}'} |D|^{-2} .$$

Thus

$$\varphi_1(\mu', \mu) = (k^{\ell} - \ell') \fint_{\mathcal{H}'} |D|^{-1} + \left(k'(\ell' - k') + \frac{1}{2} k^2 \right) \fint_{\mathcal{H}'} |D|^{-2} .$$

One has

$$k'(\ell' - k') + \frac{1}{2} k^2 = k'(\ell' - k') + \frac{1}{2} (\ell' - k')^2 = \frac{1}{2} \ell'^2 - \frac{1}{2} k'^2 ,$$

so that,

$$\varphi_1(\mu', \mu) = (k' - \ell') \fint_{\mathcal{H}'} |D|^{-1} + \frac{1}{2} (\ell'^2 - k'^2) \fint_{\mathcal{H}'} |D|^{-2} . \tag{62}$$

Now $\mu' \mu = \alpha^{*k'} \beta^0 \alpha^{k'}$, $\mu \mu' = \alpha^{*\ell'} \beta^0 \alpha^{\ell'}$ so that,

$$b\varphi_0(\mu', \mu) = \varphi_0(\mu' \mu) - \varphi_0(\mu \mu') = \varphi_0(\alpha^{*k'} \beta^0 \alpha^{k'}) - \varphi_0(\alpha^{*\ell'} \beta^0 \alpha^{\ell'}) = \rho(k') - \rho(\ell') .$$

Thus, since $\tau_1(\mu', \mu) = 0$, we get,

$$\varphi_1(\mu', \mu) = \tau_1(\mu', \mu) + (b\varphi_0)(\mu', \mu) . \tag{63}$$

Next, let us assume that $\ell = k' - \ell' > 0$. Then $\mu = \alpha^{\ell}$, $[D, \mu] = -\ell \alpha^{\ell} F$, and $\mu'[D, \mu] = -\ell \alpha^{*k'} \beta^0 \alpha^{k'} F$ so that by (57),

$$\fint \mu'[D, \mu] |D|^{-1} = -\ell \left(-\fint_{\mathcal{H}'} |D|^{-1} + k' \fint_{\mathcal{H}'} |D|^{-2} \right) ,$$

$$\fint \mu'[D, \mu] |D|^{-1} = (k' - \ell') \fint_{\mathcal{H}'} |D|^{-1} + k'(\ell' - k') \fint_{\mathcal{H}'} |D|^{-2} . \tag{64}$$

One has $\delta([D, \mu]) = \ell^2 \alpha^{\ell} F$ and,

$$\fint \mu' \delta([D, \mu]) |D|^{-2} = \ell^2 \fint \alpha^{*k'} \beta^0 \alpha^{k'} F |D|^{-1}$$

$$= -\ell^2 \fint_{\mathcal{H}'} |D|^{-2} = -(k' - \ell')^2 \fint_{\mathcal{H}'} |D|^{-2} .$$

Thus as above the coefficient of $\fint_{\mathcal{H}} |D|^{-2}$ in $\varphi_1(\mu', \mu)$ is $k'(\ell' - k') + \frac{1}{2} (k' - \ell')^2 = \frac{1}{2} \ell'^2 - \frac{1}{2} k'^2$ and,

$$\varphi_1(\mu', \mu) = (k' - \ell') \fint_{\mathcal{H}'} |D|^{-1} + \frac{1}{2} (\ell'^2 - k'^2) \fint_{\mathcal{H}'} |D|^{-2} . \tag{65}$$

Thus, as above we get,

$$\varphi_1(\mu',\mu) = \tau_1(\mu',\mu) + b\varphi_0(\mu',\mu)\,. \tag{66}$$

Before we proceed, note that (54) which defines φ_0 is only determined up to the addition of an arbitrary constant to ρ. As it turns out this constant will play a role and will be uniquely specified by equation (7) with the value $\frac{2}{3}$. Also in order to show that the above computation was largely independent of the specific numerical values of $f_{\mathcal{H}'}|D|^{-1}$ and $f_{\mathcal{H}'}|D|^{-2}$ we did not replace these expressions by their values which are,

$$\fint_{\mathcal{H}'}|D|^{-1} = -1 \qquad \fint_{\mathcal{H}'}|D|^{-2} = 2\,. \tag{67}$$

(To get (67) we use (19) and (20) and compute

$$\mathrm{Trace}_{\mathcal{H}'}(e^{-t|D|}) = \left(\sum_{k\in\mathbb{Z}} e^{-t|k|}\right)\left(\sum_1^\infty e^{-t\ell}\right)$$

$$= \left(1 + \frac{2}{e^t-1}\right)\left(\frac{1}{e^t-1}\right) \sim \frac{2}{t^2} - \frac{1}{t} + \frac{1}{3} - \frac{t}{12} + \cdots)$$

Thus (up to an additive constant), (55) gives,

$$\rho(j) = -(j+j^2) \tag{68}$$

We extend the definition of φ_0 to $\mathcal{A}$ by $\varphi_0(1) = 0$ while, as above, $\varphi_0(a)$ vanishes if the bidegree of a is $\neq (0,0)$.

7.3 Three Dimensional Components

It follows from (66) that $\psi = \varphi_1 - \tau_1 - b\varphi_0$ vanishes if one of the arguments is in $\mathcal{J}$ and thus $\psi(a_0, a_1)$ only depends on the symbols $\sigma(a_i) \in C^\infty(S^1)$ where

$$0 \longrightarrow \mathcal{J} \longrightarrow \mathcal{A} \xrightarrow{\;s\;} C^\infty(S^1) \longrightarrow 0 \tag{69}$$

is the natural exact sequence, with $\sigma(\alpha) = u$ and $\sigma(\beta) = 0$.
But the same holds for the component φ_3,

$$\varphi_3(a_0, a_1, a_2, a_3) = \fint a_0\,[D, a_1]\,[D, a_2]\,[D, a_3]\,|D|^{-3}\,. \tag{70}$$

Indeed if one of the a_j belongs to the two sided ideal $\mathcal{J}$ one is dealing with a trace class operator since $|D|^{-3}$ is trace class on the support of β. Thus $\varphi_3(a_0, a_1, a_2, a_3)$ only depends on the symbols $f_j = \sigma(a_j)$, and is given by,

$$\varphi_3(a_0, a_1, a_2, a_3) = \frac{-1}{2\pi i}\int_{S^1} f_0 f_1' f_2' f_3'\,d\theta\,, \tag{71}$$

where $f' = \frac{\partial}{\partial\theta}$. Since $F = 2P - 1$ introduces a minus sign, we use (40) and can replace $[D, a_j]$ by $-if_j'$, so that (71) follows from,

$$\oint |D|^{-3} = 1. \tag{72}$$

Thus to get the complete control of the cocycle φ it remains only to compute $\psi(\alpha^k, \alpha^{*k})$ and $\psi(\alpha^{*\ell}, \alpha^\ell)$.

Let us compute $\psi(\alpha^k, \alpha^{*k})$. One has

$$b\varphi_0(\alpha^k, \alpha^{*k}) = \varphi_0(\alpha^k \alpha^{*k}) - \varphi_0(\alpha^{*k} \alpha^k) = -\varphi_0(\alpha^{*k} \alpha^k)$$

Let $\lambda_k = \varphi_0(\alpha^{*k} \alpha^k)$, then

$$\lambda_k - \lambda_{k-1} = \varphi_0(\alpha^{*k-1}(\alpha^* \alpha - 1)\alpha^{k-1}) = -\varphi_0(\alpha^{*k-1}\beta^0\alpha^{k-1}) = -\rho(k-1)$$

since $\alpha^* \alpha - 1 = -\beta^0$. We get,

$$\lambda_k = \sum_0^{k-1}(j + j^2) = \frac{k^3}{3} - \frac{k}{3}, \tag{73}$$

and,

$$b\varphi_0(\alpha^k, \alpha^{*k}) = -\frac{k^3}{3} + \frac{k}{3} \tag{74}$$

Let us compute $\varphi_1(\alpha^k, \alpha^{*k})$. With $\varphi_1 = \varphi_1^{(0)} - \frac{1}{4}\varphi_1^{(1)} + \frac{1}{8}\varphi_1^{(2)}$, one has,

$$\varphi_1^{(0)}(\alpha^k, \alpha^{*k}) = \oint \alpha^k[D, \alpha^{*k}]|D|^{-1} = \oint k\alpha^k \alpha^{*k} F |D|^{-1} = k \oint F |D|^{-1}$$

using (40). Next,

$$\varphi_1^{(1)}(\alpha^k, \alpha^{*k}) = \oint \alpha^k \nabla([D, \alpha^{*k}])|D|^{-3} = k \oint \alpha^k \nabla(\alpha^{*k} F)|D|^{-3}.$$

But $\nabla(x) = |D|\delta(x) + \delta(x)|D| = \delta^2(x) + 2\delta(x)|D|$ and $\delta^2(\alpha^{*k} F) = k^2 \alpha^{*k} F$, thus,

$$\varphi_1^{(1)}(\alpha^k, \alpha^{*k}) = k^3 \oint F |D|^{-3} + 2k^2 \oint F |D|^{-2}$$

Finally,

$$\varphi_1^{(2)}(\alpha^k, \alpha^{*k}) = \oint \alpha^k \nabla^2([D, \alpha^{*k}])|D|^{-5}$$

$$= 4 \oint \alpha^k \delta^2([D, \alpha^{*k}])|D|^{-3} = 4k^3 \oint F |D|^{-3}.$$

Thus,

$$\varphi_1(\alpha^k, \alpha^{*k}) = k \oint F |D|^{-1} - \frac{1}{2}k^2 \oint F |D|^{-2}$$

$$+ \left(-\frac{1}{4}k^3 + \frac{1}{2}k^3\right) \oint F |D|^{-3}. \tag{75}$$

One has $\oint F|D|^{-3} = -1$ and thus the term in k^3 is $-\frac{k^3}{4}$. Thus the term in k^3 in $\psi = \varphi_1 - \tau_1 - b\varphi_0$ is, using (74),

$$\left(\frac{1}{3} - \frac{1}{4}\right) k^3 = \frac{k^3}{12}. \tag{76}$$

As we shall see now, this $\frac{1}{12}$ corresponds exactly to the coefficient $\frac{1}{12}$ in the universal index formula (theorem 6).

Indeed the term in k^3 corresponds to the cochain ψ_3 given in terms of the symbols f_0, f_1 by,

$$\psi_3(a_0, a_1) = \frac{1}{2\pi i} \int f_0 f_1''' d\theta. \tag{77}$$

Let us compute $b\psi_3$ where we only involve the symbols. One has

$$f_0 f_1 f_2''' - f_0(f_1 f_2)''' + f_2 f_0 f_1''' = f_0(-3 f_1'' f_2' - 3 f_1' f_2'')$$

thus we get,

$$b\psi_3(a_0, a_1, a_2) = \frac{1}{2\pi i} \int f_0(-3 f_1'' f_2' - 3 f_1' f_2'') \, d\theta. \tag{78}$$

We have,

$$B_0 \varphi_3(a_0, a_1, a_2) = -\frac{1}{2\pi i} \int f_0' f_1' f_2' d\theta = \frac{1}{2\pi i} \int f_0(f_1'' f_2' + f_1' f_2'') \, d\theta.$$

This is already cyclic so that,

$$B\varphi_3(a_0, a_1, a_2) = \frac{3}{2\pi i} \int f_0(f_1'' f_2' + f_1' f_2'') \, d\theta. \tag{79}$$

and we get,

$$b\psi_3 + B\varphi_3 = 0. \tag{80}$$

In fact,

$$\psi_3 = B\varphi_2 \, , \quad \varphi_3 = b\varphi_2 \, , \tag{81}$$

where φ_2 is given by,

$$\varphi_2(a_0, a_1, a_2) = \frac{-1}{2} \frac{1}{2\pi i} \int f_0 f_1' f_2'' \, d\theta. \tag{82}$$

Let us now compute $\oint F|D|^{-2}$. One has,

$$\oint |D|^{-2} = 2. \tag{83}$$

and,

$$\oint P|D|^{-2} = 1. \tag{84}$$

Thus we get, since $F = 2P - 1$,

$$\oint F |D|^{-2} = 0. \tag{85}$$

and by a similar computation,

$$\oint F |D|^{-1} = 1. \tag{86}$$

This gives,

$$\psi(\alpha^k, \alpha^{*k}) = \frac{2k}{3} + \frac{k^3}{12}. \tag{87}$$

The computation of $\psi(\alpha^{\ell*}, \alpha^\ell)$ is entirely similar and gives $\psi(\alpha^{\ell*}, \alpha^\ell) = -\frac{2\ell}{3} - \frac{\ell^3}{12}$.
We thus have $\psi = -\frac{2}{3}\psi_1 + \frac{1}{12}\psi_3$ where,

$$\psi_1(a_0, a_1) = \frac{1}{2\pi i} \int f_0 f_1' d\theta. \tag{88}$$

It just remains to see why adding a constant to ρ allows to eliminate $-\frac{2}{3}\psi_1$ from ψ. This follows from (73) and (74) i. e.

$$b\varphi_0(\alpha^k, \alpha^{*k}) = \sum_0^{k-1} \rho(j), \tag{89}$$

Thus adding $\frac{2}{3}$ to ρ gives $\psi = \frac{1}{12}\psi_3$ and ends the proof of theorem 8.
We shall now understand the conceptual meaning of the above concrete computation.

8 The η-Cochain

In this section we shall give two general formulas. The first will provide the conceptual explanation of theorem 8, and of the cochain (φ_0, φ_2) which appears there. The second will prepare for the computation of the local index formula in the general case $q \in]0, 1[$.
The explanation of theorem 8 and of the cochains,

$$\varphi_0(\alpha^{*j} f(\beta) \alpha^j) = (\frac{2}{3} - j + j^2) \frac{1}{2\pi} \int_{S^1} f \, d\theta, \tag{1}$$

$$\varphi_2(a_0, a_1, a_2) = \frac{-1}{24} \frac{1}{2\pi i} \int f_0 f_1' f_2'' \, d\theta, \quad f_j = s(a_j), \tag{2}$$

is given by the following,

Proposition 2. Let $(\mathcal{A}, \mathcal{H}, D)$ be a spectral triple with discrete simple dimension spectrum not containing 0 and upper bounded by 3. Assume that $[F, a]$ is trace class for all $a \in \mathcal{A}$. Let $\tau_1(a_0, a_1) = \mathrm{Trace}(a_0[F, a_1])$.

Then the local Chern Character (φ_1, φ_3) of $(\mathcal{A}, \mathcal{H}, D)$ is equal to $\tau_1 + (b+B)\varphi$ where (φ_0, φ_2) is the cochain given by,

$$\varphi_0(a) = \text{Trace}(F\,a\,|D|^{-s})_{s=0},$$

$$\varphi_2(a_0, a_1, a_2) = \frac{1}{24} \int a_0\,\delta(a_1)\,\delta^2(a_2)\,F\,|D|^{-3}.$$

Note that φ_0 makes sense by the absence of pole at $s = 0$, i.e. the hypothesis $0 \notin$ Dimension Spectrum. Its value for $a = 1$ coincides with the classical η-invariant ([2],[3]) and justifies the terminology of η-cochain to qualify the cochain (φ_0, φ_2). The proof of the proposition is a simple calculation based on the expansion ([16])

$$|D|^{-s}a \sim a\,|D|^{-s} - s\,\delta(a)|D|^{-s-1} + \frac{(-s)(-s-1)}{2!}\,\delta^2(a)|D|^{-s-2}$$

$$+ \frac{(-s)(-s-1)(-s-3)}{3!}\,\delta^3(a)|D|^{-s-3} + s\,o(|D|^{-s-3}), \tag{3}$$

which allows to express $b\varphi_0$ in terms of residues. More specifically one gets,

$$b\varphi_0(a_0, a_1) = -\tau_1(a_0, a_1) + \int a_0\,\delta(a_1)\,F\,|D|^{-1}$$

$$- \frac{1}{2} \int a_0\,\delta^2(a_1)\,F\,|D|^{-2} + \frac{1}{3} \int a_0\,\delta^3(a_1)\,F\,|D|^{-3} \tag{4}$$

using the hypothesis $[F, a]$ trace class for all $a \in \mathcal{A}$. This hypothesis also shows that,

$$\varphi_1(a_0, a_1) = \int a_0\,\delta(a_1)\,F\,|D|^{-1}$$

$$- \frac{1}{2} \int a_0\,\delta^2(a_1)\,F\,|D|^{-2} + \frac{1}{4} \int a_0\,\delta^3(a_1)\,F\,|D|^{-3} \tag{5}$$

Comparing (4) with (5) gives the required $\frac{1}{12}$ and allows to check that $\varphi_{odd} = \tau_1 + (b+B)\varphi_{ev}$.

Let us compute φ_{ev} in the above example. One has, as in (56),

$$\varphi_0(\alpha^{*k} e\,\alpha^k) = (\text{Trace}\,(F\,e\,(|D| + k)^{-s}))_{s=0} \tag{6}$$

Using $F = 2P - 1$ this gives,

$$\varphi_0(\alpha^{*k} e\,\alpha^k) = 2(\sum_0^\infty \frac{1}{(n+k)^s})_{s=0} - (\sum_0^\infty \frac{2n+1}{(n+k)^s})_{s=0} \tag{7}$$

One has,

$$(\sum_0^\infty \frac{1}{(n+k)^s})_{s=0} = \zeta(0) - (k-1).$$

Also,

$$\sum_{0}^{\infty} \frac{2n+1}{(n+k)^s})_{s=0} = 2\zeta(-1) + (1-2k)\zeta(0) - \sum_{1}^{k-1}(2\ell - 2k + 1)$$

$$= 2\zeta(-1) + (1-2k)\zeta(0) + (k-1)^2. \tag{8}$$

Thus we get,

$$\varphi_0(\alpha^{*k}e\,\alpha^k) = 2(\zeta(0) - (k-1)) - (2\zeta(-1) + (1-2k)\zeta(0) + (k-1)^2) \tag{9}$$

which using the values,

$$\zeta(0) = -\frac{1}{2}, \quad \zeta(-1) = -\frac{1}{12} \tag{10}$$

gives the desired result,

$$\varphi_0(\alpha^{*k}e\,\alpha^k) = \frac{2}{3} - k - k^2. \tag{11}$$

The only other non-trivial value of φ_0 is $\eta = \varphi_0(1)$, and the computation gives $\eta = \frac{1}{2}$. Finally the equality,

$$\fint a_0\,\delta(a_1)\,\delta^2(a_2)\,F\,|D|^{-3} = \frac{1}{2\pi i}\int f_0 f_1' f_2''\,d\theta, \quad f_j = \mathfrak{s}(a_j), \tag{12}$$

and the coincidence of the functional τ_1 of theorem 8 with $\mathrm{Trace}(a_0[F, a_1])$ give a perfect account of theorem 8.

In order to lighten the general computation, for $q \in]0, 1|$, we shall state a small variant of proposition 2, proved in a similar way. Given a spectral triple $(\mathcal{A}, \mathcal{H}, D)$ let us define the metric dimension $\mathrm{Dm}(P)$ of a projection P commuting with D as the lower bound of all $d \in \mathbb{R}$ such that $P(D + i)^{-1}$ is in the Schatten class L^d. We then have as above,

Proposition 3. *Let* $(\mathcal{A}, \mathcal{H}, D)$ *be a spectral triple with discrete dimension spectrum not containing 0. Assume that* $\mathrm{Dm}(\mathcal{H}) \leq 3$, *and* $\mathrm{Dm}(P) \leq 2$, $P = (1 + F)/2$, *and that* $[F, a]$ *is trace class for all* $a \in \mathcal{A}$. *Then the local Chern Character* (φ_1, φ_3) *of* $(\mathcal{A}, \mathcal{H}, D)$ *is equal to* $\psi_1 - (b + B)\varphi$ *where* ψ_1 *is the cyclic cocycle,*

$$\psi_1(a_0, a_1) = 2 \fint a_0\,\delta(a_1)\,P\,|D|^{-1} - \fint a_0\,\delta^2(a_1)\,P\,|D|^{-1}$$

and (φ_0, φ_2) *is the cochain given by,*

$$\varphi_0(a) = \mathrm{Trace}(\,a\,|D|^{-s})_{s=0},$$

$$\varphi_2(a_0, a_1, a_2) = \frac{1}{24} \fint a_0\,\delta(a_1)\,\delta^2(a_2)\,|D|^{-3}.$$

Combining Propositions 2 and 3 one obtains under the hypothesis of Proposition 3 the equality,

$$\psi_1 - \tau_1 = b(\psi_0) \tag{13}$$

where the cochain ψ_0 is given by,

$$\psi_0(a) = 2\,\mathrm{Trace}(\,a\,P\,|D|^{-s})_{s=0} \tag{14}$$

9 Pseudo-Differential Calculus and the Cosphere Bundle on $SU_q(2)$, $q \in\ ]0,1[$

In this section we shall construct the pseudo-differential calculus on $SU_q(2)$ following the general theory of [16]. We shall determine the algebra of complete symbols by computing the quotient by smoothing operators. This will give the cosphere bundle S_q^* of $SU_q(2)$ and the analogue of the geodesic flow will yield a one-parameter group of automorphisms γ_t of $C^\infty(S_q^*)$. We shall also construct the restriction morphism r to the product of two 2-disks,

$$r : C^\infty(S_q^*) \to C^\infty(D_{q+}^2 \times D_{q-}^2) \tag{1}$$

Our goal is to prepare for the computation in the next section of the dimension spectrum and of residues. Let us recall from [16] that given a spectral triple $(\mathcal{A}, \mathcal{H}, D)$ we say that an operator P in $\mathcal{H}$ is of order α when,

$$|D|^{-\alpha}\,P \in \bigcap_{n=1}^{\infty} \mathrm{Dom}\,\delta^n \tag{2}$$

where δ is the unbounded derivation given by,

$$\delta(T) = |D|\,T - T\,|D|\,. \tag{3}$$

Thus $OP^0 = \bigcap_{n=1}^{\infty} \mathrm{Dom}\,\delta^n$ is the algebra of operators of order 0 and $OP^{-\infty} = \bigcap_{k>0} OP^{-k}$ is a two sided ideal in OP^0.

We let $(\mathcal{A}, \mathcal{H}, D)$ be the spectral triple of [7] and we first determine the algebra $\mathcal{B}$ generated by the $\delta^k(a)$, $a \in \mathcal{A}$.

Recall that D is the diagonal operator in $\mathcal{H}$ given by,

$$D(e_{ij}^{(n)}) = (2\,\delta_0(n-i) - 1)\,2n\,e_{ij}^{(n)} \tag{4}$$

where $\delta_0(k) = 0$ if $k \neq 0$ and $\delta_0(0) = 1$.

By construction, the generators α, β of $\mathcal{A}$ are of the form,

$$\alpha = \alpha_+ + \alpha_-\,, \qquad \beta = \beta_+ + \beta_- \tag{5}$$

where,

$$\delta(\alpha_\pm) = \pm\alpha_\pm, \qquad \delta(\beta_\pm) = \pm\beta_\pm. \tag{6}$$

The explicit form of $\alpha_\pm$, $\beta_\pm$ is, using $\frac{n}{2}$ instead of n for the notation of the $\frac{1}{2}$ integer,

$$\alpha_\pm(e_{(i,j)}^{(n/2)}) = a_\pm(n/2,i,j)\, e_{(i-\frac{1}{2},j-\frac{1}{2})}^{\left(\frac{n\pm 1}{2}\right)} \tag{7}$$

$$\beta_\pm(e_{(i,j)}^{(n/2)}) = b_\pm(n/2,i,j)\, e_{(i+\frac{1}{2},j-\frac{1}{2})}^{\left(\frac{n\mp 1}{2}\right)} \tag{8}$$

where $a_\pm$, $b_\pm$ are as in (3) and (4) above.

Thus the algebra $\mathcal{B}$ is generated by the operators $\alpha_\pm$, $\beta_\pm$ and their adjoints.

We shall now see that, modulo the smoothing operators, we can strip the complicated formulas for the coefficients $a_\pm$, $b_\pm$ and replace them by extremely simple ones. Since we are computing *local formulas* we are indeed entitled to mod out by smoothing operators and this is exactly where great simplifications do occur.

Let us first relabel the indices i, j using,

$$x = \frac{n}{2} + i, \qquad y = \frac{n}{2} + j. \tag{9}$$

By construction x and y are *integers* which vary exactly in $\{0, 1, \ldots, n\}$.

Working modulo $OP^{-\infty}$ means that we can neglect in the formulas for $a_\pm$, $b_\pm$ any modification by a sequence of rapid decay in the set:

$$\Lambda = \{(n,x,y);\ n \in \mathbb{N},\ x,y \in \{0,\ldots,n\}\}. \tag{10}$$

Thus first, we can get rid of the denominators, since both $(1 - q^{2n})^{-1/2}$ or $(1 - q^{(2n+2)})^{-1/2}$ are equivalent to 1 and the numerators are bounded.

Next, when we rewrite the numerators in terms of the variables n, x, y we get, say for a_+, the simplified form,

$$a'_+(n,x,y) = q^{1+x+y}(1 - q^{2+2(n-x)})^{1/2}(1 - q^{2+2(n-y)})^{1/2}. \tag{11}$$

Modulo sequences of rapid decay one has,

$$q^x(1 - q^{2+2(n-x)})^{1/2} \sim q^x,$$

as one sees from the inequality $(1 - (1-u)^{1/2}) \leq u$ valid for $u \in [0,1]$, and the fact that

$$q^x\, q^{2(n-x)} \leq q^x\, q^{(n-x)} = q^n.$$

Thus we see that modulo sequences of rapid decay we can replace a'_+ by,

$$a''_+(n,x,y) = q^{1+x+y}. \tag{12}$$

To simplify formulas let us relabel the basis as,

$$f_{x,y}^{(n)} = e_{(x-n/2,y-n/2)}^{(n/2)}, \tag{13}$$

then the following operator agrees with α_+ modulo $OP^{-\infty}$,

$$\alpha'_+(f_{x,y}^{(n)}) = q^{1+x+y} f_{x,y}^{(n+1)}. \tag{14}$$

For α_- one has, as above,

$$a'_-(n, x, y) = (1 - q^{2x})^{1/2} (1 - q^{2y})^{1/2} \tag{15}$$

and the corresponding operator α'_- is,

$$\alpha'_-(f_{x,y}^{(n)}) = (1 - q^{2x})^{1/2}(1 - q^{2y})^{1/2} f_{x-1,y-1}^{(n-1)}. \tag{16}$$

Note that α'_- makes sense for $x = 0, y = 0$. For β_+ one gets,

$$b'_+(n, x, y) = -q^y(1 - q^{2+2(n-y)})^{1/2}(1 - q^{2+2x})^{1/2} \tag{17}$$

and as above we can replace it by,

$$b''_+(n, x, y) = -q^y(1 - q^{2+2x})^{1/2} \tag{18}$$

which gives,

$$\beta'_+(f_{x,y}^{(n)}) = -q^y(1 - q^{2+2x})^{1/2} f_{x+1,y}^{(n+1)}. \tag{19}$$

In a similar way one gets,

$$\beta'_-(f_{x,y}^{(n)}) = q^x(1 - q^{2y})^{1/2} f_{x,y-1}^{(n-1)} \tag{20}$$

which makes sense even for $y = 0$.

It is conspicuous in the above formulas that the new and much simpler coefficients *no longer depend upon the variable n*.

To understand these formulas we introduce the following representations $\pi_\pm$ of $\mathcal{A} = C^\infty(\mathrm{SU}_q(2))$ [1]. In both cases the Hilbert spaces are $\mathcal{H}_\pm = \ell^2(\mathbb{N})$ with basis $(\varepsilon_x)_{x \in \mathbb{N}}$ and the representations are given by,

$$\pi_\pm(\alpha)\,\varepsilon_x = (1 - q^{2x})^{1/2}\,\varepsilon_{x-1} \qquad \forall x \in \mathbb{N} \tag{21}$$

$$\pi_\pm(\beta)\,\varepsilon_x = \pm\,q^x\,\varepsilon_x \qquad \forall x \in \mathbb{N}. \tag{22}$$

With these notations, and if we ignore the n-dependence in the above formulas we have the correspondence,

$$\alpha'_+ \cong -q\,\beta^* \otimes \beta \tag{23}$$
$$\alpha'_- \cong \alpha \otimes \alpha$$
$$\beta'_+ \cong \alpha^* \otimes \beta$$
$$\beta'_- \cong \beta \otimes \alpha,$$

---

[1] see the appendix for the notation

through the representation $\pi = \pi_+ \otimes \pi_-$. Now recall that $\mathcal{A}$ is a Hopf algebra, with coproduct corresponding to matrix tensor multiplication for the following 2×2 matrix,

$$U = \begin{bmatrix} \alpha & -q\beta^* \\ \beta & \alpha^* \end{bmatrix} \tag{24}$$

which gives,

$$\Delta\alpha = \alpha \otimes \alpha - q\beta^* \otimes \beta \tag{25}$$

$$\Delta\beta = \beta \otimes \alpha + \alpha^* \otimes \beta.$$

This shows of course that $\alpha' = \alpha'_+ + \alpha'_-$ and $\beta' = \beta'_+ + \beta'_-$ provide a representation of $\mathcal{A}$ which is the tensor product in the sense of Hopf algebras of the representations π_+ and π_- of $\mathcal{A}$. However to really understand the algebra $\mathcal{B}$ modulo $OP^{-\infty}$ an its action in $\mathcal{H}$ we need to keep track of the shift of n in the formulas for $\alpha'_\pm$ and $\beta'_\pm$.

One can encode these shifts using the $\mathbb{Z}$-grading of $\mathcal{B}$ coming from the one parameter group of automorphisms $\gamma(t)$ which plays the role of the geodesic flow,

$$\gamma(t)(P) = e^{it|D|} P e^{-it|D|}. \tag{26}$$

For the corresponding $\mathbb{Z}$-grading one has,

$$\deg(\alpha_\pm) = \pm 1, \qquad \deg(\beta_\pm) = \pm 1, \tag{27}$$

which are the correct powers of the shifts of n in the above formulas for $\alpha'_\pm, \beta'_\pm$. To γ we associate the algebra morphism,

$$\gamma : \mathcal{B} \to \mathcal{B} \otimes C^\infty(S^1) = C^\infty(S^1, \mathcal{B}) \tag{28}$$

given by $\gamma(b)(t) = \gamma_t(b), \forall t \in S^1$.

Finally, note that the representations $\pi_\pm$ are not faithful on $C^\infty(SU_q(2))$ since the spectrum of β is real and positive in π_+ and real negative for π_-. We let $C^\infty(D^2_{q\pm})$ be the corresponding quotient algebras and $r_\pm$ the restriction morphisms. **Proposition**

4. *The following equalities define an algebra homomorphism ρ from $\mathcal{B}$ to*

$$C^\infty(D^2_{q+}) \otimes C^\infty(D^2_{q-}) \otimes C^\infty(S^1),$$

$$\rho(\alpha_+) = -q\beta^* \otimes \beta \otimes u, \qquad \rho(\alpha_-) = \alpha \otimes \alpha \otimes u^*,$$

$$\rho(\beta_+) = \alpha^* \otimes \beta \otimes u, \qquad \rho(\beta_-) = \beta \otimes \alpha \otimes u^*,$$

where we omitted $r_+ \otimes r_-$.

Proof. Using (28) it is enough to show that the formulas,

$$\rho_1(\alpha_+) = -q\,\pi_+(\beta^*) \otimes \pi_-(\beta), \qquad \rho_1(\alpha_-) = \pi_+(\alpha) \otimes \pi_-(\alpha)$$

$$\rho_1(\beta_+) = \pi_+(\alpha^*) \otimes \pi_-(\beta), \qquad \rho_1(\beta_-) = \pi_+(\beta) \otimes \pi_-(\alpha)$$

define a representation of $\mathcal{B}$.

But this representation is weakly contained in the natural representation of $\mathcal{B}$ in $\mathcal{H}$. To obtain ρ_1 from the latter, one just considers vectors $\varepsilon^N_{x,y}$ in $\mathcal{H}$, of the form,

$$\varepsilon^N_{x,y} = \sum h^N_{(n)} \, f^{(n)}_{x,y} \tag{29}$$

where $h^N \in \ell^2(\mathbb{N})$ corresponds to the amenability of the group $\mathbb{Z}$, i.e. to the weak containement of the trivial representation of $\mathbb{Z}$ by the regular one. Thus h^N depends on a large integer N and is $1/\sqrt{N}$ for $0 \leq n < N$ and 0 for $n \geq N$.

The almost invariance of h^N under translation of n shows that the n-dependence of the formulas (17)–(20) disappears when $N \to \infty$ and that ρ_1 is a representation of $\mathcal{B}$. Finally ρ is its amplification using (28)

Definition 2. *Let $C^\infty(S^*_q)$ be the range of ρ in $C^\infty(D^2_{q+} \times D^2_{q-} \times S^1)$.*

By construction $C^\infty(S^*_q)$ is topologically generated by $\rho(\alpha_\pm)$, $\rho(\beta_\pm)$. The NC-space S^*_q plays the role of the *cosphere bundle*. The algebra $C^\infty(S^*_q)$ is strictly contained in $C^\infty(D^2_{q+} \times D^2_{q-} \times S^1)$ since its image under $\sigma \otimes \sigma \otimes \mathrm{Id}$ is the subalgebra of $C^\infty(S^1 \times S^1 \times S^1)$ generated by $u \otimes u \otimes u^*$. Let ν_t be the S^1-action on S^*_q given by the restriction of the derivation $1 \otimes 1 \otimes \partial_u$ where $\partial_u(u) = u$. By construction,

$$\rho(\gamma(t)(P)) = \nu_t(\rho(P) \tag{30}$$

so that ν_t is the analogue of the action of the geodesic flow on the cosphere bundle. We let,

$$r : C^\infty(S^*_q) \to C^\infty(D^2_{q+} \times D^2_{q-}) \tag{31}$$

be the natural restriction morphism.

Viewing ρ as the total symbol map we shall now define a natural lifting from symbols to operators. This will only be relevant on the range of ρ but to define it we start from the representation $\pi = \pi_+ \otimes \pi_- \otimes s$ of $C^\infty(D^2_{q+}) \otimes C^\infty(D^2_{q-}) \otimes C^\infty(S^1)$ in $\ell^2(\mathbb{N}) \otimes \ell^2(\mathbb{N}) \otimes \ell^2(\mathbb{Z})$ where $s(u)$ is the shift S in $\ell^2(\mathbb{Z})$. We let Q be the orthogonal projection on the subset Λ of the basis $f^{(n)}_{x,y}$ determined by $n \geq \sup(x,y)$ and identify the range of Q with the Hilbert space $\mathcal{H}$. By definition the lifting λ is the compression,

$$\lambda(g) = Q \, \pi(g) \, Q \tag{32}$$

For g of the form $\mu \otimes u^n$, one has,

$$\lambda(g) \, f^{(\ell)}_{x,y} = \sum \mu^{(x',y')}_{(x,y)} \, f^{(\ell+n)}_{x',y'} \tag{33}$$

where $\mu^{(x',y')}_{(x,y)}$ are the matrix elements for the action of μ in $\ell^2(\mathbb{N}) \otimes \ell^2(\mathbb{N})$,

$$\mu \, \varepsilon_{x,y} = \sum \mu^{(x',y')}_{(x,y)} \, \varepsilon_{x',y'} . \tag{34}$$

It may happen in formula (33) that the indices in $f^{(\ell+n)}_{x',y'}$ do not make sense, i.e. that $f^{(\ell+n)}_{x',y'}$ does not belong to Λ. In that case the corresponding term is 0. We have now restaured the shift of n in the formulas for $\alpha'_\pm$ and $\beta'_\pm$ and get,

Lemma 1. *For any $b \in \mathcal{B}$ one has,*

$$b - \lambda(\rho(b)) \in OP^{-\infty}.$$

We refer to the appendix for the implications of this lemma. We give there another general lemma proving the stability under holomorphic functional calculus for the natural smooth algebras involved in our discussion.

10 Dimension Spectrum and Residues for $\mathrm{SU}_q(2)$, $q \in]0, 1[$

Let as above S_q^* be the cosphere bundle of $\mathrm{SU}(2)_q$, γ_t its geodesic flow and,

$$r : C^\infty(S_q^*) \to C^\infty(D_{q_+}^2 \times D_{q_-}^2) \tag{1}$$

be the natural restriction morphism.

For $C^\infty(D_{q_\pm}^2)$ we have an exact sequence of the form,

$$0 \longrightarrow \mathcal{S} \longrightarrow C^\infty(D_q^2) \overset{s}{\longrightarrow} C^\infty(S^1) \longrightarrow 0 \tag{2}$$

where the ideal $\mathcal{S}$ is isomorphic to the algebra of matrices of rapid decay. Using the representations $\pi_\pm$ of $C^\infty(D_{q_\pm}^2)$ in $\ell^2(\mathbb{N})$ with basis (ε_x), $x \in \mathbb{N}$, we define two linear functionals τ_0 and τ_1 by,

$$\tau_1(a) = \frac{1}{2\pi} \int_0^{2\pi} \sigma(a)\, d\theta \qquad \forall a \in C^\infty(D_q^2) \tag{3}$$

and

$$\tau_0(a) = \lim_{N \to \infty} \mathrm{Trace}_N(\pi(a)) - \tau_1(a)\, N \tag{4}$$

where,

$$\mathrm{Trace}_N(a) = \sum_0^N \langle a\, \varepsilon_x, \varepsilon_x \rangle. \tag{5}$$

(where we omitted $\pm$ in the above formulas). For $a \in \mathcal{S}$ one has $\sigma(a) = 0$ and $\tau_1(a) = 0$, $\tau_0(a) = \mathrm{Trace}\,(a)$. In general both τ_0 and τ_1 are invariant under the one parameter group generated by ∂_α and on the fixed points of this group, one has,

$$\tau_0(a) = \mathrm{Trace}\,(\pi(a) - \tau_1(a)\, 1) + \tau_1(a). \tag{6}$$

For all $a \in \mathcal{A}$ one has, (for all $k > 0$),

$$\mathrm{Trace}_N(\pi(a)) = \tau_1(a)\, N + \tau_0(a) + 0(N^{-k}). \tag{7}$$

We shall now prove a general formula computing residues of pseudo-differential operators in terms of their symbols,

Theorem 9.

1. *The dimension spectrum of* $\mathrm{SU}_q(2)$ *is* $\{1, 2, 3\}$.
2. *Let* $b \in \mathcal{B}$, $\rho(b) \in C^\infty(S_q^*)$ *its symbol. Then let* $\rho(b)^0$ *be the component of degree* 0 *for the geodesic flow* γ_t. *One has*,

$$\fint b\,|D|^{-3} = (\tau_1 \otimes \tau_1)(r\rho(b)^0)$$

$$\fint b\,|D|^{-2} = (\tau_1 \otimes \tau_0 + \tau_0 \otimes \tau_1)(r\rho(b)^0)$$

$$\fint b\,|D|^{-1} = (\tau_0 \otimes \tau_0)(r\rho(b)^0)\,.$$

Proof. By lemma 1, the operator $b - \lambda\rho(b)$ belongs to $OP^{-\infty}$, thus $\zeta_b(s) - \mathrm{Trace}\,(\lambda\rho(b)\,|D|^{-s})$ is a holomorphic function of $s \in \mathbb{C}$.
One has $\mathrm{Trace}\,(\lambda\rho(b)\,|D|^{-s}) = \mathrm{Trace}\,(\lambda\rho(b)^0|D|^{-s})$ and with $\rho(b)^0 = T$,

$$\mathrm{Trace}\,(\lambda(T)\,|D|^{-s}) = \sum_{n=0}^{\infty} \left(\sum_{x=0}^{n} \sum_{y=0}^{n} \langle \pi(T)\,\varepsilon_{x,y}, \varepsilon_{x,y} \rangle \right) n^{-s}\,. \tag{8}$$

Thus by (7) we get, modulo holomorphic functions of $s \in \mathbb{C}$,

$$\mathrm{Trace}\,(\lambda(T)\,|D|^{-s}) \cong (\tau_1 \otimes \tau_1)(T)\,\zeta(s-2) \tag{9}$$
$$+ (\tau_1 \otimes \tau_0 + \tau_0 \otimes \tau_1)(T)\,\zeta(s-1) + (\tau_0 \otimes \tau_0)(T)\,\zeta(s)\,.$$

This shows that $\zeta_b(s)$ extends to a meromorphic function of s with simple poles at $s \in \{1, 2, 3\}$ and gives the above values for the residues.

To show 1) we still need to adjoin $F = \mathrm{Sign}\,D$ to the algebra $\mathcal{B}$, but by [7] one has,

$$[F, a] \in OP^{-\infty} \qquad \forall a \in \mathcal{B} \tag{10}$$

so that the only elements which were not handled above are those of the form,

$$bP, \quad b \in \mathcal{B}, \quad P = \frac{1+F}{2}\,. \tag{11}$$

Thus with the above notation we still need to analyse,

$$\mathrm{Trace}\,(\lambda(T)\,P\,|D|^{-s})\,. \tag{12}$$

Since P corresponds to the subset of the basis $f_{x,y}^{(n)}$ given by $\{x = n\}$ in the above notations, the trace (12) can be expressed as,

$$\text{Trace}\,(\lambda(T)\,P\,|D|^{-s}) = \sum_{n=0}^{\infty} \sum_{y=0}^{n} \langle \pi(T)\,\varepsilon_{n,y}, \varepsilon_{n,y} \rangle\,n^{-s} \tag{13}$$

and the structure of the representation π_+ shows that the r.h.s. gives, modulo holomorphic function of $s \in \mathbb{C}$,

$$\text{Trace}\,(\lambda(T)\,P\,|D|^{-s}) \cong (\tau_1 \otimes \tau_1)(T)\,\zeta(s-1) + (\tau_1 \otimes \tau_0)(T)\,\zeta(s). \tag{14}$$

This shows (1) and also gives the two formulas,

$$\fint b\,P\,|D|^{-2} = (\tau_1 \otimes \tau_0)(\rho(b)^0) \tag{15}$$

and,

$$\fint b\,P\,|D|^{-1} = (\tau_0 \otimes \tau_0)\,\rho(b)^0 \tag{16}$$

which we shall now exploit to do the computation of the local index formula for $\text{SU}_q(2)$.

11 The Local Index Formula for $\text{SU}_q(2)$, $q \in\ ]0, 1[$

The local index formula for the spectral triple of $\text{SU}_q(2)$ uniquely determines a cyclic 1-cocycle and hence by ([10]) a corresponding one dimensional cycle. We shall first describe independently the obtained cycle since the NC-differential calculus it exhibits is of independent interest.

Let $\mathcal{A} = C^{\infty}(\text{SU}_q(2))$ and ∂ the derivation,

$$\partial = \partial_\beta - \partial_\alpha . \tag{1}$$

We extend the functional τ_0 of (4) to $\mathcal{A}$ by,

$$\tau(a) = \tau_0\,(r_-(a^{(0)})) \qquad \forall\,a \in \mathcal{A} \tag{2}$$

where $a^{(0)}$ is the component of degree 0 for ∂. By construction τ is ∂-invariant but fails to be a trace. It is the average of the transformed of $\tau_0 \circ r_-$ by the automorphism $\nu_t \in \text{Aut}(\mathcal{A})$,

$$\nu_t = \exp(it\partial) . \tag{3}$$

Thus τ fails to be a trace because τ_0 does. However we can compute the Hochschild coboundary $b\,\tau_0$, $b\,\tau_0(a_0, a_1) = \tau_0(a_0\,a_1) - \tau_0(a_1\,a_0)$. It only depends upon the symbols $\sigma(a_j) \in C^{\infty}(S^1)$ and is given by,

$$b\,\tau_0(a_0, a_1) = \frac{-1}{2\pi i} \int \sigma(a_0)\,d\sigma(a_1) . \tag{4}$$

One has

$$b\,\tau(a_0,a_1) = \frac{1}{2\pi}\int b\,\tau_0(a_0(t),a_1(t))\,\mathrm{d}t$$

where $a(t) = \nu(t)(a)$, and for homogeneous elements of $\mathcal{A}$, $b\,\tau(a_0,a_1) = 0$ unless the total degree is $(0,0)$. For such elements we thus get,

$$b\,\tau(a_0,a_1) = \frac{1}{2\pi}\int b\,\tau_0(a_0(t),a_1(t))\,\mathrm{d}t = b\,\tau_0(a_0,a_1) = -\frac{1}{2\pi i}\int \sigma(a_0)\,\mathrm{d}\sigma(a_1)$$

so that,

$$b\,\tau(a_0,a_1) = \frac{-1}{2\pi i}\int \sigma(a_0)\,\mathrm{d}\sigma(a_1) \qquad \forall\, a_j \in \mathcal{A}. \tag{5}$$

Thus, even though τ is not a trace we do control by how much it fails to be a trace and this allows us to define a *cycle* in the sense of [10] using both first and second derivatives to define the differential,

$$\mathcal{A} \xrightarrow{\ d\ } \Omega^1. \tag{6}$$

More precisely let us define the $\mathcal{A}$-bimodule Ω^1 with underlying linear space the direct sum, $\Omega^1 = \mathcal{A} \oplus \Omega^{(2)}(S^1)$ where $\Omega^{(2)}(S^1)$ is the space of differential forms $f(\theta)\,\mathrm{d}\theta^2$ of weight 2 on S^1. The bimodule structure is defined by,

$$a(\xi,f) = (a\xi,\sigma(a)f) \tag{7}$$
$$(\xi,f)b = (\xi b, -i\,\sigma(\xi)\,\sigma(b)' + f\sigma(b))$$

for $a,b \in \mathcal{A}$, $\xi \in \mathcal{A}$ and $f \in \Omega^{(2)}(S^1)$.

The differential d of (6) is then given by,

$$da = \partial a + \frac{1}{2}\sigma(a)''\,\mathrm{d}\theta^2 \tag{8}$$

as in a Taylor expansion.

The functional $\int$ is defined by,

$$\int (\xi,f) = \tau(\xi) + \frac{1}{2\pi i}\int f\,\mathrm{d}\theta. \tag{9}$$

We then have,

Proposition 5. *The triple $(\Omega, d, \int)$ is a cycle, i.e. $\Omega = \mathcal{A} \oplus \Omega^1$ equipped with d is a graded differential algebra (with $\Omega^0 = \mathcal{A}$) and the functional $\int$ is a closed graded trace on Ω.*

Proof. One checks directly that Ω^1 is an $\mathcal{A}$-bimodule so that $\Omega = \mathcal{A} \oplus \Omega^1$ is a graded algebra. The equality $\sigma(\partial a) = i\,\sigma(a)'$ together with (7) show that $d(ab) = (da)\,b + a\,db \quad \forall\, a,b \in \mathcal{A}$. It is clear also that $\int da = 0 \quad \forall\, a \in \mathcal{A}$. It remains to show that $\int$ is a (graded) trace, i.e. that $\int a\omega = \int \omega a \quad \forall\,\omega \in \Omega^1, a \in \mathcal{A}$. With $\omega = (a_1,f)$ one has

$$a\omega - \omega a = (aa_1 - a_1a, \sigma(a)f + i\sigma(a_1)\sigma(a)' - f\sigma(a)) = (aa_1 - a_1a, i\sigma(a_1)\sigma(a)').$$

Thus (5) shows that $\tau(aa_1 - a_1a) + \frac{1}{2\pi}\int i\sigma(a_1)\sigma(a)'\,d\theta = 0.$

We let χ be the cyclic 1-cocycle which is the character of the above cycle, explicitly,

$$\chi(a_0, a_1) = \int a_0\,da_1 \qquad \forall\, a_0, a_1 \in \mathcal{A}. \tag{10}$$

As above in proposition 3, we let φ be the cochain,

$$\varphi_0(a) = \mathrm{Trace}(\,a\,|D|^{-s})_{s=0},$$

$$\varphi_2(a_0, a_1, a_2) = \frac{1}{24}\int a_0\,\delta(a_1)\,\delta^2(a_2)\,|D|^{-3}.$$

Theorem 10. *The local index formula of the spectral triple $(\mathcal{A}, \mathcal{H}, D)$ is given by the cyclic cocycle χ up to the coboundary of the cochain (φ_0, φ_2).*

In other words the cocycle ψ_1 of proposition 3 is equal to χ. This follows from (15), (16).

We leave it as an exercice for the reader to compute the (non-zero) pairing between the above cyclic cocycle χ and the K-theory class of the basic unitary,

$$U = \begin{bmatrix} \alpha & -q\beta^* \\ \beta & \alpha^* \end{bmatrix} \tag{11}$$

Applying (14) we obtain the following corollary,

Corollary 2. *The character* $\mathrm{Trace}(a_0[F, a_1])$ *of the spectral triple $(\mathcal{A}, \mathcal{H}, D)$ is given by the cyclic cocycle χ up to the coboundary of the cochain ψ_0 given by $\psi_0(a) = 2\,\mathrm{Trace}(\,a\,P\,|D|^{-s})_{s=0}.$*

The cochain ψ_0 is only non-zero on elements which are functions of $\beta^*\beta$ as one sees for homogeneity reasons using the bigrading and the natural basis of $\mathcal{A}$ given by the $\alpha^k(\beta^*)^n\beta^m$. It is thus entirely determined by the values $\psi_0((\beta^*\beta)^n)$. It is an interesting problem to compute these functions of q. In order to state the result we recall that the Dedekind eta-function for the modulus q^2 is given by,

$$\eta(q^2) = q^{\frac{1}{12}}\prod_1^\infty (1 - q^{2n}). \tag{12}$$

We let G be its logarithmic derivative $q^2\partial_{q^2}\eta(q^2)$, (up to sign and after substraction of the constant term),

$$G(q^2) = \sum_1^\infty n\,q^{2n}(1 - q^{2n})^{-1}. \tag{13}$$

Theorem 11. *The functions $\frac{1}{2}\psi_0((\beta^*\beta)^r)$ of the variable q are of the form*
$q^{-2r}(q^2 R_r(q^2) - G(q^2))$ *where $R_r(q^2)$ are rational fractions of q^2 with poles only at roots of unity.*

Proof. The first step is to prove that the diagonal terms $d(n, i, j)$ of the matrix $(\beta^*\beta)^r$ fulfill the equality,

$$d(n, n, j) = \prod_{l=0}^{r-1} \frac{q^{2n+2j} - q^{4n+2+2l}}{1 - q^{4n+4+2l}}. \tag{14}$$

This follows by writing $(\beta^*\beta)^r = \beta^{*r}\beta^r = (\beta_+^* + \beta_-^*)^r(\beta_+ + \beta_-)^r$ and observing that, since $i = n$, the only term which contributes is $(\beta_+^*)^r(\beta_+)^r$.
We then change variables as above replacing n by $n/2$ and j by $-n/2 + y$, which gives for the value of $\frac{1}{2}\psi_0((\beta^*\beta)^r)$ the value at $s = 0$ of the sum,

$$Z(s) = \sum_{n=0}^{\infty} n^{-s} \sum_{y=0}^{n} \prod_{l=0}^{r-1} \frac{q^{2y} - q^{2n+2+2l}}{1 - q^{2n+4+2l}}. \tag{15}$$

(with the usual convention for $n = 0$).
To understand the appearance of $G(q^2)$ and the corresponding coefficient, let us take the constant term in

$$P(q^{2y}) = \prod_{l=0}^{r-1} \frac{q^{2y} - q^{2n+2+2l}}{1 - q^{2n+4+2l}} \tag{16}$$

viewed as a polynomial in q^{2y}, which gives, with $x = q^{2n}$,

$$R(x) = \prod_{l=0}^{r-1} \frac{(-q^{2+2l}x)}{1 - q^{4+2l}x} \tag{17}$$

The fraction $R(1/z)$ has simple distinct poles at $z = q^{4+2l}$ and vanishes at ∞, thus we can express it in the form,

$$R(1/z) = \sum_{l=0}^{r-1} \frac{\lambda_l}{z - q^{4+2l}} \tag{18}$$

Each term contributes to (15) by the value at $s = 0$ of,

$$\lambda_l \sum_{n=0}^{\infty} n^{-s}(n+1) \frac{q^{2n}}{1 - q^{2n+4+2l}}. \tag{19}$$

The value at $s = 0$ makes sense as a convergent series, and the coefficient of $G(q^2)$ is obtained by setting $n' = n + l + 2$ which gives $\lambda_l \, q^{-4-2l} \, G(q^2)$. Thus the overall coefficient for $G(q^2)$ is, using (17), (18) and the behaviour for $z = 0$,

$$\sum_{l=0}^{r-1} \frac{\lambda_l}{q^{4+2l}} = -q^{-2r}. \tag{20}$$

One has $n + 1 = n' - l - 1$ and the above terms also generate a non-zero multiple of the function,

$$G_0(q^2) = \frac{q^{2n}}{1 - q^{2n}}. \tag{21}$$

The coefficient is given by,

$$c_0 = -\sum_{l=0}^{r-1} \lambda_l (l + 1) q^{-2l-4}. \tag{22}$$

We need to show that the other terms coming from the non-constant terms in $P(q^{2y})$ exactly cancell the above multiple of $G_0(q^2)$, modulo rational functions of q^2.

Using the q^2-binomial coefficients $\binom{r}{k}_{q^2}$, one obtains, with P as in (16), that,

$$P(z) = N(z) \prod_{l=0}^{r-1} (1 - q^{2n+4+2l})^{-1} \tag{23}$$

where,

$$N(z) = \sum_{k=0}^{r} (-1)^k q^{k(k+1)} q^{2kn} \binom{r}{k}_{q^2} z^{r-k} \tag{24}$$

The constant term (in z^0) has already been taken care of, and for the others the effect of the summation $\sum_{y=0}^{n}$ in (15) is to replace z^{r-k} in the above sum by,

$$\sum_{y=0}^{n} q^{2y(r-k)} = \frac{1 - q^{2(r-k)(n+1)}}{1 - q^{2(r-k)}} \tag{25}$$

Thus the contribution of the other terms is governed by the rational fraction of $x = q^{2n}$

$$Q(x) = \left(\sum_{k=0}^{r-1} (-1)^k q^{k(k+1)} \binom{r}{k}_{q^2} \frac{x^k - x^r q^{2(r-k)}}{1 - q^{2(r-k)}}\right) \prod_{l=0}^{r-1} (1 - x \, q^{4+2l})^{-1} \tag{26}$$

The degree of the numerator is the same as the degree of the denominator, all poles are simple, and we can thus expand $Q(x)$ as,

$$Q(x) = \mu + \sum_{l=0}^{r-1} \frac{\mu_l}{1 - x \, q^{4+2l}} \tag{27}$$

The same reasoning as above shows that modulo rational functions of q^2, each term contributes to (15) by a multiple of $G_0(q^2)$, while the overall coefficient is the sum of the μ_l,

$$c_1 = Q(0) - Q(\infty) \tag{28}$$

Using (26) one obtains,

$$c_1 = (1 - q^{2r})^{-1} + (-1)^r q^{-r(r+1)} \sum_{k=0}^{r-1} (-1)^k \binom{r}{k}_{q^2} \frac{q^{k(k-1)}}{1 - q^{2(r-k)}} \tag{29}$$

To compute the λ_l one takes the residues of (18) which gives the formula,

$$\lambda_j = -(-1)^j q^{(4+j^2+j(3-2r)-3r+r^2)} \binom{r-1}{j}_{q^2} \rho(r-1)^{-1} \tag{30}$$

where,

$$\rho(r-1) = \prod_{a=1}^{r-1} (q^{2a} - 1) \tag{31}$$

This gives the following formula for the coefficient c_0,

$$c_0 = \rho(r-1)^{-1} \sum_{j=0}^{r-1} (-1)^j (j+1) q^{(j^2+j-2rj-3r+r^2)} \binom{r-1}{j}_{q^2} \tag{32}$$

The fundamental cancellation now is the identity

$$c_0 + c_1 = 0 \tag{33}$$

which is proved by differentiation of the q^2-binomial formula. The above discussion provides an explicit formula for the rational fractions R_r which allows to check that their only poles are roots of unity.

The simple expression $q^{-2r} G(q^2)$ blows up exponentially for $r \to \infty$ and if it were alone it would be impossible to extend the cochain ψ_0 from the purely algebraic to the smooth framework. However,

$$\psi_0((\beta^* \beta)^r) \to 1 + 2q^2/(q^2 - 1) \text{ when } r \to \infty.$$

Thus it is only by the virtue of the rational approximations $q^2 R_r(q^2)$ of $G(q^2)$ that the tempered behaviour of $\psi_0((\beta^* \beta)^r)$ is insured.

The list of the first $R_r(q)$ is as follows,

$$R_1[q] = \frac{3}{2(1-q)}, \qquad R_2[q] = \frac{2 + 5q - 3q^2}{2(-1+q)^2(1+q)}, \tag{34}$$

$$R_3[q] = \frac{2 + 8q + 13q^2 + 11q^3 - q^4 - 3q^5}{2(-1+q^2)^2(1+q+q^2)},$$

$$R_4[q] = \frac{2 + 10q + 24q^2 + 43q^3 + 50q^4 + 46q^5 + 24q^6 + 4q^7 - 4q^8 - 3q^9}{2(1+q^2)(-1-q+q^3+q^4)^2}.$$

Finally, note that the appearance of the function $G(q^2)$ in ψ_0 is not an artefact which could be eliminated by a better choice of cochain with the same coboundary. Indeed since

$$\alpha \alpha^* - \alpha^* \alpha = (1 - q^2) \beta^* \beta$$

the coboundary $b\psi_0(\alpha, \alpha^*)$ still involves $G(q^2)$.

12 Quantum Groups and Invariant Cyclic Cohomology

The main virtue of the above spectral triple for $SU_q(2)$ is its invariance under left translations (cf. [7]). More precisely the following equalities define an action of the envelopping algebra $\mathcal{U} = U_q(SL(2))$ on $\mathcal{H}$ which commutes with D and implements the translations on $C^\infty(SU_q(2))$,

$$k\, e_{ij}^{(n)} = q^j\, e_{ij}^{(n)} \tag{1}$$

$$e\, e_{ij}^{(n)} = q^{-n+\frac{1}{2}}(1 - q^{2(n+j+1)})^{1/2}(1 - q^{2(n-j)})^{1/2}(1 - q^2)^{-1} e_{ij+1}^{(n)} \tag{2}$$

while $f = e^*$.

With these notations one has,

$$ke = q\,ek, \quad kf = q^{-1}fk, \quad [e, f] = \frac{k^2 - k^{-2}}{q - q^{-1}}. \tag{3}$$

The vector $\Omega = e_{(0,0)}^{(0)}$ is preserved by the action and one has a natural densely defined action of $\mathcal{U}$ on $\mathcal{A} = C^\infty(SU_q(2))$ such that,

$$h(x)\Omega = h(x\Omega) \qquad \forall x \in \mathcal{A}, \ h \in \mathcal{U}. \tag{4}$$

The coproduct is given by,

$$\Delta k = k \otimes k, \quad \Delta e = k^{-1} \otimes e + e \otimes k, \quad \Delta f = k^{-1} \otimes f + f \otimes k \tag{5}$$

and the action of $\mathcal{U}$ on $\mathcal{A}$ fulfills,

$$h(xy) = \sum h_{(1)}(x)\, h_{(2)}(y) \qquad \forall x, y \in \mathcal{A}, \tag{6}$$

$\forall h \in \mathcal{U}$ with $\Delta h = \sum h_{(1)} \otimes h_{(2)}$.

On the generators $\alpha, \alpha^*, \beta, \beta^*$ of $\mathcal{A}$ one has,

$$k(\alpha) = q^{-1/2}\alpha, \quad k(\beta) = q^{-1/2}\beta, \quad e(\alpha) = q\beta^*, \tag{7}$$

$$e(\beta) = -\alpha, \quad e(\alpha^*) = 0, \quad e(\beta^*) = 0.$$

This representation of $\mathcal{U}$ in $\mathcal{H}$ generates the regular representation of the compact quantum group $SU_q(2)$ and we let $M = \mathcal{U}''$ be the von Neumann algebra it generates in $\mathcal{H}$. It is a product $M = \prod_{n \in \frac{1}{2}\mathbb{N}} M_{2n+1}(\mathbb{C})$ of matrix algebras, where $M_{2n+1}(\mathbb{C})$ acts with multiplicity $2n + 1$ in the space $\mathcal{H}_{(n)} = \{\text{span of } e_{i,j}^{(n)}\}$.

The elements of $\mathcal{U}$ are unbounded operators affiliated to M and at the qualitative level we shall leave the freedom to choose a weakly dense subalgebra $\mathcal{C}$ of M. Since all the constructions performed so far in this paper were canonically dependent on the spectral triple, the $SU_q(2)$ equivariance of $(\mathcal{A}, \mathcal{H}, D)$ should entail a corresponding *invariance* of all the objects we delt with. We shall concentrate on the cyclic cohomology aspect and show that indeed there is a fairly natural and simple notion of *invariance* fulfilled by all cochains involved in the above computation.

The main point is that we can enlarge the algebra $\mathcal{A}$ to the algebra $\mathcal{D} = \mathcal{A} \rtimes \mathcal{C}$ generated by $\mathcal{A}$ and $\mathcal{C}$, extend the cochains on $\mathcal{D}$ by similar formulas and use the commutation,

$$[D, c] = 0 \qquad \forall c \in \mathcal{C} \tag{8}$$

to conclude that the extended cochains fulfill the following key property,

Definition 3. *Let $\mathcal{D}$ be a unital algebra, $\mathcal{C} \subset \mathcal{D}$ a (unital) subalgebra and $\varphi \in C^n(\mathcal{D})$ an n-cochain. We shall say that φ is $\mathcal{C}$-constant iff both $\varphi(a^0, \ldots, a^n)$ and $(b\varphi)(a^0, \ldots, a^{n+1})$ vanish if one of the a^j, $j \geq 1$ is in $\mathcal{C}$.*

When $\mathcal{C} = \mathbb{C}$ this is a normalization condition.

When φ is $\mathcal{C}$-constant then $B_0\varphi(a^0, \ldots, a^{n-1}) = \varphi(1, a^0, \ldots, a^{n-1})$ so that $B\varphi = AB_0\varphi$ is also $\mathcal{C}$-constant. It follows that $\mathcal{C}$-constant cochains form a subcomplex of the (b, B) bicomplex of $\mathcal{D}$ and we can develop cyclic cohomology in that context, parallel to ([10],[11]). We shall denote by $HC_\mathcal{C}^*(\mathcal{D})$ the corresponding theory. In the above context we take for $\mathcal{D}$ the algebra $\mathcal{A} \rtimes \mathcal{C}$ and use the lighter notation $H\dot{C}_\mathcal{C}^*(\mathcal{A})$ for the corresponding theory.

Let us now give examples of specific cochains on $\mathcal{A} = C^\infty(\mathrm{SU}_q(2))$ which extend to $\mathcal{C}$-constant cochains on $\mathcal{A} \rtimes \mathcal{C} = \mathcal{D}$. We start with the non local form of the Chern character of the spectral triple,

$$\psi_1(a^0, a^1) = \mathrm{Trace}\,(a^0\,[F, a^1]) \qquad \forall a^0, a^1 \in \mathcal{A}. \tag{9}$$

Let us show how to extend ψ_1 to an M-constant cochain on $\mathcal{A} \rtimes M = \mathcal{D}$. An element of $\mathcal{D}$ is a finite linear combination of monomials $a_1\,m_1\,a_2\,m_2 \ldots a_\ell\,m_\ell$, where $a_j \in \mathcal{A}$, $m_\ell \in M$. But $[F, a]$ is a trace class operator for any $a \in \mathcal{A}$, while $[F, m] = 0 \quad \forall m \in M$, thus we get,

$$[F, x] \in \mathcal{L}^1 \qquad \forall x \in \mathcal{D} = \mathcal{A} \rtimes M. \tag{10}$$

We can thus define $\widetilde{\psi}_1$ as the character of the module $(\mathcal{H}, F)$ on $\mathcal{D}$, namely,

$$\widetilde{\psi}_1(x_0, x_1) = \mathrm{Trace}\,(x_0\,[F, x_1]). \tag{11}$$

It is clear that $\widetilde{\psi}_1$ is M-constant and that $b\widetilde{\psi}_1 = 0$ so that $b\widetilde{\psi}_1$ is also M-constant, $\widetilde{\psi}_1 \in HC_M^1(\mathcal{A})$. This example is quite striking in that we could extend ψ_1 to a very large algebra. Indeed if we stick to bounded operators M is the largest possible choice for $\mathcal{C}$. A similar surprising extension of a cyclic 1-cocycle in a von-Neumann algebra context already occured in the anabelian 1-traces of ([13]). As a next example let us take the functional on $\mathcal{A}$ which is the natural trace,

$$\psi_0(x) = \frac{1}{2\pi} \int \sigma(x)\,\mathrm{d}\theta \qquad \forall x \in C^\infty(\mathrm{SU}_q(2)). \tag{12}$$

When written like this, its $\mathrm{SU}_q(2)$-invariance is not clear and in fact cannot hold in the simplest sense since this would contradict the uniqueness of the Haar state on

$C^\infty(\mathrm{SU}_q(2))$. Let us however show that ψ_0 extends to an M-constant cochain (in fact an M-constant trace) on $\mathcal{D} = \mathcal{A} \rtimes M$ as above. To do this we rewrite (12) as,

$$\psi_0(x) = \mathrm{Tr}_\omega(x\,|D|^{-3}) \qquad \forall\, x \in C^\infty(\mathrm{SU}_q(2)) \tag{13}$$

where Tr_ω is the Dixmier trace (section 3 and ([29],[19])) and simply write the extension as,

$$\widetilde{\psi}_0(x) = \mathrm{Tr}_\omega(x\,|D|^{-3}). \tag{14}$$

For any monomial $\mu = a_1\, m_1 \ldots a_m\, m_m$ as above one has $[D, \mu]$ bounded and $[|D|, \mu]$ bounded. Thus it follows from the general properties of Tr_ω that,

$$\widetilde{\psi}_0(xy) = \widetilde{\psi}_0(yx) \qquad \forall\, x, y \in \mathcal{D} = \mathcal{A} \rtimes M. \tag{15}$$

This shows of course that $\widetilde{\psi}_0$ is a 0-cycle in the invariant cyclic cohomology $HC^0_M(\mathcal{A})$. After giving these simple examples it is natural to wonder wether the above notion of $\mathcal{C}$-constant cochain is restrictive enough. Here is a simple consequence of this hypothesis:

Proposition 6. Let $\mathcal{C} \subset \mathcal{D}$ be unital algebra and $\varphi \in C^n_\mathcal{C}$ be a $\mathcal{C}$-constant cochain on $\mathcal{D}$. Then for any invertible element $u \in \mathcal{C}$ one has,

$$\varphi(u\,a^0 u^{-1}, u\,a^1 u^{-1}, \ldots, u\,a^n u^{-1}) = \varphi(a^0, \ldots, a^n).$$

Proof. One has $b\varphi(a^0, u, a^1, \ldots, a^n) = 0$ so that $\varphi(a^0 u, a^1, \ldots, a^n) - \varphi(a^0, ua^1, \ldots, a^n) = 0$ since all other terms have u as an argument and hence vanish. Similarly $\varphi(a^0, \ldots, a^{j-1}u, a^j, \ldots, a^n) = \varphi(a^0, \ldots, a^{j-1}, ua^j, \ldots, a^n)$ for all $j \in \{1, \ldots, n\}$ and $\varphi(ua^0, \ldots, a^n) = \varphi(a^0, \ldots, u^n u)$. Applying these equalities yields the statement.

Let us now consider the more sophisticated cochains which appeared throughout and show how to extend them to $\mathcal{C}$-constant cochains on $\mathcal{D} = \mathcal{A} \rtimes \mathcal{C}$ for suitable algebra $\mathcal{C}$ describing the quantum group $\mathrm{SU}_q(2)$.

We first note that the action of the envelopping algebra $\mathcal{U} = U_q(\mathrm{SL}(2))$ on $\mathcal{A}$ extends to an action on the algebra of pseudo-differential operators. First it extends to $\mathcal{B}$ with the following action on the generators $\alpha_\pm, \alpha_\pm^*, \beta_\pm, \beta_\pm^*$,

$$k(\alpha_\pm) = q^{-1/2}\alpha_\pm\,, \quad k(\beta_\pm) = q^{-1/2}\beta_\pm\,, \tag{16}$$

and

$$e(\alpha_\pm) = q\beta_\mp^*\,, \ e(\beta_\pm) = -\alpha_\mp\,, \ e(\alpha_\pm^*) = 0\,, \ e(\beta_\pm^*) = 0\,. \tag{17}$$

Moreover $\mathcal{U}$ acts through the trivial representation on D, $|D|$ and F.

In fact it is important to describe the action of $\mathcal{U}$ on arbitrary pseudo-differential operators by a closed formula and this is achieved by,

Proposition 7. *The action of the generators* k, e, f *of* $\mathcal{U}$ *on pseudo-differential operators* P *is given by,* a) $k(P) = kPk^{-1}$, b) $e(P) = ePk^{-1} - qk^{-1}Pe$, c) $f(P) = fPk^{-1} - q^{-1}k^{-1}Pf$.

Proof. These formulas just describe the tensor product of the action of $\mathcal{U}$ in $\mathcal{H}$ by the contragredient representation, since the antipode S in $\mathcal{U}$ fulfills

$$S(k) = k^{-1}, \ S(e) = -qe, \ S(f) = -q^{-1}f. \tag{18}$$

One checks directly that they agree with (16) and (17) on the generators $\alpha_{\pm}, \ldots, \beta_{\pm}^*$ as well as on $D, |D|$ and F. Thus we are just using the natural implementation of the action of $\mathcal{U}$ which extends this action to operators.

The only technical difficulty is that the generators of $\mathcal{U}$ are unbounded operators in $\mathcal{H}$ so that to extend cochains to $\mathcal{A} \rtimes \mathcal{U}$ requires a little more work. In fact the only needed extension is for the residue,

$$\fint P = \text{Res}_{s=0} \text{Trace}\,(P\,|D|^{-s})\,. \tag{19}$$

Using formula (θ) we can reexpress (19) as follows,

$$\fint P = \frac{1}{2} \text{ coefficient of } \log t^{-1} \text{ in Trace}\,(Pe^{-tD^2})\,. \tag{20}$$

Thus more precisely we let $\theta_P(t) = \text{Trace}\,(Pe^{-tD^2})$ and assume that it has an asymptotic expansion for $t \to 0$ of the form

$$\theta_P(t) \sim \sum a_\alpha\, t^{-\alpha} + \lambda \log t^{-1} + a_0 + \cdots, \tag{21}$$

then the equality between (19) and (20) holds, both formulas giving $\lambda/2$. In our context we could use (20) above instead of (19) since we always controlled the size of $\zeta_b(s)$ on vertical strips to perform the inverse Mellin transform.

Let now L be an arbitrary extension of the linear form on function $f \in C^\infty \,(]0, \infty[)$ which satisfies,

$$L(f) = \frac{1}{2} \text{ coefficient of } \log t^{-1} \text{ if } f \text{ admits} \tag{22}$$

an asymptotic expansion (21).

We then extend the definition (19) by,

$$\fint_L P = L(\theta_P(t))\,. \tag{23}$$

With these notations we then have,

Proposition 8. *Let* $(k_1, \ldots, k_n)$ *be a multi-index, then the formula*

$$\widetilde{\psi}(a^0, \ldots, a^n) = \int_L a^0 [D, a^1]^{(k_1)} \ldots [D, a^n]^{(k_n)} |D|^{-n-|k|}$$

where $T^{(k)} = \delta^k(T)$, defines a $\mathcal{U}$-constant extension of the restriction ψ of $\widetilde{\psi}$ to $\mathcal{A}$ to the algebra $\mathcal{D} = \mathcal{A} \rtimes \mathcal{U}$.

Proof. In computing $b\widetilde{\psi}$ one uses the equality

$$\delta^k([D, a\,b]) = \delta^k([D, a])\,b + a\,\delta^k([D, b]) + \sum_{j=0}^{k-1} C_k^j\,\delta^j([D, a])\,\delta^{k-j}(b)$$

$$+ \sum_{\ell=1}^{k} \delta^\ell(a)\,\delta^{k-\ell}([D, b]). \qquad (24)$$

Thus in $b\widetilde{\psi}(a_0, \ldots, a_{n+1})$ the only term which does not involve a derivative of a is of the form,

$$\int a_{n+1} T|D|^{-n-|k|} - \int T|D|^{-n-|k|} a_{n+1}. \qquad (25)$$

This shows that $b\widetilde{\psi}$ vanishes if any of the $a_j \in \mathcal{U}$ for $j = 1, \ldots, n$. For $j = n + 1$, i.e. for $a_{n+1} = v \in \mathcal{U}$ one has the term (25) but since v commutes with D one has,

$$\theta_{vT} = \theta_{Tv}, \qquad (26)$$

and one gets the desired result.

This proposition shows the richness of the space of $\mathcal{U}$-constant cochains, but it does not address the more delicate issue of computing the cyclic cohomology $HC_\mathcal{U}^*(\mathcal{A})$. A much more careful choice of L would be necessary if one wanted to lift cocycles to cocycles.

We shall now show that $HC_\mathcal{U}^*(\mathcal{A})$ which obviously maps to the ordinary cyclic theory $HC^*(\mathcal{A})$,

$$HC_\mathcal{U}^*(\mathcal{A}) \xrightarrow{\rho} HC^*(\mathcal{A}), \qquad (27)$$

also maps in fact to the "twisted" cyclic cohomology $HC_{\mathrm{inv}}^*(\mathcal{A}, \theta)$ proposed in [37], where θ is the inner automorphism implemented by k^2. This, as explained in the introduction, should allow to use the latter proposal for some relevant purpose, namely as a "detector" of classes in $HC_\mathcal{U}^*(\mathcal{A})$.

To see what happens, let us start with a $\mathcal{U}$-constant 0-dimensional cochain ψ on $\mathcal{A} \rtimes \mathcal{U}$ and get an analogue of the group invariance provided by proposition 6. One has of course $\psi(kak^{-1}) = \psi(a)$ but this is not much. We would like a similar statement for the other generator e of $U_q(SL(2))$. Now by proposition 7 one has $e(a) = eak^{-1} - qk^{-1}ae$ so that $e(a)k^2 = eak - qk^{-1}aek^2$. But $\mathcal{U}$ is in the centraliser of ψ by proposition 6 and thus,

$$\psi(eak) = \psi(kea), \quad \psi(k^{-1}aek^2) = \psi(eka),$$

hence $\psi(e(a)\,k^2) = 0$. One gets in general,

$$\psi(h(a)k^2) = \varepsilon(h)\,\psi(a\,k^2) \tag{28}$$

which is the usual invariance of a linear form. More generally one has,

Proposition 9. *The equality* $\rho_\theta(\psi)(a_0,\ldots,a_n) = \psi(a_0,\ldots,a_nk^2)$ *defines a morphism,*

$$HC^*_{\mathcal{U}}(\mathcal{A}) \xrightarrow{\rho_\theta} HC^*_{\text{inv}}(\mathcal{A},\,\theta)\,.$$

where θ *is the inner automorphism implemented by* k^2.

We have seen above (11),(14) that the basic cohomology classes in the ordinary cyclic theory $HC^*(\mathcal{A})$ of $\mathcal{A}$ lift to actual cocycles in $HC^*_M(\mathcal{A})$ where M is the von-Neumann algebra bicommutant of $\mathcal{U}$. It is however not clear that they lift to $HC^*_{\mathcal{U}}(\mathcal{A})$ since the generators of $\mathcal{U}$ are unbounded operators. We can however insure that such liftings exist in the entire cyclic cohomology ([14])([57]) since the θ-summability of the spectral triple continues to hold for the algebra $\mathcal{A} \rtimes \mathcal{U}$. This point is not unrelated to the atempt by Goswami in ([31]). What we have shown here is that the local formulas work perfectly well in the context of quantum groups, and that the framework of NCG needs no change whatsoever, at least as far as $SU_q(2)$ is concerned. The only notion that requires more work is that of invariance in the q-group context.

Finally the above notion of invariant cyclic cohomology is complementary to the theory developped in ([17],[18]). In the latter the Hopf action is used to construct ordinary cyclic cocycles from twisted-traces. In the q-group situation, cocycles thus constructed from the right translations should be left-invariant in the above sense.

13 Appendix

We have not defined carefully the smooth algebras C^∞ involved in section 6. A careful definition can however be deduced from their structure and the exact sequence involving $OP^{-\infty}$ and the symbol map provided by lemma 1. What really matters is that the obtained algebras are stable under holomorphic functional calculus (h.f.c.) and we shall now provide the technical lemma which allows to check this point.

Let $(B,\mathcal{H},D)$ be a spectral triple. As above we say that an operator P in $\mathcal{H}$ is of order α when,

$$|D|^{-\alpha}\,P \in \bigcap_{n=1}^{\infty} \text{Dom}\,\delta^n \tag{1}$$

where δ is the unbounded derivation given by,

$$\delta(T) = |D|\,T - T\,|D|\,. \tag{2}$$

Thus $OP^0 = \bigcap_{n=1}^{\infty} \text{Dom}\,\delta^n$ is the algebra of operators of order 0 and $OP^{-\infty}$ is a two sided ideal in OP^0.

Let $\rho : B \to C$ a morphism of C^*-algebras, $\mathcal{C} \subset C$ be a subalgebra stable under h. f. c. and $\lambda : \mathcal{C} \to \mathcal{L}(\mathcal{H})$ be a linear map such that $\lambda(1) = 1$ and,

$$\lambda(c) \in OP^0, \ \forall c \in \mathcal{C} \tag{3}$$
$$\lambda(a\,b) - \lambda(a)\lambda(b) \in OP^{-\infty}, \ \forall a, b \in \mathcal{C}.$$

We then have the following,

Lemma 2. *Let* $\mathcal{B} = \{x \in B \,;\, x \in OP^0,\, \rho(x) \in \mathcal{C},\, x - \lambda(\rho(x)) \in OP^{-\infty}\}$. *Then* $\mathcal{B} \subset B$ *is a subalgebra stable under holomorphic functional calculus.*

Proof. Let $x \in \mathcal{B}$ be invertible in B, let us show that $x^{-1} \in \mathcal{B}$. Let $a = \rho(x)$, then since $\mathcal{C}$ is stable under h.f.c. the inverse $b = \rho(x^{-1})$ of a belongs to $\mathcal{C}$. Also since $x \in OP^0$ we have $x^{-1} \in OP^0$. Let us show that $x^{-1} - \lambda(b) \in OP^{-\infty}$. Since $a\,b = 1$ one has by (3), $\lambda(a)\lambda(b) - 1 \in OP^{-\infty}$. But $x - \lambda(a) \in OP^{-\infty}$ and $OP^{-\infty}$ is a two-sided ideal in OP^0, thus multiplying $x - \lambda(a)$ by $\lambda(b)$ on the right, we get $x\,\lambda(b) - 1 \in OP^{-\infty}$. Finally since $x^{-1} \in OP^0$ we get, multiplying $x\,\lambda(b) - 1$ on the left by x^{-1} that $\lambda(b) - x^{-1} \in OP^{-\infty}$.

References

1. M.F. ATIYAH, *Global theory of elliptic operators*. Proc. Internat. Conf. on *Functional Analysis and Related Topics* (Tokyo, 1969). University of Tokyo press, Tokyo 1970, 21-30.

2. M. F. ATIYAH, V. K. PATODI, I. M. SINGER, *Spectral Asymmetry and Riemannian Geometry*. Bull. London Math. Soc. 5 (1973), 229-234.

3. M. F. ATIYAH, H. DONNELLY, I. M. SINGER, *Eta Invariants, Signature Defects of Cusps and Values of L-functions*. Ann. of Math. 118 (1983), 131-171.

4. A.R. BERNSTEIN AND F. WATTENBERG, *Non standard measure theory*. In Applications of model theory to algebra analysis and probability, Edited by W.A.J. Luxenburg Halt, Rinehart and Winstin (1969).

5. L.G. BROWN - R.G. DOUGLAS - P.A. FILLMORE, *Extensions of C*-algebras and K-homology*. Ann. of Math., 2, 105, 1977, 265-324.

6. D. BURGHELEA, *The cyclic homology of the group rings*. Comment. Math. Helv., 60, 1985, 354-365.

7. P. S. CHAKRABORTY, A. PAL, *Equivariant Spectral triple on the Quantum SU(2)-group*, math.KT/0201004.

8. A. CONNES, *C*-algèbres et géométrie differentielle*. C.R. Acad. Sci. Paris, Ser. A-B , 290, 1980.

9. A. CONNES, *Spectral sequence and homology of currents for operator algebras*. Math. Forschungsinst. Oberwolfach Tagungsber., 41/81; Funktionalanalysis und C*-Algebren, 27-9/3-10, 1981.

10. A. CONNES, *Noncommutative differential geometry. Part I: The Chern character in K-homology*. Preprint IHES, M/82/53, 1982; *Part II: de Rham homology and noncommutative algebra*. Preprint IHES, M/83/19, 1983.

11. A. CONNES, *Noncommutative differential geometry*. Inst. Hautes Etudes Sci. Publ. Math., 62, 1985, 257-360.

70   Alain Connes

12. A. CONNES, *Cohomologie cyclique et foncteur Ext^n*. C.R. Acad. Sci. Paris, Ser. I Math, 296, 1983.
13. A. CONNES, *Cyclic cohomology and the transverse fundamental class of a foliation.* Geometric methods in operator algebras (Kyoto, 1983), pp. 52-144, Pitman Res. *Notes in Math.* **123** Longman, Harlow (1986).
14. A. CONNES, *Entire cyclic cohomology of Banach algebras and characters of θ-summable Fredholm modules.* K-theory, 1, 1988, 519-548.
15. A. CONNES - H. MOSCOVICI, *Cyclic cohomology, the Novikov conjecture and hyperbolic groups.* Topology, 29, 1990, 345-388.
16. A. CONNES AND H. MOSCOVICI, *The local index formula in noncommutative geometry,* GAFA, **5** (1995), 174-243.
17. A. CONNES - H. MOSCOVICI, *Hopf Algebras, Cyclic Cohomology and the Transverse Index Theorem.* Commun. Math. Phys.,198, 1998, 199-246.
18. A. CONNES - H. MOSCOVICI, *Cyclic Cohomology and Hopf Algebra symmetry.* Letters Math. Phys., 52, 1, 2000, 1-28, math.QA/0002125.
19. A. CONNES, *Noncommutative Geometry*, Academic Press (1994).
20. A. CONNES, *Gravity coupled with matter and foundation of noncommutative geometry.* Commun. Math. Phys., 182,1996, 155-176.
21. A. CONNES, *Noncommutative geometry and reality.* Journal of Math. Physics, 36, n.11, 1995.
22. A. CONNES, *Noncommutative Geometry: The Spectral Aspect.* Les Houches Session LXIV, Elsevier 1998, 643-685.
23. A. CONNES, G. LANDI, *Noncommutative manifolds, the instanton algebra and isospectral deformations*, math.QA/0011194.
24. A. CONNES, M. DUBOIS-VIOLETTE, *Noncommutative finite-dimensional manifolds. I. spherical manifolds and related examples.* Math QA/0107070
25. J. CUNTZ - D. QUILLEN, *Cyclic homology and singularity.* J. Amer. Math. Soc., 8, 1995, 373-442.
26. J. CUNTZ - D. QUILLEN, *Operators on noncommutative differential forms and cyclic homology.* J. Differential Geometry, to appear.
27. J. CUNTZ - D. QUILLEN, *On excision in periodic cyclic cohomology, I and II.* C. R. Acad. Sci. Paris, Ser. I Math., 317, 1993, 917-922; 318, 1994, 11-12.
28. L. DABROWSKI - G. LANDI - T. MASUDA, *Instantons on the quantum 4-spheres S_q^4,* math.QA/0012103.
29. J. DIXMIER, *Existence de traces non normales.* C.R. Acad. Sci. Paris, Ser. A-B, 262, 1966.
30. D. GOSWAMI, *Some Noncommutative Geometric Aspects of $SU_q(2)$*, math.ph/0108003.
31. D. GOSWAMI, *Twisted entire cyclic cohomology, JLO-cocycles and equivariant spectral triples*, math.ph/0204010.
32. J. M. GRACIA-BONDIA, J. C. VARILLY, H. FIGUEROA, *Elements of Noncommutative geometry.* Birkhauser, 2000.
33. A. JAFFE - A. LESNIEWSKI - K. OSTERWALDER, *Quantum K-theory: I. The Chern character.* Commun. Math. Phys., 118, 1988, 1-14.
34. M. KAROUBI, *Homologie cyclique et K-théorie*, Asterisque 149 (1987).
35. G.G. KASPAROV, *The operator K-functor and extensions of C^* algebras.* Izv. Akad. Nauk SSSR, Ser. Mat., 44, 1980, 571-636; Math. USSR Izv., 16, 1981, 513-572.
36. D. KASTLER, *The Dirac operator and gravitation.* Commun. Math. Phys., 166, 1995, 633-643.
37. J. KUSTERMANS - G.J. MURPHY - L. TUSET, *Differential calculi over quantum groups and twisted cyclic cocycles*, math.QA/0110199.

38. B. LAWSON - M.L. MICHELSON, *Spin Geometry*, Princeton 1989.
39. J.L. LODAY, *Cyclic Homology*. Springer, Berlin - Heidelberg - New York 1998.
40. S. MAJID, *Foundations of Quantum Group Theory*, Cambridge University Press (1995).
41. Y. MANIN, *Quantum groups and noncommutative geometry*. Centre Recherche Math. Univ. Montréal 1988.
42. J. MILNOR - D. STASHEFF, *Characteristic classes*. Ann. of Math. Stud., Princeton University Press, Princeton, N.J. 1974.
43. A.S. MISCENKO, C^* *algebras and K theory*. Algebraic Topology, Aarhus 1978, Lecture Notes in Math., 763, Springer-Verlag, 1979, 262-274.
44. B. RIEMANN, *Mathematical Werke*. Dover, New York 1953.
45. I.M. SINGER, *Future extensions of index theory and elliptic operators*. Ann. of Math. Studies, 70, 1971, 171-185.
46. D. SULLIVAN. *Geometric periodicity and the invariants of manifolds*. Lecture Notes in Math., 197, Springer, 1971.
47. M. WODZICKI, *Noncommutative residue, Part I. Fundamentals*, K-theory, arithmetic and geometry, Lecture Notes in Math. 1289, Springer-Berlin (1987).
48. S.L. WORONOWICZ, *Compact matrix pseudogroups*, Commun. Math. Phys. **111** (1987), 613-665.
49. S.L. WORONOWICZ, *Differential calculus on compact matrix pseudogroups (quantum groups)*, Commun. Math. Phys. **122** (1989), 125-170.

Cyclic Theory and the Bivariant Chern-Connes Character

Joachim Cuntz *

Mathematisches Institut, Universität Münster
Einsteinstr. 62, 48149 Münster
cuntz@math.uni-muenster.de

Abstract We give a survey of cyclic homology/cohomology theory including a detailed discussion of cyclic theories for various classes of topological algebras. We show how to associate cyclic classes with Fredholm modules and K-theory classes and how to construct a completely general bivariant Chern-Connes character from bivariant K-theory to bivariant cyclic theory.

1 Introduction

The two fundamental "machines" of non-commutative geometry are cyclic homology and (bivariant) topological K-theory. In the present notes we describe these two theories and their connections. Cyclic theory can be viewed as a far reaching generalization of the classical de Rham cohomology, while bivariant K-theory includes the topological K-theory of Atiyah-Hirzebruch as a very special case.

The classical commutative theories can be extended to a striking amount of generality. It is important to note however that the new theories are by no means based simply on generalizations of the existing classical constructions. In fact, the constructions are quite different and give, in the commutative case, a new approach and an unexpected interpretation of the well-known classical theories. One aspect is that some of the properties of the two theories become visible only in the non-commutative category. For instance both theories have certain universality properties in this setting.

Bivariant K-theory has first been defined and developed by Kasparov on the category of C^*-algebras (possibly with the action of a locally compact group) thereby unifying and decisively extending previous work by Atiyah-Hirzebruch, Brown-Douglas-Fillmore and others. Kasparov also applied his bivariant theory to obtain striking positive results on the Novikov conjecture. Very recently, it was discovered that in

* Research supported by the Deutsche Forschungsgemeinschaft

fact, bivariant topological K-theories can be defined on a wide variety of topological algebras ranging from discrete algebras and very general locally convex algebras to e.g. Banach algebras or C^*-algebras (possibly equipped with a group action). If E is the covariant functor from such a category of algebras given by topological K-theory or also by periodic cyclic homology, then it satisfies the following three fundamental properties:

(E1) E is diffotopy invariant, i.e., the evaluation map ev_t in any point $t \in [0, 1]$ induces an isomorphism $E(ev_t) : E(\mathfrak{A}[0, 1]) \to E(\mathfrak{A})$ for any $\mathfrak{A}$ in C. Here $\mathfrak{A}[0, 1]$ denotes the algebra of $\mathfrak{A}$-valued C^∞-functions on $[0, 1]$.

(E2) E is stable, i.e., the canonical inclusion $\iota : \mathfrak{A} \to \mathfrak{K} \hat{\otimes} \mathfrak{A}$, where $\mathfrak{K}$ denotes the algebra of infinite $\mathbb{N} \times \mathbb{N}$-matrices with rapidly decreasing coefficients, induces an isomorphism $E(\iota)$ for any $\mathfrak{A}$ in C.

(E3) E is half-exact, i.e., each extension $0 \to \mathfrak{J} \to \mathfrak{A} \to \mathfrak{B} \to 0$ in C admitting a continuous linear splitting induces a short exact sequence $E(\mathfrak{J}) \to E(\mathfrak{A}) \to E(\mathfrak{B})$

"Diffotopy", i.e., differentiable homotopy is used in (E1) for technical reasons in connection with the homotopy invariance properties of cyclic homology. For K-theory, it could also be replaced by ordinary continuous homotopy.

It turns out that the bivariant K-functor is the universal functor from the given category C of algebras into an additive category D (i.e., the morphism sets $D(\mathfrak{A}, \mathfrak{B})$ are abelian groups) satisfying these three properties 21.1.

Cyclic theory is a homology theory that has been discovered, starting from K-theory, independently by Connes and Tsygan. Connes' construction was in fact directly motivated by Kasparov's formalism for bivariant K-theory and in particular for K-homology. A crucial role is played by so-called Fredholm modules or spectral triples. Also Tsygan's work is closely related to K-theory, [43]. In fact, in his approach, the new theory was originally called "additive K-theory" and he pointed out that it is an additive version of Quillen's definition of algebraic K-theory. It was immediately realized that cyclic homology has close connections with de Rham theory, Lie algebra homology, group cohomology and index theorems.

The theory with the really good properties is the periodic theory introduced by Connes. Periodic cyclic homology HP_* satisfies the three properties (E1), (E2), (E3), [4], [18], [46], [47], [15]. Combining this fact with the universality property of bivariant K-theory leads to a multiplicative transformation (the bivariant Chern-Connes character) from bivariant K-theory to bivariant periodic cyclic theory. This transformation is a vast generalization of the classical Chern character in differential geometry. The construction of characters into cyclic theory was a motivating force for the development of the theory from the start, [4], [24].

The principal aim of this volume is to give an account of cyclic theory. Here, everything works in parallel algebraically as well as for locally convex algebras (to name just two examples think of the algebra of C^∞-functions on a smooth manifold or of algebras of pseudodifferential operators). Cyclic theory can be introduced using rather different complexes, each one of them having its own special virtues. Specifically, we will use the following complexes or bicomplexes

- The cyclic bicomplex with various realizations:

$$CC^n(A) \quad \text{with boundary operators} \quad b, b', Q, 1 - \lambda$$
$$\overline{CC}^n(\tilde{A}) \quad \text{with boundary operator} \quad B - b$$
$$\Omega(A) \quad \text{with boundary operator} \quad B - b$$

The cyclic bicomplex is well suited for the periodic theory as well as for the $\mathbb{Z}$-graded ordinary theory and for the connections between both.
- The Connes complex C_λ^n
  It has the advantage, that concrete *finite-dimensional* cocycles often arise naturally as elements of C_λ^n. The connection with Hochschild cohomology also fits naturally into this picture.
- The X-complex of any complete quasi-free extension of the given algebra A. This complex is very useful for a conceptual explanation of the properties of periodic theory, in particular for proving excision, for the connections with topological K-theory and the bivariant Chern-Connes character. It also is the natural framework for all infinite-dimensional versions of cyclic homology (analytic and entire as well as asymptotic and local theory).

In the following sections we discuss the basic properties of cyclic theory. We note however, that it is quite difficult to be exhaustive and we don't even try to give a complete account of all aspects of cyclic theory. For instance, an important notion which we don't treat is the one of a cyclic object. But also other important aspects have to be omitted. We focus on those notions and results which, we think, are most relevant for non-commutative geometry, including homotopy invariance, Morita invariance, excision but also explicit formulas for the Chern character associated to idempotents, invertibles, Fredholm modules etc..

After this we turn to a description of bivariant K-theory and to the construction of the bivariant Chern-Connes character, which generalizes the Chern character for idempotents, invertibles or Fredholm modules mentioned before.

As we pointed out already above, cyclic homology and bivariant K-theory can be defined on different categories of algebras - purely algebraically for algebras (over $\mathbb{R}$ or $\mathbb{C}$) or on categories of topological algebras like locally convex algebras, Banach algebras or C*-algebras. There are different variants of the two theories which are adapted to the different categories. In this text we treat cyclic theory in the purely algebraic case (however restricting to algebras over a field of characteristic 0) on the one hand. Concerning cyclic theory for topological algebras on the other hand, we have to make a choice. For the classical ("finite-dimensional") cyclic theories we concentrate on the category of what we call m-algebras. These are particularly nice locally convex algebras (projective limits of Banach algebras). Their advantage is that, both, ordinary cylic homology and bivariant topological K-theory make perfect sense and that the bivariant Chern-Connes-character can be constructed nicely on this category. It is important to note however that the restriction to the category of m-algebras is for convenience mainly.

Cyclic theory as well as bivariant K-theory can also be treated on many other categories of algebras. In particular, for the categories of Banach algebras or C^*-algebras there are special variants of cyclic theory, namely the entire and the local cyclic theory, which are especially designed for these categories. An interesting new feature here is the existence of *infinite-dimensional* cohomology classes. These theories, as well as a bivariant character from Kasparov's KK-theory to the local cyclic theory, are discussed in sections 22 and 23 which have been written by Ralf Meyer and Michael Puschnigg.

The text starts with a collection of examples of algebras, locally convex algebras and certain extensions of algebras. These serve as a reference for later and may be omitted at a first reading.

2 Some Examples of Algebras

We list in this section some examples of algebras and constructions on algebras that will be used later. For simplicity we always assume that the ground field is $\mathbb{C}$. Since it is important to do all constructions in the category of non-unital algebras we will frequently use the procedure of adjoining a unit to a given algebra A.

Definition 2.1. *The unitization* $\tilde{A}$ *of an algebra* A *is the vector space* $\mathbb{C} \oplus A$ *with multiplication given by*

$$(\lambda, x)(\mu, y) = (\lambda\mu, \lambda y + \mu x + xy)$$

$\tilde{A}$ *is a unital algebra and contains* A *as an ideal of codimension 1.*

2.1 Algebras of Polynomial Functions

For any algebra A, we can form the algebra $A[t]$ of polynomials with coefficients in A.

Two homomorphisms $\alpha, \beta : A \to B$ of algebras are called polynomially homotopic, if there is a homomorphism $\varphi : A \to B[t]$ such that $\varphi(x)(0) = \alpha(x), \varphi(x)(1) = \beta(x)$ for $x \in A$. Polynomial homotopy is not an equivalence relation, in general. We denote by $A[z, z^{-1}]$ or by $A(z)$ the algebra of Laurent polynomials with coefficients in A.

2.2 The Tensor Algebra

Given any vector space V, we define the (non-unital) tensor algebra over V as

$$TV = V \oplus V{\otimes}V \oplus V^{\otimes^3} \oplus \cdots$$

TV is an algebra with the product given by concatenation of tensors. We denote by $\sigma : V \to TV$ the linear map, that maps V to the first summand in TV. The map σ has the following universal property: Let $s : V \to \mathfrak{A}$ be an arbitrary linear map from V into an algebra A. Then there is a unique algebra homomorphism $\tau_s : TV \to A$ such that $\tau_s \circ \sigma = s$.

The tensor algebra is polynomially contractible, i.e., the identity map of TV is polynomially homotopic to 0. A polynomial homotopy $\varphi_t : TV \to TV$, such that $\varphi_0 = 0, \varphi_1 = \mathrm{id}$, is given by $\varphi_t = \tau_{t\sigma}, t \in [0,1]$.

2.3 The Free Product of Two Algebras

Let A and B be algebras. The algebraic free product (in the non-unital category) of A and B then is the following algebra

$$A * B = A \oplus B \oplus (A \otimes B) \oplus (B \otimes A) \oplus (A \otimes B \otimes A) \oplus \ldots$$

This direct sum is taken over all tensor products where the factors A and B alternate. The multiplication is given by concatenation of tensors but where the multiplication $A \otimes A \to A$ and $B \otimes B \to B$ is used to simplify all terms where two elements in A or two elements in B meet.

The canonical inclusions $\iota_1 : A \to A * B$ and $\iota_2 : B \to A * B$ have the following universal property: Let $\alpha : A \to E$ and $\beta : B \to E$ be two homomorphisms into an algebra E. Then there is a unique homomorphism $\alpha * \beta : A * B \to E$, such that $(\alpha * \beta) \circ \iota_1 = \alpha$ and $(\alpha * \beta) \circ \iota_2 = \beta$.

2.4 The Algebra of Finite Matrices of Arbitrary Size

For any algebra A, we can form the algebra $M_\infty(A)$ consisting of matrices $(a_{ij})_{i,j \in \mathbb{N}}$ with $a_{ij} \in A$, $i, j = 0, 1, 2 \ldots$ and only finitely many a_{ij} non-zero. Equivalently

$$M_\infty(A) = \varinjlim_n M_n(A) \tag{1}$$

where, as usual, $M_n(A)$ is embedded into $M_{n+1}(A)$ in the upper left corner.

2.5 The Algebraic Toeplitz Algebra

The algebraic Toeplitz algebra Ψ is the algebra of (row- and column-finite) $\mathbb{N} \times \mathbb{N}$-matrices generated by the matrices

$$v = \begin{pmatrix} 0 & 0 & 0 & 0 & \ldots \\ 1 & 0 & 0 & 0 & \ldots \\ 0 & 1 & 0 & 0 & \ldots \\ 0 & 0 & 1 & 0 & \ldots \\ \ldots & \ldots & \ldots & \ldots & \ldots \end{pmatrix}$$

and

$$u = \begin{pmatrix} 0 & 1 & 0 & 0 & \dots \\ 0 & 0 & 1 & 0 & \dots \\ 0 & 0 & 0 & 1 & \dots \\ 0 & 0 & 0 & 0 & \dots \\ & & \dots\dots\dots\dots & & \end{pmatrix}$$

As a vector space, the algebraic Toeplitz algebra $\mathfrak{T}$ is isomorphic to the direct sum $\Psi = M_\infty(\mathbb{C}) \oplus \mathbb{C}[z, z^{-1}]$.

Under this isomorphism v^k and u^k correspond to the elements $(0, z^k)$ and $(0, z^{-1})$, respectively, $k \in \mathbb{N}$. Moreover, $v^i(1 - vu)u^j$ corresponds to $(e^{ij}, 0)$, where e^{ij} is given by the matrix (a_{kl}) with $a_{kl} = 1$, for $k = i, l = j$, and $a_{kl} = 0$ otherwise (with the convention that $e^{ij} = 0$ for $i < 0$ or $j < 0$).

3 Locally Convex Algebras

Cyclic homology and cohomology can be defined readily for topological algebras and, in particular, locally convex algebras. Since there is a certain freedom in setting up cyclic homology in this case, we will now fix the conventions. A locally convex algebra A is, in general, an algebra with a locally convex topology for which the multiplication $A \times A \to A$ is (jointly) continuous. In the present survey we will restrict our attention however to locally convex algebras that can be represented as projective limits of Banach algebras. Most of what will be said below works for general locally convex algebras but some of the constructions (in particular the free extension) and the proofs (in particular excision for the periodic theory) become much more involved.

A locally convex algebra $\mathfrak{A}$ that can be represented as a projective limit of Banach algebras can equivalently be defined as a complete locally convex space whose topology is determined by a family $\{p_\alpha\}$ of submultiplicative seminorms, [33]. Thus for each α we have $p_\alpha(xy) \leq p_\alpha(x)p_\alpha(y)$. The algebra $\mathfrak{A}$ is then automatically a topological algebra, i.e., multiplication is (jointly) continuous. **We call such algebras m-algebras.**

The unitization $\tilde{\mathfrak{A}}$ of an m-algebra is again an m-algebra with the seminorms $\tilde{p}$ given by $\tilde{p}(\lambda 1 + x) = |\lambda| + p(x)$ for all submultiplicative seminorms p on $\mathfrak{A}$.

The direct sum $\mathfrak{A} \oplus \mathfrak{B}$ of two m-algebras is again an m-algebra with the topology defined by all seminorms $p \oplus q$ with $(p \oplus q)(x, y) = p(x) + q(y)$, where p is a continuous seminorm on $\mathfrak{A}$ and q a continuous seminorm on $\mathfrak{B}$.

Recall the definition of the projective tensor product by Grothendieck, [19], [42]. For two locally convex vector spaces V and W the projective topology on the tensor product $V \otimes W$ is given by the family of seminorms $p \otimes q$, where p is a continuous seminorm on V and q a continuous seminorm on W. Moreover $p \otimes q$ is defined by

$$p \otimes q(z) = \inf \left\{ \sum_{i=1}^n p(a_i)q(b_i) \,\middle|\, z = \sum_{i=1}^n a_i \otimes b_i, a_i \in V, b_i \in W \right\}$$

for $z \in V \otimes W$. We denote by $V \hat{\otimes} W$ the completion of $V \otimes W$ with respect to this family of seminorms. For m-algebras $\mathfrak{A}$ and $\mathfrak{B}$, the projective tensor product $\mathfrak{A} \hat{\otimes} \mathfrak{B}$ is again an m-algebra (if p and q are submultiplicative, so is $p \otimes q$).

Basically all definitions and constructions in the subsequent sections on cyclic theory will carry over to the category of m-algebras. All the definitions and results on cyclic theory of the previous sections hold not only in the algebraic setting, but also for m-algebras.

To obtain the appropriate definitions in the category of m-algebras, in general, all morphisms should be continuous, duals should be topological duals, tensor products (also in the definition of differential forms) should be completed projective tensor products (see below), extensions should admit a continuous linear splitting, there will be a natural notion of differentiable homotopy in that category and - very importantly - the tensor algebra construction gives a natural free extension for any m-algebra. We restrict to the class of m-algebras (rather than general locally convex algebras) mainly because of this natural construction of a free extension. Note however that other choices are conceivable. Note also that in the definition of the homology of a complex we divide by the image of the boundary operator and not by its closure (in many cases the image is automatically closed). We will comment on these matters along the way.

We now list some important examples of m-algebras and of constructions with m-algebras that will be used later on. This list is parallel to the list in Section 2 which describes the algebraic versions of the m-algebras in the present section.

3.1 Algebras of Differentiable Functions

Let $[a, b]$ be an interval in $\mathbb{R}$. We denote by $\mathbb{C}[a, b]$ the algebra of complex-valued $\mathcal{C}^\infty$-functions f on $[a, b]$, all of whose derivatives vanish in a and in b (while f itself may take arbitrary values in a and b).

Also the subalgebras $\mathbb{C}(a, b], \mathbb{C}[a, b)$ and $\mathbb{C}(a, b)$ of $\mathbb{C}[a, b]$, which, by definition consist of functions f, that vanish in a, in b, or in a and b, respectively, will play an important role.

The topology on these algebras is the usual Fréchet topology, which is defined by the following family of submultiplicative norms p_n:

$$p_n(f) = \|f\| + \|f'\| + \tfrac{1}{2}\|f''\| + \cdots + \tfrac{1}{n!}\|f^{(n)}\|$$

Here of course $\|g\| = \sup\{|g(t)| \mid t \in [a, b]\}$.

We note that $\mathbb{C}[a, b]$ is nuclear in the sense of Grothendieck [19] and that, for any complete locally convex space V, the space $\mathbb{C}[a, b] \hat{\otimes} V$ is isomorphic to the space of $\mathcal{C}^\infty$-functions on $[a, b]$ with values in V, whose derivatives vanish in both endpoints, [42], S 51. Exactly the same comments apply to the algebra $\mathcal{C}^\infty(M)$ of smooth functions on a compact smooth manifold where an m-algebra topology is defined locally by seminorms like the p_n above.

Given an m-algebra $\mathfrak{A}$, we write $\mathfrak{A}[a,b]$, $\mathfrak{A}[a,b)$ and $\mathfrak{A}(a,b)$ for the m-algebras $\mathfrak{A}\hat{\otimes}\mathbb{C}[a,b]$, $\mathfrak{A}\hat{\otimes}\mathbb{C}[a,b)$ and $\mathfrak{A}\hat{\otimes}\mathbb{C}(a,b)$.

Two continuous homomorphisms $\alpha, \beta : A \to B$ between complete locally convex algebras are called differentiably homotopic, or *diffotopic*, if there is a family $\varphi_t : A \to B$, $t \in [0,1]$, of continuous homomorphisms, such that $\varphi_0 = \alpha, \varphi_1 = \beta$ and such that the map $t \mapsto \varphi_t(x)$ is infinitely often differentiable for each $x \in A$. An equivalent condition is that there should be a continuous homomorphism $\varphi : A \to C^\infty([0,1])\hat{\otimes}B$ such that $\varphi(x)(0) = \alpha(x), \varphi(x)(1) = \beta(x)$ for each $x \in A$.

Let $h : [0,1] \to [0,1]$ be a monotone and bijective C^∞-map, whose restriction to $(0,1)$ gives a diffomorphism $(0,1) \to (0,1)$ and whose derivatives in 0 and 1 all vanish. Replacing φ_t by $\psi_t = \varphi_{h(t)}$ one sees that α and β are diffotopic if and only if there is a continuous homomorphism $\psi : A \to \mathbb{C}[0,1]\hat{\otimes}B$ such that $\psi(x)(0) = \alpha(x), \psi(x)(1) = \beta(x)$, $x \in A$. This shows in particular that diffotopy is an equivalence relation.

Note that this definition of differentiable homotopy applies not only to m-algebras but to arbitrary complete locally convex algebras. In particular, it applies to any algebra A over $\mathbb{C}$ if we regard it as a locally convex space with the topology given by all possible semi-norms on A. In this case $\mathbb{C}[0,1]\hat{\otimes}A$ is simply the algebraic tensor product $\mathbb{C}[0,1] \otimes A$.

3.2 The Smooth Tensor Algebra

Let V be a complete locally convex space. We define the smooth tensor algebra T^sV as the completion of the algebraic tensor algebra (see Section 2.2)

$$TV = V \oplus V{\otimes}V \oplus V^{\otimes^3} \oplus \ldots$$

with respect to the family $\{\widehat{p}\}$ of seminorms, which are given on this direct sum as

$$\widehat{p} = p \oplus p{\otimes}p \oplus p^{\otimes^3} \oplus \ldots$$

where p runs through all continuous seminorms on V. The seminorms $\widehat{p}$ are submultiplicative for the multiplication on TV. The completion T^sV therefore is an m-algebra.

In the simplest case $V = \mathbb{C}$, $T^s\mathbb{C}$ is naturally isomorphic to the algebra of holomorphic functions on the complex plane that vanish at 0 (under the isomorphism which maps a sequence (λ_n) in $T\mathbb{C}$ to the function f with $f(z) = \sum_{n=1}^{\infty} \lambda_n z^n$). The topology is the topology of uniform convergence on compact subsets.

We denote by $\sigma : V \to TV$ the map, which maps V to the first summand in TV. The map σ has the following universal property: Let $s : V \to \mathfrak{A}$ be any continuous linear map from V to an m-algebra $\mathfrak{A}$. Then there is a unique momorphism $\tau_s : TV \to \mathfrak{A}$ of m-algebras such that $\tau_s \circ \sigma = s$.

The smooth tensor algebra is differentiably contractible, i.e., the identity map on TV is diffotopic to 0. A differentiable family $\varphi_t : TV \to TV$ of homomorphisms for which $\varphi_0 = 0, \varphi_1 = $ id is given by $\varphi_t = \tau_{t\sigma}, t \in [0,1]$.

If $\mathfrak{A}$ is an m-algebra, by abuse of notation, we write $T\mathfrak{A}$ (rather than $T^s\mathfrak{A}$) for the smooth tensor algebra over $\mathfrak{A}$. Thus $T\mathfrak{A}$ is again an m-algebra.

3.3 The Free Product of Two m-Algebras

Let $\mathfrak{A}$ and $\mathfrak{B}$ be m-algebras. The (non-unital) algebraic free product $\mathfrak{A} * \mathfrak{B}$ is the direct sum over all tensor products

$$\mathfrak{A} * \mathfrak{B} = \mathfrak{A} \oplus \mathfrak{B} \oplus (\mathfrak{A} \otimes \mathfrak{B}) \oplus (\mathfrak{B} \otimes \mathfrak{A}) \oplus (\mathfrak{A} \otimes \mathfrak{B} \otimes \mathfrak{A}) \oplus \cdots$$

where the factors $\mathfrak{A}$ and $\mathfrak{B}$ alternate (see 2.3). Multiplication is, as in 2.3 concatenation of tensors combined with simplification using the multiplication in $\mathfrak{A}$ and $\mathfrak{B}$.

We denote by $\mathfrak{A}\hat{*}\mathfrak{B}$ the completion of $\mathfrak{A} * \mathfrak{B}$ with respect to the family of all seminorms of the form $p * q$, given by:

$$p * q = p \oplus q \oplus (p \otimes q) \oplus (q \otimes p) \oplus (p \otimes q \otimes p) \oplus \cdots$$

where p and q are continuous seminorms on $\mathfrak{A}$ and $\mathfrak{B}$, respectively. If p and q are submultiplivative, then so is the seminorm $p * q$ and $\mathfrak{A}\hat{*}\mathfrak{B}$ therefore is an m-algebra.

$\mathfrak{A}\hat{*}\mathfrak{B}$ is in fact the free product of $\mathfrak{A}$ and $\mathfrak{B}$ in the category of m-algebras. The canonical inclusions $\iota_1 : \mathfrak{A} \to \mathfrak{A}\hat{*}\mathfrak{B}$ and $\iota_2 : \mathfrak{B} \to \mathfrak{A}\hat{*}\mathfrak{B}$ have the following universal property: Let $\alpha : \mathfrak{A} \to \mathfrak{E}$ and $\beta : \mathfrak{B} \to \mathfrak{E}$ be two continuous homomorphisms into an m-algebra $\mathfrak{E}$. Then there is a unique continuous homomorphism $\alpha * \beta : \mathfrak{A}\hat{*}\mathfrak{B} \to \mathfrak{E}$, such that $(\alpha * \beta) \circ \iota_1 = \alpha$ and $(\alpha * \beta) \circ \iota_2 = \beta$.

If $\mathfrak{A}$ and $\mathfrak{B}$ are m-algebras, by abuse of notation, we write $\mathfrak{A} * \mathfrak{B}$ (rather than $\mathfrak{A}\hat{*}\mathfrak{B}$) for the m-algebra free product of $\mathfrak{A}$ and $\mathfrak{B}$.

3.4 The Algebra of Smooth Compact Operators

The m-algebra $\mathfrak{K}$ of "smooth compact operators" consists of all $\mathbb{N} \times \mathbb{N}$-matrices (a_{ij}) with rapidly decreasing matrix elements $a_{ij} \in \mathbb{C}$, $i,j = 0,1,2 \ldots$. The topology on $\mathfrak{K}$ is given by the family of norms $p_n, n = 0,1,2 \ldots$, which are defined by

$$p_n((a_{ij})) = \sum_{i,j} |1 + i + j|^n |a_{ij}|$$

It is easily checked that the p_n are submultiplicative and that $\mathfrak{K}$ is complete. Thus $\mathfrak{K}$ is an m-algebra. As a locally convex vector space, $\mathfrak{K}$ is isomorphic to the sequence space s and therefore is nuclear.

The map that sends $(a_{ij}) \otimes (b_{kl})$ to the $\mathbb{N}^2 \times \mathbb{N}^2$-matrix $(a_{ij}b_{kl})_{(i,k)(j,l) \in \mathbb{N}^2 \times \mathbb{N}^2}$ obviously gives an isomorphism Θ between $\mathfrak{K}\hat{\otimes}\mathfrak{K}$ and $\mathfrak{K}$ (see also [36], 2.7).

Lemma 3.4.1 *Let* $\Theta : \mathfrak{K} \to \mathfrak{K} \hat{\otimes} \mathfrak{K}$ *be as above and let* $\iota : \mathfrak{K} \to \mathfrak{K} \hat{\otimes} \mathfrak{K}$ *be the inclusion that maps* x *to* $e^{00} \otimes x$ *(where* e^{00} *is the matrix with elements* a_{ij}, *for which* $a_{ij} = 1$, *whenever* $i = j = 0$, *and* $a_{ij} = 0$ *otherwise). Then* Θ *is diffotopic to* ι. *The same holds for the corresponding maps* $\Theta' : \mathfrak{K} \to M_2(\mathfrak{K})$ *and* $\iota' : \mathfrak{K} \to M_2(\mathfrak{K})$.

Remark 3.4.2 *Let* V *be a Banach space. Then* $\mathfrak{K} \hat{\otimes} V$ *consists exactly of those matrices (or the sequences indexed by* $\mathbb{N} \times \mathbb{N}$) $(v_{ij})_{i,j \in \mathbb{N}}$, *for which the expression*

$$\bar{p}_n((v_{ij})) \underset{def}{=} \sum_{i,j} (1 + i + j)^n \|v_{ij}\|$$

is finite for all n. *The topology on* $\mathfrak{K} \hat{\otimes} V$ *is of course given exactly by the norms* $\bar{p}_n$. *To see this, consider the tensor product* α_n *of the norm* p_n *on* $\mathfrak{K}$ *with the norm* $\| \cdot \|$ *given on* V. *Then, if* x^{ij} *denotes the matrix, which has* $x \in V$ *as element on the place* i, j *and is* 0 *otherwise, we have*

$$\alpha_n(x^{ij}) = (1 + i + j)\|x\|$$

by [42], Prop. 43.1. This immediately shows that

$$\alpha_n((v_{ij})) \leq \bar{p}_n((v_{ij}))$$

for all matrices (v_{ij}) *in the algebraic tensor product* $\mathfrak{K} \otimes V$. *The converse inequality follows from the definition of the projective tensor product. Therefore, for each fixed* n, *the completion* $(\mathfrak{K} \otimes V)_{\bar{p}_n}$ *is isometrically isomorphic to* $(\mathfrak{K})_{p_n} \hat{\otimes} V$ *and consists exactly of those matrices* (v_{ij}), *for which* $\bar{p}_n((v_{ij}))$ *is finite.*

3.5 The Schatten Ideals $\ell^p(H)$

Let H be an infinite-dimensional separable Hilbert space (i.e., $H \cong \ell^2(\mathbb{N})$), and $(\xi_n)_{n \in \mathbb{N}}$ an orthonormal basis. Let also $\mathcal{L}(H)$ denote the algebra of bounded operators on H. For each positive x in $\mathcal{L}(H)$, we can define the trace $Tr(x)$ by

$$Tr(x) = \sum_n (x\xi_n \mid \xi_n)$$

This expression is ≥ 0 and may be infinite. It does not depend on the choice of the orthonormal basis. Moreover $Tr(x^*x) = Tr(xx^*)$ holds for all x.

Definition 3.5.1 *Let* $p \geq 1$. *We define* $\ell^p(H)$ *to be the set of all operators* $x \in \mathcal{L}(H)$ *for which*

$$Tr\,|x|^p < \infty$$

where $|x| = \sqrt{x^*x}$.

Proposition 3.5.2 *For each* $p \geq 1$, $\ell^p(H)$ *is an ideal in* $\mathcal{L}(H)$ *which is contained in the ideal* $\mathcal{K}(H)$ *of compact operators on* H. *It is a Banach algebra with respect to the norm*

$$\|x\|_p = (Tr\,|x|^p)^{1/p}$$

Moreover $\ell^p(H) \subset \ell^q(H)$ *for* $p \leq q$

3.6 The Smooth Toeplitz Algebra

The elements of the algebra $C^\infty S^1$ can be written as power series in the generator z (defined by $z(t) = t, t \in S^1 \subset \mathbb{C}$). The coefficients of these power series are rapidly decreasing. Thus

$$C^\infty(S^1) = \left\{ \sum_{k \in \mathbb{Z}} a_k z^k \mid \sum_{k \in \mathbb{Z}} |a_k| |k|^n < \infty \text{ for all } n \in \mathbb{N} \right\}$$

Submultiplicative norms that determine the topology are given by

$$q_n \left(\sum a_k z^k \right) = \sum |1 + k|^n |a_k|$$

As a topological vector space, the smooth Toeplitz algebra $\mathfrak{T}$ may be defined as the direct sum $\mathfrak{T} = \mathfrak{K} \oplus C^\infty(S^1)$.

To describe the multiplication in $\mathfrak{T}$, we write v_k for the element $(0, z^k)$ in $\mathfrak{T}$ and simply x for the element $(x, 0)$ with $x \in \mathfrak{K}$. Moreover, e^{ij} denotes the element of $\mathfrak{T}$, which is determined by the matrix (a_{kl}), where $a_{kl} = 1$, whenever $k = i, l = j$, and $a_{kl} = 0$ otherwise (with the convention that $e^{ij} = 0$, if $i < 0$ oder $j < 0$). Multiplication in $\mathfrak{T}$ then is determined by the following rules:

$$e^{ij} e^{kl} = \delta_{jk} e^{il}$$

$$v_k e^{ij} = e^{(i+k),j} \qquad e^{ij} v_k = e^{i,(j-k)}$$

$(i, j, k \in \mathbb{Z})$; and

$$v_k v_{-l} = \begin{cases} v_{k-l}(1 - E_{l-1}) & l > 0 \\ \\ v_{k-l} & l \leq 0 \end{cases}$$

where $E_l = e^{00} + e^{11} + \cdots + e^{ll}$. If p_n are the norms on $\mathfrak{K}$ defined in 3.4 and q_n the norms on $C^\infty(S^1)$ defined above, it is easy to see that each norm $p_n \oplus q_m$ is submultiplicative on $\mathfrak{T} = \mathfrak{K} \oplus C^\infty(S^1)$ with the multiplication introduced above. Obviously, $\mathfrak{K}$ is a closed ideal in $\mathfrak{T}$ and the quotient $\mathfrak{T}/\mathfrak{K}$ is $C^\infty(S^1)$.

4 Standard Extensions of a Given Algebra

Extensions play a fundamental role in all aspects of non-commutative geometry. In this section we describe standard extensions that can be associated to any given algebra. All of these extensions exist in a purely algebraic setting, but also in the category of m-algebras. We treat both cases in parallel.

4.1 The Suspension Extension

This is the fundamental extension of algebraic topology. It has the form

$$0 \to \mathbb{C}(0,1) \to \mathbb{C}[0,1) \to \mathbb{C} \to 0$$

or more generally

$$0 \to \mathfrak{A}(0,1) \to \mathfrak{A}[0,1) \to \mathfrak{A} \to 0$$

with an arbitrary m-algebra $\mathfrak{A}$.

Since cyclic homology is homotopy invariant only for differentiable homotopies (not for continuous ones) we have to work with algebras of differentiable functions. Recall that $\mathbb{C}(0,1)$ and $\mathbb{C}[0,1)$ denote the algebras of C^∞- functions on the interval $[0,1]$, whose derivatives to arbitrary order vanish in 0 and 1, or or only in 1, respectively, and that the algebra $\mathbb{C}[0,1)$ is differentiably contractible.

The algebraic suspension extension for an algebra A is

$$0 \to t(1-t)A[t] \to tA[t] \to A \to 0$$

4.2 The Free Extension

For any algebra there exists a natural extension by the free algebra TA—the nonunital tensor algebra over A.

$$0 \to JA \to TA \to A \to 0. \tag{1}$$

This extension is universal in the sense that, given any extension $0 \to I \to E \to A \to 0$ of A, there is a morphism of extensions

$$
\begin{array}{ccccccccc}
0 \to & JA & \to TA \to & A & \to 0 \\
 & \downarrow \gamma_A & \downarrow & \downarrow id & \\
0 \to & I & \to E \to & A & \to 0
\end{array}
$$

(the map $TA \to E$ is obtained by choosing a linear splitting $s : A \to E$ in the given extension). The map $JA \to I$ in this commutative diagram is called the *classifying map* of the extension $0 \to I \to E \to A \to 0$.

More generally, any n-step extension of the form

$$0 \to I \to E_1 \to E_2 \to \ldots \to E_n \to A \to 0$$

can be compared with the free n-step extension

$$0 \to J^n A \to T(J^{n-1}A) \to T(J^{n-2}A) \to \ldots \to TA \to A \to 0$$

and thus has a classifying map $J^n A \to I$ (which is unique up to homotopy) (here $J^n A$ is defined recursively by $J^n A = J(J^{n-1}A)$).

For any m-algebra $\mathfrak{A}$, the (smooth) tensor algebra $T\mathfrak{A}$ has been defined in 3.2.

Recall that the linear map $\sigma : V \to TV$ that sends V to the first summand in $T\mathfrak{A}$ has the following universal property:

Let $s : \mathfrak{A} \to \mathfrak{B}$ be an arbitrary continuous linear map from $\mathfrak{A}$ to an m-Algebra $\mathfrak{B}$. Then there exists a unique homomorphism $\tau_s : T\mathfrak{A} \to \mathfrak{B}$ of m-algebras with the property that $\tau_s \circ \sigma = s$. Thus exactly as in the algebraic case we get an extension

$$0 \to J\mathfrak{A} \to T\mathfrak{A} \to \mathfrak{A} \to 0. \tag{2}$$

by defining $J\mathfrak{A}$ as the kernel of the natural map $T\mathfrak{A} \to \mathfrak{A}$. This extension has the property that, given any extension $0 \to \mathfrak{J} \to \mathfrak{E} \to \mathfrak{A} \to 0$ of $\mathfrak{A}$ **with a continuous linear splitting** $s : \mathfrak{A} \to \mathfrak{E}$, there is a morphism of extensions

$$
\begin{array}{ccccccccc}
0 \to & J\mathfrak{A} & \to & T\mathfrak{A} \to & \mathfrak{A} & \to 0 \\
& \downarrow \gamma_A & & \downarrow & \downarrow id \\
0 \to & \mathfrak{J} & \to & \mathfrak{E} \to & \mathfrak{A} & \to 0
\end{array}
$$

The map $J\mathfrak{A} \to \mathfrak{J}$ in this commutative diagram is called the *classifying map* of the extension $0 \to \mathfrak{J} \to \mathfrak{E} \to \mathfrak{A} \to 0$.

More generally, any n-step extension of the form

$$0 \to \mathfrak{J} \to \mathfrak{E}_1 \to \mathfrak{E}_2 \to \ldots \to \mathfrak{E}_n \to \mathfrak{A} \to 0$$

has a classifying map $J^n\mathfrak{A} \to \mathfrak{J}$ (which is unique up to homotopy) (here $J^n\mathfrak{A}$ is defined recursively by $J^n\mathfrak{A} = J(J^{n-1}\mathfrak{A})$).

4.3 The Universal Two-Fold Trivial Extension

With any m-algebra $\mathfrak{A}$ we associate as in [9] the m-algebra $Q\mathfrak{A} = \mathfrak{A} * \mathfrak{A}$. We denote by ι and $\bar{\iota}$ the two canonical inclusions of $\mathfrak{A}$ into $Q\mathfrak{A}$. The algebra $Q\mathfrak{A}$ is, in a natural way, $\mathbb{Z}/2$-graded by the involutive automorphism τ which exchanges $\iota(\mathfrak{A})$ and $\bar{\iota}(\mathfrak{A})$.

The ideal $q\mathfrak{A}$ in $Q\mathfrak{A}$ is by definition the kernel of the homomorphism $\pi = id * id : \mathfrak{A} * \mathfrak{A} \to \mathfrak{A}$. The extension

$$0 \to q\mathfrak{A} \to Q\mathfrak{A} \xrightarrow{\pi} \mathfrak{A} \to 0 \tag{3}$$

admits two natural splittings via the algebra homomorphisms ι and $\bar{\iota}$. It has the following universal property: Let

$$0 \to \mathfrak{E}_0 \to \mathfrak{E}_1 \to \mathfrak{A} \to 0 \tag{4}$$

be an extension with two splittings $\alpha, \bar{\alpha} : \mathfrak{A} \to \mathfrak{E}_1$, which are continuous algebra homomorphisms. Then there is a morphism of extensions (i.e., a commutative diagram) as follows:

$$
\begin{array}{ccccccccc}
0 \to & q\mathfrak{A} & \to & Q\mathfrak{A} & \to & \mathfrak{A} & \to 0 \\
& \downarrow \alpha*\bar{\alpha} & & \downarrow \alpha*\bar{\alpha} & & \downarrow id \\
0 \to & \mathfrak{E}_0 & \to & \mathfrak{E}_1 & \to & \mathfrak{A} & \to 0
\end{array}
$$

By construction, this morphism sends ι and $\bar{\iota}$ to α and $\bar{\alpha}$.

The same construction can of course be performed purely algebraically and we obtain, for any algebra A, an extension

$$0 \to qA \to QA \to A \to 0$$

where $QA = A * A$. This extension has the same universal property in the category of algebras without topology.

4.4 The Toeplitz Extension

The smooth Toeplitz algebra $\mathfrak{T}$ was introduced in 3.6. By definition, $\mathfrak{T}$ contains the algebra $\mathfrak{K}$ of smooth compact operators as a closed ideal and we obtain the following extension (Toeplitz extension)

$$0 \to \mathfrak{K} \to \mathfrak{T} \xrightarrow{\pi} \mathcal{C}^\infty(S^1) \to 0$$

which, of course, admits a continuous linear splitting.

Let $\kappa : \mathfrak{T} \to \mathbb{C}$ be the canonical homomorphism, which maps v_1 and v_{-1} to 1 (with the notation of 3.6) and put $\mathfrak{T}_0 = \mathrm{Ker}\,\kappa$. By restriction we obtain the following extension

$$0 \to \mathfrak{K} \to \mathfrak{T}_0 \to \mathcal{C}_0^\infty(S^1 \backslash 1) \to 0$$

The algebraic Toeplitz extension has the following form

$$0 \to M_\infty(\mathbb{C}) \to \Psi \to \mathbb{C}[t, t^{-1}] \to 0$$

where Ψ is the algebraic Toeplitz algebra, 2.5

5 Preliminaries on Homological Algebra

A complex is a $\mathbb{Z}$-graded vector space C with an endomorphism ∂ of degree 1 such that $\partial^2 = 0$. The map δ is called the boundary operator of C. We will also use $\mathbb{Z}/2$-graded complexes ("supercomplexes"). They can be viewed as periodic ordinary complexes.

Definition 5.1. *The homology $H(C)$ of a complex C is the quotient space*

$$H(C) = \mathrm{Ker}\,\partial/\mathrm{Im}\,\partial$$

$H(C)$ *is a graded vector space with* $H_n(C) = (\mathrm{Ker}\,\partial \cap C_n)/\partial(C_{n+1})$.

A linear map between two complexes C and D is said to have degree k if it maps C_n to D_{n+k} for all n. The Hom-complex $\mathrm{Hom}(C, D)$ is the (graded) vector space consisting of linear maps φ from the vector space C to the vector space D which are linear combinations of such homogeneous maps. The boundary map on $\mathrm{Hom}(C, D)$ is given by the graded commutator

$$\partial(\varphi) = \partial_Y \circ \varphi - (-1)^{\deg} \varphi \circ \partial_X .$$

with the boundary maps. A map $\varphi : C \to D$ of degree 0 is called a morphism of complexes if $\partial \varphi = 0$, i.e., if it commutes with the boundary operators. Any such morphism induces a morphism

$$\varphi_* : H(C) \longrightarrow H(D)$$

Two morphisms in $\mathrm{Hom}(C, D)$ are said to be (chain) homotopic if they differ by a boundary in $\mathrm{Hom}(X, Y)$. Homotopic morphisms induce the same maps on homology. Two complexes are called homotopy equivalent if there exist

$$\varphi \in \mathrm{Hom}(C, D), \ \partial \varphi = 0$$

and

$$\psi \in \mathrm{Hom}(C, D), \ \partial \psi = 0$$

$\deg \varphi = \deg \psi = 0$ such that $\varphi \psi$ and $\psi \varphi$ are homotopic to id. Finally, the complex C will be said to be contractible if it is homotopy equivalent to 0 (this is the case iff there exists a map $s : C \to C$ of degree 1, such that $\mathrm{id}_C = \partial s + s \partial$). The homology of a contractible complex is 0.

A sequence

$$\cdots \xrightarrow{\alpha_{n+2}} G_{n+1} \xrightarrow{\alpha_{n+1}} G_n \xrightarrow{\alpha_n} G_{n-1} \xrightarrow{\alpha_{n-1}} \cdots$$

of morphisms between vector spaces or abelian groups is called exact, if, for each n, $\mathrm{Ker}\,\alpha_n = \mathrm{Im}\,\alpha_{n+1}$.

Proposition 5.1. *Let*

$$0 \longrightarrow C \xrightarrow{\iota} D \xrightarrow{\pi} E \longrightarrow 0$$

be an extension of complexes, i.e., an exact sequence of (graded) vector spaces, where the maps ι and π are morphisms of complexes. Then there is a map δ : $H(E) \to H(C)$ of degree 1 and an exact sequence of the form

$$H(D)$$
$$\iota_* \nearrow \qquad \searrow \pi_*$$
$$H(C) \qquad \xleftarrow{\ \delta\ } \qquad H(E)$$

This can be written as a long exact sequence of the form

$$\xrightarrow{\pi_*} H_{n+1}E \xrightarrow{\delta} H_nC \xrightarrow{\iota_*} H_nD \xrightarrow{\pi_*} H_nE \xrightarrow{\delta} H_{n-1}C \xrightarrow{\iota_*} H_{n-1}D \xrightarrow{\pi_*}$$

Proof. Given $[x]$ in $H(E)$ with $\partial(x) = 0$, put $\delta([x]) = \partial(z)$, where $z \in D$ is such that $\pi(z) = x$.

Definition 5.2. *A bicomplex is a* $\mathbb{Z} \times \mathbb{Z}$*-graded vector space or, equivalently, a family* $(C_{pq})_{p,q \in \mathbb{Z}}$ *of vector spaces with two endomorphisms* ∂_h *and* ∂_v *of degree* $(-1, 0)$ *and* $(0, -1)$, *respectively, such that*

$$\partial_h^2 = \partial_v^2 = \partial_h \partial_v + \partial_v \partial_h = 0$$

∂_h *and* ∂_v *are called the horizontal and vertical boundary maps of C.*
The total complex for C is the graded vector space D whose component of degree n is given by the infinite product

$$D_n = \prod_{p+q=n} C_{pq}$$

(The letter D stands for diagonals). With the boundary operator $\partial = \partial_h + \partial_v$ *it becomes a complex.*

Remark 5.1. Instead of the direct product, one can also take the direct sum in the definition of the total complex. Both variants will be used in the sequel.

For each p, the family $(C_{pq})_{q \in \mathbb{Z}}$ defines a complex with boundary operator ∂_v. We call its homology the vertical homology. The horizontal homology is defined similarly for each q. We will often consider bicomplexes where $C_{pq} = 0$ for $p < 0$ or $q < 0$. The following proposition is then often useful.

Proposition 5.2. *Let* $C = C_{pq}$ *and* $E = E_{pq}$ *be bicomplexes with* $C_{pq} = 0$ *and* $E_{pq} = 0$ *for* $p < 0$ *or* $q < 0$, *and let D and F denote the total complexes of C and E, respectively. Assume that* $\alpha, \beta : C \rightarrow E$ *are two morphisms of bicomplexes which, for each p, induce the same map on the vertical homology of C and E, respectively. Then the induced maps* $\alpha_*, \beta_* : H(D) \rightarrow H(F)$ *coincide.*

6 Definition of Cyclic Homology/Cohomology Using the Cyclic Bicomplex and the Connes Complex

In this section, we introduce cyclic homology using, what are probably the most fundamental operators in cyclic theory (b, b', λ, Q), and we explain the long exact sequence connecting cyclic homology with Hochschild homology. The important properties of cyclic theory will be discussed only much later.

The cyclic bicomplex $CC_{pq}(A)$, $p, q \in \mathbb{N}$, for an algebra A is the following

$$
\begin{array}{ccccccc}
\downarrow & & \downarrow & & \downarrow & & \\
A^{\otimes 3} & \xleftarrow{1-\lambda} & A^{\otimes 3} & \xleftarrow{Q} & A^{\otimes 3} & \xleftarrow{1-\lambda} & \cdots \\
\downarrow b & & \downarrow -b' & & \downarrow b & & \\
A^{\otimes 2} & \xleftarrow{1-\lambda} & A^{\otimes 2} & \xleftarrow{Q} & A^{\otimes 2} & \xleftarrow{1-\lambda} & \cdots \\
\downarrow b & & \downarrow -b' & & \downarrow b & & \\
A & \xleftarrow{1-\lambda} & A & \xleftarrow{Q} & A & \xleftarrow{1-\lambda} & \cdots
\end{array}
$$

i.e.,

$$CC_{pq}(A) = A^{\otimes(q+1)} \qquad p, q \geq 0$$

In this section and later, bicomplexes are understood to continue with 0 outside the represented region (thus here $C_{pq} = 0$ for $p < 0$ or $q < 0$). The map b' is given by

$$b'(x_0 \otimes x_1 \otimes \ldots \otimes x_n) = \sum_{j=0}^{n-1} (-1)^j x_0 \otimes \ldots \otimes x_j x_{j+1} \otimes \ldots \otimes x_n$$

while b is the usual Hochschild operator

$$b(x_0 \otimes \ldots \otimes x_n) = b'(x_0 \otimes \ldots \otimes x_n) + (-1)^n x_n x_1 \otimes x_0 \otimes \ldots \otimes x_{n-1}$$

Finally, λ is Connes' signed cyclic permutation operator

$$\lambda(x_0 \otimes \ldots \otimes x_n) = (-1)^n x_n \otimes x_0 \otimes \ldots \otimes x_{n-1}$$

and $Q = \sum_{j=0}^n \lambda^j$ is averaging with respect to signed cyclic permutations, thus

$$Q(x_0 \otimes \ldots \otimes x_n) = \sum_{j=0}^{n} (-1)^{jn} x_j \otimes \ldots \otimes x_n \otimes x_0 \otimes \ldots \otimes x_{j-1}$$

One has $b^2 = 0, b'^2 = 0$ and $(1 - \lambda)Q = Q(1 - \lambda) = 0$. Moreover, $CC_{pq}(A)$ is in fact a bicomplex, due to the fundamental identities

$$Qb = b'Q$$

$$(1 - \lambda)b' = b(1 - \lambda)$$

The rows of the cyclic bicomplex $CC_{pq}(A)$ are obviously exact as long as $p > 0$.

Definition 6.1. $HC_n A$ *is the homology of the total complex of the cyclic bicomplex* $CC_{pq}(A)$, *i.e.,* $HC_n A$ *is the homology of the complex*

$$\ldots \longrightarrow D_n \xrightarrow{\partial} D_{n-1} \xrightarrow{\partial} \ldots \xrightarrow{\partial} D_1 \xrightarrow{\partial} D_0 \longrightarrow 0$$

where

$$D_n = CC_{0,n} \oplus CC_{1,(n-1)} \oplus \ldots \oplus CC_{n,0}$$

$$= A^{\otimes(n+1)} \oplus A^{\otimes n} \oplus \ldots \oplus A$$

are the diagonals in $CC_{pq}(A)$ *and* ∂ *is the sum of the respective boundary operators* $b, b', 1 - \lambda, Q$.

Consider the bicomplex $CC_*^{Hochschild}$ consisting of the first two columns of CC_*:

$$
\begin{array}{ccc}
\downarrow & & \downarrow \\
A^{\otimes 3} & \xleftarrow{1-\lambda} & A^{\otimes 3} \\
\downarrow b & & \downarrow -b' \\
A^{\otimes 2} & \xleftarrow{1-\lambda} & A^{\otimes 2} \\
\downarrow b & & \downarrow -b' \\
A & \xleftarrow{1-\lambda} & A
\end{array}
\qquad (CC_*^{Hochschild})
$$

and its total complex

$$
\ldots \xrightarrow{\partial'} D_n^{Hochschild} \xrightarrow{\partial'} D_{n-1}^{Hochschild} \xrightarrow{\partial'} \ldots \xrightarrow{\partial'} D_0^{Hochschild} \longrightarrow 0
$$

where

$$
D_n^{Hochschild} = A^{\otimes(n+1)} \oplus A^{\otimes n}, \quad n \geq 1, \qquad D_0^{Hochschild} = A
$$

$$
\text{and} \quad \partial' = \begin{pmatrix} b & (1-\lambda) \\ 0 & -b' \end{pmatrix}
$$

Definition 6.2. *For any (possibly nonunital) algebra A, the Hochschild homology $HH_*(A)$ is defined as the homology of the complex*

$$
\ldots \xrightarrow{\partial'} D_n^{Hochschild} \xrightarrow{\partial'} D_{n-1}^{Hochschild} \xrightarrow{\partial'} \ldots \xrightarrow{\partial'} D_0^{Hochschild} \longrightarrow 0
$$

If A has a unit, then the second column in $CC_*^{Hochschild}$ is obviously exact and can therefore be disregarded for the computation of $HH_*(A)$. We therefore find in this case the usual definition of the Hochschild homology of A (with coefficients in A), namely as the homology of the complex

$$
\ldots \xrightarrow{b} A^{\otimes 3} \xrightarrow{b} A^{\otimes 2} \xrightarrow{b} A \longrightarrow 0
$$

More generally, if this holds, we call A H-unital, see [46].

Definition 6.3. *The algebra A is called H-unital, if the second column in $CC_*^{Hochschild}$, i.e., the complex*

$$
\ldots \xrightarrow{b'} A^{\otimes 3} \xrightarrow{b'} A^{\otimes 2} \xrightarrow{b'} A \longrightarrow 0
$$

is exact.

For any H-unital algebra A, the Hochschild homology $HH(A)$ is the homology of the complex

$$
\ldots \xrightarrow{b} A^{\otimes 3} \xrightarrow{b} A^{\otimes 2} \xrightarrow{b} A \longrightarrow 0
$$

If A is unital, then the Hochschild homology of A has a well known interpretation as a derived functor. In particular it can be computed from any projective A-bimodule resolution of the A-bimodule A. This method applies in many cases.

Theorem 6.1. *There is a long exact sequence (Connes's SBI-sequence) of the form*

$$\longrightarrow HC_nA \xrightarrow{S} HC_{n-2}A \xrightarrow{B} HH_{n-1}A \xrightarrow{I} HC_{n-1}A \xrightarrow{S} HC_{n-3}A \xrightarrow{B}$$

$$\cdots \xrightarrow{I} HC_1A \xrightarrow{S} HC_{-1}A \longrightarrow HH_0A \xrightarrow{I} HC_0A \longrightarrow 0$$

Proof. This is the long exact sequence associated to the obvious exact sequence of complexes

$$0 \longrightarrow D_n^{Hochschild} \longrightarrow D_n \longrightarrow D_{n-2} \longrightarrow 0$$

The operators I and S in the above sequence are simply induced by the inclusion $D_n^{Hochschild} \to D_n$ and by the projection $D_n \to D_{n-2}$, while the operator B is given by the boundary map. Note that $HC_{-1}A = 0$ and that $I : HH_0A \xrightarrow{\cong} HC_0A$ is an isomorphism. If the Hochschild homology of A is known, then the *SBI*-sequence allows in many cases to determine the cyclic homology groups for A recursively.

Consider the following sub-bicomplex $CC_*^{trivial}(A)$ of $CC_{pq}(A)$

$$\begin{array}{ccccc}
\downarrow & & \downarrow & & \downarrow \\
(1-\lambda)A^{\otimes 3} & \xleftarrow{1-\lambda} & A^{\otimes 3} & \xleftarrow{Q} & A^{\otimes 3} \xleftarrow{1-\lambda} \cdots \\
\downarrow b & & \downarrow -b' & & \downarrow b \\
(1-\lambda)A^{\otimes 2} & \xleftarrow{1-\lambda} & A^{\otimes 2} & \xleftarrow{Q} & A^{\otimes 2} \xleftarrow{1-\lambda} \cdots \qquad (CC_*^{trivial}(A)) \\
\downarrow b & & \downarrow -b' & & \downarrow b \\
(1-\lambda)A & \xleftarrow{1-\lambda} & A & \xleftarrow{Q} & A \xleftarrow{1-\lambda}
\end{array}$$

Since the rows in $CC_*^{trivial}(A)$ are exact, its total complex $D^{trivial}$ has no homology. We denote by $C^\lambda(A)$ the quotient complex $D/D^{trivial}$. Explicitly

$$C_n^\lambda(A) = D_n/D_n^{trivial} = A^{\otimes(n+1)}/(1-\lambda)A^{\otimes(n+1)}$$

Since $H_n(D^{trivial}) = 0$, we immediately get

Proposition 6.1. *For any algebra A, the cyclic homology $HC_n(A)$ can be computed as the homology of the complex*

$$\cdots \longrightarrow C_n^\lambda(A) \xrightarrow{b} C_{n-1}^\lambda(A) \xrightarrow{b} \cdots \longrightarrow C_0^\lambda(A) \longrightarrow 0$$

This is Connes' original definition of cyclic homology.

The two-sided (or periodic) cyclic complex $CC_*^{twosided}$ continues CC_* to the left. It is defined by $CC_{pq}^{twosided}(A) = A^{\otimes(q+1)}$ $p,q \in \mathbb{Z}, q \geq 0$ with boundary operators as follows

$$\downarrow \qquad \downarrow \qquad \downarrow$$

$$\ldots \xleftarrow{Q} A^{\otimes 3} \xleftarrow{1-\lambda} A^{\otimes 3} \xleftarrow{Q} A^{\otimes 3} \xleftarrow{1-\lambda} \ldots$$

$$\downarrow b \qquad \downarrow -b' \qquad \downarrow b$$

$$\ldots \xleftarrow{Q} A^{\otimes 2} \xleftarrow{1-\lambda} A^{\otimes 2} \xleftarrow{Q} A^{\otimes 2} \xleftarrow{1-\lambda} \ldots \qquad (CC_{pq}^{twosided}(A))$$

$$\downarrow b \qquad \downarrow -b' \qquad \downarrow b$$

$$\ldots \xleftarrow{Q} A \xleftarrow{1-\lambda} A \xleftarrow{Q} A \xleftarrow{1-\lambda} \ldots$$

The left cyclic complex CC_*^{left} is the left-hand side of $CC_*^{twosided}$

$$\downarrow \qquad \downarrow \qquad \downarrow$$

$$\ldots \xleftarrow{1-\lambda} A^{\otimes 3} \xleftarrow{Q} A^{\otimes 3} \xleftarrow{1-\lambda} A^{\otimes 3}$$

$$\downarrow -b' \qquad \downarrow b \qquad \downarrow -b'$$

$$\ldots \xleftarrow{1-\lambda} A^{\otimes 2} \xleftarrow{Q} A^{\otimes 2} \xleftarrow{1-\lambda} A^{\otimes 2} \qquad (CC_*^{left})$$

$$\downarrow -b' \qquad \downarrow b \qquad \downarrow -b'$$

$$\ldots \xleftarrow{1-\lambda} A \xleftarrow{Q} A \xleftarrow{1-\lambda} A$$

Consider the total complexes $D^{left}, D^{twosided}, D$ of the bicomplexes $CC_{pq}^{left}(A), CC_{pq}^{twosided}(A)$ and $CC_{pq}(A)$, respectively, where here we use direct sums rather than direct products for the definition of the diagonals in the total complex. We get an exact sequence of complexes

$$0 \longrightarrow D^{left} \longrightarrow D^{twosided} \longrightarrow D \longrightarrow 0$$

Since the rows of $CC_{pq}^{twosided}(A)$ are exact, a standard staircase argument shows that $D^{twosided}$ has no homology (this argument would not work for the direct product total complex). Therefore

$$HC_n(A) = H_n(D) \cong H_{n-1}(D^{left})$$

whence the following alternative description of cyclic homology

Proposition 6.2. *For any algebra A, the cyclic homology $HC_n(A)$ can be computed as*

$$HC_n(A) = H_{n-1}(D^{left})$$

where D^{left} is the total complex of $CC_{pq}^{left}(A)$.

By using the dual complexes, the analogue of any statement in this section is also valid for cohomology in place of homology.

In particular, cyclic cohomology can be defined using the dual $C_\lambda^n A = (C_n^\lambda A)'$ of $C_n^\lambda A$, i.e.,

$$C_\lambda^n A = \left\{ f \;\middle|\; \begin{array}{l} f \text{ is an } (n+1)\text{-linear functional on } A^{n+1} \text{ such that} \\ f(x_n, x_0, x_1, \ldots, x_{n-1}) = (-1)^n f(x_0, x_1, \ldots, x_n) \end{array} \right\}$$

The boundary operator on $C_\lambda^n A$ is the restriction of the Hochschild operator b.

A **cyclic n-cocycle** then is an $(n+1)$-linear functional such that

$$f(x_n, x_0, x_1, \ldots, x_{n-1}) = (-1)^n f(x_0, x_1, \ldots, x_n)$$

and

$$f(x_0 x_1, x_2, \ldots, x_{n+1}) - f(x_0, x_1 x_2, \ldots, x_{n+1}) + \ldots$$

$$+ (-1)^n f(x_0, \ldots, x_n x_{n+1}) + (-1)^{n+1} f(x_{n+1} x_0, x_1, \ldots, x_n) = 0$$

This is the original definition of a cyclic cocycle due to A. Connes. It also is the form in which cyclic cocycles appear often in practice.

7 The Algebra ΩA of Abstract Differential Forms over A and Its Operators

The boundary operators in the cyclic bicomplex take a simpler form and can be "explained" if interpreted as operators on the algebra ΩA of abstract differential forms over A. Another deeper and even more conceptual explanation will be presented in Section 10.1.

Given an algebra A, we denote by ΩA the universal algebra generated by $x \in A$ with relations of A and symbols $dx, x \in A$, where dx is linear in x and satisfies $d(xy) = xdy + d(x)y$. We do not impose $d1 = 0$, i.e., if A has a unit, $d1 \neq 0$. ΩA is a direct sum of subspaces $\Omega^n A$ generated by linear combinations of $x_0 dx_1 \ldots dx_n$, and $dx_1 \ldots dx_n$, $x_j \in A$. This decomposition makes ΩA into a graded algebra. As usual, we write $\deg(\omega) = n$ if $\omega \in \Omega^n A$.

As a vector space, for $n \geq 1$,

$$\Omega^n A \cong \widetilde{A} \otimes A^{\otimes n} \cong A^{\otimes(n+1)} \oplus A^{\otimes n} \tag{1}$$

(where $\widetilde{A}$ is A with a unit adjoined and $1 \otimes x_1 \otimes \cdots \otimes x_n$ corresponds to $dx_1 \ldots dx_n$). The operator d is defined on ΩA by

$$d(x_0 dx_1 \ldots dx_n) = dx_0 dx_1 \ldots dx_n$$
$$d(dx_1 \ldots dx_n) = 0 \tag{2}$$

The operator b is defined by

$$b(\omega dx) = (-1)^{\deg \omega}[\omega, x]$$
$$b(dx) = 0, \, b(x) = 0, \qquad x \in A, \, \omega \in \Omega A \tag{3}$$

Then clearly $d^2 = 0$ and one easily computes that also $b^2 = 0$.
Under the isomorphism in equation (1) d becomes

$$d(x_0 \otimes \ldots \otimes x_n) = 1 \otimes x_0 \otimes \ldots \otimes x_n \quad x_0 \in A$$

$$d(1 \otimes x_1 \otimes \ldots \otimes x_n) = 0$$

while b corresponds to the usual Hochschild operator

$$b(\widetilde{x_0} \otimes x_1 \otimes \ldots \otimes x_n) =$$

$$\widetilde{x_0}x_1 \otimes \ldots \otimes x_n + \sum_{j=2}^{n}(-1)^{j-1}\widetilde{x_0} \otimes \ldots \otimes x_{j-1}x_j \otimes \ldots \otimes x_n$$

$$+ (-1)^n x_n\widetilde{x_0} \otimes x_1 \otimes \ldots \otimes x_{n-1}, \qquad \widetilde{x_0} \in \tilde{A}, \ x_1, \ldots, x_n \in A$$

Another important natural operator is the degree (or number) operator:

$$N(\omega) = \deg(\omega)\omega \tag{4}$$

We also define the Karoubi operator κ on ΩA by

$$\kappa = 1 - (db + bd) \tag{5}$$

Explicitly, κ is given by

$$\kappa(\omega dx) = (-1)^{deg\omega} dx\omega$$

The operator κ satisfies $(\kappa^n - 1)(\kappa^{n+1} - 1) = 0$ on Ω^n. Therefore, by linear algebra, there is a projection operator P on ΩA corresponding to the generalized eigenspace for 1 of the operator $\kappa = 1 - (db + bd)$.

Lemma 7.1. Let $L = (Nd)b + b(Nd)$. Then $\Omega A = Ker\,L \oplus Im\,L$ and P is exactly the projection onto $Ker\,L$ in this splittting.

Proof. This follows from the identity

$$L = (\kappa - 1)^2(\kappa^{n-1} + 2\kappa^{n-2} + 3\kappa^{n-3} + \ldots + (n-1)\kappa + n)$$

The operator L thus behaves like a "selfadjoint" operator. It can be viewed as an abstract Laplace operator on the algebra of abstract differential forms ΩA. The elements in the image of P are then "abstract harmonic forms".

By construction, P commutes with b, d, N. Thus setting $B = NPd$ one finds $Bb + bB = PL = 0$ and $B^2 = 0$.
Explicitly, B is given on $\omega \in \Omega^n$ by the formula

$$B(\omega) = \sum_{j=0}^{n} \kappa^j d\omega$$

Under the isomorphism in equation (1), this corresponds to:

$$B(x_0 dx_1 \ldots dx_n) = dx_0 dx_1 \ldots dx_n + (-1)^n dx_n dx_0 \ldots dx_{n-1}$$
$$+ \ldots + (-1)^{nn} dx_1 \ldots dx_n dx_0$$

The preceding identities show that we obtain a bicomplex - the (B, b)-bicomplex - in the following way

$$
\begin{array}{ccccccc}
\downarrow b & & \downarrow b & & \downarrow b & & \downarrow b \\
\Omega^3 A & \xleftarrow{B} & \Omega^2 A & \xleftarrow{B} & \Omega^1 A & \xleftarrow{B} & \Omega^0 A \\
\downarrow b & & \downarrow b & & \downarrow b & & \\
\Omega^2 A & \xleftarrow{B} & \Omega^1 A & \xleftarrow{B} & \Omega^0 A & & \\
\downarrow b & & \downarrow b & & & & \\
\Omega^1 A & \xleftarrow{B} & \Omega^0 A & & & & \\
\downarrow b & & & & & & \\
\Omega^0 A & & & & & &
\end{array}
\tag{6}
$$

One can rewrite the (B, b)-bicomplex (6) using the isomorphism $\Omega^n A \cong \Lambda^{\otimes(n+1)} \oplus A^{\otimes n}$ in equation (1). An easy computation shows that the operator $b : \Omega^n A \to \Omega^{n-1} A$ corresponds under this isomorphism to the operator $A^{\otimes(n+1)} \oplus A^{\otimes n} \to A^{\otimes n} \oplus A^{\otimes(n-1)}$ which is given by the matrix

$$\begin{pmatrix} b & (1 - \lambda) \\ 0 & -b' \end{pmatrix}$$

where b', b and λ are the operators discussed in Section 6.
Similarly, the operator $B : \Omega^n A \to \Omega^{n+1} A$ corresponds to the operator $A^{\otimes(n+1)} \oplus A^{\otimes n} \to A^{\oplus(n+2)} \oplus A^{\otimes(n+1)}$ given by the 2×2-matrix

$$\begin{pmatrix} 0 & 0 \\ Q & 0 \end{pmatrix}$$

where Q is as in Section 6.
This shows that the (B, b)-bicomplex is just another way of writing the cyclic bi-complex, i.e., the total complex D^Ω for the (B, b)-bicomplex is exactly isomorphic to the total complex D for the cyclic bicomplex. As a consequence we have

Proposition 7.1. $HC_n A$ *is the homology of the complex*

$$\longrightarrow D_n^\Omega \xrightarrow{B'-b} D_{n-1}^\Omega \xrightarrow{B'-b} \ldots \xrightarrow{B'-b} D_1^\Omega \xrightarrow{B'-b} D_0^\Omega \longrightarrow 0$$

where

$$D_{2n}^\Omega = \Omega^0 A \oplus \Omega^2 A \oplus \ldots \oplus \Omega^{2n} A$$
$$D_{2n+1}^\Omega = \Omega^1 A \oplus \Omega^3 A \oplus \ldots \oplus \Omega^{2n+1} A$$

and B' is the truncated B-operator, i.e., $B' = B$ on the components $\Omega^k A$ of D_n^Ω, except on the highest component $\Omega^n A$, where it is 0.
$HH_n(A)$ is the homology of the complex

$$\longrightarrow \Omega^n A \xrightarrow{b} \Omega^{n-1} A \xrightarrow{b} \ldots \xrightarrow{b} \Omega^1 A \xrightarrow{b} \Omega^0 A \longrightarrow 0$$

Remark 7.1. Assume that A has a unit 1. We may introduce in ΩA the additional relation $d(1) = 0$, i.e., divide by the ideal M generated by $d(1)$ (this is equivalent to introducing the relation $1 \cdot \omega = \omega$ for all ω in ΩA). We denote the quotient by $\overline{\Omega} A$. Now M is a graded subspace, invariant under b and B, and its homology with respect to b is trivial. The preceding proposition is thus still valid if we use $\overline{\Omega} A$ in place of ΩA. The convention $d(1) = 0$ is often used (implicitly) in the literature. In some cases it simplifies computations considerably. There are however situations where one cannot reduce the computations to the unital situation (in particular this is true for the excision problem).

If $\mathfrak{A}$ is an m-algebra, then $\Omega\mathfrak{A}$ is defined using the completed projective tensor product

$$\Omega\mathfrak{A} = \widetilde{\mathfrak{A}} \hat{\otimes} \mathfrak{A}^{\hat{\otimes} n} \cong \mathfrak{A}^{\hat{\otimes}(n+1)} \oplus \mathfrak{A}^{\hat{\otimes} n}$$

The operators on $\Omega\mathfrak{A}$ considered above are of course continuous and extend to the completed tensor products.

8 Periodic Cyclic Homology and the Bivariant Theory

Let A be an algebra. We denote by $\widehat{\Omega} A$ the infinite product

$$\widehat{\Omega} A = \prod_n \Omega^n A$$

and by $\widehat{\Omega}^{ev} A$, $\widehat{\Omega}^{odd} A$ its even and odd part, respectively. $\widehat{\Omega} A$ may be viewed as the (periodic) total complex for the bicomplex

$$
\begin{array}{ccccccc}
\downarrow{-b} & & \downarrow{-b} & & \downarrow{-b} & & \downarrow{-b} \\
\longleftarrow \Omega^3 A & \xleftarrow{B} & \Omega^2 A & \xleftarrow{B} & \Omega^1 A & \xleftarrow{B} & \Omega^0 A \\
\downarrow{-b} & & \downarrow{-b} & & \downarrow{-b} & & \\
\longleftarrow \Omega^2 A & \xleftarrow{B} & \Omega^1 A & \xleftarrow{B} & \Omega^0 A & & \\
\downarrow{-b} & & \downarrow{-b} & & & & \\
\longleftarrow \Omega^1 A & \xleftarrow{B} & \Omega^0 A & & & & \\
\downarrow{-b} & & & & & & \\
\longleftarrow \Omega^0 A & & & & & &
\end{array}
$$

Similarly, the (continuous) dual $(\widehat{\Omega}A)'$ of $\widehat{\Omega}A$ is

$$(\widehat{\Omega}A)' = \bigoplus_n (\Omega^n A)'$$

Definition 8.1. *The periodic cyclic homology* $HP_*(A)$, $* = 0, 1$, *is defined as the homology of the* $\mathbb{Z}/2$-*graded complex*

$$\widehat{\Omega}^{ev} A \quad \overset{B-b}{\underset{B-b}{\rightleftarrows}} \quad \widehat{\Omega}^{odd} A$$

and the periodic cyclic cohomology $HP^*(A)$, $* = 0, 1$, *is defined as the homology of the* $\mathbb{Z}/2$-*graded complex*

$$(\widehat{\Omega}^{ev} A)' \quad \overset{B-b}{\underset{B-b}{\rightleftarrows}} \quad (\widehat{\Omega}^{odd} A)'$$

Now by definition, S is the projection

$$D_n^{\Omega} = \Omega^n \oplus \Omega^{n-2} \oplus \dots \quad \longrightarrow \quad D_{n-2}^{\Omega} = \Omega^{n-2} \oplus \Omega^{n-4} \oplus \dots$$

where D_n^{Ω} is as in 7.1. Therefore we get

$$\widehat{\Omega}A = \varprojlim_S (D_{2n}^{\Omega} \oplus D_{2n+1}^{\Omega})$$

and

$$(\widehat{\Omega}A)' = \varinjlim_{S'} (D_{2n}^{\Omega} \oplus D_{2n+1}^{\Omega})'$$

We deduce

Proposition 8.1. *For any algebra A and $* = 0, 1$ one has*

$$HP^*(A) = \varinjlim_S HC^{2n+*} A$$

and an exact sequence

$$0 \to \varprojlim_S {}^1 HC_{2n+*+1}A \to HP_*A \to \varprojlim_S HC_{2n+*}A \to 0$$

(where as usual $\varprojlim_S {}^1 HC_{2n+*+1}A$ *is defined as*

$(\prod_n HC_{2n+*+1}A)/((1 - s)(\prod_n HC_{2n+*+1}A))$, *$s$ being the shift on the infinite product).*

Now $\widehat{\Omega}A$ is in a natural way a complete metric space (with the metric induced by the filtration on ΩA—the distance of families (x_n) and (y_n) in $\prod \Omega^n A$ is $\leq 2^{-k}$ if the k first x_i and y_i agree). We call this the filtration topology and denote by

$\mathrm{Hom}(\widehat{\Omega}A, \widehat{\Omega}B)$ the set of continous linear maps $\widehat{\Omega}A \to \widehat{\Omega}B$. It can also be described as

$$\mathrm{Hom}(\widehat{\Omega}A, \widehat{\Omega}B) = \varprojlim_m (\varinjlim_n \mathrm{Hom}(\bigoplus_{i \leq n} \Omega^i A, \bigoplus_{j \leq m} \Omega^j B).$$

It is a $\mathbb{Z}/2$-graded complex with boundary map

$$\partial \varphi = \varphi \circ \partial - (-1)^{deg(\varphi)} \partial \circ \varphi$$

where $\partial = B - b$. If A and B are themselves topological algebra (m-algebras) of course one takes, in the definition of Hom above, continuous (for the locally convex vector space topology) linear maps from $\Omega^i A$ to $\Omega^j B$.

Definition 8.2. *Let A and B be algebras. Then the bivariant periodic cyclic homology $HP_*(A, B)$ is defined as the homology of the Hom-complex*

$$HP_*(A, B) = H_*(\mathrm{Hom}(\widehat{\Omega}A, \widehat{\Omega}B)) \quad * = 0, 1$$

It is not difficult to see that the $\mathbb{Z}/2$-graded complex $\widehat{\Omega}\mathbb{C}$ is (continuously with respect to the filtration topology) homotopy equivalent to the trivial complex

$$\mathbb{C} \; \underset{\longleftarrow}{\overset{\longrightarrow}{}} \; 0$$

Therefore

$$HP_*(\mathbb{C}, B) = HP_*(B) \quad \text{and} \quad HP_*(A, \mathbb{C}) = HP^*(A)$$

There is an obvious product

$$HP_i(A_1, A_2) \times HP_j(A_2, A_3) \to HP_{i+j}(A_1, A_3)$$

induced by the composition of elements in $\mathrm{Hom}(\widehat{\Omega}A_1, \widehat{\Omega}A_2)$ and $\mathrm{Hom}(\widehat{\Omega}A_2, \widehat{\Omega}A_3)$, which we denote by $(x, y) \mapsto x \cdot y$. In particular, $HP_0(A, A)$ is a unital ring with unit 1_A given by the identity map on $\widehat{\Omega}A$.

An element $\alpha \in HP_*(A, B)$ is called invertible if there exists $\beta \in HP_*(B, A)$ such that $\alpha \cdot \beta = 1_A \in HP_0(A, A)$ and $\beta \cdot \alpha = 1_B \in HP_0(B, B)$. An invertible element of degree 0, i.e., in $HP_0(A, B)$ will also be called an HP-equivalence. Such an HP-equivalence exists in $HP_0(A, B)$ if and only if the supercomplexes $\widehat{\Omega}A$ and $\widehat{\Omega}B$ are continuously homotopy equivalent. Multiplication by an invertible element α on the left or on the right induces natural isomorphisms $HP_*(B, D) \cong HP_*(A, D)$ and $HP_*(D, A) \cong HP_*(D, B)$ for any algebra D.

One can also define a $\mathbb{Z}$-graded version of the bivariant cyclic theory, [23], as follows. Say that a linear map $\alpha : \widehat{\Omega}A \to \widehat{\Omega}B$, continuous for the filtration topology, is of order $\leq k$ if $\alpha(F^n \widehat{\Omega}A) \subset F^{n-k} \widehat{\Omega}B$ for $n \geq k$, where $F^n \widehat{\Omega}A$ is the infinite product of $b(\Omega^{n+1})$, Ω^{n+1}, Ω^{n+2}, ... We denote by $\mathrm{Hom}^k(\widehat{\Omega}A, \widehat{\Omega}B)$ the set of all maps of order $\leq k$. This is, for each k, a subcomplex of the $\mathbb{Z}/2$-graded complex $\mathrm{Hom}(\widehat{\Omega}A, \widehat{\Omega}B)$. We can define

$$HC_n(A, B) = H_i(\mathrm{Hom}^n(\widehat{\Omega}A, \widehat{\Omega}B)) \text{ where } i \in \{0, 1\}, i \equiv n \bmod 2$$

The bivariant theory $HC_n(A, B)$ has a product $HC_n(A, B) \times HC_m(B, C) \longrightarrow HC_{n+m}(A, C)$ and satisfies

$$HC_n(\mathbb{C}, B) = HC_n(B) \quad \text{and} \quad HC_n(A, \mathbb{C}) = HC^n(A)$$

In general, there exist elements in $HP_*(A, B)$ which are not in the range of the natural map $HC_{2n+*}(A, B) \to HP_*(A, B)$ for any n, [14].

9 Mixed Complexes

In this section we present still another construction of cyclic theory which is sometimes convenient for explicit computations of cyclic homology in terms of Hochschild homology.

Following [26], a *mixed complex* is a $\mathbb{Z}$-graded complex $\bigoplus_n M_n$, bounded below, whose differential (usually denoted b) has degree -1, which is equipped with an operator B of degree $+1$ satisfying $[b, B] = B^2 = 0$. The homology of M as a complex, i.e., with respect to b, is by definition the *Hochschild homology* of M:

$$HH_n M = H_n M$$

Associated to M is the $\mathbb{Z}/2$-graded complex given by

$$\widehat{M} = \prod_n M_n$$

equipped with the usual even-odd grading and the differential $B - b$. *The periodic cyclic homology* is then the homology of this supercomplex:

$$HP_\nu M = H_\nu \widehat{M}, \qquad \nu \in \mathbb{Z}/2$$

The *cyclic homology* of M is defined as

$$HC_n M = H_n(D)$$

where D is the complex given by

$$D_n = \bigoplus_{p \geq 0} M_{n-2p}$$

with boundary operator $B' - b$, where $B' : M_n \to M_{n-1}$ is, as in 7.1, the truncated B-operator.

We have seen that, for any algebra A, there is a natural mixed complex, given by

$(\Omega(A), b, B)$ and the Hochschild, cyclic and periodic cyclic homology of A is by definition the Hochschild, cyclic and periodic cyclic homology of this mixed complex.

The following simple observation is useful for computations of cyclic homology invariants.

Proposition 9.1. *Let* $\varphi : (M, b, B) \rightarrow (M', b, B)$ *be a morphism of mixed complexes (i.e., φ maps M_n to M'_n for all n and φ commutes with the b- and B-operators). If φ is a quasi-isomorphism from (M, b) to (M', b), then φ induces isomorphisms on the cyclic and periodic cyclic homology of M and M'.*

Proof. This follows from the analogue of 6.1 and 8.1 for mixed complexes.

For instance, the cyclic homology of smooth (in the sense of Grothendieck) algebras, of the locally convex algebra of smooth functions on a compact differentiable manifold or of the irrational rotation algebra (noncommutative torus) can be computed that way. In these cases, the Hochschild homology $HH(A)$ can be computed and there is an operator $\dot{B}$ of degree 1 on $HH(A)$ as well as a surjective map of mixed complexes

$$(\Omega(A), b, B) \rightarrow (HH(A), 0, \dot{B})$$

see [4], [31]. In the case of a smooth algebra or of the smooth functions on a compact manifold, $HH(A)$ is simply the algebra of ordinary classical differential forms and the operator $\dot{B}$ is the the classical de Rham operator d.

10 The X-Complex Description of Cyclic Homology

The simple X-complex provides a useful alternative description of cyclic theory. It has been introduced first by Quillen [40] in connection with cyclic homology for coalgebras and then been used systematically in [13], [14], [15]. This different approach to cyclic theory is particularly useful for the study of periodic theories and cyclic theories for topological algebras, for proving the excision theorem, and for understanding the connection with topological K-theory.

Definition 10.1. *Let A be an algebra. The X-complex $X(A)$ is the $\mathbb{Z}/2$-graded complex*

$$A \underset{b}{\overset{\natural d}{\rightleftarrows}} \Omega^1 A_\natural$$

where $\Omega^1 A_\natural = \Omega^1 A / b(\Omega^2 A) = \Omega^1 A / [A, \Omega^1 A]$ and $\natural : \Omega^1 A \rightarrow \Omega^1 A_\natural$ is the canonical quotient map.

Recall that, from the definition of b, its image exactly consists of all commutators, thus $b(\Omega^2 A) = [A, \Omega^1 A]$.

The X-complex is constructed from the lowest level of the (B, b)-bicomplex

$$\downarrow b \qquad\qquad \downarrow b \qquad\qquad \downarrow b$$

$$\longleftarrow \Omega^2 A \xleftarrow{\ B\ } \boxed{\Omega^1 A} \xleftarrow{\ B=d\ } \boxed{\Omega^0 A}$$

$$\downarrow b \qquad\qquad \downarrow b$$

$$\longleftarrow \boxed{\Omega^1 A} \xleftarrow{\ B=d\ } \boxed{\Omega^0 A}$$

$$\downarrow b$$

$$\longleftarrow \boxed{\Omega^0 A}$$

To be precise, the X-complex is the total complex of the quotient of the bicomplex

$$\downarrow b \qquad\quad \downarrow b \qquad\quad \downarrow b \qquad\quad \downarrow b$$

$$\longleftarrow \Omega^3 A \xleftarrow{\ B\ } \Omega^2 A \xleftarrow{\ B\ } \Omega^1 A \xleftarrow{\ B\ } \Omega^0 A$$

$$\downarrow b \qquad\quad \downarrow b \qquad\quad \downarrow b$$

$$\longleftarrow \Omega^2 A \xleftarrow{\ B\ } \Omega^1 A \xleftarrow{\ B\ } \Omega^0 A$$

$$\downarrow b \qquad\quad \downarrow b$$

$$\longleftarrow \Omega^1 A \xleftarrow{\ B\ } \Omega^0 A$$

$$\downarrow b$$

$$\longleftarrow \Omega^0 A$$

by the sub-bicomplex

$$\downarrow b \qquad\qquad \downarrow b \qquad\qquad \downarrow b$$

$$\longleftarrow \Omega^3 A \xleftarrow{\ B\ } \Omega^2 A \xleftarrow{\ B\ } b(\Omega^2 A)$$

$$\downarrow b \qquad\qquad \downarrow b$$

$$\longleftarrow \Omega^2 A \xleftarrow{\ B\ } b(\Omega^2 A) \tag{1}$$

$$\downarrow b$$

$$\longleftarrow b(\Omega^2 A)$$

where $\Omega^0 A$ is replaced by 0 and $\Omega^1 A$ is replaced by $b(\Omega^2 A)$. It is periodic of period two and specifically designed to compute the periodic cyclic theory for a special class of "smooth" algebras - the quasi-free algebras, because for quasi-free algebras the subcomplex by which we divide is contractible.

Proposition-Definition 10.1 *Let A be an algebra. The following conditions are equivalent:*

(1) Let $0 \to N \to S \xrightarrow{q} B \to 0$ be an extension of algebras where the ideal N is nilpotent (i.e., $N^k = \{0\}$ for some $k \geq 1$) and $A \xrightarrow{\alpha} B$ a homomorphism. Then there exists a homomorphism $A \xrightarrow{\alpha'} S$ such that $q \circ \alpha' = \alpha$.

(2) A has cohomological dimension ≤ 1 with respect to Hochschild cohomology.

(3) $\Omega^1 A$ is a projective bimodule over A.

The algebra A is called <u>quasi-free</u> if these equivalent conditions are satisfied.

The list of equivalent conditions characterizing quasi-freeness can be extended. In particular, (2) is equivalent to

(2)' There exists a linear map $\varphi : A \to \Omega^2 A$ such that

$$\varphi(xy) = x\varphi(y) + \varphi(x)y - dxdy \tag{2}$$

(i.e., the universal Hochschild 2-cocycle $c : A \times A \to \Omega^2 A$ defined by $c(x, y) = dxdy$ is the boundary of φ: $c = b\varphi$).

and (3) is equivalent to

(3)' There exists a linear map $\nabla : \Omega^1 A \to \Omega^2 A$ (a "right connection") satisfying $\nabla(a\omega) = a\nabla\omega$

$$\nabla(\omega a) = (\nabla\omega)a + \omega da \qquad (a \in A, \omega \in \Omega^2 A)$$

The equivalence between (2)' and (3)' is based on the correspondence between φ and ∇ given by the equality $\nabla(xdy) = x\varphi(y)$. For explicit computations the property (2) of quasi-free algebras often is particularly useful.

Proposition 10.1. *Let A be quasi-free and $HX_*(A), * = 0, 1$ be the homology of the complex $X(A)$. Then $HX_*(A) = HP_*A$.*

Proof. The proof is based on a contraction of the infinite product $(b\Omega^2 A, \Omega^2 A, \Omega^3 A, \ldots)$ inside $\hat{\Omega} A$ obtained from the connection ∇. Or, equivalently on showing that the the total complex of the sub-bicomplex (1) is contractible.

The preceding characterizations of quasi-free algebras carry over directly to the category of m-algebras. In condition (1) of 10.1 N^k has to be interpreted as the image of the n-fold projective tensor power of N under the multiplication map and the nilpotent extension has to be assumed to admit a continuous linear splitting. The most important examples of quasi-free algebras are free algebras. In particular, for any vector space V, the (non-unital) tensor algebra

$$TV = V \oplus V^{\otimes 2} \oplus V^{\otimes 3} \oplus \ldots$$

is quasi-free. It obviously satisfies condition (1) in 10.1. In fact, it admits liftings in any (not necessarily nilpotent) extension. The map φ of (2) is determined by the condition that $\varphi(x) = 0$ for all generators $x \in V$. If $\mathfrak{A}$ is an m-algebra, then the tensor algebra $T\mathfrak{A}$ in the category of m-algebras is quasi-free. Other examples of quasifree algebras include $\mathbb{C}$, $M_n(\mathbb{C})$ or $\mathbb{C}[z, z^{-1}]$. One obtains from this

Corollary 10.1.

$$HP_0(\mathbb{C}) = \mathbb{C} \qquad HP_1(\mathbb{C}) = 0$$

$$HP_0(\mathbb{C}[z, z^{-1}]) = \mathbb{C} \; HP_1(\mathbb{C}[z, z^{-1}]) = \mathbb{C}$$

Definition 10.2. *Let T be an algebra and J an ideal in T. The J-adic filtration on $X(T)$ is defined by*

$$F^{2n+1}X(T) = J^{n+1} \oplus \natural(J^{n+1}dT + J^n dJ)$$

$$F^{2n}X(T) = (J^{n+1} + [J^n, T]) \oplus \natural(J^n dT)$$

Remark 10.1. This defines a decreasing filtration which is closely related to the inverse system $X(T/J^n)$. In fact. let F^n be the J-adic filtration of $X(T)$ as above and G^n the kernel of the quotient map $X(T) \to X(T/J^n)$. Then

$$G^{n+1} \subset F^{2n+1} \subset F^{2n} \subset G^n \subset \dots$$

Therefore $\widehat{X}(T) = X(\widehat{T})$ where

$$\widehat{T} = \varprojlim_n T/J^n \qquad \text{and} \qquad \widehat{X}(T) = \varprojlim_n X(T)/F^n$$

The connection between the X-complex for free-algebras and the cyclic bicomplex is based on the following results.

Proposition 10.2. *Let A be an algebra.*

(1) The following defines an associative product (Fedosov product) on ΩA:

$$\omega_1 \circ \omega_2 = \omega_1 \omega_2 + (-1)^{deg(\omega_1)} d\omega_1 d\omega_2$$

where $\omega_1 \omega_2$ is the ordinary product of ω_1, ω_2 in ΩA. Clearly, the even forms $\Omega^{ev}A$ form a subalgebra of $(\Omega A, \circ)$ for $\circ$.

(2) Let $p : A \to TA$ be the canonical linear inclusion of A into the tensor algebra over A and, for $x, y \in A$, $\omega(x, y) = p(xy) - pxpy$. Then the map

$$\alpha : x_0 dx_1 dx_2 \dots dx_{2n-1} dx_{x_{2n}} \mapsto x_0 \omega(x_1, x_2) \dots \omega(x_{2n-1}, x_{2n})$$

defines an isomorphism of algebras $(\Omega^{ev}A, \circ) \xrightarrow{\alpha} TA$.

(3) The map $\omega dx \mapsto \natural(\alpha(\omega)d(px))$ (where α and p are as in (2)) defines a linear isomorphism $\Omega^{odd}A \xrightarrow{\alpha'} \Omega^1 TA_\natural$.

(4) The isomorphism $\alpha \oplus \alpha' : \Omega A \to X(TA)$ described in (2) and (3) identifies the natural filtration $F^n \Omega A = b\Omega^{n+1}A \oplus \Omega^{n+1}A \oplus \Omega^{n+2}A \oplus \dots$ on $\Omega(A)$ with the JA-adic filtration

$$F^{2n+1}X(TA) = (JA)^{n+1} \oplus \natural((JA)^{n+1}d(TA) + (JA)^n d(JA))$$

$$F^{2n}X(TA) = ((JA)^{n+1} + [(JA)^n, TA]) \oplus \natural((JA)^n d(TA))$$

on $X(TA)$, see 10.2, where JA is the ideal in TA given as the kernel of the natural homomorphism $TA \to A$.

Remark 10.2. The full algebra $(\Omega A, \circ)$ with Fedosov product is isomorphic to the algebra $QA = A * A$, 4.3. The isomorphism sends $x + dx$ to $\iota(x)$ and $x - dx$ to $\bar{\iota}(x)$. Thus, in particular, the algebra TA is isomorphic to the even part, for the natural $\mathbb{Z}/2$-grading of QA.

As a consequence of the preceding proposition, the X-complex $X(TA)$: $TA \overset{\natural d}{\underset{b}{\rightleftarrows}}$
$\Omega^1(TA)_\natural$ is isomorphic to a complex of the form $\Omega^{ev}A \overset{\delta}{\underset{\beta}{\rightleftarrows}} \Omega^{odd}A$ where the maps δ and β have to be determined.

This is the content of the following key theorem.

Theorem 10.2. *Let δ, β be as above and $\Omega A \overset{\kappa}{\to} \Omega A$ the Karoubi operator. Then*

(a)

$$\beta(\omega) = b\omega - (1 + \kappa)d\omega, \qquad \omega \in \Omega^{odd}A$$

$$\delta(\omega) = \sum_{j=0}^{2n+1} \kappa^j d\omega - \sum_{j=0}^{n-1} \kappa^{2j} b\omega, \, \omega \in \Omega^{2n}A$$

(b) Let P be the spectral projection to 1 for κ (see 7.1). P commutes with b, B and β, δ and

$$P\beta = P(b - \tfrac{2}{N}B)$$

$$P\delta = P(B - b\tfrac{N}{2})$$

(b) follows immediately from (a) by using the equality $P\kappa = P$ and $B = NPd$.

The complex $(1 - P)\widehat{\Omega}A$ is contractible, both with the boundary operator $B - b$ and with β and δ. The same holds for the complexes in the associated filtration. For a precise statement we refer to [14], S6.

We still denote by P the projection operator corresponding to P under the isomorphism $\Omega A \cong X(TA)$. Let c be the scaling operator on ΩA which is multiplication by c_q on Ω^q where $c_{2n} = c_{2n+1} = (-1)^n n!$. It follows from 10.2 that $c : PX(TA) \to P\Omega A$ is an isomorphism identifying the differential $\natural d$ and $\bar{b}$ on $PX(TA)$ with the differential $B - b$ on $P\Omega A$. It obviously is compatible with the filtration F^n on both sides described in 10.2(4).

This proves the following theorem

Theorem 10.3. *Let A be an algebra*

(a) For each n we have

$$HC_n(A) = H_i\Big(X(TA) / F^n X(TA)\Big)$$

where $H_i, i = 0, 1$ is the homology of the $\mathbb{Z}/2$-graded complex ("supercomplex") $X(TA)/F^n X(TA)$ and $i \equiv n \bmod 2$.

(b) Let $\widehat{\Omega}A = \varprojlim_{n} \Omega A / F^n \Omega A$ and $\widehat{X}(TA) = \varprojlim_{n} X(TA)/F^n X(TA)$ be the completions. Then

$$HP_i A = H_i(\widehat{\Omega}A) \cong H_i(\widehat{X}(TA))$$

where $i = 0, 1$ and the differentials on $\widehat{\Omega}A$ and $\widehat{X}(TA)$ are given by $B - b$ and by $(\natural d, \dot{b})$, respectively.

Corollary 10.2. *For each n we have*

$$HC_n(A) = H_{i+1}(F^n X(TA)) \qquad HC^n(A) = H_{i+1}((F^n X(TA))')$$

where $H_i, i = 0, 1$ is the homology of the $\mathbb{Z}/2$-graded complex $F^n X(TA)$ and its dual $(F^n X(TA))'$ while $i \equiv n \bmod 2$. In particular, any trace τ on $(JA)^n$ defines an element in $HC^{2n-1}A$. This element is represented by the cyclic cocycle f_{2n-1} given by

$$f_{2n-1}(x_0, x_1, \ldots x_{2n-1}) = \tau(\omega(x_0, x_1)\omega(x_2, x_3) \ldots \omega(x_{2n-2}, x_{2n-1}))$$

$$-\tau(\omega(x_{2n-1}, x_0)\omega(x_1, x_2) \ldots \omega(x_{2n-3}, x_{2n-2}))$$

where $\omega(x, y) = p(xy) - p(x)p(y)$ for the canonical linear inclusion $p : A \to TA$.

Proof. The first assertion follows from 10.3 and the exact sequence of complexes $0 \to F^n X(TA) \to X(TA) \to X(TA)/(F^n X(TA)) \to 0$ together with the fact that the homology of $X(TA)$ is trivial.

The second assertion is an easy consequence of the fact that $(JA)^n$ is exactly the even part of $F^{2n-1} X(TA)$.

The preceding corollary is a translation of 6.2. There is a $\mathbb{Z}/2$-graded version using supertraces. A supertrace on a $\mathbb{Z}/2$-graded algebra with grading automorphism $x \to \bar{x}$ is by definition a linear functional f such that $f(xy) = (-1)^{\deg x \deg y} f(yx)$. f is called odd if $f(\bar{x}) = -f(x)$.

Proposition 10.3. *Let qA be the canonical ideal in the free product $QA = A * A$ defined in 4.3. Any odd supertrace τ on $(qA)^{2n+1}$ defines an element of HC^{2n} represented by the cyclic cocycle*

$$f(x_0, x_1, \ldots, x_{2n}) = \tau(q(x_0)q(x_1) \ldots q(x_{2n}))$$

Theorem 10.4. *Let $0 \to J \to T \to A \to 0$ be any extension of A where T is quasi-free. Then the complex $X(T)$ is homotopy equivalent to $X(TA)$ by a chain map respecting the filtrations induced by J and JA as in 10.2. Therefore $HC_n(A)$ and $HP_{0/1}A$ can be computed from $X(T)$. Explicitly*

$$HC_n(A) = H_i\big(X(T)/F^n X(T)\big) \quad \text{for } i \equiv n \bmod 2$$

$$HP_i A = H_i(\widehat{X}(T)) \quad \text{for } i = 0, 1$$

Proof. This follows from the fact that the X-complex is homotopy invariant for quasi-free algebras (see 12.1) and the fact that T and TA are homotopy equivalent by a homomorphism that respects the filtrations of T and TA by the powers of the ideals J and JA.

Remark 10.3. (a) The operator $S : HC_{n+2} \to HC_n$ is induced by the natural map $\Omega/F^{n+2}\Omega \to \Omega/F^n\Omega$ or by the map $X/F^{n+2}X \to X/F^nX$.
(b) Also the Hochschild homology $HH_n(A)$ can be read off from the filtered complex $X(T)$, if T is any quasi-free extension of A. In fact, $HH_n(A)$ is the homology of the $\mathbb{Z}/2$-graded complex $F^{n-1}XT/F^nXT$ in dimension $i \in \{0,1\}$ with $i \equiv n$ mod 2.

Corollary 10.3. *(Goodwillie's theorem). Let I be an ideal in the algebra A that is nilpotent, i.e., $I^k = 0$ for some k. Then the quotient map $A \to A/I$ induces an invertible element in $HP_*(A, A/I)$. In particular $HP_*(A) \cong HP_*(A/I)$.*

Proof. We can choose TA as a quasi-free extension for both A and A/I. Since I is nilpotent, the resulting filtrations on $X(TA)$ are continuous with respect to each other.

All constructions in this section also work in the category of m-algebras. If $\mathfrak{J}$ is an m-algebra then, by definition, $\mathfrak{J}^2$ is the image of the multiplication map $\mathfrak{J}\hat{\otimes}\mathfrak{J} \to \mathfrak{J}$. Note the following important property of ideals in quasi-free m-algebras:

If $\mathfrak{J}$ is the ideal in an extension

$$0 \longrightarrow \mathfrak{J} \longrightarrow \mathfrak{A} \longrightarrow \mathfrak{A}/\mathfrak{J} \longrightarrow 0$$

of m-algebras with a continuous splitting and $\mathfrak{A}$ is quasi-free, then also $\mathfrak{J}^n$ is an ideal which is a direct summand in $\mathfrak{A}$ as a topological vector space for each n. For instance $(J\mathfrak{A})^n$ is a direct summand in $T\mathfrak{A}$ for each n.

Remark 10.4. The fact that cyclic and periodic cyclic homology can be reduced to the homology of the X-complex applied to a quasi-free algebra, means that the properties of the theory can be checked by computations that only involve the cyclic homology of a quasi-free algebra *in dimensions 0 and 1*! This fact is essential for establishing some of the most important properties of the theory.

11 Cyclic Homology as Non-commutative de Rham Theory

For smooth commutative algebras the cyclic homology/cohomology can be determined by computing first the Hochschild homology and then using Connes' SBI-sequence. This applies in essentially the same way to the coordinate ring of a smooth variety or to the Fréchet algebra of smooth functions on a compact C^∞-manifold.

Definition 11.1. *A commutative algebra A is smooth if any homomorphism $\alpha : A \to B/N$ where B is a commutative algebra and N an ideal in B with $N^2 = 0$, can be lifted to a homomorphism $\widehat{\alpha} : A \to B$ such that $\pi \circ \widehat{\alpha} = \alpha$ for the quotient map $\pi : B \to B/N$.*

There are various other characterizations of smoothness. Note that, in particular, the coordinate ring of a smooth algebraic variety is smooth.

For a unital commutative algebra we define $\Omega^*_{comm}A$ as the largest graded commutative quotient of ΩA, i.e., as ΩA divided by the ideal generated by all differences $xdy - d(y)x$ and $dxdy + dydx$—with the additional relation $d(1) = 0$. It also can be defined as the exterior algebra over A on the A-bimodule $\Omega^1 A$, with additional relation $d(1) = 0$.

The Hochschild homology of a smooth algebra is determined by the following theorem.

Theorem 11.1. *(Hochschild-Kostant-Rosenberg) If A is a smooth unital commutative algebra then the natural map $\Omega^*_{comm}A \to HH_*(A)$ is an isomorphism (of graded algebras).*

The de Rham cohomology $H^{dR}A$ of A is by definition the homology of the complex

$$\cdots \to \Omega^k_{comm}A \xrightarrow{d} \Omega^{k+1}_{comm}A \to \cdots$$

For $* = 0, 1$ we write $H^{dR}_0 A$ for the direct sum of all de Rham homology groups in even dimensions and $H^{dR}_1 A$ for the sum of all de Rham homology groups in odd dimensions.

We have

Theorem 11.2. *Let A be a unital smooth algebra of homological dimension n. Then*

(a) $HP_*A \cong H^{dR}_*A$

(a) If $k > n$, then $HC_k A \cong H^{dR}_*A$ for $* \equiv k \mod 2$.

Proof. The quotient map $\Omega^n A \to \Omega^n_{comm}A$ induces a natural map from the (B,b)-bicomplex for A to the bicomplex

$$\downarrow 0 \qquad\qquad \downarrow 0 \qquad\qquad \downarrow 0$$

$$\Omega^2_{comm}A \xleftarrow{d} \Omega^1_{comm}A \xleftarrow{d} \Omega^0_{comm}A$$

$$\downarrow 0 \qquad\qquad \downarrow 0$$

$$\Omega^1_{comm}A \xleftarrow{d} \Omega^0_{comm}A$$

$$\downarrow 0$$

$$\Omega^0_{comm}A$$

By the Hochschild-Kostant-Rosenberg theorem this map is a quasi-isomorphism on the columns. Thus the theorem follows for instance from the spectral sequence comparison theorem (or also from Connes's SBI-sequence).

Remark 11.1. If A is the coordinate ring for a not necessarily smooth variety V then HP_*A is isomorphic to the infinitesimal cohomology of V in the sense of Grothendieck which is in this case the correct replacement for de Rham theory.

In the case of the Fréchet algebra $C^\infty M$ of course we use the cyclic homology for locally convex algebras.

Theorem 11.3. *(Connes) Let $C^\infty M$ be the algebra of smooth functions on a smooth compact manifold M and $C^\infty(\Lambda T^*M)$ be the algebra of differential forms on M. Then the natural map $C^\infty(\Lambda^k T^*M) \rightarrow HH_k(C^\infty M)$ is an isomorphism for each k.*

Theorem 11.4. *Let M be a smooth compact manifold of dimension n. Then*

(a) $HP_*(C^\infty M) \cong H_*^{dR}(C^\infty M)$
(a) If $k > n$, then $HC_k(C^\infty M) \cong H_*^{dR}(C^\infty M)$ for $* \equiv k \bmod 2$.

All results in this section are of course also valid in cohomology in place of homology.

Remark 11.2. Karoubi has defined de Rham homology for an arbitrary algebra A as the homology of the complex given by the A-bimodules $\Omega^n(A)/[\Omega^*(A), \Omega^*(A)]$, where the quotients are by graded commutators, and has shown that it is isomorphic to the image of $HC_{n+2}(A)$ in $HC_n(A)$ under the operator S, see [24].

There is an even more significant and important analogy between cyclic homology and Grothendieck's notion of infinitesimal homology which describes the topology of non-smooth varieties in algebraic geometry. In this analogy, quasi-free algebras play the role of smooth varieties in the non-commutative category and the X-complex corresponds to the de Rham complex.

In fact, in algebraic geometry, the infinitesimal cohomology is defined by writing the coordinate ring A of the variety as a quotient of the coordinate ring of a smooth variety T and by taking the de Rham cohomology of the completion $\widehat{T}$. It can be shown that this does not depend on the choice of the embedding into a smooth variety.

The periodic cyclic homology of an algebra A is obtained by writing A as a quotient of a quasi-free algebra T and taking the X-complex-homology of the completion $\widehat{T}$. This does not depend on the choice of the quasi-free extension T, see 10.3.

12 Homotopy Invariance for Cyclic Theory

The important basic features of cyclic homology are homotopy invariance (under differentiable homotopies), Morita invariance, and excision. These properties hold very generally for the periodic theory HP—but only in a very limited sense for the non-periodic theory HC.

In this section we discuss homotopy invariance. The first (and main) step consists in showing that derivations act trivially on HP (or on HC after stabilizing by the

S-operator). This can be done by guessing the correct homotopy formula on the complex $(\Omega^n(A), B - b)$ computing the cyclic homology $HC_*(A)$, or, in the X-complex-picture, by using the fact that the X-complex is naturally homotopy invariant for quasi-free algebras.

Given an algebra B, we denote by $B[0, 1]$ the algebra $B \otimes C^\infty[0, 1]$ where $C^\infty[0, 1]$ stands for the algebra of $\mathbb{C}$-valued C^∞-functions on $[0, 1]$ (i.e., restrictions of functions in $C^\infty(\mathbb{R})$ to $[0, 1]$). **When we work in the category of m-algebras, we use the completed projective tensor product. In this case, $B[0, 1]$ can be identified with the algebra of B-valued C^∞-functions on $[0, 1]$.**

By a differentiable homotopy from an algebra A to an algebra B we mean a homomorphism α from A to the algebra $B[0, 1]$. Given such a family α, evaluation for each $t \in [0, 1]$ defines a homomorphism $\alpha_t : A \to B$. We say that two homomorphisms $\beta_0, \beta_1 : A \to B$ are differentiably homotopic if there exists a differentiable homotopy α such that $\alpha_0 = \beta_0, \alpha_1 = \beta_1$.

Let $u : A \to B[0, 1]$ be a differentiable family of homomorphisms between algebras A and B. For each fixed $t \in [0, 1]$ let $u_t : A \to B$ be the morphism given by u at the point t and $\dot{u}_t$ the derivative, with respect to t, at the point t, which is a linear map $A \to B$ and behaves like a derivation with respect to u_t, i.e.,

$$\dot{u}_t(xy) = u_t(x)\dot{u}_t(y) + \dot{u}_t(x)u_t(y),$$

$x, y \in A$.

The Lie derivative

$$L_t = L(u_t, \dot{u}_t) : X(A) \to X(B)$$

is defined by

$$L_t(x) = \dot{u}_t(x), \quad L_t(\natural(xdy)) = \natural(\dot{u}_t(x)d(u_ty) + u_t(x)d(\dot{u}_ty)),$$

$x, y \in A$. Then L_t is a morphism of supercomplexes, see [14], beginning of S7.

Proposition 12.1. *Let $u : A \to B[0, 1]$ be as above and assume that A is quasi-free. Then, for each $t \in [0, 1]$, there is a map of supercomplexes of odd degree $h_t : X(A) \to X(B)$ such that $L_t = \partial h_t + h_t\partial$ where ∂ is the boundary operator in the X-complex.*

Proof. Let φ be the fundamental map $A \to \Omega^2 A$ for the quasi-free algebra A, see (2). Put

$$h_t(x) = \natural i\varphi(x)$$
$$h_t(\natural(xdy)) = i(xdy + b(x\varphi(y))$$

$x, y \in A$, where

$$i(x_0dx_1 \ldots dx_k) = u_t(x_0)\dot{u}_t(x_1)du_t(x_2) \ldots du_t(x_k).$$

One computes that $L = \partial h + h\partial$, see [14].

By integrating h_t with respect to t, see [14], 8.1, we obtain an operator $H = \int_0^1 h_t dt$ and

Proposition 12.2. *Let* $u : A \to B[0,1]$ *be as above and* A *quasi-free. There is an odd map of supercomplexes* $H : X(A) \to X(B)$ *such that* $u_{1*} - u_{0*} = \partial H + H\partial$ *for the maps* $u_{t*} : X(A) \to X(B)$ *induced by* u_t, $t = 0, 1$.

Assume now that A and B are written as quotients of free algebras T and R, respectively:

$$A \cong T/I \qquad B \cong R/J$$

and that $u : A \to B[0,1]$ is a homotopy as before. Since T is free, u induces a homomorphism $w : T \to R[0,1]$ which makes the following diagram commutative:

$$
\begin{array}{ccc}
T & \xrightarrow{w} & R[0,1] \\
\downarrow & & \downarrow \\
A & \xrightarrow{u} & B[0,1]
\end{array}
$$

Necessarily, w maps I to $J[0,1]$ and thus each w_t maps I to J. Therefore the map induced by w_t on the X-complexes maps the subspace F^n in the filtration of $X(T)$ induced by I, see 10.2, to the corresponding subspace F^n in the filtration of $X(R)$ induced by J. The formulas used in the definition of $L(u_t, \dot{u}_t)$ and H show that these maps send F^n to F^{n-1}. As a consequence, we get

Theorem 12.1. *Let* $u : A \to B[0,1]$ *be a differentiable homotopy and* $u_t : A \to B$ *the corresponding family of homomorphisms.*

(a) *The element* $HP_0(u_t) \in HP_0(A, B)$ *defined by* u_t *does not depend on* t. *Thus, in particular, the maps induced by* u_t *on* HP_* *and on* HP^* *are constant.*

(b) *Let* $(u_t)_*$ *be the map induced by* u_t *on* HC_n *and* $S : HC_{n-2} \to HC_n$ *the S-operator. Then* $(u_t)_* \circ S : HC_{n-2}(A) \to HC_n(B)$ *is constant.*

13 Morita Invariance for Periodic Cyclic Theory

Theorem 13.1. *Let* E *and* F *be linear subspaces of the algebra* A *and* $A(EF)$, $A(FE)$ *the subalgebras of* A *generated by the spaces of products* EF *and* FE. *Then there is an invertible element in* $HP_0(A(EF), A(FE))$

This theorem can be proved using the following quasi-free extensions of $A(EF)$ and $A(FE)$:

$$0 \to K(EF) \to T(E \otimes F) \xrightarrow{\pi_1} A(EF) \to 0$$

$$0 \to K(FE) \to T(F \otimes E) \xrightarrow{\pi_2} A(FE) \to 0$$

Here, π_1 and π_2 are the homomorphisms induced by the linear maps $E \otimes F \ni x \otimes y \mapsto xy \in A(EF)$, $F \otimes E \ni y \otimes x \mapsto yx \in A(FE)$ and $K(EF), K(FE)$ are the kernels of π_1, π_2.

The invertible element in $HP_0(A(EF), A(FE))$ is induced by the linear isomorphism

$$\lambda : X(T(E \otimes F)) \to X(T(F \otimes E))$$

which is simply cyclic permutation on tensor products

$$\lambda(x_1 \otimes y_1 \otimes x_2 \otimes y_2 \otimes \ldots \otimes x_k \otimes y_k)$$
$$= y_k \otimes x_1 \otimes y_1 \otimes x_2 \otimes \ldots \otimes x_k$$

$$\lambda(\natural(x_1 \otimes y_1 \otimes \ldots \otimes x_{k-1} \otimes y_{k-1}d(x_k \otimes y_k)))$$
$$= \natural(y_k \otimes x_1 \otimes y_1 \otimes \ldots \otimes x_{k-1}d(y_{k-1} \otimes x_k))$$

$$x_1, \ldots, x_k \in E; \; y_1, \ldots, y_k \in F$$

It is trivially checked that λ commutes with the boundary operators $\natural d$ and $\dot{b}$ of the X-complex, using the definition of these operators.

One can also show that $\lambda : X(T(E \otimes F)) \to X(T(F \otimes E))$ is "continuous" with respect to the natural filtration, see 10.2, i.e., that for each n there is k, such that λ maps F^k to F^n, see [12].

A Morita context

$$\begin{pmatrix} A & X \\ Y & B \end{pmatrix}$$

for two algebras A and B is by definition an algebra with a splitting into four linear subspaces such that when the elements are written as 2×2 matrices the multiplication is consistent with matrix multiplication.

Corollary 13.1. *Let A and B be algebras which are related by a Morita context as above and assume that $A = XY$, $B = YX$. Then there is an invertible element in $HP_0(A, B)$.*

The results of this section and of the following one are also valid for m-agebras replacing $A(EF)$, $A(FE)$, XY and YX by the closed subalgebras generated by the respective products.

14 Morita Invariance for the Non-periodic Theory

In the non-periodic theory, Morita invariance only holds for unital or at least H-unital algebras, [47].

Theorem 14.1. *Let E and F be linear subspaces of the algebra A and $A(EF)$, $A(FE)$ the subalgebras of A generated by the spaces of products EF and FE. Assume that $A(EF)$, $A(FE)$ are both H-unital (see 6.3). Then there is an invertible element in $HC^0(A(EF), A(FE))$. This element induces isomorphisms*

$$HC^n(A(EF)) \cong HC^n(A(FE))$$

for all n.

Corollary 14.1. *Let A and B be H-unital algebras. Let E be a $A - B$-bimodule and F be a $B - A$-bimodule such that*

$$E \otimes_B F \cong A \qquad F \otimes_A E \cong B$$

Then there is an invertible element in $HC^0(A, B)$ inducing isomorphisms

$$HC^n(A) \cong HC^n(B)$$

for all n.

15 Excision for Periodic Cyclic Theory

Theorem 15.1. *Let $0 \to S \to P \to Q \to 0$ be an extension of algebras and A an algebra. There are two natural six-term exact sequences*

$$
\begin{array}{ccccc}
HP_0(A, S) & \to & HP_0(A, P) & \to & HP_0(A, Q) \\
\uparrow & & & & \downarrow \\
HP_1(A, Q) & \leftarrow & HP_1(A, P) & \leftarrow & HP_1(A, S)
\end{array}
$$

and

$$
\begin{array}{ccccc}
HP_0(S, A) & \leftarrow & HP_0(P, A) & \leftarrow & HP_0(Q, A) \\
\downarrow & & & & \uparrow \\
HP_1(Q, A) & \to & HP_1(P, A) & \to & HP_1(S, A)
\end{array}
$$

where the horizontal arrows are induced by the maps in the given extension (a description of the vertical arrows will be given below).

In the category of m-algebras, this theorem is valid for extensions admitting a continuous linear splitting, see [10]. We include a brief discussion of the vertical arrows in 15.1 (see [15] and also [27], III 2.2). Given an extension of algebras $0 \to S \to P \to Q \to 0$ let δ, δ' be the connecting maps in the following two six-term exact sequences.

$$\longrightarrow HP_*(P, S) \longrightarrow HP_*(S, S) \overset{\delta}{\longrightarrow} HP_{*+1}(Q, S) \longrightarrow$$

$$\longrightarrow HP_*(Q, P) \longrightarrow HP_*(Q, Q) \overset{\delta'}{\longrightarrow} HP_{*+1}(Q, S) \longrightarrow$$

Let 1_S and 1_Q denote the class of id in $HP_0(S, S)$ and $HP_0(Q, Q)$ respectively.

Proposition 15.1. *One has $\delta(1_S) = -\delta'(1_Q)$.*

Theorem 15.2. *Let $E : 0 \to S \to P \to Q \to 0$ be an extension of algebras. We write $\mathrm{ch}(E)$ for the element $\delta'(1_Q) = -\delta(1_S) \in HP_1(Q, S)$. The connecting map in the first exact sequence in Theorem 5.3 acts on $HP_j(A, Q)$ by multiplication by $(-1)^j \mathrm{ch}E$ on the right, while the connecting map in the second exact sequence in Theorem 5.3 is given by multiplication by $-\mathrm{ch}(E)$ on the left.*

16 Excision for the Non-periodic Theory

Let I be a (non-unital) algebra and $(\Omega^n(I), b)$ and $(D_n^\Omega(I), B' - b)$ be the standard complexes computing Hochschild and cyclic homology of I, respectively, see 7.1. If $0 \to I \to A \to B \to 0$ with I as ideal, we let $(\Omega^n(I : A), b)$ and $(D_n^\Omega(I : A)$ denote the kernels of the maps $\Omega^n(A) \to \Omega^n(B)$ and $D_n^\Omega(A) \to D_n^\Omega(B)$, respectively, so that we have exact sequences of complexes:

$$0 \to \Omega^n(I : A) \to \Omega^n(A) \to \Omega^n(B) \to 0$$

$$0 \to D_n^\Omega(I : A) \to D_n^\Omega(A) \to D_n^\Omega(B) \to 0$$

We say that I is *excisive* in Hochschild homology (respectively in cyclic homology), if the inclusion $\Omega^n(I) \to \Omega^n(I : A)$ (respectively $D_n^\Omega(I) \to D_n^\Omega(I : A)$) is a homotopy equivalence of complexes. If this is the case then obviously the exact sequences of complexes above induce long exact sequences in Hochschild (respectively cyclic) homology.

Theorem 16.1. *([46], [47]) Let I be a (non-unital) algebra. The following conditions are equivalent:*

(a) I is H-unital, (see 6.3).

(b) I is excisive in Hochschild homology, i.e., any extension of the form $0 \to I \to A \to B \to 0$ induces a long exact sequence in Hochschild homology

$$\ldots HH_n(I) \to HH_n(A) \to HH_n(B) \to HH_{n-1}(I)$$

$$\to HH_{n-1}(A) \to HH_{n-1}(B) \to \ldots$$

(c) I is excisive in cyclic homology, i.e., any extension of the form $0 \to I \to A \to B \to 0$ induces a long exact sequence in cyclic homology

$$\ldots HC_n(I) \to HC_n(A) \to HC_n(B) \to HC_{n-1}(I)$$

$$\to HC_{n-1}(A) \to HC_{n-1}(B) \to \ldots$$

17 Cyclic Homology for Schatten Ideals

The cyclic homology of the m-algebra $\ell^p(H)$, see 3.5 can — somewhat unexpectedly — be computed using the excision property of periodic cyclic homology.

Proposition 17.1. *Let $\mathfrak{J}$ and $\mathfrak{A}$ be m-algebras. We assume that there are continuous homomorphisms $\alpha : \mathfrak{J} \to \mathfrak{A}$ and $\mu : \mathfrak{A} \hat\otimes \mathfrak{A} \to \mathfrak{J}$ with the following properties:*

(a) $\alpha \circ \mu$ is the multiplication on $\mathfrak{A}$.

(b) $\mu \circ (\alpha \otimes \alpha)$ is the multiplication on $\mathfrak{J}$.

(thus, in particular $\alpha(\mathfrak{J})$ is an ideal in $\mathfrak{A}$ with $\mathfrak{A}^2 \subset \alpha(\mathfrak{J})$). Then α induces an invertible element $HP_0(\alpha)$ in $HP_0(\mathfrak{J}, \mathfrak{A})$.

Proof. This follows by applying the excision theorem in HP for m-algebras to the following extension.

$$0 \to \mathfrak{J}(0,1) \to \mathfrak{J}(0,1) + \mathfrak{A}t \to \mathfrak{A} \to 0 \tag{1}$$

The m-algebra $\mathfrak{J}(0,1) + \mathfrak{A}t$ is defined in the following way. As a locally convex vector space it simply is the direct sum of $\mathfrak{J}(0,1)$ and $\mathfrak{A}$. The symbol t denotes the identity function on $[0,1]$. The elements of $\mathfrak{A}t$ are regarded as functions on $[0,1]$ with values in $\mathfrak{A}$, which are multiples of t by elements of $\mathfrak{A}$. The multiplication on the first summand is the one of $\mathfrak{J}(0,1)$. The product of a function f in $\mathfrak{J}(0,1)$ by an element $xt \in \mathfrak{A}t$ is $\mu(\alpha(f) \otimes xt)$ (we extend μ and α canonically to functions). The product of xt and yt in the second factor is defined as $\mu(x \otimes y)(t^2 - t) + \alpha\mu(x \otimes y)t$, where the first summand is in $\mathfrak{J}(0,1)$ and the second one in $\mathfrak{A}t$. The extension constructed that way obviously has a continuous linear splitting and defines an element u in $HP_1(\mathfrak{A}, \mathfrak{J}(0,1)) \cong HP_0(\mathfrak{A}, \mathfrak{J})$. This element is inverse to $HP_0(\alpha)$.

We now apply this result to the Schatten ideals $\ell^p = \ell^p(H)$, $p \geq 1$. More generally, consider the case where $\mathfrak{J} = \ell^p \hat{\otimes} \mathfrak{B}$ and $\mathfrak{A} = \ell^q \hat{\otimes} \mathfrak{B}$ for an arbitrary m-algebra $\mathfrak{B}$ and $p \leq q \leq 2p$. The maps α and μ are obtained from the continuous inclusion $\ell^p \to \ell^q$ and from the multiplication map $\ell^q \hat{\otimes} \ell^q \to \ell^p$.

Corollary 17.1. *For any m-algebra $\mathfrak{B}$, $\ell^p \hat{\otimes} \mathfrak{B}$ is equivalent in HP_* to $\ell^q \hat{\otimes} \mathfrak{B}$, i.e., the map $\ell^p \hat{\otimes} \mathfrak{B} \to \ell^q \hat{\otimes} \mathfrak{B}$ induced by the inclusion $\ell^p \to \ell^q$ gives an invertible element in $HP_0(\ell^p \hat{\otimes} \mathfrak{B}, \ell^q \hat{\otimes} \mathfrak{B})$ for $q \geq p$.*

Proposition 17.2. *The inclusion map $\mathbb{C} \to \ell^1(H)$ (which maps $1 \in \mathbb{C}$ to a one-dimensional projector in ℓ^1) and the canonical trace $Tr : \ell^1(H) \to \mathbb{C}$ define elements in $HP_0(\mathbb{C}, \ell^1(H))$ and $HP_0(\ell^1(H), \mathbb{C})$, respectively, which are inverse to each other. Thus in particular $HP_*(\ell^1(H)) \cong HP_*(\mathbb{C})$.*

Corollary 17.2. *The inclusion map $\mathbb{C} \to \ell^p(H)$ defines an invertible element in $HP_0(\mathbb{C}, \ell^p(H))$ for each $p \geq 1$. Thus in particular $HP_*(\ell^p(H)) \cong HP_*(\mathbb{C})$.*

Remark 17.1. Assume (for simplicity) that p is an odd integer. The generator of $HP^0(\ell^p(H)) = \varinjlim HC^{2n}(\ell^p(H)$ is represented by the cyclic cocycle f in C_λ^{p-1} given by

$$f(x_0, x_1, \ldots, x_{p-1}) = \mathrm{Tr}\ (x_0 x_1 \ldots x_{p-1})$$

It follows from the SBI-sequence that the class of f is not in the image of the S-operator and thus that $p - 1$ is the smallest integer k for which the map $HC^k(\ell^p(H)) \to HP^0(\ell^p(H))$ is non-trivial.

18 The Chern Character for K-Theory Classes Given by Idempotents and Invertibles

With homotopy classes of idempotents and invertibles in a matrix algebra over A one can associate periodic cyclic homology classes as follows.

Let A be unital, e an idempotent in a matrix algebra $M_n(A)$ and $[e]$ its homotopy class. e may be thought of as a homomorphism $\mathbb{C} \to M_n(A)$. It induces a map $HP_0\mathbb{C} \to HP_0(M_n(A)) \cong HP_0A$. We denote by $ch([e])$ the image of the generator of $HP_*\mathbb{C} \cong \mathbb{C}$ (see 10.1) under this map.

Similarly, an invertible element u in $M_n(A)$ may be thought of as a map from the algebra $\mathbb{C}[z, z^{-1}]$ to $M_n(A)$ mapping z to u. This induces a map $HP_1\mathbb{C}[z, z^{-1}] \to HP_1(M_n(A)) \cong HP_1A$. We denote by $ch([u])$ the image in HP_1A of the generator of $HP_1\mathbb{C}[z, z^{-1}] \cong \mathbb{C}$ (see 10.1) under this map.

Remark 18.1. It is very easily seen that the classes $ch([e])$ and $ch([u])$ are additive for the natural K-theory addition. Therefore one may view these classes as defining homomorphisms $ch : K_*^{alg}A \to HP_*A$, $* = 0, 1$ from the algebraic K-theory to HP_* (which in fact can be extended to the higher-dimensional algebraic K-theory in a rather straightforward way). This homomorphism even passes to the quotient of algebraic K-theory by homotopy. Since there is no good definition of topological K-theory for arbitrary m-algebras based only on idempotents and invertibles in matrix algebras, it is not clear, a priori, how to interpret this as a map from topological K-theory though.

To obtain explicit formulas for the cycles $ch([e])$ and $ch([u])$ in the periodic cyclic complex it is very convenient to use the description of $HP_*(A)$ as the homology of the completed X-complex $\widehat{X}(TA)$ and its homotopy equivalence with the (B, b)-bicomplex $\widehat{\Omega}A$.

The idempotent e admits a canonical lifting to an idempotent $\hat{e}$ in each nilpotent extension, and therefore in the projective limit $M_n(\widehat{T}A)$ of nilpotent extensions of $M_n(A)$. This is easily seen putting $f = 2e - 1$ in the unitization and defining a lift $\hat{f}$ for f by the formula $\hat{f} = s(f)(s(f)^2)^{-1/2}$, given by a power series in the nilpotent variable $1 - s(f)^2$ (s denotes the canonical linear lift $M_n(A) \to M_n(TA)$). Under the isomorphism $M_n(\widehat{T}A) \cong M_n(\widehat{\Omega}^{ev}A)$ with Fedosov product, 10.2, $\hat{e}$ is given by the following formula

$$\hat{e} = e + \sum_{n \geq 1} \frac{(2n)!}{(n!)^2} \left(e - \frac{1}{2}\right)(de)^{2n}$$

where de is to be considered as the matrix (de_{ij}) if $e = (e_{ij})$. Now, $Tr(\hat{e})$ is an even cycle in the complex $\widehat{X}(TA) = X(\widehat{T}A)$, representing $ch([e])$ (since the generator of $HP_0(\mathbb{C})$ is represented by 1 in $X(\mathbb{C})$). Under the identification of $\widehat{X}(TA)$ with $\widehat{\Omega}A$, this becomes

$$ch([e]) = Tr(e) + \sum_{n \geq 1}(-1)^n \frac{(2n)!}{n!} Tr\left(\left(e - \frac{1}{2}\right)(de)^{2n}\right)$$

as an element of the complex $\widehat{\Omega}A$ with boundary operator $B - b$.

Let further u be an invertible element in the unitization $M_n(A)^{\sim}$ and consider the reduced X-complex $\overline{X}(M_n(A)^{\sim})$ where we introduce the relation $d(1) = 0$ for the adjoined unit 1. By a similar argument as before, the element $\natural(u^{-1}du)$ can be lifted to a cycle $\natural(\hat{u}^{-1}d\hat{u})$ in $\overline{X}(M_n(\widehat{T}A)^\sim)$. The trace on M_n induces a map $Tr: \overline{X}(M_n(\widehat{T}A)^\sim) \to \overline{X}(\widehat{T}A^\sim)$ and $TR(\natural(\hat{u}^{-1}d\hat{u}))$ will represent $ch([u])$ (the generator of $HP_1(\mathbb{C}[z,z^{-1}])$ is represented by $\natural(z^{-1}dz)$ in $X(\mathbb{C}[z,z^{-1}])$). Then $ch([u])$ corresponds to

$$ch([u]) = \sum_{n \geq 0} n! \, Tr((u^{-1}-1)d(u-1)\,(d(u^{-1}-1)d(u-1))^n)$$

19 Cyclic Cocycles Associated with Fredholm Modules

The concept of a Fredholm module over an algebra A has been introduced by Atiyah and Kasparov as a framework for K-homology. The bivariant version (over two algebras A and B) is due to Kasparov and is usually called a Kasparov module. It describes an extension of B by an algebra Morita equivalent to A which admits an inverse (i.e, there exists another extension such that its sum with the given extension is equivalent to a trivial extension). The notion of an unbounded Fredholm module or a spectral triple plays a fundamental role in Alain Connes's "dictionary" for noncommutative geometry. We treat in this section only the simplest but typical case of a "bounded" Fredholm module.

Given a Hilbert space H, we denote by $\mathcal{L}(H)$ the algebra of bounded operators on H and by $\ell^p(H)$ the Schatten ideal of p-summable operators in $\mathcal{L}(H)$.

Definition 19.1. *Let A be an algebra*

(a) A p-summable odd Fredholm module over A is a representation $\pi: A \to \mathcal{L}(H)$ together with an operator $F \in \mathcal{L}(H)$ such that $F = F^*$, $F^2 = 1$ and such that the commutators $[F, \pi(x)]$, $x \in A$ are all in $\ell^p(H)$.

(b) A p-summable even Fredholm module over A is a representation $\pi: A \to \mathcal{L}(H)$ together with an operator $F \in \mathcal{L}(H)$ such that $F = F^*$, $F^2 = 1$ and $[F, \pi(x)] \in \ell^p(H) \in$, $x \in A$ and a grading operator γ on H, $\gamma = \gamma^*$, $\gamma^2 = 1$, such that $\gamma F \gamma = -F$ and $\gamma \pi(x) \gamma = \pi(x)$, $x \in A$ (i.e., F has degree 1, while $\pi(A)$ has degree 0 for the $\mathbb{Z}/2$-grading induced by γ).

An odd p-summable Fredholm module determines an extension of the following form

$$0 \longrightarrow \ell^{p/2}(H) \longrightarrow (\ell^{p/2}(H) + \sigma(A)) \longrightarrow A_0 \longrightarrow 0 \tag{1}$$

where $\sigma(x) = P\pi(x)P$, $P = 1/2(F + 1)$ the spectral projection for 1 and $A_0 = A/\mathrm{Ker}(\sigma)$. Note that we can work with $p/2$ rather than p, since $\sigma(xy) - \sigma(x)\sigma(y) = (1/2)P[F,x][F,y]P \in \ell^{p/2}(H)$.

We know from 17.1 that $HP^0(\ell^{2n+1}(H))$ is one-dimensional with generator the image $\dot{g}$ in HP^0 of the cyclic cocycle $g(x_0, \ldots, x_{2n}) = \mathrm{Tr}(x_0 x_1 \ldots x_{2n})$. By excision for HP^*, the image of $\dot{g}$ under the boundary map in the long exact sequence associated to the extension (1) gives an element $ch(A, F)$ in $HP^1(A)$ which is called the character of the Fredholm module (A, F). A more refined version of this argument is the following. By 4.2, the extension (1) determines a classifying map $\gamma : JA \to \ell^p(H)$ which maps the p-th power $(JA)^p$ of JA to the p-th power of $\ell^p(H)$, i.e., to $\ell^1(H)$. Composition of γ with the trace on $\ell^1(H)$ gives a trace τ on $(JA)^p$. Now, by 10.2, any trace on $(JA)^p$ determines for each n with $2n + 2 \geq p$ cyclic cocycles $f_{2n+1} \in C_\lambda^{2n+1}(A)$ which are related by the S-operator. According to 10.2 the explicit formula for f_{2n+1} is

$$f_{2n+1}(x_0, x_1, \ldots x_{2n+1}) = \mathrm{Tr}(\omega(x_0, x_1)\omega(x_2, x_3) \ldots \omega(x_{2n}, x_{2n+1}))$$

$$-\mathrm{Tr}(\omega(x_{2n+1}, x_0)\omega(x_1, x_2) \ldots \omega(x_{2n-1}, x_{2n}))$$

where $\omega(x, y) = \sigma(xy) - \sigma(x)\sigma(y) = P\pi(xy)P - P\pi x P\pi y P$. A simple computation shows that this formula can be rewritten in terms of commutators $[F, \cdot]$ as follows

$$f_{2n+1}(x_0, x_1, \ldots x_{2n+1}) = \mathrm{Tr}\left(F([F, x_0][F, x_1] \ldots [F, x_{2n+1}])\right)$$

It follows from the proof of excision in HP^* that $ch(A, F)$ is the image of $\frac{(-1)^n}{n!} f_{2n+1}$ under the map $HC^{2n+1} \to HP^1$. There is a simple index theorem based on this cocycle. Recall that a Fredholm operator on a Hilbert space is a bounded operator which is invertible modulo compact operators and that the index of a Fredholm operator L is defined as $\mathrm{Index}\, L = \dim \mathrm{Ker}\, L - \dim \mathrm{Coker}\, L$.

Theorem 19.1. *(Connes) Let A be unital (and unitally represented), $u \in A$ invertible and (A, F) a p-summable Fredholm module over A. Then the $u_P = P\pi(u)P$ is a Fredholm operator on the Hilbert space PH. Its index is given by*

$$\mathrm{ind}\, u_P = \langle ch(A, F) | ch(u) \rangle$$
$$= f_{2n+1}(u^{-1}, u, u^{-1}, \ldots, u)$$

With obvious modifications, the theorem is still valid for invertibles in a matrix algebra over A.

An even p-summable Fredholm module determines an extension of the form

$$0 \longrightarrow \ell^p(H) \longrightarrow (\ell^p(H) + \pi(A)) \longrightarrow A_0 \longrightarrow 0$$

with two different splittings $\pi, \bar{\pi} : A \to (\ell^p(H) + \pi(A))$ where $\bar{\pi}(x) = F\pi(x)F$. Using excision for HP we see that $HP^0(\ell^p(H) + \pi(A)) \cong HP^0(\ell^p(H)) \oplus HP^0(A)$. Denote by $ch(A, F)$ the image of 1 under the map

$$\mathbb{C} = HP^0(\ell^p(H)) \quad \xrightarrow{\pi^* - \bar{\pi}^*} \quad HP^0(A)$$

Again there is an explicit formula for a cyclic cocycle representing $ch(A, F)$. In fact, the universality property of the extension

$$0 \longrightarrow qA \longrightarrow QA \longrightarrow A \longrightarrow 0$$

shows that there is a natural homomorphism $qA \to \ell^p(H)$. Composing again with the supertrace on $\ell^1(H)$ (recall that H is graded) we obtain an odd supertrace on $(qA)^p$ and thus by 10.3 cyclic cocycles determined for each n with $2n + 1 \geq p$ by the formula

$$f_{2n}(x_0, x_1, \ldots x_{2n}) = \mathrm{Tr}(\gamma q(x_0) q(x_1) \ldots q(x_{2n}))$$

where $q(x) = \pi(x) - \bar{\pi}(x)$ and γ is the grading automorphism on H. Again, one may rewrite this as

$$f_{2n}(x_0, x_1, \ldots x_{2n}) = \mathrm{Tr}\left(\gamma F([F, x_0][F, x_1] \ldots [F, x_{2n}])\right)$$

The class $ch(A, F)$ is represented by $\frac{(-1)^n n!}{(2n)!} f_{2n}$

Theorem 19.2. *(Connes) Let A be unital, $e \in A$ idempotent and (A, F) a p-summable Fredholm module over A. Then the $F_e = \pi(e)F\pi(e)$ is a Fredholm operator from the Hilbert space $\pi(e)H_+$ to the Hilbert space $\pi(e)H_-$, where $H = H_+ \oplus H_-$ is the eigenspace-decomposition of H with respect to the grading operator γ. Its index is given by*

$$ind\, F_e = \langle ch(A, F) | ch(e) \rangle$$
$$= f_{2n}(e, e, \ldots, e)$$

20 Bivariant K-Theory for Locally Convex Algebras

A natural category on which topological K-theory can be defined with good properties, is the one of Banach algebras. It has been studied with particular intensity on the category of C*-algebras. On the category of C*-algebras Kasparov, [25], extended K-theory to a bivariant theory KK in such a way that, for a C*-algebra A, one has $K_*(A) = KK_*(\mathbb{C}, A)$, $* = 0, 1$. A closely related similar bivariant theory on the category of C*-algebras is the E-theory of Connes-Higson, [8]. It is an alternative extension of K-theory to a bivariant theory. In particular it also satisfies $K_*(A) = E_*(\mathbb{C}, A)$, $* = 0, 1$.

Now it is a well-known fact that cyclic homology is degenerate on Banach or C*-algebras. Thus, for instance, if we define cyclic and periodic cyclic homology for a locally convex algebra as in Section 8, we obtain for the algebra $\mathcal{C}(X)$ of continuous functions on the compact space X

$$HP^0(\mathcal{C}(X)) = \{ \text{bounded complex measures on} X \}$$
$$HP^1(\mathcal{C}(X)) = 0$$

It is possible to define cyclic homology theories (even in the bivariant setting) with good properties on the category of C*-algebras (or Banach algebras) and to define a bivariant transformation from $KK(A, B)$ or from $E(A, B)$ into this theory, [37], [38], [39].

On the other hand, it is in fact also possible to construct a bivariant topological K-theory on the category of locally convex algebras and then to define a bivariant Chern-Connes character from this theory into the ordinary bivariant cyclic theory $HP_*(A, B)$, as we discussed it in Section 8. We will describe the construction in this section. As before, to simplify constructions, we work within the category of m-algebras (projective limits of Banach algebras), rather than with general locally convex algebras. Since cyclic theory is homotopy invariant only for differentiable homotopies (called diffotopies below), we have to set up the theory in such a way that it uses only diffotopies in place of general homotopies.

For an m-algebra $\mathfrak{A}$ we write $\mathfrak{A}[a, b]$, $\mathfrak{A}[a, b)$ and $\mathfrak{A}(a, b)$ for the m-algebras $\mathfrak{A}\hat{\otimes}\mathbb{C}[a, b]$, $\mathfrak{A}\hat{\otimes}\mathbb{C}[a, b)$ and $\mathfrak{A}\hat{\otimes}\mathbb{C}(a, b)$, see 3.1 .

Definition 20.1. *(see 3.1) Two continuous homomorphisms $\alpha, \beta : \mathfrak{A} \to \mathfrak{B}$ between two m-algebras are called differentiably homotopic or* **diffotopic**, *if there is a family $\varphi_t : \mathfrak{A} \to \mathfrak{B}$, $t \in [0, 1]$ of continuous homomorphisms, such that $\varphi_0 = \alpha, \varphi_1 = \beta$ and such that the map $t \mapsto \varphi_t(x)$ is infinitely differentiable for each $x \in \mathfrak{A}$.*

Let $h : [0, 1] \to [0, 1]$ be a monotone and bijective C^∞-map, whose restriction to $(0, 1)$ is a diffeomorphism from $(0, 1) \to (0, 1)$ and whose derivatives in 0 and 1 all vanish. Replacing φ_t by $\psi_t = \varphi_{h(t)}$ one sees that α and β are diffotopic if and only if there is a continuous homomorphism $\psi : \mathfrak{A} \to \mathbb{C}[0, 1]\hat{\otimes}\mathfrak{B}$ with the property that $\psi(x)(0) = \alpha(x)$, $\psi(x)(1) = \beta(x)$ for each $x \in \mathfrak{A}$. This shows that diffotopy is an equivalence relation.

Given a homomorphism α we denote by $\langle\alpha\rangle$ its diffotopy class and given two m-algebras $\mathfrak{A}$ and $\mathfrak{B}$, we denote by $\langle\mathfrak{A}, \mathfrak{B}\rangle$ the set of diffotopy classes of homomorphisms from $\mathfrak{A}$ to $\mathfrak{B}$.

Definition 20.2. *Let $\mathfrak{A}$ and $\mathfrak{B}$ be m-algebras. For any homomorphism $\varphi : \mathfrak{A} \to \mathfrak{B}$ of m-algebras, we denote by $\langle\varphi\rangle$ the equivalence class of φ with respect to diffotopy and we set*

$$\langle\mathfrak{A}, \mathfrak{B}\rangle = \{\langle\varphi\rangle|\, \varphi \text{ is a continuous homomorphism } \mathfrak{A} \to \mathfrak{B}\}$$

Let $\mathfrak{K}$ be the algebra of smooth compact operators defined in 3.4. For two continuous homomorphisms $\alpha, \beta : \mathfrak{A} \to \mathfrak{K}\hat{\otimes}\mathfrak{B}$ we define the direct sum $\alpha \oplus \beta$ as

$$\alpha \oplus \beta = \begin{pmatrix} \alpha & 0 \\ 0 & \beta \end{pmatrix} : \mathfrak{A} \longrightarrow M_2(\mathfrak{K}\hat{\otimes}\mathfrak{B}) \cong \mathfrak{K}\hat{\otimes}\mathfrak{B}$$

With the addition defined by $\langle\alpha\rangle + \langle\beta\rangle = \langle\alpha \oplus \beta\rangle$ the set $\langle\mathfrak{A}, \mathfrak{K}\hat{\otimes}\mathfrak{B}\rangle$ of diffotopy classes of homomorphisms from $\mathfrak{A}$ to $\mathfrak{K}\hat{\otimes}\mathfrak{B}$ is an abelian semigroup with 0-element $\langle0\rangle$.

Definition 20.3. *(see 4.2) Given an m-algebra $\mathfrak{A}$ we define the ideal $J\mathfrak{A}$ in the standard free extension as in (1). $J\mathfrak{A}$ is, for each m-algebra $\mathfrak{A}$, again an m-algebra. By* *iteration we can therefore form* $J^2\mathfrak{A} = J(J\mathfrak{A}), \ldots, J^n\mathfrak{A} = J(J^{n-1}\mathfrak{A})$. *For any extension of* $\mathfrak{A}$ *of length* n *there is a classifying map from* $J^n\mathfrak{A}$ *into the first ideal in the* n*-step extension.*

Consider the set $H_k = \langle J^k\mathfrak{A}, \mathfrak{K}\hat{\otimes}\mathfrak{B}\rangle$, where $H_0 = \langle \mathfrak{A}, \mathfrak{K}\hat{\otimes}\mathfrak{B}\rangle$. Each H_k is an abelian semigroup with 0-element for the K-theory addition defined in 20.2.

To define a map $S : H_k \rightarrow H_{k+2}$, we use the classifying map $\varepsilon : J^2\mathfrak{A} \rightarrow \mathfrak{K}\hat{\otimes}\mathfrak{A}$ for the two-step extension which is obtained by composing the Toeplitz with the suspension extension, see 4.4 and 4.1.

Definition 20.4. *Let* $\varphi : J\mathfrak{A} \rightarrow C^\infty(S^1)\hat{\otimes}\mathfrak{A}$ *be the composition of the classifying map* $J\mathfrak{A} \rightarrow \mathfrak{A}(0,1)$, *belonging to the extension*

$$0 \longrightarrow \mathfrak{A}(0,1) \longrightarrow \mathfrak{A}[0,1) \longrightarrow \mathfrak{A} \longrightarrow 0$$

with the inclusion map $\mathfrak{A}(0,1) \longrightarrow C^\infty(S^1)\hat{\otimes}\mathfrak{A}$. *The map*

$$\varepsilon : J^2\mathfrak{A} \longrightarrow \mathfrak{K}\hat{\otimes}\mathfrak{A}$$

is defined as the composition $\varepsilon = \gamma \circ J(\varphi)$ *with the classifying map* γ *for the extension*

$$0 \longrightarrow \mathfrak{K}\hat{\otimes}\mathfrak{A} \longrightarrow \mathfrak{T}\hat{\otimes}\mathfrak{A} \overset{s}{\longrightarrow} C^\infty(S^1)\hat{\otimes}\mathfrak{A} \longrightarrow 0$$

For $\langle\alpha\rangle \in H_k$, $\alpha : J^k\mathfrak{A} \longrightarrow \mathfrak{K}\hat{\otimes}\mathfrak{B}$, let $S\langle\alpha\rangle = \langle(\mathrm{id}_{\mathfrak{K}}\otimes\alpha)\circ\varepsilon\rangle$. Here $\varepsilon : J^{k+2}\mathfrak{A} \longrightarrow \mathfrak{K}\otimes J^k\mathfrak{A}$ is the ε-map for $J^k\mathfrak{A}$. Let $\varepsilon_- : J^{k+2}\mathfrak{A} \longrightarrow \mathfrak{K}\otimes J^k\mathfrak{A}$ be the map which is obtained by replacing, in the definition of ε, the Toeplitz extension by the inverse Toeplitz extension. The sum $\varepsilon\oplus\varepsilon_-$ is then diffotopic to 0. Therefore $S\langle\alpha\rangle + S_-\langle\alpha\rangle = 0$, putting $S_-\langle\alpha\rangle = \langle(\mathrm{id}_{\mathfrak{K}}\otimes\alpha)\circ\varepsilon_-\rangle$.

Definition 20.5. *Let* $\mathfrak{A}$ *and* $\mathfrak{B}$ *be* m*-algebras and* $* = 0$ *or* 1. *We define*

$$kk_*(\mathfrak{A},\mathfrak{B}) = \varinjlim_k H_{2k+*} = \varinjlim_k \langle J^{2k+*}\mathfrak{A}, \mathfrak{K}\hat{\otimes}\mathfrak{B}\rangle$$

The preceding discussion shows that $kk_*(\mathfrak{A},\mathfrak{B})$ is not only an abelian semigroup, but even an abelian group (every element admits an inverse).

Theorem 20.1. *(a) There is an associative product*

$$kk_i(\mathfrak{A},\mathfrak{B}) \times kk_j(\mathfrak{B},\mathfrak{C}) \longrightarrow kk_{i+j}(\mathfrak{A},\mathfrak{C})$$

($i, j \in \mathbb{Z}/2$; $\mathfrak{A}, \mathfrak{B}$ *and* $\mathfrak{C}$ *m-algebras), which is additive in both variables.*
(b) There is a bilinear, graded commutative, exterior product

$$kk_i(\mathfrak{A}_1,\mathfrak{A}_2) \times kk_j(\mathfrak{B}_1,\mathfrak{B}_2) \longrightarrow kk_{i+j}(\mathfrak{A}_1\hat{\otimes}\mathfrak{A}_2, \mathfrak{B}_1\hat{\otimes}\mathfrak{B}_2)$$

(c) Each homomorphism $\varphi : \mathfrak{A} \rightarrow \mathfrak{B}$ *defines an element* $kk(\varphi)$ *in the group* $kk_0(\mathfrak{A},\mathfrak{B})$. *If* $\psi : \mathfrak{B} \rightarrow \mathfrak{C}$, *is another homomorphism, then*

$$kk(\psi \circ \varphi) = kk(\varphi) \cdot kk(\psi)$$

$kk_*(\mathfrak{A}, \mathfrak{B})$ is a contravariant functor in $\mathfrak{A}$ and a covariant functor in $\mathfrak{B}$. If $\alpha : \mathfrak{A}' \to \mathfrak{A}$ and $\beta : \mathfrak{B} \to \mathfrak{B}'$ are homomorphisms, then the induced maps, in the first and second variable of kk_*, are given by left multiplication by $kk(\alpha)$ and right multiplication by $kk(\beta)$.

(d) $kk_*(\mathfrak{A}, \mathfrak{A})$ is, for each m-algebra $\mathfrak{A}$, a $\mathbb{Z}/2$-graded ring with unit element $kk(\mathrm{id}_{\mathfrak{A}})$.

(e) The functor kk_* is invariant under diffotopies in both variables.

(f) The canonical inclusion $\iota : \mathfrak{A} \to \mathfrak{K} \hat{\otimes} \mathfrak{A}$ defines an invertible element in $kk_0(\mathfrak{A}, \mathfrak{K} \hat{\otimes} \mathfrak{A})$. In particular, $kk_*(\mathfrak{A}, \mathfrak{B}) \cong kk_*(\mathfrak{K} \hat{\otimes} \mathfrak{A}, \mathfrak{B})$ and $kk_*(\mathfrak{B}, \mathfrak{A}) \cong kk_*(\mathfrak{B}, \mathfrak{K} \hat{\otimes} \mathfrak{A})$ for each m-algebra $\mathfrak{B}$.

(g) (Bott periodicity) The suspension extension gives rise to classifying maps $J\mathfrak{A} \to \mathfrak{A}(0, 1)$ and $J^2\mathfrak{A} \to \mathfrak{A}(0, 1)^2$ (where $\mathfrak{A}(0, 1)^2 = \mathfrak{A}(0, 1)(0, 1)$). The elements in $kk_0(J\mathfrak{A}, \mathfrak{A}(0, 1))$, in $kk_1(\mathfrak{A}, \mathfrak{A}(0, 1))$ and in $kk_0(\mathfrak{A}, \mathfrak{A}(0, 1)^2)$ defined by these homomorphisms are invertible.

(h) Let $\mathfrak{D}$ be any m-algebra. Every extension admitting a continuous linear split

$$E : 0 \to \mathfrak{J} \xrightarrow{i} \mathfrak{A} \xrightarrow{q} \mathfrak{B} \to 0$$

induces exact sequences in $kk(\mathfrak{D}, \cdot)$ and $kk(\cdot, \mathfrak{D})$ of the following form:

$$
\begin{array}{ccccc}
kk_0(\mathfrak{D}, \mathfrak{J}) & \xrightarrow{\cdot kk(i)} & kk_0(\mathfrak{D}, \mathfrak{A}) & \xrightarrow{\cdot kk(q)} & kk_0(\mathfrak{D}, \mathfrak{B}) \\
\uparrow & & & & \downarrow \\
kk_1(\mathfrak{D}, \mathfrak{B}) & \xleftarrow{\cdot kk(q)} & kk_1(\mathfrak{D}, \mathfrak{A}) & \xleftarrow{\cdot kk(i)} & kk_1(\mathfrak{D}, \mathfrak{J})
\end{array}
\tag{1}
$$

and

$$
\begin{array}{ccccc}
kk_0(\mathfrak{J}, \mathfrak{D}) & \xleftarrow{kk(q)\cdot} & kk_0(\mathfrak{A}, \mathfrak{D}) & \xleftarrow{kk(i)\cdot} & kk_0(\mathfrak{B}, \mathfrak{D}) \\
\downarrow & & & & \uparrow \\
kk_1(\mathfrak{B}, \mathfrak{D}) & \xrightarrow{kk(i)\cdot} & kk_1(\mathfrak{A}, \mathfrak{D}) & \xrightarrow{kk(q)\cdot} & kk_1(\mathfrak{J}, \mathfrak{D})
\end{array}
\tag{2}
$$

The given extension E defines a classifying map $J\mathfrak{B} \to \mathfrak{J}$ and thus an element of $kk_1(\mathfrak{J}, \mathfrak{B})$, which we denote by $kk(E)$. The vertical arrows in (1) and (2) are (up to a sign) given by right and left multiplication, respectively, by this class $kk(E)$.

Finally, we have to compare kk with the ordinary topological K-theory. For a Fréchet m-algebra $\mathfrak{A}$ one can define the following abelian groups:

$$
K_0(\mathfrak{A}) = \big\{ \langle e \rangle \,\big|\, e \text{ is an idempotent element in}
$$
$$
M_2(\mathfrak{K} \hat{\otimes} \mathfrak{A}^\sim) \text{ such that } e - \begin{pmatrix} 1 & 0 \\ 0 & 0 \end{pmatrix} \in M_2(\mathfrak{K} \hat{\otimes} \mathfrak{A}) \big\}
\tag{3}
$$
$$
K_1(\mathfrak{A}) = K_0(\mathfrak{A}(0, 1))
$$

Here, as usual $\mathfrak{K} \hat{\otimes} \mathfrak{A}^\sim$ denotes the algebra obtained by adjoining a unit to $\mathfrak{K} \hat{\otimes} \mathfrak{A}$.

Phillips shows in [36] that this defines a topological K-theory with the usual properties (homotopoy invariance, stability, Bott periodicity, six-term exact sequences) on the category of Fréchet m-algebras and in addition gives the usual well-known K-theory if $\mathfrak{A}$ is a Banach algebra. We have

Theorem 20.2. *For every Fréchet m-algebra $\mathfrak{A}$, the groups $kk_*(\mathbb{C}, \mathfrak{A})$ and $K_* \mathfrak{A}$ are naturally isomorphic.*

Remark 20.1. V.Lafforgue, [29] has introduced a bivariant K-theory KK^{Ban} which is defined on the category of Banach algebras. It does not have a product but it has very good properties for very general Morita equivalences of Banach algebras. Lafforgue uses his theory to obtain striking positive results on the Baum-Connes conjecture. For Banach algebras, it is possible to construct an analogue kk_*^{Ban} of kk_* on the category of Banach algebras using continuous homotopies rather than differentiable one (this is straightforward) and to show that there is a natural transformation $KK_*^{Ban} \to kk_*^{Ban}$.

21 The Bivariant Chern-Connes Character

In this section we describe the construction of a bivariant multiplicative transformation from kk_* to the bivariant theory HP_* on the category of m-algebras.

Consider a covariant functor E from the category of m-algebras to the category of abelian groups which satisfies the following conditions:

(E1) E is diffotopy invariant, i.e., the evaluation map ev_t in any point $t \in [0,1]$ induces an isomorphism $E(ev_t) : E(\mathfrak{A}[0,1]) \to E(\mathfrak{A})$

(E2) E is stable, i.e., the canonical inclusion $\iota : \mathfrak{A} \to \mathfrak{K} \hat{\otimes} \mathfrak{A}$ induces an isomorphism $E(\iota)$.

(E3) E is half-exact, i.e., each extension $0 \to \mathfrak{J} \to \mathfrak{A} \to \mathfrak{B} \to 0$ admitting a continuous linear splitting induces a short exact sequence $E(\mathfrak{J}) \to E(\mathfrak{A}) \to E(\mathfrak{B})$

(The same conditions can of course be formulated analogously for a contravariant functor E.) We note that a standard construction from algebraic topology, using property (E1) and mapping cones, permits to extend the short exact sequence in (E3) to an infinite long exact sequence of the form

$$\cdots \to E(\mathfrak{B}(0,1)^2) \to E(\mathfrak{J}(0,1)) \to E(\mathfrak{A}(0,1))$$
$$\to E(\mathfrak{B}(0,1)) \to E(\mathfrak{J}) \to E(\mathfrak{A}) \to E(\mathfrak{B})$$

see, e.g., [25] or [1].

Theorem 21.1. *Let E be a covariant functor with the properties (E1), (E2), (E3). Then we can associate in a unique way with each $h \in kk_0(\mathfrak{A}, \mathfrak{B})$ a morphism of abelian groups $E(h) : E(\mathfrak{A}) \to E(\mathfrak{B})$, such that $E(h_1 \cdot h_2) = E(h_2) \circ E(h_1)$ for the product $h_1 \cdot h_2$ of $h_1 \in kk_0(\mathfrak{A}, \mathfrak{B})$ and $h_2 \in kk_0(\mathfrak{B}, \mathfrak{C})$ and such that $E(kk(\alpha)) = E(\alpha)$ for each morphism $\alpha : \mathfrak{A} \to \mathfrak{B}$ of m-algebras.*
An analogous statement holds for contravariant functors.

Proof. Let h be represented by $\eta : J^{2n}\mathfrak{A} \to \hat{\mathfrak{K}}\hat{\otimes}\mathfrak{B}$. We set

$$E(h) = E(\iota)^{-1}E(\eta)E(\varepsilon^n)^{-1}E(\iota)$$

It is clear that $E(h)$ is well-defined and that $E(kk(\alpha)) = E(\alpha)$.

The preceding result can be interpreted differently, see also [20], [1]. For this, note first that kk_0 can be regarded as a category, whose objects are the m-algebras, and where the morphism set between $\mathfrak{A}$ and $\mathfrak{B}$ is given by $kk_0(\mathfrak{A}, \mathfrak{B})$. This category is additive in the sense that the morphism set between two objects forms an abelian group and that the product of morphisms is bilinear.

We denote the natural functor from the category of m-algebras to the category kk_0, which is the identity on objects, also by kk_0.

Corollary 21.1. *Let F be a functor from the category of m-algebras to an additive category C, such that $F(\beta \circ \alpha) = F(\alpha) \cdot F(\beta)$, for any two homomorphisms α : $A_1 \to A_2$ and $\beta : A_2 \to A_3$ between m-algebras.*

We assume that for each $\mathfrak{B}$, the contravariant functor $C(F(\,\cdot\,), F(\mathfrak{B}))$ and the covariant functor $C(F(\mathfrak{B}), F(\,\cdot\,))$ on the category of m-algebras satisfy the properties (E1), (E2), (E3). Then there is a unique covariant functor F' from the category kk_0 to C, such that $F = F' \circ kk_0$.

Remark 21.1. Property (E3) implies that any such functor F' is automatically additive:

$$F'(h + g) = F'(h) + F'(g)$$

As a consequence of the preceding corollary we get a bilinear multiplicative transformation from kk_0 to HP_0—the bivariant Chern-Connes character.

Corollary 21.2. *There is a unique (covariant) functor $ch : kk_0 \to HP_0$, such that $ch(kk_0(\alpha)) = HP_0(\alpha) \in HP_0(\mathfrak{A}, \mathfrak{B})$ for every morphism $\alpha : \mathfrak{A} \to \mathfrak{B}$ of m-algebras.*

Proof. This follows from 21.1, since HP_0 satisfies properties (E1), (E2) and (E3).

The Chern-Connes-character ch is compatible with the exterior product on kk_0 as well, see 20.1(b), and the corresponding product on HP_0, see [15], p.86.

It remains to extend ch to a multiplicative transformation from the $\mathbb{Z}/2$-graded theory kk_* to HP_* and to study the compatibilty with the boundary maps in the long exact sequences associated to an extension for kk and HP.

For the special case of the free extension

$$E_u : \qquad 0 \to J\mathfrak{A} \to T\mathfrak{A} \to \mathfrak{A} \to 0$$

we put, using the notation in 15.2,

$$x_{\mathfrak{A}} = ch(E_u) = \delta'(1_{\mathfrak{A}}) = -\delta(1_{J\mathfrak{A}}) \in HP_1(\mathfrak{A}, J\mathfrak{A})$$

The boundary maps δ and δ' in the long exact sequences associated to E_u are given by left and right multiplication by $x_{\mathfrak{A}}$, see 15.2.

Let now u be an element in $kk_1(\mathfrak{A}, \mathfrak{B})$. By definition we have $kk_1(\mathfrak{A}, \mathfrak{B}) = kk_0(J\mathfrak{A}, \mathfrak{B})$. Let u_0 be the element in $kk_0(J\mathfrak{A}, \mathfrak{B})$ that corresponds to u under this isomorphism. We set

$$ch(u) = \sqrt{2\pi i}\, x_{\mathfrak{A}} \cdot ch(u_0) \qquad \in HP_1(\mathfrak{A}, \mathfrak{B})$$

Theorem 21.2. *The thus defined Chern-Connes character $ch : kk_* \to HP_*$ is multiplicative, i.e., for $u \in kk_i(\mathfrak{A}, \mathfrak{B})$ and $v \in kk_j(\mathfrak{B}, \mathfrak{C})$ we have*

$$ch(u \cdot v) = ch(u) \cdot ch(v)$$

For m-algebras, the bivariant character ch constructed here is a far reaching generalization of the Chern character from K-Theory and K-homology in Sections 18 and 19.

If e is an idempotent in $M_n\mathfrak{A}$, then e can be viewed as a homomorphism from $\mathbb{C}$ to $\mathfrak{K} \otimes \mathfrak{A}$, thus as an element h_e of $kk_0(\mathbb{C}, \mathfrak{A})$. It is not difficult to see that for the class $ch([e])$ constructed in section 18 one has

$$ch([e]) = ch(h_e)$$

Similarly any invertible element u in $M_n\mathfrak{A}^\sim$ defines an element h_u in $kk_1(\mathbb{C}, \mathfrak{A})$ and we have $ch([u]) = ch(h_u)$

Similarly, any p-summable Fredholm module over $\mathfrak{A}$ gives an element of $kk_*(\mathfrak{A}, \ell^p(H))$.

Corollary 17.1 shows that $\ell^p \hat{\otimes} \mathfrak{B}$ and $\ell^1 \hat{\otimes} \mathfrak{B}$ are equivalent in HP_0 for each $p \geq 1$. On the other hand, it is easy to see that $\ell^1 \hat{\otimes} \mathfrak{B}$ is in HP_0 equivalent to $\mathfrak{B}$, see for instance [17]. We thus obtain

Corollary 21.3. *The m-algebra $\ell^p \hat{\otimes} \mathfrak{B}$ is equivalent, in HP_0, to $\mathfrak{B}$ for each $p \geq 1$. The Chern-Connes-character gives a transformation $ch^{(p)} : kk_*(\mathfrak{A}, \ell^p \hat{\otimes} \mathfrak{B}) \to HP_*(\mathfrak{A}, \mathfrak{B})$ such that*

$$ch^{(p)}(x \cdot kk(\iota^{(p)})) = ch(x) \qquad \text{für } x \in kk_*(\mathfrak{A}, \mathfrak{B})$$

where $\iota^{(p)}$ denotes the canonical inclusion $\mathfrak{B} \to \ell^p \hat{\otimes} \mathfrak{B}$.

Comparing the functorial properties, one sees that this transformation generalizes the construction of cyclic cohomology classes associated to p-summable Fredholm modules in Section 19. It also generalizes previous constructions of a bivariant Chern-Connes character in special cases [27], [45], [34], [35].

22 Entire Cyclic Cohomology

In the study of discrete subgroups of Lie groups and of quantum field theory, one is naturally lead to consider Fredholm modules that are not finitely summable but only θ-summable:

Definition 22.1. *A Fredholm module* (F, π, H) *is called* θ-*summable iff the (graded) commutators* $[F, \pi(a)]$ *are contained in the operator ideal*

$$\mathrm{Li}^{1/2}(H) := \{T \in \mathcal{L}(H) \mid \mu_n(T) = O\big((\ln n)^{-1/2}\big) \text{ for } n \to \infty\}.$$

Here $\mu_n(T)$ *denotes the* nth *singular value of* T.

In periodic cyclic cohomology, one can only express the Chern character of a finitely summable Fredholm module. Entire cyclic cohomology was introduced to handle θ-summable Fredholm modules as well [5].

Definition 22.2. *Let* A *be a Banach algebra. Let* K *be the* $\mathbb{Z}/2$-*graded vector space of all linear maps* $\varphi: \Omega A \to \mathbb{C}$ *that satisfy the following growth condition: For all bounded subsets* $S \subseteq A$, *there is a constant* C_S *such that*

$$|\varphi(a_0 da_1 \ldots da_n)| \leq C_S, \qquad |\varphi(da_1 \ldots da_n)| \leq C_S \tag{1}$$

for all $a_0, \ldots, a_n \in S$. *We equip* K *with the boundary* $\partial^*: K \to K$ *induced from the boundary maps of* $X(TA)$. *That is,* $\partial^*(\varphi) := \varphi \circ \beta$ *for even* φ *and* $\partial^*(\varphi) := -\varphi \circ \delta$ *for odd* φ. *The* entire cyclic cohomology $HE^*(A)$ *is defined to be the homology of the complex* K.

If we use $B \pm b$ instead of the X-complex boundary, we have to modify the growth condition above, replacing (1) by

$$[n/2]! \cdot |\varphi(a_0 da_1 \ldots da_n)| \leq C_S, \quad [n/2]! \cdot |\varphi(da_1 \ldots da_n)| \leq C_S, \tag{2}$$

where $[n/2] = k$ if $n = 2k$ or $n = 2k + 1$. This yields a chain homotopic complex [32].

Definition 22.2 can be carried over literally to locally convex algebras with separately continuous multiplication. However, since we only use the bounded subsets of A, it is more appropriate to define entire cyclic cohomology for *complete (convex) bornological algebras*. These are algebras equipped with a collection S of subsets satisfying some reasonable axioms like

$$S, T \in \mathcal{S}, \ \lambda \in \mathbb{C} \Longrightarrow S \cdot T, \ S + T, \ \lambda \cdot S \in \mathcal{S}$$

Convexity means that if $S \in \mathcal{S}$, so is the absolutely convex hull of S. If $S \subseteq A$ is absolutely convex, then the linear span A_S of S carries a unique semi-norm with unit ball S. We call S *completant* iff A_S is a Banach space. We call A *separated* iff A_S is a normed space for all S and we call A *complete* iff each $S \in \mathcal{S}$ is contained in a completant absolutely convex set $T \in \mathcal{S}$. A complete bornological vector space is automatically separated. In the following, we will drop the words convex and separated from our notation and assume all bornological vector spaces to be convex and separated. The collection $\mathcal{S}$ is called the *bornology* of A. We call $S \subseteq A$ *small* iff $S \in \mathcal{S}$. See [21] for an introduction to bornological vector spaces.

Here are some important examples of bornologies. If V is just a vector space, we can equip it with the finest possible bornology: We let $S \in \mathcal{S}$ iff S is a bounded subset of a finite dimensional vector subspace of V. If V is a (quasi)complete locally convex vector space, then the collection of bounded subsets of V is a complete bornology. Another reasonable complete bornology is the collection of all precompact or relatively compact subsets. (If V is (quasi)complete, then a subset is precompact iff it is relatively compact iff it is contained in a compact subset.) Actually, the precompact bornology tends to be better behaved than the bounded one. For local cyclic cohomology, it is essential to use the precompact bornology.

Let A be a bornological algebra. Then there is an obvious bornology on ΩA such that the space K of Definition 22.2 is equal to the space of bounded linear functionals on ΩA. Namely, the bornology generated by the sets $\bigcup_{n\geq 0} S(dS)^n \cup (dS)^{n+1}$. That is, a subset of ΩA is small if and only if it is contained in the absolutely convex hull of a set of the form $\bigcup_{n\geq 0} S(dS)^n \cup (dS)^{n+1}$. The operators b, d, B, κ and the multiplication of differential forms are bounded with respect to this bornology. Identifying TA and $X(TA)$ with $\Omega^{\mathrm{ev}}A$ and ΩA as in Proposition 10.2, we obtain canonical bornologies on TA and $X(TA)$. We always equip $\Omega A \cong X(TA)$ and TA with these bornologies.

If V is a bornological vector space, we write V^c for its *completion*. Moreover, we write $\mathcal{L}(X, Y)$ for the space of bounded linear maps between two bornological vector spaces X and Y.

Definition 22.3. *Let A and B be bornological algebras. We define* entire cyclic homology *and* bivariant entire cyclic homology *by*

$$HE_*(B) := H_*\big(X(TB)^c\big),$$
$$HE_*(A, B) := H_*\Big(\mathcal{L}\big(X(TA)^c, X(TB)^c\big)\Big).$$

It is necessary to complete $X(TB)$ because the complex $X(TB)$ itself is always acyclic. We have $HE_*(A, \mathbb{C}) \cong HE^*(A)$ and $HE_*(\mathbb{C}, A) \cong HE_*(A)$ because the complex $X(T\mathbb{C})^c$ is chain homotopic to the complex $\mathbb{C}$ with $\mathbb{C}$ in degree 0 and 0 in degree 1.

The composition of chain maps gives rise to an associative *composition product* in bivariant entire cyclic homology. A bounded homomorphism $f : A \to B$ induces a bounded chain map $f_* : X(TA)^c \to X(TB)^c$. Thus we obtain an element $[f_*] \in HE_0(A, B)$. Composition with $[f_*]$ yields the functoriality of the entire cyclic theories.

Since $X(TA)^c$ is contained in the complex $\hat{X}(TA)$ defining the periodic cyclic theories, there are natural transformations $HP^*(A) \to HE^*(A)$ and $HE_*(A) \to HP_*(A)$.

Entire cyclic (co)homology has the same stability, homotopy invariance, and exactness properties as periodic cyclic (co)homology [32].

To formulate homotopy invariance and stability, we need the completed bornological tensor product $\hat{\otimes}$. It is defined by a universal property analogous to the universal property of the projective tensor product: Bounded linear maps $V \hat{\otimes} W \to X$ correspond bijectively to bounded bilinear maps $V \times W \to X$. For Banach spaces equipped with the bounded bornology, $V \hat{\otimes} W$ is nothing but the projective tensor product, again equipped with the bounded bornology. If V and W are Fréchet spaces equipped with the precompact bornology, then $V \hat{\otimes} W$ will be isomorphic to the projective tensor product $V \hat{\otimes}_\pi W$ equipped with the precompact bornology [32]. A slightly weaker statement holds if V and W carry the bounded bornology.

For a complete bornological algebra, we define $B[0,1] := \mathbb{C}[0,1] \hat{\otimes} B$, where $\mathbb{C}[0,1]$ carries the bounded bornology (which is equal to the precompact bornology). This gives rise to a notion of smooth homotopy. Entire cyclic (co)homology is smoothly homotopy invariant, that is, Theorem 12.1.(a) holds with HE instead of HP. Similarly, entire cyclic homology is stable with respect to the trace class operators $\ell^1(H)$. That is, the canonical embedding $A \to A \hat{\otimes} \ell^1(H)$ is invertible in HE_0. These two assertions can be proved as for the periodic theory. The excision theorem also holds [39], [32]:

Theorem 22.1. *Bivariant entire cyclic homology satisfies excision in both variables for extensions with a bounded linear section.*

That is, if $0 \to S \to P \to Q \to 0$ is an extension of complete bornological algebras with a bounded linear section and if A is another complete bornological algebra, then there are two natural six-term exact sequences as in Theorem 15.1, only with HE instead of HP. However, the proof of Theorem 22.1 is more difficult than the proof of Theorem 15.1. We will explain the strategy of the proof below.

The formulas for the Chern character in K-theory in Section 18 actually yield elements of $X(TA)^c$. Hence these formulas define a Chern character

$$\text{ch}\colon K_*(A) \to HE_*(A).$$

The Chern character $K_*(A) \to HP_*(A)$ is the composition of this character with the natural map $HE_*(A) \to HP_*(A)$.

There are two apparently different definitions for the Chern character of a θ - summable Fredholm module, due to Connes [5] and Jaffe-Lesniewski-Osterwalder [57], respectively. It is an important fact that both approaches yield homologous cocycles [6]. In [57], an entire cochain is written down explicitly and it is checked that it is a cocycle. However, it is not clear why this particular formula should be the right one. Connes's construction is more conceptual. Presumably, it can be explained using the excision theorem, as in Section 19. However, Connes's construction is technically quite involved, so that we cannot discuss it here.

Both constructions of the Chern character work with *unbounded* Fredholm modules in the following sense.

Definition 22.4. *Let A be a bornological algebra. Let H be a Hilbert space, let $\pi\colon A \to \mathcal{L}(H)$ be a bounded representation of A, and let D be an unbounded self-adjoint operator on H. We call the triple (D, π, H) an unbounded odd Fredholm module over A iff $(1 + D^2)^{-1}$ is compact, $D\pi(a)$ is densely defined for all $a \in A$, $[D, \pi(a)]$ extends to a bounded operator on H for all $a \in A$, and the map $A \ni a \mapsto [D, \pi(a)] \in \mathcal{L}(H)$ is bounded.*
If H is a graded Hilbert space, $\pi(a)$ has degree 0 for all $a \in A$, and D has degree 1, we call the triple (D, π, H) an unbounded even Fredholm module.
We call (D, π, H) θ-summable iff $\exp(-D^2) \in \ell^1(H)$.

In applications, it frequently happens that $\exp(-\beta D^2) \in \ell^1(H)$ for all $\beta > 0$. This stronger assumption is made in [5] and [57], although it is not necessary to obtain the Chern character.

Given an unbounded Fredholm module (D, π, H), we get an associated bounded Fredholm module as follows. If D is injective, we may put $F := \operatorname{sign} D$. Otherwise, we first modify (D, π, H) so that D becomes injective and then put $F := \operatorname{sign} D$. In the odd case, we may simply add to D the projection onto the kernel of D. In the even case, this does not work because the resulting operator will no longer have degree 1. In fact, there may be no injective self-adjoint degree 1 operator on $\ker D$. Therefore, we replace H by the direct sum $H \oplus \ker D$ and extend $\pi(a)$ and D to be 0 on the summand $\ker D$. We equip the summand $\ker D$ with the opposite grading. This insures that the operator

$$\begin{pmatrix} 0 & 1 \\ 1 & 0 \end{pmatrix}$$

on $\ker D \oplus \ker D$ has degree 1. Adding this operator to $D \oplus 0$, we get an unbounded Fredholm module $(D', \pi \oplus 0, H \oplus \ker D)$ with $\ker D' = \{0\}$.

If we start with a θ-summable unbounded Fredholm module in the sense of Definition 22.4, the resulting bounded Fredholm module will be θ-summable as well [41]. Conversely, if A is countably generated, then any bounded θ-summable Fredholm module arises in this way from a suitable unbounded θ-summable Fredholm module [7]. This construction only yields $\exp(-D^2) \in \ell^1(H)$, not $\exp(-\beta D^2) \in \ell^1(H)$ for all $\beta > 0$.

Let (D, π, H) be an unbounded, even, θ-summable Fredholm module with grading operator γ. The associated *JLO-cocycle* is defined as follows. Let

$$\Delta_k := \{(s_0, \ldots, s_k) \in [0,1]^{k+1} \mid s_0 + \cdots + s_k = 1\}$$

be the standard k-simplex and let $\int_{\Delta_k}$ denote integration over Δ_k with respect to the usual Lebesgue measure $ds_0 \ldots ds_{k-1}$. The JLO-cocycle is defined by the formula

$$\tau(a_0 da_1 \ldots da_{2n})$$
$$:= \int_{\Delta_{2n}} \operatorname{tr}\left(\gamma a_0 e^{-s_0 D^2}[D, a_1]e^{-s_1 D^2} \cdots [D, a_{2n}]e^{-s_{2n} D^2}\right) \quad (3)$$

for $n \geq 0$. An involved computation shows that $\tau \circ (B + b) = 0$ and that τ satisfies the growth condition (2). Hence τ defines an element of $HE^0(A)$. For odd Fredholm modules, almost the same formula works, see [7]. As expected, the Chern character for Fredhom modules is compatible with the Chern character in K-theory $\operatorname{ch}: K_*(A) \to HE_*(A)$. That is, the analogues of Theorem 19.1 and 19.2 hold:

Theorem 22.2. *Let $x \in K_*(A)$ and let y be a θ-summable Fredholm module with the same parity as x. Then the pairing between the Chern characters $\langle \operatorname{ch} x, \operatorname{ch} y \rangle$ equals the index pairing between K-theory and K-homology.*

Viewing entire cyclic cohomology as a version of periodic cyclic cohomology that contains, in addition, certain infinite dimensional cycles, we expect that $HE^*(A) =$

$HP^*(A)$ for "finite dimensional" algebras. A positive result in this direction is the following [28]:

Theorem 22.3. *Let A be a Banach algebra of finite Hochschild homological dimension. Then $HE^*(A) \cong HP^*(A)$.*

However, there are quasi-free Fréchet algebras for which entire and periodic cyclic cohomology differ.

Entire and periodic cyclic cohomology have many properties in common. An explanation for this is that we can describe both $HP^0(A)$ and $HE^0(A)$ using traces on "nilpotent" extensions of A. For the purposes of periodic cyclic cohomology, a nilpotent extension of A is an extension of algebras $N \to E \to A$ in which the kernel N is nilpotent in the sense that $N^n = 0$ for some $n \in \mathbb{N}$. A trace on E gives rise to an element of $HP^0(E)$ and hence of $HP^0(A)$ because Goodwillie's theorem asserts that $HP^0(A) \cong HP^0(E)$. Moreover, all elements of $HP^0(A)$ arise in this way. It is also possible to describe when two traces give rise to the same element in $HP^0(A)$, but we will not explain this here.

We get entire cyclic cohomology by weakening the notion of nilpotence.

Definition 22.5. *A bornological algebra A is called* analytically nilpotent *iff $S^\infty :=$ $\bigcup_{n \geq 1} S^n$ is small for all $S \in \mathcal{S}(A)$.*

An analytically nilpotent extension *of A is an extension $N \to E \to A$ of complete bornological algebras with a bounded linear section and analytically nilpotent kernel N.*

Let A be a complete bornological algebra that is analytically nilpotent and let $a \in A$. If the power series $\sum c_j z^j$ has non-zero radius of convergence, then the series $\sum c_j a^j$ converges in A. This means that all elements of A have "spectral radius zero". A typical example of an analytically nilpotent algebra is JA or its completion $(JA)^c$. Indeed, the bornology on TA is the coarsest one for which JA is analytically nilpotent and the canonical linear map $A \to TA$ is bounded.

Consequently, the extension $(JA)^c \to (TA)^c \to A$ is universal among the analytically nilpotent extensions in the following sense: If $N \to E \to A$ is an analytically nilpotent extension of A, then there is a morphism of extensions from $(TA)^c \to A$ to $E \to A$, that is, a commutative diagram

$$
\begin{array}{ccccc}
(JA)^c & \longrightarrow & (TA)^c & \longrightarrow & A \\
\downarrow & & \downarrow & & \| \\
N & \longrightarrow & E & \longrightarrow & A
\end{array}
$$

In addition, this morphism is unique up to smooth homotopy.

Definition 22.6. *Let $l : A \to B$ be a linear map between bornological algebras. For $S \in \mathcal{S}(A)$, let*

$$\omega_l(S, S) := \{l(xy) - l(x)l(y) \mid x, y \in S\} \subseteq B.$$

We say that l is almost multiplicative *or that l has* analytically nilpotent curvature *iff $\bigcup_{n \geq 1} \omega_l(S, S)^n$ is small for all $S \in \mathcal{S}(A)$.*

It is an important fact that almost multiplicative linear maps form a category, that is, compositions of almost multiplicative linear maps are again almost multiplicative. This is closely related to the observation that the class of analytically nilpotent algebras is closed under extensions. That is, if $A \to B \to C$ is an extension with analytically nilpotent A and C, then B is analytically nilpotent as well. Working with almost multiplicative linear maps, it is easy to obtain the following analogue of Goodwillie's theorem:

Theorem 22.4. *Let $N \to E \xrightarrow{p} A$ be an extension of complete bornological algebras with a bounded linear section. Suppose that N is analytically nilpotent. Then $[p_*] \in HE_0(E, A)$ is invertible. Therefore, $p^* \colon HE^0(A) \to HE^0(E)$ is an isomorphism.*

Another consequence of the composability of almost multiplicative maps is that any extension $N \to E \to (TA)^c$ with analytically nilpotent kernel N splits by a bounded homomorphism. Thus $(TA)^c$ is analytically quasi-free:

Definition 22.7. *A complete bornological algebra R is called* analytically quasi-free *iff any analytically nilpotent extension $N \to E \to R$ splits by a bounded homomorphism.*

Analytical quasi-freeness is a much stronger condition than quasi-freeness and is not related to homological algebra. Examples of analytically quasi-free algebras besides $(TA)^c$ are $\mathbb{C}$ and the algebra of Laurent series $\mathbb{C}[u, u^{-1}]$, equipped with the finest possible bornology.

Theorem 22.5. *Let $N \to R \to A$ be an extension of complete bornological algebras with a bounded linear section. If N is analytically nilpotent and R is analytically quasi-free, then the complexes $X(R)^c$ and $X(TA)^c$ are homotopy equivalent via bounded chain maps and homotopies.*

We now sketch the proof of the Excision Theorem 22.1 in [32]. Let $S \to P \to Q$ be an extension with bounded linear section. The basic idea of the proof is to play with the *left* ideal $L \subseteq TP$ that is generated by $S \subseteq TP$. It turns out that while the two-sided ideal generated by S does not have good properties, the ideal L behaves very nicely. First, there is an explicit free L-bimodule resolution of $\tilde{L}$ of length 1 related to the embedding $L \subseteq TP$. It follows that L is quasi-free. An analysis of this resolution shows that the complex $X(L)$ is homotopy equivalent to the kernel of the canonical map $X(TP) \to X(TQ)$. Secondly, L is not just quasi-free but analytically quasi-free. This is more difficult to show. One can construct a bounded homomorphism $L \to TL$ using the universal property of TP. The kernel of the canonical projection $L \to S$ is contained in JP and therefore analytically nilpotent. Hence Theorem 22.5 shows that $X(TS)$ and $X(L)$ are homotopy equivalent. Since $X(L)$ is homotopy equivalent to the kernel of the canonical map $X(TP) \to X(TQ)$, the Excision Theorem follows.

23 Local Cyclic Cohomology

Neither periodic nor entire cyclic cohomology yield good results for big algebras like C^*-algebras. If A is a nuclear C^*-algebra or, more generally, an amenable Banach algebra, then $HE^1(A) = HP^1(A) = 0$ and $HE^0(A) \cong HP^0(A) \cong HH^0(A)$ is the space of traces on A [28]. To obtain reasonable results for C^*-algebras as well, one has to use local cyclic cohomology [38].

The definition of local cyclic cohomology is based on entire cyclic cohomology. The first modification with respect to the entire theory is that we equip a Banach (or Fréchet) algebra A with the bornology of all *precompact* subsets. This change of bornology can still be handled within the framework of section 22. The second modification is that we view a bornological vector space as an inductive system and replace the ordinary cohomology by a "local" cohomology that is defined in the setting of inductive systems.

Let X be a bornological vector space. Then the absolutely convex small subsets of X form a directed set with respect to inclusion. Recall that if $S \subseteq X$ is an absolutely convex small subset, then its linear span X_S carries a canonical norm. The map $S \mapsto X_S$ defines an inductive system of normed spaces with the additional property that all the structure maps are injective. Its limit (in the category of bornological vector spaces) is isomorphic to X. Thus any bornological vector space can be written in a canonical way as an inductive limit of normed spaces. This construction gives rise to an equivalence between the category of bornological vector spaces and bounded linear maps and the category of inductive systems of normed spaces with *injective* structure maps and morphisms of inductive systems [32].

If we apply this construction to a $\mathbb{Z}/2$-graded complex of bornological vector spaces, the resulting inductive system is (isomorphic to) an inductive system of $\mathbb{Z}/2$-graded complexes of normed spaces. In the following, we view $X(TA)$ as an inductive system of $\mathbb{Z}/2$-graded complexes of normed spaces in this way. *We still write $X(TA)$ for this inductive system to simplify notation.*

The space of bounded linear functionals on an inductive system $(C_i)_{i \in I}$ is equal to the projective limit of the dual spaces C_i'. The projective limit functor has the following undesirable property: Even if all the complexes C_i are contractible, it may happen that the cohomology of the inductive limit is non-zero. We would like to have another cohomology functor H_{loc}^* that is *local* in the following sense: if $H^*(C_i) = 0$ for all $i \in I$, then $H_{\text{loc}}^*((C_i)_{i \in I}) = 0$ as well.

To define H_{loc}^*, we observe the following. The category $\text{Ho}(\text{Ind})$ of inductive systems of $\mathbb{Z}/2$-graded complexes of normed spaces with homotopy classes of chain maps as morphisms is a triangulated category. The subclass $\mathcal{N}$ of inductive systems of *contractible* complexes is a "null system" in $\text{Ho}(\text{Ind})$. Thus the corresponding quotient category $\text{Ho}(\text{Ind})/\mathcal{N}$ is again a triangulated category. The decisive property of this quotient category is that in it all elements of $\mathcal{N}$ become isomorphic to the zero object.

In the following definition, we view $\mathbb{C}$ as a constant inductive system of complexes with zero boundary as usual.

Definition 23.1. *Let $K = (K_i)_{i \in I}$ and $L = (L_j)_{j \in J}$ be two inductive systems of complexes. Define $H_*^{\text{loc}}(K, L)$ to be the space of morphisms $K \to L$ in the category $\text{Ho}(\text{Ind})/\mathcal{N}$. Let*

$$H_{\text{loc}}^*(K) := H_*^{\text{loc}}(K, \mathbb{C}), \qquad H_*^{\text{loc}}(L) := H_*^{\text{loc}}(\mathbb{C}, L).$$

One can compute $H_*^{\mathrm{loc}}(K, L)$ via an appropriate projective resolution of K (see [38]). An analysis of this resolution yields that $H_*^{\mathrm{loc}}(K, L)$ is a limit of a spectral sequence whose E^2-term involves the homologies $H_*\big(\mathcal{L}(K_i, L_j)\big)$ and the derived functors of the projective limit functor $\varprojlim$. For countable inductive systems, the derived functors $R^p \varprojlim$ with $p \geq 2$ vanish. Hence the spectral sequence degenerates to a Milnor $\varprojlim^1$-exact sequence

$$0 \longrightarrow \varprojlim_I^1 \varinjlim_J H_{*-1}\big(\mathcal{L}(K_i, L_j)\big) \longrightarrow H_*^{\mathrm{loc}}(K, L)$$
$$\longrightarrow \varprojlim_I \varinjlim_J H_*\big(\mathcal{L}(K_i, L_j)\big) \longrightarrow 0$$

if I is countable. In the local theory we have this exact sequence for arbitrary countable inductive systems (K_i). In particular, for $K = \mathbb{C}$ we obtain

$$H_*^{\mathrm{loc}}(L) = \varinjlim H_*(L_j).$$

Since the inductive limit functor is exact, $H_*^{\mathrm{loc}}(L)$ is equal to the homology of the inductive limit of L.

The completion of an inductive system of normed spaces $(X_i)_{i \in I}$ is defined entry-wise: $(X_i)^{\mathrm{c}} := (X_i^{\mathrm{c}})_{i \in I}$.

Definition 23.2. *Let A and B be complete bornological algebras. We define* local cyclic cohomology, local cyclic homology, *and* bivariant local cyclic homology *by*

$$HE^*_{\mathrm{loc}}(A) := H^*_{\mathrm{loc}}(X(TA)), \qquad HE_*^{\mathrm{loc}}(B) := H_*^{\mathrm{loc}}(X(TB)^{\mathrm{c}}),$$
$$HE_*^{\mathrm{loc}}(A, B) := H_*^{\mathrm{loc}}(X(TA)^{\mathrm{c}}, X(TB)^{\mathrm{c}}) \cong H_*^{\mathrm{loc}}(X(TA), X(TB)^{\mathrm{c}}).$$

As usual, we have $HE_*^{\mathrm{loc}}(A, \mathbb{C}) \cong HE^*_{\mathrm{loc}}(A)$ and $HE_*^{\mathrm{loc}}(\mathbb{C}, A) \cong HE_*^{\mathrm{loc}}(A)$. The composition of morphisms gives rise to a product on HE_*^{loc}.

If we view a bornological vector space X as an inductive system (X_i) with injective structure maps, then the structure maps of the completion $(X_i)^{\mathrm{c}}$ need not be injective any more, so that $(X_i)^{\mathrm{c}}$ may differ from the inductive system that corresponds to the bornological vector space X^{c}. Hence we have to be careful in what category we form the completion $X(TB)^{\mathrm{c}}$. For the local theory, it is crucial to take the completion in the category of inductive systems. A consequence is that, in general, $HE_*^{\mathrm{loc}}(B)$ may be different from $HE_*(B)$. However, in applications it often turns out that the two completions coincide. This holds, for instance, if B is a Fréchet algebra equipped with the precompact bornology or if B is nuclear in the sense of Grothendieck. In these two cases we have $HE_*(B) \cong HE_*^{\mathrm{loc}}(B)$ and a natural map $HE_*(A, B) \to HE_*^{\mathrm{loc}}(A, B)$.

The main advantage of local cyclic cohomology is that it behaves well when passing to "smooth" subalgebras.

Definition 23.3. *Let A be a complete bornological algebra with bornology $\mathcal{S}(A)$, let B be a Banach algebra with closed unit ball U, and let $j \colon A \to B$ be an injective bounded homomorphism with dense image. We call A a* smooth subalgebra *of B iff $S^{\infty} := \bigcup S^n \in \mathcal{S}(A)$ whenever $S \in \mathcal{S}(A)$ and $j(S) \subseteq rU$ for some $r < 1$.*

Let $A \subseteq B$ be a smooth subalgebra. Then an element $a \in A$ that is invertible in B is already invertible in A. Hence A is closed under holomorphic functional calculus. In the setting of topological algebras, smooth subalgebras are defined in [2] using norms with special properties. If $A \subseteq B$ is a smooth subalgebra in the sense of [2], then A equipped with the precompact bornology is a smooth subalgebra in the sense of Definition 23.3.

Whereas the inclusion of a smooth subalgebra $A \to B$ induces an isomorphism $K_*(A) \cong K_*(B)$ in K-theory, the periodic or entire cyclic theories of A and B may differ drastically. However, local theory behaves like K-theory in this situation:

Theorem 23.1. *Let $A \subseteq B$ be a smooth subalgebra. Assume that A and B are both endowed with the bornology of precompact subsets. Let $j: A \to B$ be the inclusion and let $j_* \in HE_0^{\mathrm{loc}}(A, B)$ be the corresponding element in the bivariant local cyclic homology.*

If B has Grothendieck's approximation property, then j_* is invertible. More generally, j_* is invertible if the identity map on B can be approximated uniformly on small subsets of B by maps of the form $j \circ l$ with $l: B \to A$ bounded and linear.

Bivariant local cyclic homology is exact for extensions with a bounded linear section, homotopy invariant for smooth homotopies and stable with respect to tensor products with the trace class operators $\ell^1(H)$. The proofs are the same as for entire cyclic cohomology. Using Theorem 23.1, we can strengthen these properties considerably:

Theorem 23.2. *Let A be a C^*-algebra. The functors $B \mapsto HE_*^{\mathrm{loc}}(A, B)$ and $B \mapsto HE_*^{\mathrm{loc}}(B, A)$ are split exact, stable homotopy functors on the category of C^*-algebras.*

For separable C^*-algebras, there is a natural bivariant Chern character

$$\mathrm{ch}: KK_*(A, B) \to HE_*^{\mathrm{loc}}(A, B).$$

The Chern character is multiplicative with respect to the Kasparov product on the left and the composition product on the right hand side.

If both A and B are in the bootstrap class satisfying the universal coefficient theorem, then there is a natural isomorphism

$$HE_*^{\mathrm{loc}}(A, B) \cong \mathrm{Hom}(K_*(A) \otimes_{\mathbb{Z}} \mathbb{C}, K_*(B) \otimes_{\mathbb{Z}} \mathbb{C}).$$

Proof. Since $C^\infty([0, 1], A) \subseteq C([0, 1], A)$ is a smooth subalgebra, Theorem 23.1 and smooth homotopy invariance imply continuous homotopy invariance. The subalgebra $M_\infty(A) = M_\infty \hat{\otimes} A$ defined as in (1) is a smooth subalgebra of the C^*-algebraic stabilization of A. Hence Theorem 23.1 and stability with respect to M_∞ imply C^*-algebraic stability.

The existence of the bivariant Chern character follows from these homological properties by the universal property of Kasparov's KK-theory.

The last assertion is trivial for $A = B = \mathbb{C}$. The class of C^*-algebras for which it holds is closed under KK-equivalence, inductive limits and extensions with completely positive section. Hence it contains all nuclear C^*-algebras satisfying the universal coefficient theorem (see [1]).

Examples of C^*-algebras satisfying the universal coefficient theorem are commutative C^*-algebras and the C^*-algebras of amenable groupoids [44].

Assume that Γ is a discrete torsion free group that is the fundamental group of a manifold M of non-positive sectional curvature. Then M is a model for $B\Gamma$, so that $\mathbb{C}[\Gamma]$ has finite homological dimension. We have

$$HE_*^{\mathrm{loc}}\big(\ell^1(\Gamma)\big) \cong HP_*\big(\mathbb{C}[\Gamma]\big) \cong H_*(\Gamma,\mathbb{C}).$$

However, the local cyclic homology of $C_{\mathrm{red}}^*\Gamma$ is still unknown.

References

1. B.Blackadar: K-theory for Operator Algebras, Springer-Verlag, Heidelberg / Berlin / New York / Tokyo, 1986.
2. B. Blackadar and J. Cuntz. Differential Banach algebra norms and smooth subalgebras of C^*-algebras. *J. Operator Theory*, 26(2):255–282, 1991.
3. L.G.Brown, R.G.Douglas und P.Fillmore: Extensions of C*-algebras and K-homology, Ann. of Math. **105** (1977), 265-324.
4. A. Connes, Non-commutative differential geometry. Publ. Math. I.H.E.S. **62** (1985), 257-360.
5. A. Connes. Entire cyclic cohomology of Banach algebras and characters of θ-summable Fredholm modules. *K-Theory*, 1(6):519–548, 1988.
6. A. Connes. On the Chern character of θ summable Fredholm modules. *Comm. Math. Phys.*, 139(1):171–181, 1991.
7. A. Connes, Non-commutative Geometry, Academic Press, London-Sydney-Tokyo-Toronto, 1994.
8. A. Connes und N.Higson: Déformations, morphismes asymptotiques et K-théorie bivariante, C.R.Acad.Sci. Paris Ser. I, Math, **311** (1990), 101-106.
9. J. Cuntz: A new look at KK-theory, K-theory **1** (1987), 31-52.
10. J. Cuntz: Excision in bivariant cyclic theory for topological algebras, Proc. Workshop on Cyclic Cohomology and Noncommutative Geometry, Fields Institute, wird erscheinen.
11. J. Cuntz: Bivariante K-theorie für lokalkonvexe Algebren und der bivariante Chern-Connes-Charakter, Docum. Math. J. DMV 2 (1997), 139-182, http://www.mathematik.uni-bielefeld.de/documenta
12. J. Cuntz: Morita invariance in cyclic homology for nonunital algebras, K-theory **15** (1998), 301 - 305.
13. J.Cuntz and D. Quillen, Algebra extensions and nonsingularity, J.Amer. Math. Soc. **8** (1995), 251-289
14. J. Cuntz and D. Quillen: Cyclic homology and nonsingularity, J. Amer. Math. Soc. **8** (1995), 373-442.
15. J. Cuntz and D. Quillen: Excision in bivariant periodic cyclic cohomology, Invent. math. **127** (1997), 67-98.
16. J. Cuntz and D. Quillen, Operators on noncommutative differential forms and cyclic homology, in "Geometry, Topology and Physics; for Raoul Bott", International Press, Cambridge MA, 1995.
17. G.A.Elliott, T.Natsume and R.Nest: Cyclic cohomology for one-parameter smooth crossed products, Acta Math. **160** (1988), 285-305.

18. T.G. Goodwillie, Cyclic homology and the free loopspace. Topology **24** (1985), 187-215.
19. A.Grothendieck: Produits tensoriels topologiques et espaces nucléaires, Memoirs of the Amer. Math. Soc. **16** (1955).
20. N.Higson: On Kasparov theory, M.A. thesis, Dalhousie University, 1983.
21. H. Hogbe-Nlend. *Bornologies and functional analysis*. North-Holland Publishing Co., Amsterdam, 1977.
22. A. Jaffe, A. Lesniewski, and K. Osterwalder. Quantum K-theory. I. The Chern character. *Comm. Math. Phys.*, 118(1):1–14, 1988.
23. J.D.S. Jones and C. Kassel, Bivariant cyclic theory, K-theory **3** (1989), 339-365.
24. M.Karoubi, Homologie cyclique et K-théorie, Astérisque **149** (1987).
25. G. G. Kasparov, The operator K-functor and extensions of C^*-algebras, Izv. Akad. Nauk. SSSR, Ser. Mat. **44** (1980), 571-636
26. C. Kassel, Cyclic homology, comodules and mixed complexes, J. of Algebra **107** (1987), 195 - 216.
27. C. Kassel, Caractère de Chern bivariant. K-Theory **3** (1989), 367-400.
28. M. Khalkhali. Algebraic connections, universal bimodules and entire cyclic cohomology. *Comm. Math. Phys.*, 161(3):433–446, 1994.
29. V.Lafforgue: K-théorie bivariante pour les algèbres de Banach et conjecture de Baum-Connes, thesis, Paris 1999.
30. J.-L.Loday, Cyclic Homology, Grundlehren **301**, Springer Verlag, 1992.
31. J.-L.Loday and D. Quillen, Cyclic homology and the Lie algebra homology of matrices, Comment. Math. Helvetici **59** (1984), 565 - 591.
32. R. Meyer. Analytic cyclic cohomology. Thesis. math.KT/9906205, 1999.
33. E. A. Michael, Locally multiplicatively-convex topological algebras, Memoirs of the AMS, number **11**, 1952.
34. V. Nistor, A bivariant Chern character for p-summable quasihomomorphisms. K-Theory **5** (1991), 193-211.
35. V. Nistor, A bivariant Chern-Connes character, Ann. of Math. **138** (1993), 555-590.
36. C.Phillips: K-theory for Fréchet algebras, International Journal of Mathematics vol. **2** no. 1 (1991), 77 - 129.
37. M.Puschnigg: Asymptotic cyclic cohomology, Lecture Notes in Mathematics **1642**, Springer 1996.
38. M.Puschnigg: Cyclic homology theories for topological algebras, preprint.
39. M.Puschnigg: to appear in Inv. Math.
40. D. Quillen, Algebra cochains and cyclic cohomology, Publ. Math. IHES, **68** (1988), 139 - 174.
41. E. Schrohe, M. Walze, and J.-M. Warzecha. Construction de triplets spectraux à partir de modules de Fredholm. *C. R. Acad. Sci. Paris Sér. I Math.*, 326(10):1195–1199, 1998.
42. F.Treves: Topological vector spaces, distributions and kernels, Academic Press, New York, London, 1967.
43. F.Tsygan: The homology of matrix Lie algebras over rings and the Hochschild homology (in Russian), Uspekhi Mat. Nauk 38 (1983), 217-218 - Russ. Math. Survey 38 no. 2 (1983), 198-199.
44. J.-L. Tu. La conjecture de Baum-Connes pour les feuilletages moyennables. *K-Theory*, 17(3):215–264, 1999.
45. X. Wang, A bivariant Chern character, II, Can. J. Math. **35** (1992), 1-36.
46. M.Wodzicki, The long exact sequence in cyclic homology associated with an extension of algebras, C.R. Acad. Sci. Paris **306** (1988), 399-403.
47. M.Wodzicki, Excision in cyclic homology and in rational algebraic K-theory, Ann. of Math. **129** (1989), 591-639.

Group C*-Algebras and K-Theory

Nigel Higson[1] and Erik Guentner[2]

[1] Department of Mathematics, Pennsylvania State University
   University Park, PA 16802
   higson@psu.edu
[2] Department of Mathematical Sciences University of Hawaii, Manoa
   2565 McCarthy Mall, Keller 401A Honolulu, HI 96822
   erik@math.hawaii.edu

Preface

These notes are about the formulation of the Baum-Connes conjecture in operator algebra theory and the proofs of some cases of it. They are aimed at readers who have some prior familiarity with K-theory for C^*-algebras (up to and including the Bott Periodicity theorem). I hope the notes will be suitable for a second course in operator K-theory.

The lectures begin by reviewing K-theory and the Bott periodicity theorem. Much of the Baum-Connes theory has to do with broadening the periodicity theorem in one way or another, and for this reason quite some time is spent formulating and proving the theorem in a way which is suited to later extensions. Following that, the lectures turn to the machinery of bivariant K-theory and the formulation of the Baum-Connes conjecture. The main objective of the notes is reached in Lecture 4, where the conjecture is proved for groups which act properly and isometrically on affine Euclidean spaces. The remaining lectures deal with partial results which are important in applications and with counterxamples to various overly optimistic strengthenings of the conjecture.

Despite their length the notes are not complete in every detail, and the reader will have to turn to the references, or his own inner resources, to fill some gaps. In addition the lectures contain no discussion of applications or connections to geometry, topology and harmonic analysis, nor do they cover the remarkable work of Vincent Lafforgue. For the former see [7]; for the latter see [62, 44].

The notes are based on joint work carried out over a period of many years now with many people: Paul Baum, Alain Connes, Erik Guentner, Gennadi Kasparov, Vincent Lafforgue, John Roe, Georges Skandalis and Jody Trout. It is a pleasure to thank them all. I am especially grateful to Erik Guentner for writing the first draft of these notes and for his valuable assistance throughout their creation. Both authors were partially supported by NSF grants during the preparation of this paper.

Nigel Higson

1 K-Theory

In the first three lectures we shall be developing machinery needed to formulate the Baum-Connes conjecture and prove some cases of it. We shall presume some prior familiarity with C^*-algebra K-theory, but we shall also develop a 'spectral' picture of K-theory from scratch. In Lecture 1 we shall prove the Bott periodicity theorem in C^*-algebra K-theory in a way which will be suited to generalization in subsequent lectures.

1.1 Review of K-Theory

We begin by briefly reviewing the rudiments of C^*-algebra K-theory, up to and including the Bott periodicity theorem. As the reader knows, C^*-algebra K-theory is a development of the topological K-theory of Atiyah and Hirzebruch [4]. But the basic definition is completely algebraic in nature:

Definition 1.1. *Let A be a ring with a multiplicative unit. The group $K_0(A)$ is the abelian group generated by the set of isomorphism classes of finitely generated and projective (unital, right) A-modules, subject to the relations $[E] + [F] = [E \oplus F]$.*

Remark 1.1. Functional analysts usually prefer to formulate the basic definition in terms of equivalence classes idempotents in the matrix rings $M_n(A)$. This is because in several contexts idempotents arise more naturally than modules. We shall use both definitions below, bearing in mind that they are related by associating to an idempotent $P \in M_n(A)$ the projective module $E = PA^n$.

The group $K_0(A)$ is functorial in A since associated to a ring homomorphism $A \to B$ there is an induction operation on modules, $E \mapsto E \otimes_A B$.

Most of the elementary algebraic theory of the functor $K_0(A)$ is a consequence of a structure theorem involving pull-back diagrams like this one:

$$
\begin{array}{ccc}
A & \xrightarrow{q_1} & A_1 \\
{\scriptstyle q_2}\downarrow & & \downarrow{\scriptstyle p_1} \\
A_2 & \xrightarrow{p_2} & B
\end{array}
\qquad A = \{(a_1, a_2) \in A_1 \oplus A_2 \,|\, p_1(a_1) = p_2(a_2)\}.
$$

Theorem 1.1. *Assume that in the above diagram at least one of the two homomorphisms into B is surjective. If E_1 and E_2 are finitely generated and projective modules over A_1 and A_2, and if $F: p_{1*}E_1 \to p_{2*}E_2$ is an isomorphism of B modules, then the A-module*

$$
E = \{\, (e_1, e_2) \in E_1 \times E_2 \,|\, F(e_1 \otimes 1) = e_2 \otimes 1 \,\}
$$

is finitely generated and projective. Moreover, up to isomorphism, every finitely generated and projective module over A has this form. □

This is proved in the first few pages of Milnor's algebraic K-theory book [49]. The theorem describes projective modules over A in terms of projective modules over A_1, projective modules over A_2, and invertible maps between projective modules over B. It leads very naturally to the definition of a group $K_1(B)$ in terms of invertible matrices, but at this point the purely algebraic and the C^*-algebraic theories diverge, as a result of an important homotopy invariance principle.

Definition 1.2. *Let A be a C^*-algebra. Denote by $A[0,1]$ the C^*-algebra of continuous functions from the unit interval $[0,1]$ into A.*

We shall similarly denote by $A(X)$ the C^*-algebra of continuous functions from a compact space X into a C^*-algebra A.

Theorem 1.2. *Let A be a C^*-algebra with unit. If E is a finitely generated and projective module over $A[0,1]$ then the induced modules over A obtained by evaluation at $0 \in [0,1]$ and $1 \in [0,1]$ are isomorphic to one another.* □

As a result, K-theory is a *homotopy functor* in the sense of the following definition:

Definition 1.3. *A homotopy of $*$-homomorphisms between C^*-algebras is a family of homomorphisms $\varphi_t \colon A \to B$ $(t \in [0,1])$, for which the maps $t \mapsto \varphi_t(a)$ are continuous, for all $a \in A$. A functor F on the category of C^*-algebras is a homotopy functor if all the homomorphisms φ_t in any homotopy induce one and the same map $F(\varphi_t) \colon F(A) \to F(B)$.*

We shall now define the K-theory group $K_1(A)$.

Definition 1.4. *Let A be a C^*-algebra with unit. Denote by $M_n(A)$ the C^*-algebra of $n \times n$ matrices with entries in A and denote by $GL_n(A)$ the group of invertible elements in $M_n(A)$. View $GL_n(A)$ as a subgroup of each $GL_{n+k}(A)$ via the embeddings*

$$A \mapsto \begin{pmatrix} A & 0 \\ 0 & I \end{pmatrix}.$$

Denote by $K_1(A)$ the direct limit of the component groups $\pi_0(GL_n(A))$:

$$K_1(A) = \varinjlim \pi_0(GL_n(A)).$$

Remark 1.2. This is a group, thanks to the group structure in $GL_n(A)$, and in fact an abelian group since $\begin{pmatrix} XY & 0 \\ 0 & I \end{pmatrix}$ is homotopic to $\begin{pmatrix} X & 0 \\ 0 & Y \end{pmatrix}$, and hence to $\begin{pmatrix} YX & 0 \\ 0 & I \end{pmatrix}$.

Returning to our pullback diagram and Theorem 1.1, it is now straightforward to derive all but the dotted part of the following six-term 'Mayer-Vietoris' exact sequence of K-theory groups:

$$K_0(A) \longrightarrow K_0(A_1) \oplus K_0(A_2) \longrightarrow K_0(B) \tag{1}$$

$$K_1(B) \longleftarrow K_1(A_1) \oplus K_1(A_2) \longleftarrow K_1(A).$$

The diagram is completed (along the dotted arrow) as follows. Consider first the pullback diagram

$$
\begin{array}{ccc}
A(S^1) & \xrightarrow{q_1} & A_1(S^1) \\
{\scriptstyle q_2}\downarrow & & \downarrow{\scriptstyle p_1} \\
A_2(S^1) & \xrightarrow[p_2]{} & B(S^1)
\end{array}
$$

involving algebras of functions on the circle S^1. The Mayer-Vietoris sequence associated to it,

$$
K_0(A(S^1)) \longrightarrow K_0(A_1(S^1)) \oplus K_0(A_2(S^1)) \longrightarrow K_0(B(S^1)) \qquad (2)
$$

$$
K_1(B(S^1)) \longleftarrow K_1(A_1(S^1)) \oplus K_1(A_2(S^1)) \longleftarrow K_1(A(S^1)),
$$

maps to the Mayer-Vietoris sequence (1) via the operation ε of evaluation at $1 \in S^1$, and in fact this map is the projection onto a direct summand since ε has a one-sided inverse consisting of the inclusion of the constant functions into the various algebras of functions on S^1. The complementary summands are computed using the following two results:

Theorem 1.3. *Let A be a C^*-algebra. The kernel of the evaluation homomorphism*

$$
\varepsilon \colon K_0(A(S^1)) \to K_0(A)
$$

is naturally isomorphic to $K_1(A)$. $\quad\square$

This is a simple application of the partial Mayer-Vietoris sequence (think of $A(S^1)$ as assembled by a pullback operation from two copies of $A[0,1]$).

Theorem 1.4. *Let A be a C^*-algebra. The kernel of the evaluation homomorphism*

$$
\varepsilon \colon K_1(A(S^1)) \to K_1(A)
$$

is naturally isomorphic to $K_0(A)$.

This is much harder; it is one formulation of the Bott periodicity theorem. But granting ourselves the result for a moment, we can complete the diagram (1) by the simple device of viewing its horizontal reflection (with the K_1-groups on the top) as a direct summand of the diagram (2). The required connecting map $\partial \colon K_0(B) \to K_1(A)$ appears as a direct summand of the connecting map $\partial \colon K_1(B(S^1)) \to K_0(A(S^1))$.

The full Mayer-Vietoris sequence is a powerful computational tool, especially for commutative algebras. For example it implies that the functors $X \mapsto K_j(A(X))$ constitute a cohomology theory on compact spaces (as in algebraic topology). A simple consequence is the formula

$$K_0(A(S^2)) \cong K_0(A) \oplus K_0(A)$$

which is a perhaps more familiar formulation of Bott periodicity.

Let us conclude our review of K-theory with a quick look at the proof of Theorem 1.4. The launching point is the definition of a map

$$\beta : K_0(A) \to K_1(A(S^1))$$

by associating to the class of an idempotent $P \in M_n(A)$ the element

$$u_P(z) = zP + (1 - P) \tag{3}$$

in $GL_n(A(S^1))$. The following argument (due to Atiyah and Bott [6]) then shows that this *Bott homomorphism* is an isomorphism onto the kernel of the evaluation map $\varepsilon : K_1(A(S^1)) \to K_1(A)$. The key step is to show β is surjective; the proof in injectivity is a minor elaboration of the surjectivity argument[3] and we shall not comment on it further.

By an approximation argument involving trigonometric polynomials the proof of surjectivity quickly reduces to showing that a polynomial loop of invertible matrices

$$u(z) = b_0 + zb_1 + \cdots + z^m b_m, \quad b_j \in M_n(A),$$

which defines a element of the kernel of the evaluation map must lie in the image of β. By elementary row operations, the loop $u(z)$ is equivalent to the 'linear' loop

$$v(z) = \begin{pmatrix} b_0 & b_1 & \ldots & b_{m-1} & b_m \\ -z & 1 & \ldots & 0 & 0 \\ \multicolumn{5}{c}{\dotfill} \\ 0 & 0 & \ldots & -z & 1 \end{pmatrix} = Az + B,$$

for suitable matrices A and B. Evaluating at $z = 1$ and bearing in mind that v is in the kernel of the evaluation map we see that $A + B$ is path connected to I (in some suitable $GL_N(A)$) and so v is equivalent to

$$w(z) = (A + B)^{-1}(Az + B) = Cz + (I - C).$$

The final step of the argument is for our purposes the most interesting, since in involves in a crucial way the spectral theory of elements in C^*-algebras. Since $w(z)$ is invertible for all $z \in S^1$ the spectrum of C contains no element on the line $\mathrm{Re}(z) = \frac{1}{2}$ in $\mathbb{C}$. If P denotes the idempotent associated to the part of the spectrum of C to the right of this line (obtained from the Riesz functional calculus) then $w(z)$ is homotopic to the path

$$u_P(z) = Pz + (1 - P),$$

the K-theory class of which is of course in the image of β. This concludes the proof.

In the following sections we shall recast the definition of K-theory and the proof of Bott periodicity in a way which brings spectral theory very much to prominence. As we shall eventually see, this is an important first step toward our principal goal of computing K-theory for group C^*-algebras.

---

[3] As Shmuel Weinberger puts it, uniqueness is a relative form of existence.

1.2 Graded C*-Algebras

To proceed further with K-theory we shall find it convenient to work with *graded* C^*-algebras, which are defined as follows.

Definition 1.5. *Let A be a C^*-algebra. A grading on A is a $*$-automorphism α of A satisfying $\alpha^2 = 1$. Equivalently, a grading is a decomposition of A as a direct sum of two $*$-linear subspaces, $A = A_0 \oplus A_1$, with the property that $A_i A_j \subset A_{i+j}$, where $i, j \in \mathbb{Z}/2$. Elements of A_0 (for which $\alpha(a) = a$) are said to be of* even *grading-degree while elements of A_1 (for which $\alpha(a) = -a$) are of* odd *grading-degree.*

Example 1.1. The *trivial grading* on A is defined by the $*$-automorphism $\alpha = \mathrm{id}$, or equivalently by setting $A_0 = A$ and $A_1 = 0$.

In fact, we shall require only a very small collection of non-trivially graded C^*-algebras, among which the following two are the most important.

Example 1.2. Let $\mathcal{H}$ be a graded Hilbert space; that is, a Hilbert space equipped with an orthogonal decomposition $\mathcal{H} = \mathcal{H}_0 \oplus \mathcal{H}_1$. The C^*-algebras $\mathcal{K}(\mathcal{H})$ of compact operators and $\mathcal{B}(\mathcal{H})$ of bounded operators on $\mathcal{H}$ are graded. To describe the grading, think of an operator T on $\mathcal{H}$ as a 2×2 matrix of operators. We declare the diagonal matrices to be even and the off-diagonal ones to be odd.

Example 1.3. Let $\mathcal{S} = C_0(\mathbb{R})$, the C^*-algebra of continuous, complex-valued functions on $\mathbb{R}$ which vanish at infinity, and define a grading on $\mathcal{S}$ by the decomposition

$$\mathcal{S} = C_0(\mathbb{R}) = \{ \text{ even functions } \} \oplus \{ \text{ odd functions } \}.$$

The grading operator is the automorphism $f(x) \mapsto f(-x)$.

Warning: In K-theory it is customary to introduce the C^*-algebra $C_0(\mathbb{R})$ in connection with the operation of 'suspension'. But in what follows the algebra $\mathcal{S}$ will play a quite different role.

Definition 1.6. *A graded C^*-algebra A is inner-graded if there exists a self-adjoint unitary ε in the multiplier algebra of A which implements the grading automorphism α on A:*

$$\alpha(a) = \varepsilon a \varepsilon, \quad \text{for all } a \in A.$$

Examples 1.5 *The trivial grading on a C^*-algebra A is inner: take $\varepsilon = 1$. In addition the gradings on $\mathcal{K}(\mathcal{H})$ and $\mathcal{B}(\mathcal{H})$ are inner: take ε to be the operator which is $+I$ on H_0 and $-I$ on H_1. However the grading on $\mathcal{S}$ is not inner.*

All the fundamental constructions on C^*-algebras have graded counterparts, and we shall require below some familiarity with the notion of tensor product for graded C^*-algebras. As is the case with ungraded C^*-algebras, tensor products of graded C^*-algebras are defined as completions of the algebraic graded tensor product. And as is the case in the ungraded world, there is not usually a unique such completion.

Let us introduce the symbol ∂a defined by

$$\partial a = \begin{cases} 0, & \text{if } a \in A_0 \\ 1, & \text{if } a \in A_1. \end{cases}$$

An element $a \in A$ is *homogeneous* if $a \in A_0$ or $a \in A_1$. Keep in mind that ∂a is defined only when a is a homogeneous element.

Definition 1.7. *Let A and B be graded C^*-algebras. Let $A \widehat{\odot} B$ be the algebraic tensor product of the linear spaces underlying A and B. Define a multiplication, involution and grading on $A \widehat{\odot} B$ by means of the following formulas involving elementary tensors:*

$$(a_1 \widehat{\odot} b_1)(a_2 \widehat{\odot} b_2) = (-1)^{\partial b_1 \partial a_2} a_1 a_2 \widehat{\odot} b_1 b_2,$$

$$(a \widehat{\odot} b)^* = (-1)^{\partial a \partial b} a^* \widehat{\odot} b^*$$

$$\partial(a \widehat{\odot} b) = \partial a + \partial b, \quad (mod\ 2),$$

for all homogeneous elements $a, a_1, a_2 \in A$ and $b, b_1, b_2 \in B$. (The multiplication and involution are extended by linearity to all of $A \widehat{\odot} B$.)

The construction of $A \widehat{\odot} B$ satisfies the usual associativity and commutativity rules but with occasional twists. For example, an isomorphism $A \widehat{\odot} B \to B \widehat{\odot} A$ is defined by

$$a \widehat{\odot} b \longmapsto (-1)^{\partial a \partial b} b \widehat{\odot} a. \tag{4}$$

Definition 1.8. *The graded commutator of elements in a graded C^*-algebra is given by the formula*

$$[a, b] = ab - (-1)^{\partial a \partial b} ba,$$

on homogeneous elements (this is extended by linearity to all elements).

Lemma 1.1. *If C is a graded C^*-algebra and if $\varphi: A \to C$ and $\psi: B \to C$ are graded $*$-homomorphisms[4] whose images graded-commute (meaning that all graded commutators $[\varphi(a), \psi(b)]$ are zero) then there is a unique graded $*$-homomorphism from $A \widehat{\odot} B$ into C which maps $a \widehat{\odot} b$ to $\varphi(a)\psi(b)$.* □

Example 1.4. Let $\mathcal{H}$ be a graded Hilbert space and denote by $\mathcal{H} \widehat{\otimes} \mathcal{H}$ the ordinary Hilbert space tensor product, but considered as a graded Hilbert space. The construction of the lemma produces a graded $*$-homomorphism from the tensor product algebra $\mathcal{B}(\mathcal{H}) \widehat{\odot} \mathcal{B}(\mathcal{H})$ into $\mathcal{B}(\mathcal{H} \otimes \mathcal{H})$ which takes the homogeneous elementary tensor $S \widehat{\odot} T$ to the operator

$$v \otimes w \mapsto Sv \otimes (-1)^{\partial v \partial T} Tw.$$

---

[4] A $*$-homomorphism is *graded*, or *grading-preserving*, if it maps homogeneous elements to homogeneous elements of the same grading-degree.

Definition 1.9. *Let A and B be graded C^*-algebras and let $A \odot B$ be their algebraic tensor product. The maximal graded tensor product, which we will denote by $A \widehat{\otimes} B$, or occasionally by $A \widehat{\otimes}_{max} B$, is the completion of $A \odot B$ in the norm*

$$\| \sum a_i \widehat{\odot} b_i \| = \sup \| \sum_i \varphi(a_1)\psi(b_i) \|,$$

where the supremum is taken over graded-commuting pairs of graded $*$-homomorphisms, mapping A and B into a common third graded C^*-algebra C.

Warning: Our use of the undecorated symbol $\widehat{\otimes}$ to denote the *maximal* tensor product (as opposed to the *minimal* one, which we shall define in a moment) runs counter to ordinary C^*-algebra usage. In situations where the choice of tensor product really is crucial we shall try to write $\widehat{\otimes}_{max}$.

Remark 1.3. It is clear from the definition that the tensor product $\widehat{\otimes}$ is functorial: if $\varphi : A \to C$ and $\psi : B \to D$ are graded $*$-homomorphisms then there is a unique graded $*$-homomorphism $\varphi \widehat{\otimes} \psi : A \widehat{\otimes} B \to C \widehat{\otimes} D$ mapping $a \widehat{\otimes} b$ to $\varphi(a) \widehat{\otimes} \psi(b)$, for all $a \in A$ and $b \in B$.

Example 1.5. If one of A or B is inner-graded then the ungraded C^*-algebra underlying the graded tensor product $A \widehat{\otimes} B$ is isomorphic to the usual tensor product of the ungraded C^*-algebras underlying A and B. If say A is inner-graded then the isomorphism $A \widehat{\otimes} B \to A \otimes B$ is defined by

$$a \widehat{\otimes} b \longmapsto a\varepsilon^{\partial b} \otimes b.$$

We also note that the graded tensor product of two inner-graded C^*-algebras is itself inner-graded. Indeed

$$\varepsilon_A \widehat{\otimes} \varepsilon_B \in \mathcal{M}(A \widehat{\otimes} B) = \mathcal{M}(A \otimes B).$$

For the most part we shall use the maximal tensor product of graded C^*-algebras, but occasionally we shall work with the following 'minimal' product:

Definition 1.10. *Let A and B be graded C^*-algebras and let $A \odot B$ be their algebraic tensor product. The minimal graded tensor product of A and B is the completion of $A \odot B$ in the representation obtained by first faithfully representing A and B as graded subalgebras of $\mathcal{B}(\mathcal{H})$, and then mapping $\mathcal{B}(\mathcal{H}) \odot \mathcal{B}(\mathcal{H})$ to $\mathcal{B}(\mathcal{H} \widehat{\otimes} \mathcal{H})$ as above.*

The minimal tensor product is also functorial, but from our point of view it has some serious shortcomings. These will be explained in the next lecture.

Exercise 1.6 *Show that the minimal and maximal completions of $A \odot \mathcal{K}(\mathcal{H})$ and $S \odot A$ are the same.*

Exercise 1.1. Describe the tensor product C^*-algebra $S \widehat{\otimes} S$ (note that although S itself is a commutative C^*-algebra, the tensor product $S \widehat{\otimes} S$ is not).

Exercise 1.7 *Show that $\mathcal{K}(\mathcal{H}) \widehat{\otimes} \mathcal{K}(\mathcal{H}') \cong \mathcal{K}(\mathcal{H} \widehat{\otimes} \mathcal{H}')$.*

1.3 Amplification

The graded C^*-algebra $\mathcal{S} = C_0(\mathbb{R})$ will play a special role for us. Using it we shall enrich, or 'amplify', the category of graded C^*-algebras and $*$-homomorphisms.

To do so we introduce two $*$-homomorphisms, as follows:

$$\eta \colon \mathcal{S} \to \mathbb{C} \quad \text{and} \quad \Delta \colon \mathcal{S} \to \mathcal{S} \widehat{\otimes} \mathcal{S}.$$

The first is defined by $\eta(f) = f(0)$. In the world of ungraded C^*-algebras and K-theory η is not so interesting since it is homotopic to the zero $*$-homomorphism. But as $*$-homomorphism of graded C^*-algebras η is definitely non-trivial, even at the level of K-theory (which we will come to in the next section). The defining formula for Δ,

$$\Delta \colon f(X) \mapsto f(X \widehat{\otimes} 1 + 1 \widehat{\otimes} X),$$

is explained as follows. Denote by $\mathcal{S}_R$ the quotient of $\mathcal{S}$ consisting of functions on the interval $[-R, R]$ (the quotient map is the operation of restriction of functions) and denote by $X_R \in \mathcal{S}_R$ the function $x \mapsto x$. If $f \in \mathcal{S}$ then we can apply the functional calculus to the self-adjoint element $X_R \widehat{\otimes} 1 + 1 \widehat{\otimes} X_R \in \mathcal{S}_R \widehat{\otimes} \mathcal{S}_R$ to obtain an element $f(X_R \widehat{\otimes} 1 + 1 \widehat{\otimes} X_R) \in \mathcal{S}_R \widehat{\otimes} \mathcal{S}_R$.

Lemma 1.2. *There is a unique graded $*$-homomorphism $\Delta \colon \mathcal{S} \to \mathcal{S} \widehat{\otimes} \mathcal{S}$ whose composition with the quotient map $\mathcal{S} \widehat{\otimes} \mathcal{S} \to \mathcal{S}_R \widehat{\otimes} \mathcal{S}_R$ is the $*$-homomorphism*

$$\Delta \colon f \mapsto f(X_R \widehat{\otimes} 1 + 1 \widehat{\otimes} X_R),$$

for every $R > 0$. $\square$

Exercise 1.8 *Show that the intersection of the kernels of the maps $\mathcal{S} \widehat{\otimes} \mathcal{S} \to \mathcal{S}_R \widehat{\otimes} \mathcal{S}_R$ is zero. This proves the uniqueness part of the Lemma.*

Remark 1.4. If the self-adjoint homogeneous elements u and v in $\mathcal{S}$ are defined by

$$u(x) = e^{-x^2}, \quad \text{and} \quad v(x) = xe^{-x^2}.$$

then

$$\Delta(u) = u \widehat{\otimes} u \quad \text{and} \quad \Delta(v) = u \widehat{\otimes} v + v \widehat{\otimes} u.$$

Since u and v generate the C^*-algebra $\mathcal{S}$, formulas involving Δ and η can often be verified by checking them on u and v.

Remark 1.5. Another approach to the definition of Δ is to use the theory of unbounded multipliers. See the short appendix to this lecture.

The $*$-homomorphisms η and Δ provide $\mathcal{S}$ with a sort of coalgebra structure: the diagrams

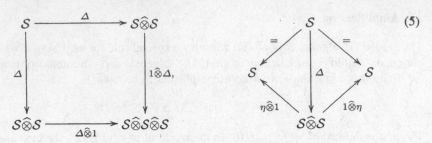

(5)

commute, as is easily verified by considering the elements u and $v \in S$.

Definition 1.11. *Let A be a graded C^*-algebra. The amplification of A is the graded tensor product $SA = S\widehat{\otimes}A$.*

Definition 1.12. *The* amplified category of graded C^*-algebras *is the category whose objects are the graded C^*-algebras and for which the morphisms from A to B are the graded $*$-homomorphisms from SA to B. Composition of morphisms $\varphi\colon A \to B$ and $\psi\colon B \to C$ in the amplified category is given by the following composition of $*$-homomorphisms:*

$$SA \xrightarrow{\Delta\widehat{\otimes}1} S^2 A \xrightarrow{S(\varphi)} SB \xrightarrow{\psi} C.$$

Exercise 1.2. Using (5) verify that the composition law is associative and that the $*$-homomorphisms $SA \to A$ obtained by taking the tensor product of the augmentation $\eta\colon S \to \mathbb{C}$ with the identity map on A serve as identity morphisms for this composition law.

Remark 1.6. Most features of the category of graded C^*-algebras pass to the amplified category. One example is the tensor product operation: given amplified morphisms from $\varphi_1\colon A_1 \to B_1$ and $\varphi_2\colon A_2 \to B_2$ there is a tensor product morphism from $A_1\widehat{\otimes}A_2$ to $B_1\widehat{\otimes}B_2$ (in other words a $*$-homomorphism from $S(A_1\widehat{\otimes}A_2)$ into $B_1\widehat{\otimes}B_2$) defined by the composition of $*$-homomorphisms

$$\mathcal{S}(A_1\widehat{\otimes}A_2) = \mathcal{S}\widehat{\otimes}A_1\widehat{\otimes}A_2 \xrightarrow{\Delta\widehat{\otimes}1\widehat{\otimes}1} \mathcal{S}^2\widehat{\otimes}A_1\widehat{\otimes}A_2 \cong \mathcal{S}A_1\widehat{\otimes}\mathcal{S}A_2 \xrightarrow{\varphi_1\widehat{\otimes}\varphi_2} B_1\widehat{\otimes}B_2$$

(the formula incorporates the transposition isomorphism (4)).

Exercise 1.3. Show that the tensor product is functorial (compatible with composition) and associative.

1.4 Stabilization

A second means of enriching the notion of $*$-homomorphism is the process of *stabilization*. This is of course very familiar in K-theory: stabilization means replacing a C^*-algebra A with $A\widehat{\otimes}\mathcal{K}(\mathcal{H})$, its tensor product with the C^*-algebra of compact operators.

If A is a trivially graded C^*-algebra with unit then each projection p in $A \otimes \mathcal{K}(\mathcal{H})$ determines a projective module over A (namely $p(A \otimes \mathcal{K}(\mathcal{H}))$ with the obvious right action of A) and in fact the set of isomorphism classes of finitely generated A-modules is identified in this way with the set of homotopy classes of projections in $A \otimes \mathcal{K}(\mathcal{H})$. For this reason stabilization is a central idea in K-theory.

Let us now return to the graded situation. There are $*$-homomorphisms

$$\mathbb{C} \to \mathcal{K}(\mathcal{H}) \quad \text{and} \quad \mathcal{K}(\mathcal{H}) \widehat{\otimes} \mathcal{K}(\mathcal{H}) \to \mathcal{K}(\mathcal{H})$$

defined by mapping $\lambda \in \mathbb{C}$ to λe, where e is the projection onto a one-dimensional, grading-degree zero subspace of $\mathcal{H}$, and by identifying $\mathcal{H} \widehat{\otimes} \mathcal{H}$ with $\mathcal{H}$ by a grading-degree zero unitary isomorphism. These play a role similar to the maps η and Δ introduced in the previous section. There is no canonical choice of the projection e or the isomorphism $\mathcal{H} \widehat{\otimes} \mathcal{H} \cong \mathcal{H}$, and for this reason we cannot 'stabilize' the category of C^*-algebras in quite the way we amplified it in the previous section. But at the level of homotopy the situation is better:

Lemma 1.3. *Let $\mathcal{H}$ and $\mathcal{H}'$ be graded Hilbert spaces. Any two grading-preserving isometries from $\mathcal{H}$ into $\mathcal{H}'$ induce graded $*$-homomorphisms from $\mathcal{K}(\mathcal{H})$ to $\mathcal{K}(\mathcal{H}')$ which are homotopic through graded $*$-homomorphisms.* $\square$

As as result there are canonical, up to homotopy, maps $\mathbb{C} \to \mathcal{K}(\mathcal{H})$ and $\mathcal{K}(\mathcal{H}) \widehat{\otimes} \mathcal{K}(\mathcal{H}) \to \mathcal{K}(\mathcal{H})$. We could therefore create a stabilized homotopy category, in which the morphisms from A to B are the homotopy classes of graded $*$-homomorphisms from A to $B \widehat{\otimes} \mathcal{K}(\mathcal{H})$. We could even stabilize and amplify simultaneously, and create the category in which the morphisms between C^*-algebras A and B are the homotopy classes of graded $*$-homomorphisms from SA to $B \widehat{\otimes} \mathcal{K}(\mathcal{H})$. We won't exactly do this, but the reader will notice echoes of this construction in the following sections.

1.5 A Spectral Picture of K-Theory

We are going provide a 'spectral' description of K-theory which is well adapted to Fredholm index theory and to an eventual bivariant generalization. Actually our definition is a back formation from the bivariant theory described in [13, 14, 27] (it is also closely related to various other approaches to K-theory).

For the rest of this section we shall fix a graded Hilbert space $\mathcal{H}$ whose even and odd grading-degree parts are both countably infinite-dimensional. Unless explicitly noted otherwise we shall be working with graded C^*-algebras and grading-preserving $*$-homomorphisms between them.

Definition 1.13. *We shall denote by $[A, B]$ the set of homotopy classes of grading-preserving $*$-homomorphisms between the graded C^*-algebras A and B.*

With this notation in hand, our description of K-theory is quite simple:

Definition 1.14. *If A is a graded C^*-algebra then we define*

$$K(A) = [\mathcal{S}, A\widehat{\otimes}\mathcal{K}(\mathcal{H})].$$

For the moment $K(A)$ is just a set, although we will soon give it the structure of an abelian group. But first let us give two examples of classes in $K(A)$ to help justify the definition.

Example 1.6. Take $A = \mathbb{C}$. Let D be an unbounded self-adjoint operator on the graded Hilbert space $\mathcal{H}$ of the form

$$D = \begin{pmatrix} 0 & D_- \\ D_+ & 0 \end{pmatrix}$$

(in other words D is a grading-degree one operator) and assume that D has compact resolvent. (For example, D might be a Dirac-type operator on a compact manifold.) The functional calculus

$$\psi_D \colon f \mapsto f(D)$$

defines a graded $*$-homomorphism $\psi_D \colon \mathcal{S} \to \mathcal{K}(\mathcal{H})$ and hence a class in $K(\mathbb{C})$.

Example 1.7. Suppose that A is unital and trivially graded, so that the K-theory group $K_0(A)$ of Section 1.1 can be described in terms of equivalence classes of projections in $A\widehat{\otimes}\mathcal{K}(\mathcal{H})$. If p_0, p_1 are two such projections, acting on the even and odd parts of the graded Hilbert space $\mathcal{H} = \mathcal{H}_0 \oplus \mathcal{H}_1$, then the formula

$$\psi_p \colon f \longmapsto \begin{pmatrix} f(0)p_0 & 0 \\ 0 & f(0)p_1 \end{pmatrix}$$

defines a grading preserving $*$-homomorphism from $\mathcal{S}$ to $A\widehat{\otimes}\mathcal{K}(\mathcal{H})$.

The second example is related to the first as follows: if D is a self-adjoint, grading-degree one, compact resolvent operator on $\mathcal{H}$ then the family

$$\psi_s \colon f \mapsto f(s^{-1}D), \qquad s \in [0,1]$$

is a homotopy from the $*$-homomorphism ψ_D at $s = 1$ to the $*$-homomorphism ψ_p at $s = 0$, where $p = p_0 \oplus p_1$ is the projection onto the kernel of D.

Before reading any further the reader may enjoy solving the following problem.

Exercise 1.9 *Prove that $K(\mathbb{C}) \cong \mathbb{Z}$ in such a way that to the class of the $*$-homomorphism ψ_D of Example 1.6 is associated the Fredholm index of D_+.*

Let us turn now to the operation of addition on $K(A)$. This is given by the direct sum operation which associates to a pair of $*$-homomorphisms ψ_1 and ψ_2 the $*$-homomorphism

$$\psi_1 \oplus \psi_2 \colon \mathcal{S} \to A\widehat{\otimes}\mathcal{K}(\mathcal{H} \oplus \mathcal{H}).$$

(One identifies $\mathcal{H} \oplus \mathcal{H}$ with $\mathcal{H}$ by some degree zero unitary isomorphism to complete the definition; at the level of homotopy any two such identifications are equivalent.)

The zero element is the class of the zero homomorphism. To prove the existence of additive inverses it is convenient to make the following preliminary observation which will be important for other purposes as well. The proof is a simple exercise with the functional calculus.

Lemma 1.4. *Let D be any graded C^*-algebra and let $\psi: S \to D$ be a grading-preserving $*$-homomorphism. Adjoin units to S and D, extend ψ, and form the unitary element*

$$U_\psi = \psi\left(\frac{x-i}{x+i}\right)$$

in the unitalization of D. The correspondence $\varphi \leftrightarrow U_\psi$ is a bijection between the set of $*$-homomorphisms $\psi: S \to D$ and the set of unitary elements U in the unitalization of D which are equal to 1 modulo D and which are mapped to their adjoints by the grading automorphism: $\alpha(U) = U^*$. $\square$

Definition 1.15. *If D is a graded C^*-algebra then by a* Cayley *transform for D we shall mean a unitary in the unitalization of D which is equal to the identity, modulo D, and which is switched to its adjoint by the grading automorphism.*

Returning to the question of additive inverses in $K(A)$, if U is the Cayley transform of ψ then it is tempting to say that the additive inverse to ψ should be represented by the Cayley transform U^*. But this is not quite right; we must also view U^* as a Cayley transform for $A\widehat{\otimes}\mathcal{K}(\mathcal{H}^{\mathrm{opp}})$, where $\mathcal{H}^{\mathrm{opp}}$ is the Hilbert space $\mathcal{H}$ but with the grading reversed. The rotation homotopy

$$\begin{pmatrix} \cos(t)U & \sin(t)I \\ -\sin(t)I & \cos(t)U^* \end{pmatrix}$$

is then a path of Cayley transforms for $A\widehat{\otimes}\mathcal{K}(\mathcal{H} \oplus \mathcal{H}^{\mathrm{opp}})$ connecting $\begin{pmatrix} U & 0 \\ 0 & U^* \end{pmatrix}$ to $\begin{pmatrix} 0 & I \\ -I & 0 \end{pmatrix}$, which is in turn connected to the identity.

Remark 1.7. In terms of $*$-homomorphisms rather than Cayley transforms, the additive inverse of ψ is represented by the $*$-homomorphism

$$\psi^{\mathrm{opp}} = \psi \circ \alpha: S \to A\widehat{\otimes}\mathcal{K}(\mathcal{H}^{\mathrm{opp}})$$

obtained by composing ψ with the grading automorphism on S and also reversing the grading on the Hilbert space $\mathcal{H}$.

Remark 1.8. In the next lecture we shall give an account of additive inverses using the comultiplication map Δ we introduced in the previous section.

Proposition 1.10 *On the category of trivially graded and unital C^*-algebras the functor $K(A)$ defined in this section is naturally isomorphic to the K-theory functor $K_0(A)$ introduced at the beginning of this lecture.*

Proof. We have already seen that $K(A)$ is the group of path components of the space of Cayley transforms for $A \otimes \mathcal{K}(\mathcal{H})$ (we can dispense with the graded tensor product here since A is trivially graded). If $\varepsilon = \left(\begin{smallmatrix} I & 0 \\ 0 & -I \end{smallmatrix}\right)$ is the grading operator and if U is a Cayley transform then εU is a self-adjoint unitary whose $+1$ spectral projection,

$$P = \tfrac{1}{2}(\varepsilon U + 1),$$

is equal to the $+1$ spectral projection $P_\varepsilon = \left(\begin{smallmatrix} I & 0 \\ 0 & 0 \end{smallmatrix}\right)$ of ε, modulo $A \otimes \mathcal{K}(\mathcal{H})$. Conversely if P is a projection which is equal to P_ε modulo $A \otimes \mathcal{K}(\mathcal{H})$ then the formula

$$U = \varepsilon(2P - I)$$

defines a Cayley transform for $A \otimes \mathcal{K}(\mathcal{H})$. We therefore have a new description of the new $K(A)$, as the group of path components of the projections which are equal to P_ε, modulo $A \otimes \mathcal{K}(\mathcal{H})$. We leave it to the reader to determine that the formula $[P] \mapsto [P] - [P_\varepsilon]$ is an isomorphism from this new component space to the usual $K_0(A)$ (the argument involves the familiar stability property of K-theory).

Exercise 1.11 *Denote by $\mathcal{C}_1$ the C^*-algebra $\mathbb{C} \oplus \mathbb{C}$ with grading operator $\lambda_1 \oplus \lambda_2 \mapsto \lambda_2 \oplus \lambda_1$ (this is an example of a Clifford algebra — see Section 1.11). Show that if A is trivially graded and unital then $K(A \widehat{\otimes} \mathcal{C}_1) \cong K_1(A)$.*

Exercise 1.4. Show that if a graded C^*-algebra B is the closure of the union of a direct system of graded C^*-subalgebras B_α then the natural map

$$\varinjlim K(B_\alpha) \longrightarrow K(B)$$

is an isomorphism. (*Hint:* Show that every Cayley transform for $B \widehat{\otimes} \mathcal{K}(\mathcal{H})$ is a limit of Cayley transforms for the subalgebras $B_\alpha \widehat{\otimes} \mathcal{K}(\mathcal{H})$.)

1.6 Long Exact Sequences

Although it is not absolutely necessary we shall invoke some ideas of elementary homotopy theory to construct the K-theory long exact sequences. For this purpose let us introduce the following space:

Definition 1.16. *Let A be a graded C^*-algebra. Denote by $\mathbb{K}(A)$ the space of all graded $*$-homomorphisms from S into $A \widehat{\otimes} K(\mathcal{H})$, equipped with the topology of pointwise convergence (so that $\psi_\alpha \to \psi$ iff $\psi_\alpha(f) \to \psi(f)$ in the norm topology, for every $f \in S$). Thus:*
$$\mathbb{K}(A) = \mathrm{Map}(S, A \widehat{\otimes} K(\mathcal{H})).$$

Remark 1.9. As it happens, the space $\mathbb{K}(A)$ is a spectrum in the sense of homotopy theory—see for example [1]—but we shall not need the homotopy-theoretic notion of spectrum in these lectures.

The space $\mathbb{K}(A)$ has a natural base-point, namely the zero homomorphism from $\mathcal{S}$ into $A \widehat{\otimes} \mathcal{K}(\mathcal{H})$. It also has a more or less natural 'direct sum' operation

$$\mathbb{K}(A) \times \mathbb{K}(A) \to \mathbb{K}(A)$$

which associates to a pair of $*$-homomorphisms ψ_1 and ψ_2 the $*$-homomorphism $\varphi_1 \oplus \varphi_2$ into $A \widehat{\otimes} \mathcal{K}(\mathcal{H} \oplus \mathcal{H})$. (One identifies $\mathcal{H} \oplus \mathcal{H}$ with $\mathcal{H}$ by some degree zero unitary isomorphism to complete the definition; at the level of homotopy any two such identifications are equivalent.) It is of course this operation which gives the addition operation on the groups $K(A) = \pi_0(\mathbb{K}(A))$. By a general principle in homotopy theory the direct sum operation agrees with the group operations on the higher homotopy groups $\pi_n(\mathbb{K}(A))$, for $n \geq 1$.

As for the higher groups $\pi_n(\mathbb{K}(A))$, they may be identified as follows. There is an obvious homeomorphism of spaces

$$\mathbb{K}(C_0(\mathbb{R}^n) \otimes A) \cong \Omega^n \mathbb{K}(A).$$

Indeed by evaluation at points of $\mathbb{R}^n$ we obtain from an element of $\mathbb{K}(C_0(\mathbb{R}^n) \otimes A)$ a map from $\mathbb{R}^n$ to $\mathbb{K}(A)$ which converges to the zero homomorphism at infinity, or in other words a pointed map from the one-point compactification S^n of $\mathbb{R}^n$ into $\mathbb{K}(A)$, which is to say an element of $\Omega^n \mathbb{K}(A)$. It follows that

$$\pi_n(\mathbb{K}(A)) = \pi_0(\Omega^n \mathbb{K}(A)) \cong K(C_0(\mathbb{R}^n) \otimes A).$$

Definition 1.17. *Let A be a graded C^*-algebra. The higher K-theory groups of A are the homotopy groups of the space $\mathbb{K}(A)$:*

$$K_n(A) = \pi_n(\mathbb{K}(A)), \qquad n \geq 0$$

The space $\mathbb{K}(A)$, and therefore also the groups $K_n(A)$, are clearly functorial in A. They are well adapted to the construction of long exact sequences, as the following computation shows:

Lemma 1.5. *If $A \to B$ is a surjective homomorphism of graded C^*-algebras then the induced map from $\mathbb{K}(A)$ to $\mathbb{K}(B)$ is a fibration.*

Recall that a map $X \to Y$ is a (Serre) fibration if for every map from a cube (of any finite dimension) into Y, and for every lifting to X of the restriction of f to a face of the cube, there is an extension to a lifting defined on the whole cube.

Proof. Think of $\mathbb{K}(A)$ as the space of Cayley transforms for $A \widehat{\otimes} \mathcal{K}(\mathcal{H})$, and thus as a space of unitary elements. The proof that the map $\mathbb{K}(A) \to \mathbb{K}(B)$ is a fibration is then only a small modification of the usual proof that the map of unitary groups corresponding to a surjection of C^*-algebras is a fibration.

The fiber of the map $\mathbb{K}(A) \to \mathbb{K}(B)$ (meaning the inverse image of the base-point) is of course $\mathbb{K}(J)$ where the ideal J is the kernel of the surjection. So elementary homotopy theory now provides us with long exact sequences

$$\cdots \longrightarrow K_{n+1}(A) \longrightarrow K_{n+1}(B) \longrightarrow K_n(J) \longrightarrow K_n(B) \longrightarrow \cdots$$

(ending at $K(B)$) as well as Mayer-Vietoris sequences

$$\cdots \longrightarrow K_{n+1}(B) \longrightarrow K_n(A) \longrightarrow K_n(A_1) \oplus K_n(A_2) \longrightarrow K_n(B) \longrightarrow \cdots$$

associated to pullback squares of the sort we considered in the first part of this lecture.

1.7 Products

A key feature of our spectral picture of K-theory is that it is very well adapted to *products*. Recall that in the realm of ungraded C^*-algebras there is a product operation

$$K_0(A) \otimes K_0(B) \to K_0(A \otimes B)$$

defined for unital C^*-algebras by the prescription $[p] \otimes [q] = [p \otimes q]$. This is the first in a sequence of more and more complicated, and more and more powerful, product operations, which culminates with the famous Kasparov product in bivariant K-theory.

In our spectral picture the product is defined using the 'comultiplication' map Δ that we introduced during our discussion of graded C^*-algebras. Using Δ we obtain a map of spaces

$$\mathbb{K}(A) \times \mathbb{K}(B) \to \mathbb{K}(A \widehat{\otimes} B)$$

by associating to a pair (ψ_A, ψ_B) the composition

$$\mathcal{S} \xrightarrow{\Delta} \mathcal{S} \widehat{\otimes} \mathcal{S} \xrightarrow{\psi_A \widehat{\otimes} \psi_B} (A \widehat{\otimes} \mathcal{K}(\mathcal{H})) \widehat{\otimes} (B \widehat{\otimes} \mathcal{K}(\mathcal{H})) \cong A \widehat{\otimes} B \widehat{\otimes} \mathcal{K}(\mathcal{H})$$

(in the last step we employ a transposition isomorphism and we also pick an isomorphism $\mathcal{H} \widehat{\otimes} \mathcal{H} \cong \mathcal{H}$). Taking homotopy groups we obtain pairings

$$K_i(A) \otimes K_j(B) \to K_{i+j}(A \widehat{\otimes} B),$$

as required.

Example 1.8. Suppose that $A = B = \mathbb{C}$ and that ψ_1 and ψ_2 are the functional calculus homomorphisms associated to self-adjoint operators D_1 and D_2, as in Example 1.6. Then the product of ψ_1 and ψ_2 is the functional calculus homomorphism for the self-adjoint operator[5] $D_1 \widehat{\otimes} I + I \widehat{\otimes} D_2$. This type of formula is familiar from index theory; in fact it is the standard construction of an operator whose Fredholm index is the *product* of the indices of D_1 and D_2. It is this example which dictates our use of the comultiplication Δ.

The various features of the product are summarized in the following two results.

---

[5] To be accurate, the formula defines an essentially self-adjoint operator defined on the algebraic tensor product of the domains of D_1 and D_2.

Proposition 1.1. *The K-theory product has the following properties:*

(a) *It is associative.*

(b) *It is commutative, in the sense that if $x \in K(A)$ and $y \in K(B)$, and if $\tau: A \widehat{\otimes} B \to B \widehat{\otimes} A$ is the transposition isomorphism, then $\tau_*(x \times y) = y \times x$.*

(c) *It is functorial, in the sense that if $\varphi: A \to A'$ and $\psi: B \to B'$ are graded $*$-homomorphisms then $(\varphi \widehat{\otimes} \psi)_*(x \times y) = \varphi_*(x) \times \psi_*(y)$.* $\square$

Remark 1.10. In item (b), if we take $x \in K_i(A)$ and $y \in K_j(B)$ then the appropriate formula is $\tau_*(x \times y) = (-1)^{ij} y \times x$.

Proposition 1.2. *Denote by $1 \in K(\mathbb{C})$ the class of the homomorphism which maps the element $f \in S$ to the element $f(0)P \in \mathcal{K}(\mathcal{H})$, where P is the orthogonal projection onto a one-dimensional, grading-degree zero subspace of $\mathcal{H}$. If A is any graded C^*-algebra and if $x \in K(B)$ then under the isomorphism $\mathbb{C} \widehat{\otimes} B \cong B$ the class $1 \times x$ corresponds to x.* $\square$

1.8 Asymptotic Morphisms

We are now going to introduce a concept which can be used as a tool to compute K-theory for C^*-algebras. Other tools are available (for example Kasparov's theory or the theory of C^*-algebra extensions) but we shall work almost exclusively with asymptotic morphisms in these lectures.

Definition 1.18. *Let A and B be graded C^*-algebras. An asymptotic morphism from A to B is a family of functions $\varphi_t : A \to B$, $t \in [1, \infty)$ satisfying the continuity condition that for all $a \in A$*

$$t \longmapsto \varphi_t(a) : [1, \infty) \to B \text{ is bounded and continuous}$$

and the asymptotic conditions that for all $a, a_1, a_2 \in A$ and $\lambda \in \mathbb{C}$

$$\left. \begin{array}{l} \varphi_t(a_1 a_2) - \varphi_t(a_1)\varphi_t(a_2) \\ \varphi_t(a_1 + a_2) - \varphi_t(a_1) - \varphi_t(a_2) \\ \varphi_t(\lambda a) - \lambda \varphi_t(a) \\ \varphi_t(a^*) - \varphi_t(a)^* \end{array} \right\} \to 0, \quad as\ t \to \infty.$$

If A and B are graded we shall require that in addition

$$\alpha(\varphi_t(a)) - \varphi_t(\alpha(a)) \to 0 \quad as\ t \to \infty,$$

where α denotes the grading automorphism. We shall denote an asymptotic morphism with a dashed arrow, thus: $\varphi : A \dashrightarrow B$.

In short, an asymptotic morphism is a one-parameter family of maps from A to B which are asymptotically $*$-homomorphisms.

We shall postpone for a little while the presentation of nontrivial examples of asymptotic morphisms (the main ones are given in Sections 1.12 and 2.6). As for trivial examples, observe that each $*$-homomorphism from A to B can be viewed as a (constant) asymptotic morphism from A to B.

It is usually convenient to work with equivalence classes of asymptotic morphisms, as follows:

Definition 1.19. *Two asymptotic morphisms* $\varphi^1, \varphi^2 : A \dashrightarrow B$ *are (asymptotically) equivalent if for all* $a \in A$

$$\lim_{t \to \infty} \left\| \varphi_t^1(a) - \varphi_t^2(a) \right\| = 0.$$

Up to equivalence, an asymptotic morphism $\varphi \colon A \dashrightarrow B$ is exactly the same thing as a $*$-homomorphism from A into the following *asymptotic algebra* associated to B.

Definition 1.20. *Let* B *be a graded* C^*-algebra. *Denote by* $\mathrm{i}(B)$ *the* C^*-algebra of *bounded, continuous functions from* $[1, \infty)$ *into* B, *and denote by* $\mathrm{i}_0(B)$ *the ideal comprised of functions which vanish at infinity. The asymptotic* C^*-algebra of B is *the quotient* C^*-algebra

$$\mathfrak{A}(B) = \mathrm{i}(B)/\mathrm{i}_0(B).$$

If $\varphi \colon A \to \mathfrak{A}(B)$ is a $*$-homomorphism then by composing φ with a set-theoretic section of the quotient mapping from $\mathrm{i}(B)$ to $\mathfrak{A}(B)$ we obtain an asymptotic morphism from A to B; its equivalence class is independent of the choice of section. Conversely an asymptotic morphism can be viewed as a function from A into $\mathrm{i}(B)$, and by composing with the quotient map into $\mathfrak{A}(B)$ we obtain a $*$-homomorphism from A to $\mathfrak{A}(B)$ which depends only on the asymptotic equivalence class of the asymptotic morphism.

Suppose now that we are given an asymptotic morphism

$$\varphi \colon A \widehat{\otimes} \mathcal{K}(\mathcal{H}) \dashrightarrow B \widehat{\otimes} \mathcal{K}(\mathcal{H}).$$

If $\psi \colon \mathcal{S} \to A \widehat{\otimes} \mathcal{K}(\mathcal{H})$ is a graded $*$-homomorphism then the composition

$$\mathcal{S} \xrightarrow{\ \psi\ } A \widehat{\otimes} \mathcal{K}(\mathcal{H}) - \overset{\varphi}{\ } \!\! \twoheadrightarrow B \widehat{\otimes} \mathcal{K}(\mathcal{H}) \tag{6}$$

is an asymptotic morphism from $\mathcal{S}$ into $B \widehat{\otimes} \mathcal{K}(\mathcal{H})$.

Lemma 1.6. *Every asymptotic morphism from* $\mathcal{S}$ *into a graded* C^*-algebra D is *asymptotic to a family of graded* $*$-homomorphisms from $\mathcal{S}$ to D.

Proof. We saw previously that a $*$-homomorphism from $\mathcal{S}$ to D is the same thing as Cayley transform for D — a unitary in the unitalization of D (equal to 1 modulo D) which is switched to its adjoint by the grading automorphism. In the same way, by making use of the asymptotic algbra $\mathfrak{A}(D)$ we see that an asymptotic morphism from $\mathcal{S}$ to D is the same thing, up to equivalence, as a norm continuous family of elements X_t in the unitalization, equal to 1 modulo D, which are asymptotically unitary and asymptotically switched to their adjoints by the grading automorphism. But such an 'asymptotic Cayley transform' family can, for large t, be altered to produce a family of *actual* Cayley transforms: first replace X_t by

$$Y_t = \frac{1}{2}(X_t + \alpha(X_t^*))$$

(this ensures that the grading automorphism switches the element and its adjoint) and then unitarize by forming

$$U_t = Y_t(Y_t^*Y_t)^{-\frac{1}{2}}$$

(note that Y_t is invertible for large t). Since X_t and U_t are asymptotic we have shown that every asymptotic morphism from $\mathcal{S}$ into a C^*-algebra is asymptotic to a family of $*$-homomorphisms (corresponding to U_t), as required.

Definition 1.21. *Two asymptotic morphisms φ^0 and φ^1 from A to B are* homotopic *if there is an asymptotic morphism φ from A to $B[0,1]$ from which φ^0 and φ^1 can be recovered by evaluation at $0, 1 \in [0, 1]$. Homotopy is an equivalence relation and we shall use the notation*

$$[\![A, B]\!] = \{ \text{ homotopy classes of asymptotic morphisms from } A \text{ to } B \} .$$

There is a natural map from $[A, B]$ into $[\![A, B]\!]$ since each $*$-homomorphism can be regarded as a constant asymptotic morphism. It follows easily from the previous lemma that:

Proposition 1.3. *If D is any graded C^*-algebra then the natural map*

$$[\mathcal{S}, D] \longrightarrow [\![\mathcal{S}, D]\!]$$

is an isomorphism.   □

Returning to the composition (6), it gives rise to the following diagram:

$$[\![\mathcal{S}, A\widehat{\otimes}\mathcal{K}(\mathcal{H})]\!] \xrightarrow{\text{composition with } \varphi} [\![\mathcal{S}, B\widehat{\otimes}\mathcal{K}(\mathcal{H})]\!]$$

$$\Big\uparrow \cong$$

$$[\mathcal{S}, B\widehat{\otimes}\mathcal{K}(\mathcal{H})]$$

We arrive at the following conclusion: composition with $\varphi \colon A\widehat{\otimes}\mathcal{K}(\mathcal{H}) \dashrightarrow B\widehat{\otimes}\mathcal{K}(\mathcal{H})$ induces a homomorphism $\varphi_* \colon K(A) \to K(B)$.

1.9 Asymptotic Morphisms and Tensor Products

The construction of maps $\varphi_* \colon K(A) \to K(B)$ from asymptotic morphisms has several elaborations which are quite important. They rely on the following observation:

Lemma 1.7. *Let D be a C^*-algebra and let $\varphi\colon A \dashrightarrow B$ be an asymptotic morphism between C^*-algebras. There is an asymptotic morphism $\varphi\widehat{\otimes}1\colon A\widehat{\otimes}D \dashrightarrow B\widehat{\otimes}D$ such that, on elementary tensors,*

$$(\varphi\widehat{\otimes}1)_t\colon a\widehat{\otimes}d \mapsto \varphi_t(a)\widehat{\otimes}d.$$

Moreover this formula determines $\varphi\widehat{\otimes}1$ uniquely, up to asymptotic equivalence.

Proof. Assume for simplicity that B and D are unital (the general case, which can be attacked by adjoining units, is left to the reader). There are graded $*$-homomorphisms from A and D into the asymptotic algebra $\mathfrak{A}(B\widehat{\otimes}D)$, determined by the formulas $a \mapsto \varphi_t(a)\widehat{\otimes}1$ and $d \mapsto 1\widehat{\otimes}d$. They graded commute and so determine a homomorphism $\varphi\widehat{\otimes}1\colon A\widehat{\otimes}D \to \mathfrak{A}(B\widehat{\otimes}D)$. This in turn determines an asymptotic morphism $\varphi\widehat{\otimes}1\colon A\widehat{\otimes}D \dashrightarrow B\widehat{\otimes}D$, as required. Two asymptotic morphisms which are asymptotic on the elementary tensors $a\widehat{\otimes}d$ determine $*$-homomorphisms into $\mathfrak{A}(B\widehat{\otimes}D)$ which are equal on elementary tensors, and hence equal everywhere. From this it follows that the two asymptotic morphisms are equivalent.

Remark 1.11. It is clear from the argument that it is crucial here to use the *maximal* tensor product.

Here then are the promised elaborations:

(a) An asymptotic morphism $\varphi\colon A \dashrightarrow B$ determines an asymptotic morphism from $A\widehat{\otimes}\mathcal{K}(\mathcal{H})$ to $B\widehat{\otimes}\mathcal{K}(\mathcal{H})$ by tensor product, and hence a K-theory map $\varphi_*\colon K(A) \to K(B)$.

(b) An asymptotic morphism $\varphi\colon A \dashrightarrow B\widehat{\otimes}\mathcal{K}(\mathcal{H})$ determines an asymptotic morphism from $A\widehat{\otimes}\mathcal{K}(\mathcal{H})$ to $B\widehat{\otimes}\mathcal{K}(\mathcal{H})\widehat{\otimes}\mathcal{K}(H)$ by tensor product. After identifying $\mathcal{K}(\mathcal{H})\widehat{\otimes}\mathcal{K}(\mathcal{H})$ with $\mathcal{K}(\mathcal{H})$ we can apply the construction of the previous section to obtain a map $\varphi_*\colon K(A) \to K(B)$.

(c) An asymptotic morphism $\varphi\colon S\widehat{\otimes}A \dashrightarrow B$ determines an asymptotic morphism from $S\widehat{\otimes}A\widehat{\otimes}\mathcal{K}(\mathcal{H})$ to $B\widehat{\otimes}\mathcal{K}(\mathcal{H})\widehat{\otimes}\mathcal{K}(H)$ by tensor product. If $\psi\colon S \to A\widehat{\otimes}\mathcal{K}(\mathcal{H})$ represents a class in $K(A)$ then by forming the composition

$$S \xrightarrow{\;\Delta\;} S\widehat{\otimes}S \xrightarrow{\;1\widehat{\otimes}\psi\;} S\widehat{\otimes}A\widehat{\otimes}\mathcal{K}(\mathcal{H}) \xdashrightarrow{\;\varphi\widehat{\otimes}1\;} B\widehat{\otimes}\mathcal{K}(\mathcal{H})$$

we obtain a class in $K(B)$, and we obtain a K-theory map $\varphi_*\colon K(A) \to K(B)$.

(d) Combining (b) and (c), an asymptotic morphism $\varphi\colon S\widehat{\otimes}A \to B\widehat{\otimes}\mathcal{K}(\mathcal{H})$ determines a K-theory map $\varphi_*\colon K(A) \to K(B)$.

1.10 Bott Periodicity in the Spectral Picture

We are going to formulate and prove the Bott periodicity theorem using the spectral picture of K-theory, products, and a line of argument which is due to Atiyah [5]. In the course of doing so we shall introduce many of the ideas which will feature in our later discussion of the Baum-Connes conjecture.

In this section present an abstract outline of the argument; in the next three sections we shall fill in the details using the theory of Clifford algebras to construct suitable K-theory classes and asymptotic morphisms.

Definition 1.22. *Let us say that a graded C^*-algebra B has the* rotation property *if the automorphism $b_1 \widehat{\otimes} b_2 \mapsto (-1)^{\partial b_1 \partial b_2} b_2 \widehat{\otimes} b_1$ which interchanges the two factors in the tensor product $B \widehat{\otimes} B$ is homotopic to a tensor product $*$-homomorphism $1 \otimes \iota \colon B \widehat{\otimes} B \to B \widehat{\otimes} B$.*

Example 1.9. The trivially graded C^*-algebra $B = C_0(\mathbb{R}^{2n})$ has this property (with $\iota = 1$).[6]

Theorem 1.12. *Let B be a graded C^*-algebra with the rotation property. Suppose there exists a class $b \in K(B)$ and an asymptotic morphism*

$$\alpha \colon S \widehat{\otimes} B \to \mathcal{K}(\mathcal{H})$$

with the property that the induced K-theory homomorphism $\alpha_* \colon K(B) \to K(\mathbb{C})$ maps b to 1. Then for every C^*-algebra A the maps

$$\alpha_* \colon K(A \widehat{\otimes} B) \to K(A) \quad and \quad \beta_* \colon K(A) \to K(A \widehat{\otimes} B)$$

induced by α and by multiplication by the K-theory class b are inverse to one another.

Proof. From our definitions it is clear that the diagram

$$
\begin{array}{ccc}
K(C) \otimes K(A\widehat{\otimes}B) & \xrightarrow{\ K\text{-theory product}\ } & K(C\widehat{\otimes}A\widehat{\otimes}B) \\
{\scriptstyle 1\otimes\alpha_*}\big\downarrow & & \big\downarrow{\scriptstyle \alpha_*} \\
K(C) \otimes K(A) & \xrightarrow[\ K\text{-theory product}\ ]{} & K(C\widehat{\otimes}A)
\end{array}
$$

commutes. Let us express this by saying that the maps $\alpha_* \colon K(A \widehat{\otimes} B) \to K(A)$ are *multiplicative*. It follows directly from the multiplicative property that α_* is left-inverse to the map $\beta_* \colon K(A) \to K(A \widehat{\otimes} B)$:

$$\alpha_*(\beta_*(x)) = \alpha_*(x \times b) = x \times \alpha_*(b) = x \times 1 = x.$$

To prove that α_* is also left-inverse to β_* we introduce the isomorphisms

$$\sigma \colon A \widehat{\otimes} B \to B \widehat{\otimes} A$$

and

$$\tau \colon B \widehat{\otimes} A \widehat{\otimes} B \to B \widehat{\otimes} A \widehat{\otimes} B$$

which interchange the first and last factors in the tensor products. Note that

---

[6] So does $B = C_0(\mathbb{R}^{2n+1})$, but Theorem 1.12 does not apply in the odd-dimensional case.

$$\sigma_*(y) \times z = \tau_*(z \times y), \qquad \forall y \in K(A \widehat{\otimes} B), z \in K(B).$$

Since B has the rotation property, τ is homotopic to the tensor product $\iota \widehat{\otimes} 1 \widehat{\otimes} 1$, where ι is as in Definition 1.22. Therefore, setting $z = b$ above, we get

$$\sigma_*(y) \times b = \tau_*(b \times y) = \iota_*(b) \times y.$$

Applying α_* we deduce that

$$\sigma_*(y) = \alpha_*(\sigma_*(y) \times b) = \alpha_*(\iota_*(b) \times y) = \iota_*(b) \times \alpha_*(y)$$

(the first and last inequalities follow from the multiplicative property of α_*). Applying another flip isomorphism we conclude that $y = \alpha_*(y) \times \iota_*(b)$. This shows that multiplication by $\iota_*(b)$ is left-inverse to α_*. Therefore α_*, being both left and right invertible, is invertible. Moreover the left inverse β_* is necessarily a two-sided inverse.

Remark 1.12. It follows that $\iota_*(b) = b$. This fact can be checked in the example presented in the next section.

1.11 Clifford Algebras

We begin by venturing a bit further into the realm of graded C^*-algebras. We are going to introduce the (complex) *Clifford algebras*, which are a familiar presence in K-theory and index theory.

Definition 1.23. *Let V be a finite-dimensional Euclidean vector space (that is, a real vector space equipped with a positive-definite inner product). The complex Clifford algebra of V is the graded complex C^*-algebra generated by a linear copy of V, whose elements are self-adjoint and of grading-degree one, subject to the relations $v^2 = \|v\|^2 \cdot 1$ for every $v \in V$.*

Remark 1.13. The Clifford algebra can be concretely constructed from the complexified tensor algebra $T(V)$ be dividing $T(V)$ the ideal generated by the elements $v \otimes v - \|v\|^2 \cdot 1$.

It follows immediately from the definition that if $e_1, \ldots, e_n$ is an orthonormal basis for V then regarded as members of $\mathrm{Cliff}(V)$ these elements satisfy the relations

$$e_j^2 = 1 \quad \text{and} \quad e_i e_j + e_j e_i = 0 \quad \text{if } i \neq j.$$

The monomials $e_{i_1} \cdots e_{i_p}$, where $1 \leq i_1 < \cdots < i_p \leq n$ span $\mathrm{Cliff}(V)$ as a complex linear space. In fact these monomials constitute a basis for $\mathrm{Cliff}(V)$. The monomial $e_{i_1} \cdots e_{i_p}$ has grading-degree p (mod 2).

Example 1.10. The C^*-algebra $\mathrm{Cliff}(\mathbb{R})$ is isomorphic to $\mathbb{C} \oplus \mathbb{C}$, with e_1 corresponding to $(1, -1)$. The grading automorphism transposes the two copies of $\mathbb{C}$.

Example 1.11. The C^*-algebra $\mathrm{Cliff}(\mathbb{R}^2)$ is isomorphic to $M_2(\mathbb{C})$ in such a way that

$$e_1 = \begin{pmatrix} 0 & 1 \\ 1 & 0 \end{pmatrix} \quad \text{and} \quad e_2 = \begin{pmatrix} 0 & i \\ -i & 0 \end{pmatrix}.$$

The (inner) grading is given by the grading operator $\varepsilon = ie_1 e_2 = \begin{pmatrix} 1 & 0 \\ 0 & -1 \end{pmatrix}$.

Remark 1.14. More generally, each even Clifford algebra $\mathrm{Cliff}(\mathbb{R}^{2k})$ is a matrix algebra $M_{2^k}(\mathbb{C})$, graded by $\varepsilon = i^k e_1 \ldots e_{2k} = \begin{pmatrix} I & 0 \\ 0 & -I \end{pmatrix}$; each odd Clifford algebra $\mathrm{Cliff}(\mathbb{R}^{2k+1})$ is a direct sum $M_{2^k}(\mathbb{C}) \oplus M_{2^k}(\mathbb{C})$, graded by the automorphism which switches the summands.

Definition 1.24. *Let V by a finite-dimensional Euclidean vector space. Denote by $\mathcal{C}(V)$ the graded C^*-algebra of continuous functions, vanishing at infinity, from V into $\mathrm{Cliff}(V)$. (The grading on $\mathcal{C}(V)$ comes from $\mathrm{Cliff}(V)$ alone—thus for example an even function is a function which takes values in the even part of $\mathrm{Cliff}(V)$.)*

Example 1.12. Thus $\mathcal{C}(\mathbb{R}^1)$ is isomorphic to $C_0(\mathbb{R}) \oplus C_0(\mathbb{R})$ (and the grading automorphism switches the summands) while the C^*-algebra $\mathcal{C}(\mathbb{R}^2)$ is isomorphic to $M_2(C_0(\mathbb{R}^2))$, graded by $\begin{pmatrix} I & 0 \\ 0 & -I \end{pmatrix}$.

Suppose now that V and W are finite-dimensional Euclidean vector spaces. Each of V and W is of course a subspace of $V \oplus W$, and there are corresponding inclusions of $\mathrm{Cliff}(V)$ and $\mathrm{Cliff}(W)$ into $\mathrm{Cliff}(V \oplus W)$. They determine a $*$-isomorphism

$$\mathrm{Cliff}(V) \widehat{\otimes} \mathrm{Cliff}(W) \cong \mathrm{Cliff}(V \oplus W)$$

(this can be checked either by computing with the standard linear bases for the Clifford algebras, or by checking that that the tensor product $\mathrm{Cliff}(V) \widehat{\otimes} \mathrm{Cliff}(W)$ has the defining property of the Clifford algebra $\mathrm{Cliff}(V \oplus W)$).

Proposition 1.13 *Let V and W be finite-dimensional Euclidean spaces. The map $f_1 \widehat{\otimes} f_2 \mapsto f$, where $f(v + w) = f_1(v) f_2(w)$ determines an isomorphism of graded C^*-algebras*

$$\mathcal{C}(V \oplus W) \cong \mathcal{C}(V) \widehat{\otimes} \mathcal{C}(W)$$

Proof. This follows easily by combining the isomorphism $\mathrm{Cliff}(V) \widehat{\otimes} \mathrm{Cliff}(W) \cong \mathrm{Cliff}(V \widehat{\otimes} W)$ above with the isomorphism $C_0(V) \otimes C_0(W) \cong C_0(V \oplus W)$.

Proposition 1.4. *Let V be a finite-dimensional Euclidean vector space. The C^*-algebra $\mathcal{C}(V)$ has the rotation property.*

Proof. Let $g: W_1 \to W_2$ be an isometric isomorphism of finite-dimensional Euclidean vector spaces. There is a corresponding $*$-isomorphism $g_*: \mathrm{Cliff}(W_1) \to \mathrm{Cliff}(W_2)$ and also a $*$-isomorphism

$$g_{**}: \mathcal{C}(W_1) \to \mathcal{C}(W_2)$$

defined by $(g_{**}f)(w_2) = g_*(f(g^{-1}w_2))$. Under the isomorphism

$$\mathcal{C}(V)\widehat{\otimes}\mathcal{C}(V) \cong \mathcal{C}(V \oplus V)$$

of Proposition 1.13 the flip isomorphism on the tensor product corresponds to the $*$-automorphism τ_{**} of $\mathcal{C}(V \oplus V)$ associated to the map τ which exchanges the two copies of V in the direct sum $V \oplus V$. But τ is homotopic, through isometric isomorphisms of $V \oplus V$, to the map $(v_1, v_2) \mapsto (v_1, -v_2)$, and so τ_{**} is homotopic to $1\widehat{\otimes}\iota_{**}$, where $\iota: V \to V$ is multiplication by -1.

Of course, as we noted earlier, the algebra $C_0(V)$ has the rotation property too. The virtue of dealing with $\mathcal{C}(V)$ rather than the plainer object $C_0(V)$ is that with Clifford algebras to hand we can present in a very concise fashion the following important element of the group $K(\mathcal{C}(V))$.

Definition 1.25. *Denote by* $C: V \to \mathrm{Cliff}(V)$ *the function* $C(v) = v$ *which includes* V *as a real linear subspace of self-adjoint elements in* $\mathrm{Cliff}(V)$.

This is a continuous function on V into $\mathrm{Cliff}(V)$, but

$$C(v)^2 = \|v\|^2 \cdot 1$$

so C does not vanish at infinity (far from it) and it is therefore not an element of $\mathcal{C}(V)$. However if $f \in \mathcal{S}$ then the function $f(C)$ defined by

$$v \mapsto f(C(v)), \qquad v \in V,$$

where f is applied to the element $C(v) \in \mathrm{Cliff}(V)$ in the sense of the functional calculus, *does* belong to $\mathcal{C}(V)$ and the assignment $\beta: f \mapsto f(C)$ is a $*$-homomorphism from $\mathcal{S}$ to $\mathcal{C}(V)$.

Definition 1.26. *The* Bott element $b \in K(\mathcal{C}(V))$ *is the K-theory class of the $*$-homomorphism* $\beta: \mathcal{S} \to \mathcal{C}(V)$ *defined by* $\beta: f \mapsto f(C)$.

Remark 1.15. The function C is an example of an *unbounded multiplier* of the C^*-algebra $\mathcal{C}(V)$. See the appendix.

Example 1.13. Bearing in mind the isomorphisms of Examples 1.10 and 1.11, we have

$$C(x) = (x, -x), \qquad x \in \mathbb{R}^1$$

and

$$C(z) = \begin{pmatrix} 0 & z \\ \bar{z} & 0 \end{pmatrix}, \qquad z \in \mathbb{C} \cong \mathbb{R}^2.$$

We can now formulate the Bott periodicity theorem.

Theorem 1.14. *For every graded C^*-algebra A and every finite-dimensional Euclidean space V the* Bott map

$$\beta: K(A) \to K(A\widehat{\otimes}\mathcal{C}(V)),$$

defined by $\beta(x) = x \times b$, *is an isomorphism of abelian groups.*

We shall prove the theorem in the next two sections by constructing a suitable asymptotic morphism α and proving that $\alpha_*(b) = 1$.

Remark 1.16. To relate the above theorem to more familiar formulations of Bott periodicity we note, as we did earlier, that if $n = 2k$ is even then the Clifford algebra C_n is isomorphic to $M_{2^k}(C_0(\mathbb{R}^n))$, from which it follows that if A is trivially graded then

$$K(A\hat{\otimes}\mathcal{C}(\mathbb{R}^{2k})) \cong K(A \otimes C_0(\mathbb{R}^{2k})).$$

The 'graded' theorem above therefore implies the more familiar isomorphism

$$K(A \otimes C_0(\mathbb{R}^{2k})) \cong K(A).$$

1.12 The Dirac Operator

We are going to construct an asymptotic morphism as in the following result. (The actual proof of the theorem will be carried out in the next section.)

Theorem 1.15. *There exists an asymptotic morphism*

$$\alpha \colon S\hat{\otimes}\mathcal{C}(V) \dashrightarrow \mathcal{K}(\mathcal{H})$$

for which the induced homomorphism $\alpha \colon K(\mathcal{C}(V)) \to K(\mathbb{C})$ *maps the Bott element* $b \in K(\mathcal{C}(V))$ *to* $1 \in K(\mathbb{C})$.

Definition 1.27. *Let V be a finite-dimensional Euclidean vector space. Let us provide the finite-dimensional linear space underlying the algebra* $\mathrm{Cliff}(V)$ *with the Hilbert space structure for which the monomials* $e_{i_1}\cdots e_{i_p}$ *(associated to an orthonormal basis of V) are orthonormal. The Hilbert space structure so obtained is independent of the choice of $e_1, \ldots, e_n$. Denote by $\mathcal{H}(V)$ the infinite-dimensional complex Hilbert space of square-integrable $\mathrm{Cliff}(V)$ valued functions on V. Thus:*

$$\mathcal{H}(V) = L^2(V, \mathrm{Cliff}(V)).$$

The Hilbert space $\mathcal{H}(V)$ is a graded Hilbert space, with grading inherited from $\mathrm{Cliff}(V)$.

Definition 1.28. *Let V be a finite-dimensional Euclidean vector space and let $e, f \in V$. Define linear operators on the finite-dimensional graded Hilbert space underlying* $\mathrm{Cliff}(V)$ *by the formulas*

$$e(x) = e \cdot x$$
$$\widehat{f}(x) = (-1)^{\partial x} x \cdot f.$$

Observe that the operator $e \colon \mathrm{Cliff}(V) \to \mathrm{Cliff}(V)$ is self-adjoint while the operator $\widehat{f} \colon \mathrm{Cliff}(V) \to \mathrm{Cliff}(V)$ is skew-adjoint.

Exercise 1.5. Let $e_1, \ldots, e_n$ be an orthonormal basis for V. Show that if $i_1 < \cdots < i_p$ then the 'number' operator

$$N = \sum_{i=1}^{n} \hat{e}_i e_i$$

maps the monomial $e_{i_1} \cdots e_{i_p}$ in $\mathrm{Cliff}(V)$ to $(2p - n)e_{i_1} \cdots e_{i_p}$.

Definition 1.29. *Let V be a finite-dimensional Euclidean vector space. Denote by $\mathfrak{s}(V)$ the dense subspace of $\mathcal{H}(V)$ comprised of Schwartz-class $\mathrm{Cliff}(V)$-valued functions:*

$$\mathfrak{s}(V) = \textit{Schwartz-class } \mathrm{Cliff}(V)\textit{-valued functions.}$$

The Dirac operator of V is the unbounded operator D on $\mathcal{H}(V)$, with domain $\mathfrak{s}(V)$, defined by

$$(Df)(v) = \sum_{1}^{n} \hat{e}_i \left(\frac{\partial f}{\partial x_i}(v) \right),$$

where $e_1, \ldots, e_n$ is an orthonormal basis of V and $x_1, \ldots, x_n$ are the corresponding coordinates on V.

Since the individual $\hat{e}_i$ are skew-adjoint and since they commute with the partial derivatives we see that D is formally self-adjoint on $\mathfrak{s}(V)$.

Lemma 1.8. *Let V be a finite-dimensional Euclidean vector space. The Dirac operator on V is essentially self-adjoint. If $f \in \mathcal{S}$, if $h \in \mathcal{C}(V)$ and if M_h is the operator of pointwise multiplication by h on the Hilbert space $\mathcal{H}(V)$, then the product $f(D)M_h$ is a compact operator on $\mathcal{H}(V)$.*

Proof. The operator D is a constant coefficient operator acting on a Schwartz space of vector valued functions on $V \cong \mathbb{R}^n$. It has the form $D = \sum_{i=1}^{n} E_i \frac{\partial}{\partial x_i}$, where the matrices E_i are skew adjoint. Under the Fourier transform (a unitary isomorphism) D corresponds to the multiplication operator $\hat{D} = \sqrt{-1} \sum_{i=1}^{n} E_i \xi_i$, and from this we see that $\hat{D}$, and hence D, is essentially self-adjoint. Moreover from the formula

$$\hat{D}^2 = \left(\sqrt{-1} \sum_{i=1}^{n} E_i \xi_i \right)^2 = \|\xi\|^2,$$

for all $\xi \in \mathbb{R}^n$, it follows that if say $f(x) = e^{-ax^2}$ then $f(\hat{D})$ is pointwise multiplication by $e^{-\|\xi\|^2}$, and therefore the inverse Fourier transform $f(D)$ is convolution by $e^{-\frac{1}{4}\|x\|^2}$ (give or take a constant). It follows that $h \in \mathcal{C}(V)$ is compactly supported then $f(D)M_h$ is a Hilbert-Schmidt operator, and is therefore compact. The lemma follows from this since the set of $f \in \mathcal{S}$ for which $f(D)M_h$ is compact, for all h, is an ideal in $\mathcal{S}$, while the function e^{-x^2} generates $\mathcal{S}$ as an ideal.

We are almost ready to define our asymptotic morphism α.

Definition 1.30. *Let V be a finite-dimensional Euclidean space. If $h \in C(V)$ and if $t \in [1, \infty)$ then denote by $h_t \in C(V)$ the function $h_t(v) = h(t^{-1}v)$.*

Lemma 1.9. *Let V be a finite-dimensional Euclidean space with Dirac operator D. For every $f \in S$ and $h \in C(V)$ we have*

$$\lim_{t \to \infty} \left\| [f(t^{-1}D), M_{h_t}] \right\| = 0,$$

where $M_{h_t} \in B(\mathcal{H}(V))$ is the operator of pointwise multiplication by h_t and $f(t^{-1}D)$ is defined using the functional calculus of unbounded operators.

Remark 1.17. The commutator $[,]$ here is the graded commutator of Definition 1.8.

Proof. By an approximation argument involving the Stone-Weierstrass theorem it suffices to consider the cases where $f(x) = (x \pm i)^{-1}$ and where h is smooth and compactly supported. We compute

$$[(t^{-1}D \pm iI)^{-1}, M_{h_t}] = t^{-1}(t^{-1}D \pm iI)^{-1}[M_{h_t}, D](t^{-1}D \pm iI)^{-1},$$

which has norm bounded by $t^{-1} \| [M_{h_t}, D] \|$. But the commutator of M_{h_t} with D is the operator of pointwise multiplication by (minus) the function

$$v \mapsto t^{-1} \sum_{i=1}^{n} \hat{e}_i \left(\frac{\partial h}{\partial x_i}(t^{-1}v) \right).$$

So its norm is $\mathcal{O}(t^{-1})$, and the proof is complete.

Proposition 1.5. *There is, up to equivalence, a unique asymptotic morphism*

$$\alpha_t : S \widehat{\otimes} C(V) \to \mathcal{K}(\mathcal{H}(V))$$

for which, on elementary tensors,

$$\alpha_t(f \widehat{\otimes} h) = f(t^{-1}D)M_{h_t}.$$

Proof. For $t \in [1, \infty)$ define a linear map $\alpha_t : S \widehat{\odot} C(V) \to B(\mathcal{H}(V))$ by the formula

$$\alpha_t(f \widehat{\otimes} h) = f(t^{-1}D)M_{h_t}.$$

Lemma 1.9 shows that the maps α_t define a homomorphism from $S \widehat{\odot} C(V)$ into $\mathfrak{A}(B(\mathcal{H}(V)))$. By the universal property of the tensor product $\widehat{\otimes}$ this extends to a $*$-homomorphism defined on $S \widehat{\otimes} C(V)$. Now, although neither of the operators $f(t^{-1}D)$ or M_{h_t} are compact it follows from elementary elliptic operator theory that their product is compact. So our $*$-homomorphism actually maps $S \widehat{\otimes} C(V)$ into the subalgebra $\mathfrak{A}(\mathcal{K}(\mathcal{H}(V))) \subseteq \mathfrak{A}(B(\mathcal{H}(V)))$. Therefore we obtain an asymptotic morphism as required.

Remark 1.18. The presence of h_t, instead of the plainer h, in the definition of α is not at this stage very important. The 't' could be removed without any problem. But later on it will turn out to have been convenient to have used h_t.

Exercise 1.6. Show that if J is an ideal in a C^*-algebra A then there is a short exact sequence of asymptotic algebras

$$0 \longrightarrow \mathfrak{A}(J) \longrightarrow \mathfrak{A}(A) \longrightarrow \mathfrak{A}(A/J) \longrightarrow 0.$$

1.13 The Harmonic Oscillator

In this section we shall verify that $\alpha_*(b) = 1$, which will complete the proof of the Bott periodicity theorem. Actually we shall make a more refined computation which will be required later on.

We begin by taking a second look at the basic construction of Section 1.11.

Definition 1.31. *Let V be a finite-dimensional Euclidean vector space. The Clifford operator is the unbounded operator on $\mathcal{H}(V)$, with domain the Schwartz space $\mathfrak{s}(V)$, which is given by the formula*

$$(Cf)(v) = \sum_{i=1}^{n} x_i e_i(f(v)),$$

where x_i are the coordinates on V dual to the orthonormal basis e_i of V (the definition of C is independent of the choice of basis).

The Clifford operator is essentially self-adjoint on the domain $\mathfrak{s}(V)$. So if $f \in \mathcal{S}$ we may form the bounded operator $f(C) \in \mathcal{B}(\mathcal{H}(V))$ by the functional calculus.

Lemma 1.10. *Let V be a finite-dimensional Euclidean vector space and let $\beta \colon \mathcal{S} \to \mathcal{C}(V)$ be the homomorphism of Definition 1.26. If $\mathcal{C}(V)$ is represented on the Hilbert space $\mathcal{H}(V)$ by pointwise multiplication operators then the composition*

$$\mathcal{S} \xrightarrow{\ \beta\ } \mathcal{C}(V) \xrightarrow{\ M\ } \mathcal{B}(\mathcal{H}(V))$$

maps $f \in \mathcal{S}$ to $f(C) \in \mathcal{B}(\mathcal{H}(V))$. □

We shall compute the compostion $\alpha_*(b)$ by analyzing the following operator:

Definition 1.32. *Let V be a finite-dimensional Euclidean vector space. Define an unbounded operator B on $\mathcal{H}(V)$, with domain $\mathfrak{s}(V)$, by the formula*

$$(Bf)(v) = \sum_{1}^{n} x_i e_i(f(v)) + \sum_{1}^{n} \widehat{e}_i \left(\frac{\partial f}{\partial x_i}(v) \right).$$

Thus $B = C + D$, where C is the Clifford operator and D is the Dirac operator.

Example 1.14. Suppose $V = \mathbb{R}$. Then

$$B = \begin{pmatrix} 0 & x - d/dx \\ x + d/dx & 0 \end{pmatrix},$$

if we identify $\mathcal{H}(V)$ with $L^2(\mathbb{R}) \oplus L^2(\mathbb{R})$ in the way suggested by Example 1.10.

Observe that the operator B maps the Schwartz space $\mathfrak{s}(V)$ into itself. So the operator $H = B^2$ is defined on $\mathfrak{s}(V)$.

Proposition 1.16 *Let V be a finite-dimensional euclidean vector space of dimension n, let $B = C + D$ as above. There exists within $\mathfrak{s}(V)$ an orthonormal basis for $\mathcal{H}(V)$ consisting of eigenvectors for B^2 such that*

(a) *the eigenvalues are nonnegative integers, and each eigenvalue occurs with finite multiplicity, and*

(b) *the eigenvalue 0 occurs precisely once and the corresponding eigenfunction is $\exp(-\frac{1}{2}\|v\|^2)$*

Proof. Let us consider the case $V = \mathbb{R}$ first. Here,

$$B^2 = \begin{pmatrix} x^2 - \frac{d^2}{dx^2} - 1 & 0 \\ 0 & x^2 - \frac{d^2}{dx^2} + 1 \end{pmatrix},$$

and so it suffices to prove that within the Schwartz subspace of $L^2(\mathbb{R})$ there is an orthonormal basis of eigenfunctions for the operator

$$H = x^2 - \frac{d^2}{dx^2},$$

for which the eigenvalues are positive integers (with finite multiplicities) and for which the eigenvalue 1 appears with multiplicity one. This is a well-known computation, and is done as follows. Define $K = x + \frac{d}{dx}$ and $L = x - \frac{d}{dx}$, and let $f_1(x) = e^{-\frac{1}{2}x^2}$. Observe that

$$H = KL - I = LK + I$$

and that $Kf_1 = 0$, so that $Hf_1 = f_1$. It follows that $HL = LH + 2L$ and $HL^n = L^n H + 2nL^n$. So if we define $f_{n+1} = L^n f_1$ then $Hf_{n+1} = (2n+1)f_{n+1}$. The functions f_{n+1} are orthogonal (being eigenfunctions of the symmetric operator H with distinct eigenvalues), nonzero, and they span $L^2(\mathbb{R})$ (since, by induction, f_{n+1} is a polynomial of degree n times f_1). So after L^2-normalization we obtain the required basis.

The general case follows from the (purely algebraic) calculation

$$B^2 = C^2 + D^2 + N = \sum_{i=1}^{n} x_i^2 + \sum_{i=1}^{n} -\frac{\partial^2}{\partial x_i^2} + (2p - n) \quad \text{on } \mathcal{H}_p(V),$$

where N is the number operator introduced in Exercise 1.5 and $\mathcal{H}_p(V)$ denotes the subspace of $\mathcal{H}(V)$ comprised of functions $V \to \mathrm{Cliff}(V)$ whose values are combinations of the degree p monomials $e_{i_1} \cdots e_{i_p}$. From this an eigenbasis for B^2 may be found by separation of variables.

We shall use the following consequences of this computation:

Corollary 1.1. *Let V be a finite-dimensional Euclidean vector space. Let $B = B_V$ be the Bott-Dirac operator of V, considered as an unbounded operator on $\mathcal{H}(V)$ with domain $\mathfrak{s}(V)$. Then*

(a) B is essentially self-adjoint

(b) B has compact resolvent.

(c) The kernel of B is one-dimensional and is generated by the function $\exp(-\|v\|^2)$.

Theorem 1.17. Let V be a finite-dimensional Euclidean vector space. The composition

$$S \xrightarrow{\Delta} S\widehat{\otimes}S \xrightarrow{1\widehat{\otimes}\beta} S\widehat{\otimes}\mathcal{C}(V) - \xrightarrow{\alpha} \mathcal{K}(\mathcal{H}(V))$$

is asymptotically equivalent to the asymptotic morphism $\gamma \colon S \dashrightarrow \mathcal{K}(\mathcal{H})$ defined by

$$\gamma_t(f) = f(t^{-1}B) \qquad (t \geq 1).$$

The idea of the proof is to check the equivalence of the asymptotic morphisms $\alpha \circ \beta$ and γ on the generators

$$u(x) = e^{-x^2}, \quad \text{and} \quad v(x) = xe^{-x^2}.$$

of the C^*-algebra S. Since for example

$$\gamma_t(u) = e^{-tH} \quad \text{and} \quad \alpha_t(\beta(u)) = e^{-tD^2}e^{-tC^2}$$

(the latter thanks to Lemma 1.10) we shall need to know that e^{-tH} is asymptotic to $e^{-tD^2}e^{-tC^2}$. For this purpose we invoke Mehler's formula:

Proposition 1.6 (Mehler's Formula). Let V be a finite-dimensional Euclidean space and let C and D be the Clifford and Dirac operators for V. The operators D^2, C^2 and $C^2 + D^2$ are essentially self-adjoint on the Schwartz space $s(V)$, and if $s > 0$ then

$$e^{-s(C^2+D^2)} = e^{-\frac{1}{2}s_1C^2}e^{-s_2D^2}e^{-\frac{1}{2}s_1C^2},$$

where $s_1 = (\cosh(2s) - 1)/\sinh(2s)$ and $s_2 = \sinh(2s)/2$. In addition,

$$e^{-s(C^2+D^2)} = e^{-\frac{1}{2}s_1D^2}e^{-s_2C^2}e^{-\frac{1}{2}s_1D^2},$$

for the same s_1 and s_2. □

See for example [16]. Note that the second identity follows from the first upon taking the Fourier transform on $L^2(\mathbb{R})$, which interchanges the operators D^2 and C^2.

Lemma 1.11. If X is any unbounded self-adjoint operator then there are asymptotic equivalences

$$e^{-\frac{1}{2}\tau_1X^2} \sim e^{-\frac{1}{2}t^{-2}X^2}, \qquad e^{-\tau_2X^2} \sim e^{-t^{-2}X^2}$$

and

$$t^{-1}Xe^{-\frac{1}{2}\tau_1X^2} \sim t^{-1}Xe^{-\frac{1}{2}t^{-2}X^2}, \qquad t^{-1}Xe^{-\tau_2X^2} \sim t^{-1}Xe^{-t^{-2}X^2}$$

where $\tau_1 = (\cosh(2t^{-2}) - 1)/\sinh(2t^{-2})$ and $\tau_2 = \sinh(2t^{-2})/2$.

Remark 1.19. By 'asymptotic equivalence' we mean here that the differences between the left and right hand sides in the above relations all converge to zero, in the operator norm, as t tends to infinity.

Proof (Proof of the Lemma). By the spectral theorem it suffices to consider the same problem with the self-adjoint operator X replaced by a real variable x and the operator norm replaced by the supremum norm on $C_0(\mathbb{R})$. The lemma is then a simple calculus exercise, based on the Taylor series $\tau_1, \tau_2 = t^{-2} + o(t^{-2})$.

Lemma 1.12. *If $f, g \in \mathcal{S} = C_0(\mathbb{R})$ then*

$$\lim_{t \to \infty} \left\| [f(t^{-1}C), g(t^{-1}D)] \right\| = 0.$$

Proof. For any fixed $f \in \mathcal{S}$, the set of $g \in \mathcal{S}$ for which the lemma holds is a C^*-subalgebra of $C_0(\mathbb{R})$. So by the Stone-Weierstrass theorem it suffices to prove the lemma when g is one of the resolvent functions $(x \pm i)^{-1}$. It furthermore suffices to consider the case where f is a smooth and compactly supported function. In this case we have

$$\left\| [f(t^{-1}C), (t^{-1} \pm i)^{-1}] \right\| \leq \left\| [f(t^{-1}C), t^{-1}D] \right\|$$

by the commutator identity for resolvents. But then

$$\left\| [f(t^{-1}C), t^{-1}D] \right\| \leq t^{-2} \cdot \text{constant} \cdot \left\| \text{grad}(f(C)) \right\|.$$

This proves the lemma.

Proof (Proof of Theorem 1.17). Denote by $N \colon \mathcal{H}(V) \to \mathcal{H}(V)$ the 'number operator' which multiplies the degree p component of $\mathcal{H}(V)$ by $2n - p$. We observed in the proof of Proposition 1.16 that

$$B^2 = C^2 + D^2 + N,$$

and let us observe now that the operator N commutes with C^2 and D^2. As a result,

$$e^{t^{-2}B^2} = e^{t^{-2}(C^2+D^2)}e^{t^{-2}N}$$

and therefore, by Mehler's formula,

$$e^{-t^{-2}B^2} = e^{-\frac{1}{2}\tau_1 C^2}e^{-\tau_2 D^2}e^{-\frac{1}{2}\tau_1 C^2}e^{-t^{-2}N},$$

It follows from Lemma 1.11 that

$$e^{-t^{-2}B^2} \sim e^{\frac{1}{2}t^{-2}C^2}e^{-t^{-2}D^2}e^{-\frac{1}{2}t^{-2}C^2}e^{-t^{-2}N},$$

and hence from Lemma 1.12 that

$$e^{-t^{-2}B^2} \sim e^{-t^{-2}C^2}e^{-t^{-2}D^2}$$

(since the operator N is bounded the operators $e^{-t^{-2}N}$ converge in norm to the identity operator). Now the homomorphism $\beta\colon \mathcal{S} \to \mathcal{S}\widehat{\otimes}\mathcal{C}(V)$ maps $u(x) = e^{-x^2}$ to $u\widehat{\otimes}u(C)$, and applying α_t we obtain

$$\alpha_t(\beta(u)) = u(t^{-1}C)u(t^{-1}D) = e^{-t^{-2}C^2}e^{t^{-2}D^2},$$

as we noted earlier. But $\gamma_t(u) = e^{-t^{-2}B^2}$, and so we have shown that $\alpha_t(\beta(u))$ and $\gamma_t(u)$ are asymptotic to one another. A similar computation shows that if $v(x) = xe^{-x^2}$ then $\alpha_t(\beta(v))$ and $\gamma_t(v)$ are asymptotic to one another. Since u and v generate $\mathcal{S}$, this completes the proof.

Corollary 1.2. *The homomorphism* $\alpha_*\colon K(\mathcal{C}(V)) \to K(\mathbb{C})$ *maps the element* $b \in K(\mathcal{C}(V))$ *to the element* $1 \in K(\mathbb{C})$.

Proof. The class $\alpha_*(b)$ is represented by the composition of the $*$-homomorphism β with the asymptotic morphism α. By Theorem 1.17, this composition is asymptotic to the asymptotic morphism

$$\gamma_t(f) = f(t^{-1}B).$$

But each map γ_t is actually a $*$-homomorphism, and so the asymptotic morphism γ is homotopic to the single $*$-homomorphism $f \mapsto f(B)$. Now denote by p the projection onto the kernel of B. The formula

$$f \longmapsto \begin{cases} f(s^{-1}D), & \text{if } s \in (0,1] \\ \begin{pmatrix} f(0)p & 0 \\ 0 & 0 \end{pmatrix}, & \text{if } s = 0, \end{cases}$$

defines a homotopy proving that $\alpha_*(b) = 1$.

Appendix: Unbounded Multipliers

Any C^*-algebra A may be regarded as a right Hilbert module over itself (see the book [45] for an introduction to Hilbert modules). An *unbounded (essentially self-adjoint) multiplier* of A is then an essentially self-adjoint operator on the Hilbert module A, in the sense of the following definition:

Definition 1.33. *(Compare [45, Chapter 9].) Let A be a C^*-algebra and let $\mathcal{E}$ be a Hilbert A-module. An essentially self-adjoint operator on $\mathcal{E}$ is an A-linear map T from a dense A-submodule $\mathcal{E}_T \subseteq \mathcal{E}$ into $\mathcal{E}$ with the following properties:*

(a) $\langle Tv, w \rangle = \langle v, Tw \rangle$, *for all* $v, w \in \mathcal{E}_T$.
(b) *The operator* $I + T^2$ *is densely defined and has dense range.*

If T is essentially self-adjoint then the closure of T (the graph of which is the closure of the graph of T) is self-adjoint and regular, which means that the operators

$(\overline{T} \pm iI)$ are bijections from the domain of $\overline{T}$ to $\mathcal{E}$, and that the inverses $(\overline{T} \pm iI)^{-1}$ are adjoints of one another. See [45, Chapter 9] again.

If T is essentially self-adjoint then there is a functional calculus $*$-homomorphism from $\mathcal{S} = C_0(\mathbb{R})$ into the bounded, adjoinable operators on $\mathcal{E}$. It maps $(x \pm i)^{-1}$ to $(\overline{T} \pm iI)^{-1}$.

In the case where $\mathcal{E} = A$, if the densely defined operators $(T \pm iI)^{-1}$ are given by right multiplication with elements of A, then the functional calculus homomorphism maps $\mathcal{S}$ into A (acting on A as right multiplication operators). If A is graded, if the domain A_T of T is graded, and if T has odd grading-degree (as a map from the graded space A_T into the graded space A) then the functional calculus homomorphism is a graded $*$-homomorphism.

Example 1.15. If $A = \mathcal{S}$ then the operator $X: f(x) \mapsto xf(x)$, defined on say the compactly supported functions, is essentially self-adjoint.

Lemma 1.13. *If X_1 is an essentially self-adjoint multiplier of A_1 and if X_2 is essentially self-adjoint multiplier of A_2, then $X_1 \widehat{\otimes} 1 + 1 \widehat{\otimes} X_2$, with domain $A_{X_1} \widehat{\odot} A_{X_2}$, is an essentially self-adjoint multiplier of $A_1 \widehat{\otimes} A_2$.* □

Example 1.16. Using the lemma we can define $\Delta: \mathcal{S} \to \mathcal{S} \widehat{\otimes} \mathcal{S}$ by $\Delta(f) = f(X \widehat{\otimes} 1 + 1 \widehat{\otimes} X)$.

2 Bivariant K-Theory

We saw in the last section that asymptotic morphisms between C^*-algebras determine maps between K-theory groups. In this lecture we shall organize homotopy classes of asymptotic morphisms into a *bivariant* version of K-theory, whose purpose is to streamline the computation of K-theory groups via asymptotic morphisms. In doing so we shall be following the lead of Kasparov (see [39, 37, 38]), although the theory we obtain, called E-theory [13, 14, 27], will in fact be a minor modification of Kasparov's KK-theory.

2.1 The E-Theory Groups

Definition 2.1. *Let A and B be separable, graded C^*-algebras. We shall denote by $E(A, B)$ the set of homotopy classes of asymptotic morphisms from $\mathcal{S} \widehat{\otimes} A \widehat{\otimes} \mathcal{K}(\mathcal{H})$ to $B \widehat{\otimes} \mathcal{K}(\mathcal{H})$. Thus:*

$$E(A, B) = [\![\mathcal{S} \widehat{\otimes} A \widehat{\otimes} \mathcal{K}(\mathcal{H}), B \widehat{\otimes} \mathcal{K}(\mathcal{H})]\!].$$

Example 2.1. Each $*$-homomorphism φ from A to B, or more generally from $\mathcal{S} \widehat{\otimes} A \widehat{\otimes} \mathcal{K}(\mathcal{H})$ to $B \widehat{\otimes} \mathcal{K}(\mathcal{H})$, determines an element of $E(A, B)$. This element depends only on the homotopy class of φ, and will be denoted $[\varphi] \in E(A, B)$.

The sets $E(A, B)$ come equipped with an operation of addition, given by direct sum of asymptotic morphisms, and the zero asymptotic morphism provides a zero element for this addition.

Lemma 2.1. *The abelian monoids $E(A, B)$ are in fact abelian groups.*

Proof. Let $\varphi\colon S\hat{\otimes}A\hat{\otimes}\mathcal{K}(\mathcal{H}) \dashrightarrow B\hat{\otimes}\mathcal{K}(\mathcal{H})$ be an asymptotic morphism. Define an asymptotic morphism

$$\varphi^{\mathrm{opp}}\colon S\hat{\otimes}A\hat{\otimes}\mathcal{K}(\mathcal{H}) \dashrightarrow B\hat{\otimes}\mathcal{K}(\mathcal{H}^{\mathrm{opp}})$$

by the formula $\varphi_t^{\mathrm{opp}}(x) = \varphi_t(\alpha(x))$, where α is the grading automorphism. We shall show that φ^{opp} defines an additive inverse to φ in $E(A, B)$.

For a fixed scalar $s \geq 0$ the formula

$$\Phi_t^s\colon f\hat{\otimes}x \mapsto f\begin{pmatrix} 0 & s \\ s & 0 \end{pmatrix}\begin{pmatrix} \varphi_t(x) & 0 \\ 0 & \varphi_t^{\mathrm{opp}}(x) \end{pmatrix} \qquad f \in S, \quad x \in S\hat{\otimes}A\hat{\otimes}\mathcal{K}(\mathcal{H})$$

defines an asymptotic morphism Φ^s from $S\hat{\otimes}S\hat{\otimes}A\hat{\otimes}\mathcal{K}(\mathcal{H})$ into $B\hat{\otimes}\mathcal{K}(\mathcal{H} \oplus \mathcal{H}^{\mathrm{opp}})$. By composing Φ^s with the comultiplication $\Delta\colon S \to S\hat{\otimes}S$ we obtain asymptotic morphisms

$$S\hat{\otimes}A\hat{\otimes}\mathcal{K}(\mathcal{H}) \xrightarrow{\Delta\hat{\otimes}1} S\hat{\otimes}S\hat{\otimes}A\hat{\otimes}\mathcal{K}(\mathcal{H}) \xrightarrow{\Phi^s} B\hat{\otimes}\mathcal{K}(\mathcal{H} \oplus \mathcal{H}^{\mathrm{opp}})$$

which constitute a homotopy (parametrized by $s \in [0, \infty]$) connecting $\varphi \oplus \varphi^{\mathrm{opp}}$ to 0.

Remark 2.1. The above argument provides another proof that the K-theory groups described in the last lecture are in fact groups.

If e is a rank-one projection in $\mathcal{K}(\mathcal{H})$ then by composing asymptotic morphisms with the $*$-homomorphism which maps the element $f\hat{\otimes}a \in S\hat{\otimes}A$ to the element $f\hat{\otimes}a\hat{\otimes}e \in S\hat{\otimes}A\hat{\otimes}\mathcal{K}(\mathcal{H})$ we obtain a map (of sets, or in fact abelian groups)

$$[\![S\hat{\otimes}A\hat{\otimes}\mathcal{K}(\mathcal{H}), B\hat{\otimes}\mathcal{K}(\mathcal{H})]\!] \longrightarrow [\![S\hat{\otimes}A, B\hat{\otimes}\mathcal{K}(\mathcal{H})]\!].$$

Lemma 2.2. *The above map is a bijection.*

Proof. The inverse is given by tensor product with the identity on $\mathcal{K}(\mathcal{H})$. Details are left to the reader as an exercise.

The groups $E(A, B)$ are contravariantly functorial in A and covariantly functorial in B on the category of graded C^*-algebras.

Proposition 2.1. *The functor $E(\mathbb{C}, B)$ on the category of graded C^*-algebras is naturally isomorphic to $K(B)$.*

Proof. This follows from Proposition 1.3 and Lemma 2.2.

2.2 Composition of Asymptotic Morphisms

The main feature of E-theory is the existence of a bilinear 'composition law'

$$E(A, B) \otimes E(B, C) \to E(A, C)$$

which is associative in the sense that the two possible iterated pairings

$$E(A, B) \otimes E(B, C) \otimes E(C, D) \to E(A, D)$$

are equal, and which gathers the E-theory groups together into an additive category (the objects are separable graded C^*-algebras, the morphisms from A to B are the elements of the abelian group $E(A, B)$, and the above pairing is the composition law).

The E-theory category plays an important role in the computation of C^*-algebra K-theory groups, as follows. To compute the K-theory of a C^*-algebra A one can, on occasion, find a C^*-algebra B and elements of $E(A, B)$ and $E(B, A)$ whose compositions are the identity morphisms in $E(A, A)$ and $E(B, B)$. Composition with these two elements of $E(A, B)$ and $E(B, A)$ now gives a pair of mutually inverse maps between $E(\mathbb{C}, A)$ and $E(\mathbb{C}, B)$. But as we noted in the last section $E(\mathbb{C}, A)$ and $E(\mathbb{C}, B)$ are the K-theory groups $K(A)$ and $K(B)$. It therefore follows that $K(A) \cong K(B)$. Therefore, assuming that $K(B)$ can be computed, so can $K(A)$. This is the main strategy for computing the K-theory of group C^*-algebras.

In this section and the next we shall lay the groundwork for the construction of the composition pairing. The following sequence of definitions and lemmas presents a reasonably conceptual approach to the problem. The proofs are all very simple, and by and large they are omitted. Details can be found in the monograph [27].

We begin by repeating a definition from the last lecture.

Definition 2.2. *Let B be a graded C^*-algebra. Denote by $\mathrm{i}(B)$ the C^*-algebra of bounded, continuous functions from $[1, \infty)$ into B, and denote by $\mathrm{i}_0(B)$ the ideal comprised of functions which vanish at infinity. The asymptotic C^*-algebra of B is the quotient C^*-algebra*

$$\mathfrak{A}(B) = \mathrm{i}(B)/\mathrm{i}_0(B).$$

Observe (as we did in the last section) that an asymptotic morphism $\varphi \colon A \dashrightarrow B$ defines a $*$-homomorphism $\varphi \colon A \to \mathfrak{A}(B)$ in the obvious manner and that two asymptotic morphism from A to B define the same $*$-homomorphism from A to $\mathfrak{A}(B)$ precisely when they are asymptotically equivalent.

The asymptotic algebra construction $B \mapsto \mathfrak{A}(B)$ is a functor, since a $*$-homomorphism from B to C induces a $*$-homomorphism from $\mathfrak{A}(B)$ to $\mathfrak{A}(C)$ by composition.

Definition 2.3. *The asymptotic functors $\mathfrak{A}^0, \mathfrak{A}^1, \ldots$ are defined by $\mathfrak{A}^0(B) = B$ and*

$$\mathfrak{A}^n(B) = \mathfrak{A}(\mathfrak{A}^{n-1}(B)).$$

Two $*$-homomorphisms $\varphi^0, \varphi^1 \colon A \to \mathfrak{A}^n(B)$ are n-homotopic if there exists an $*$-homomorphism $\Phi \colon A \to \mathfrak{A}^n(B[0,1])$ from which the $*$-homomorphisms φ^0 and φ^1 are recovered as the compositions

$$A \xrightarrow{\hspace{3cm}} \mathfrak{A}^n(B[0,1]) \xrightarrow{\text{evaluate at } 0, \, 1} \mathfrak{A}^n(B).$$

Lemma 2.3. [27, Proposition 2.3] *The relation of n-homotopy is an equivalence relation on the set of $*$-homomorphisms from A to $\mathfrak{A}^n(B)$.* $\square$

Definition 2.4. *Let A and B be graded C^*-algebras. Denote by $[\![A, B]\!]_n$ the set of n-homotopy classes of $*$-homomorphisms from A to $\mathfrak{A}^n(B)$:*

$$[\![A, B]\!]_n = \{ \ n\text{-Homotopy classes of } *\text{-homomorphisms from } A \text{ to } \mathfrak{A}^n(B) \ \}.$$

Example 2.2. Observe that $[\![A, B]\!]_0$ is the set of homotopy classes of $*$ - homomorphisms and $[\![A, B]\!]_1$ is the set of homotopy classes of asymptotic morphisms.

Remark 2.2. The relation of n-homotopy is *not* the same thing as homotopy: homotopic $*$-homomorphisms into $\mathfrak{A}^n(B)$ are n-homotopic, but not *vice-versa*, in general.

There is a natural transformation of functors, from $\mathfrak{A}^n(B)$ to $\mathfrak{A}^{n+1}(B)$, defined by including $\mathfrak{A}^n(B)$ as constant functions in $\mathfrak{A}^{n+1}(B) = \mathfrak{A}(\mathfrak{A}^n(B))$. A second and different natural transformation from $\mathfrak{A}^n(B)$ to $\mathfrak{A}^{n+1}(B)$ may be defined by including B into $\mathfrak{A}(B)$ as constant functions, and then applying the functor $\mathfrak{A}^n$ to this inclusion. Both natural transformations are compatible with homotopy in the sense that they define maps

$$[\![A, B]\!]_n \longrightarrow [\![A, B]\!]_{n+1}.$$

Lemma 2.4. [27, Proposition 2.8] *The above natural transformations define the same map $[\![A, B]\!]_n \longrightarrow [\![A, B]\!]_{n+1}$* $\square$.

With the above maps the sets $[\![A, B]\!]_n$ are organized into a directed system

$$[\![A, B]\!]_1 \to [\![A, B]\!]_2 \to [\![A, B]\!]_3 \to \cdots$$

Definition 2.5. *Let A and B be graded C^*-algebras. Denote by $[\![A, B]\!]_\infty$ the direct limit of the above directed system.*

Proposition 2.2. [27, Proposition 2.12] *Let $\varphi \colon A \to \mathfrak{A}^n(B)$ and $\psi \colon B \to \mathfrak{A}^m(C)$ be $*$-homomorphisms. The class of the composite $*$-homomorphism*

$$A \xrightarrow{\varphi} \mathfrak{A}^n(B) \xrightarrow{\mathfrak{A}^n(\psi)} \mathfrak{A}^{n+m}(C).$$

in the set $[\![A, C]\!]_\infty$ depends only on the classes of φ and ψ in the sets $[\![A, B]\!]_\infty$ and $[\![B, C]\!]_\infty$. The composition law

$$[\![A, B]\!]_\infty \times [\![B, C]\!]_\infty \to [\![A, C]\!]_\infty$$

so defined is associative. $\square$

Exercise 2.1. Show that the identity *-homomorphism from A to A determines an element of $[\![A, A]\!]_\infty$ which serves as an identity morphism for the above composition law.

Thanks to Proposition 2.2 and the exercise we obtain a category:

Definition 2.6. *The asymptotic category is the category whose objects are the graded C^*-algebras, whose are elements of the sets $[\![A, B]\!]_\infty$, and whose composition law is the process described in Proposition 2.2.*

Observe that there is a functor from the category of graded C^*-algebras and *-homomorphisms into the asymptotic category (which is the identity on objects and which assigns to a *-homomorphism $\varphi\colon A \to B$ its class in $[\![A, B]\!]_\infty$).

Exercise 2.2. Show that K-theory, thought of as a functor from graded C^*-algebras to abelian groups, factors through the asymptotic category.

2.3 Operations

We want to define tensor products, amplifications and other operations on the asymptotic category. For this purpose we introduce the following definitions.

Definition 2.7. *Let F be a functor from the category of graded C^*-algebras to itself. If B is a graded C^*-algebra and if $f \in F(B[0,1])$ then define a function $\hat{f}$ from $[0, 1]$ into $F(B)$ by assigning to $t \in [0,1]$ the image of f under the homomorphism $F(\varepsilon_t)\colon F(B[0,1]) \to F(B)$, where ε_t is evaluation at t. The functor F is continuous if for every B and every $f \in F(B[0,1])$ the function $\hat{f}$ is continuous.*

Example 2.3. The tensor product functors $A \mapsto A\widehat{\otimes}B$ (for both the minimal and maximal tensor product) are continuous.

Definition 2.8. *A functor F from the category of graded C^*-algebras to itself is exact if for every short exact sequence*

$$0 \longrightarrow J \longrightarrow A \longrightarrow A/J \longrightarrow 0$$

the induced sequence

$$0 \longrightarrow F(J) \longrightarrow F(A) \longrightarrow F(A/J) \longrightarrow 0$$

is also exact.

Exercise 2.3. The *maximal* tensor product functor $A \mapsto A\widehat{\otimes}_{max}B$ is exact.

Remark 2.3. In contrast the *minimal* tensor product functor $A \mapsto A\widehat{\otimes}_{min}B$ is not exact for every B. See [66] for examples (and also Lecture 6).

If F is a continuous functor then the construction of $\hat{f}$ from f described in Definition 2.7 determines a natural transformation

$$F(B[0,1]) \to F(B)[0,1].$$

The same process also determines natural transformations

$$F(\mathrm{i}(B)) \to \mathrm{i}(F(B) \quad \text{and} \quad F(\mathrm{i}_0(B)) \to \mathrm{i}_0(F(B))$$

(recall that $\mathrm{i}(B)$ is the C^*-algebra of bounded and continuous functions from $[1, \infty)$ into B and $\mathrm{i}_0(B)$ is the ideal of functions vanishing at infinity). So if F is in addition an exact functor then we obtain an induced map from $F(\mathfrak{A}(B))$ into $\mathfrak{A}(F(B))$, as indicated in the following diagram:

$$
\begin{array}{ccccccccc}
0 & \longrightarrow & F(\mathrm{i}_0(B)) & \longrightarrow & F(\mathrm{i}(B)) & \longrightarrow & F(\mathfrak{A}(B)) & \longrightarrow & 0 \\
 & & \downarrow & & \downarrow & & \downarrow & & \\
0 & \longrightarrow & \mathrm{i}_0(F(B)) & \longrightarrow & \mathrm{i}(F(B)) & \longrightarrow & \mathfrak{A}(F(B)) & \longrightarrow & 0.
\end{array}
$$

Proposition 2.3. [27, Theorem 3.5] *Let F be a continuous and exact functor on the category of graded C^*-algebras. The process which assigns to each $*$-homomorphism $\varphi\colon A \to \mathfrak{A}^n(B)$ the composition*

$$F(A) \xrightarrow{F(\varphi)} F(\mathfrak{A}^n(B)) \longrightarrow \mathfrak{A}^n(F(B))$$

defines a functor on the asymptotic category. $\square$

Applying this to the (maximal) tensor product functors we obtain the following result.

Proposition 2.1 [27, Theorem 4.6] *There is a functorial tensor product $\widehat{\otimes}_{max}$ on the asymptotic category.* $\square$

With a tensor product operation in hand we can construct an amplified asymptotic category in the same way we constructed the amplification of the category of C^*-algebras and $*$-homomorphisms in Definition 1.12.

Definition 2.9. *The amplified asymptotic category is the category whose objects are the graded C^*-algebras and for which the morphisms from A to B are the elements of $[\![S \widehat{\otimes} A, B]\!]_\infty$. Composition of morphisms $\varphi\colon A \to B$ and $\psi\colon B \to C$ in the amplified asymptotic category is given by the following composition of morphisms in the asymptotic category:*

$$S\widehat{\otimes}A \xrightarrow{\Delta\widehat{\otimes}1} S\widehat{\otimes}S\widehat{\otimes}A \xrightarrow{1\widehat{\otimes}\varphi} S\widehat{\otimes}B \xrightarrow{\psi} C.$$

2.4 The E-Theory Category

The main technical theorem in E-theory is the following:

Theorem 2.2. [27, Theorem 2.16] *Let A and B be graded C^*-algebras and assume that A is separable. The natural map of $[\![A, B]\!]_1$ into the direct limit $[\![A, B]\!]_\infty$ is a bijection. Thus every morphism from A to B in the asymptotic category is represented by a unique homotopy class of asymptotic morphisms from A to B.*   □

Unlike the results of the previous two sections, this is a little delicate. We refer the reader to [27] for details.

It follows from Theorem 2.2 and Definition 2.1 that the group $E(A, B)$ (for A separable) may be identified with the set of morphisms in the amplified asymptotic category from $A \widehat{\otimes} \mathcal{K}(\mathcal{H})$ to $B \widehat{\otimes} \mathcal{K}(\mathcal{H})$. As a result we obtain a pairing

$$E(A, B) \otimes E(B, C) \longrightarrow E(A, C)$$

from the composition law in the asymptotic category. We have now reached the main objective of the lecture:

Theorem 2.3. *The E-theory groups $E(A, B)$ are the morphism groups in an additive category $\mathbf{E}$ whose objects are the separable graded C^*-algebras. There is a functor from the homotopy category of graded separable C^*-algebras and graded $*$-homomorphisms into $\mathbf{E}$ which is the identity on objects.*   □

Remark 2.4. If $\varphi \colon A \to B$ is a $*$-homomorphism and if $\psi \colon B \dashrightarrow C$ is an asymptotic morphism then φ and ψ determine elements $[\varphi] \in E(A, B)$ and $[\psi] \in E(B, C)$. In addition the (naive) composition $\psi \circ \varphi$ is an asymptotic morphism from A to C, and so defines an element $[\psi \circ \varphi] \in E(A, C)$. We have that $[\psi \circ \varphi] = [\psi] \circ [\varphi]$. The same applies to compositions of $*$-homomorphisms and asymptotic morphisms the other way round, and also to compositions in the amplified category.

The tensor product functor on the asymptotic category extends to the amplified asymptotic category (compare Remark 1.6), and we obtain a tensor product in E-theory:

Theorem 2.4. *There is a functorial tensor product $\widehat{\otimes}_{max}$ on the E-theory category which is compatible with the tensor product on C^*-algebras via the functor from the category of graded separable C^*-algebras and graded $*$-homomorphisms into the E-theory category.*   □

The minimal tensor product does not carry over to E-theory, but we have at least a partial result. First, here is some standard C^*-algebra terminology.

Definition 2.10. *A (graded) C^*-algebra B is* exact *if, for every short exact sequence of graded C^*-algebras*

$$0 \longrightarrow J \longrightarrow A \longrightarrow A/J \longrightarrow 0$$

the sequence of minimal tensor products

$$0 \longrightarrow J\widehat{\otimes}_{min}B \longrightarrow A\widehat{\otimes}_{min}B \longrightarrow A/J\widehat{\otimes}_{min}B \longrightarrow 0$$

is exact.

In other words, B is exact if and only if the functor $A \mapsto A\widehat{\otimes}_{min}B$ is exact.

Theorem 2.5. *Let B be a separable, graded and exact C^*-algebra. There is a functor $A \mapsto A\widehat{\otimes}_{min}B$ on the E-theory category. In particular, if A_1 and A_2 are isomorphic in the E-theory category then $A_1\widehat{\otimes}_{min}B$ and $A_2\widehat{\otimes}_{min}B$ are isomorphic there too.* □

We shall return to the topic of minimal tensor products in Lecture 6.

2.5 Bott Periodicity

Our proof of Bott periodicity in Lecture 1 may be recast as a computation in E-theory, as follows.

Definition 2.11. *Let V be a finite-dimensional Euclidean vector space. Denote by $\beta \in E(\mathbb{C}, \mathcal{C}(V))$ the E-theory class of the $*$-homomorphism $\beta \colon S \to \mathcal{C}(V)$ introduced in Definition 1.26. Denote by $\alpha \in E(\mathcal{C}(V), \mathbb{C})$ the E-theory class of the asymptotic morphism $\alpha \colon S\widehat{\otimes}\mathcal{C}(V) \dashrightarrow \mathcal{K}(\mathcal{H}(V))$ introduced in Proposition 1.5.*

Proposition 2.4. *The composition*

$$\mathbb{C} \xrightarrow{\ \beta\ } \mathcal{C}(V) \xrightarrow{\ \alpha\ } \mathbb{C}$$

in the E-theory category is the identity morphism $\mathbb{C} \to \mathbb{C}$.

Proof. This follows from Remark 2.4 and Theorem 1.17, as in the proof of Corollary 1.2.

A small variation on the rotation argument we discussed in Section 1.10 now proves the following basic result:

Theorem 2.6. *The morphisms $\alpha \colon \mathcal{C}(V) \to \mathbb{C}$ and $\beta \colon \mathbb{C} \to \mathcal{C}(V)$ in the E-theory category are mutual inverses.* □

2.6 Excision

The purpose of this section is to discuss the construction of 6-term exact sequences in E-theory. First, we need a simple definition.

Definition 2.12. *Let A be a C^*-algebra. The suspension of A is the C^*-algebra*

$$\Sigma A = \{\, f \in A[0,1] : f(0) = f(1) = 0 \,\}.$$

In other words ΣA is the tensor product of A with $\Sigma = C_0(0,1)$. If A is graded then so is ΣA (the algebra Σ itself is given the trivial grading).

Theorem 2.7. *The suspension map*

$$E(A, B) \longrightarrow E(\Sigma A, \Sigma B)$$

is an isomorphism. Moreover there are natural isomorphisms

$$E(A, B) \cong E(\Sigma^2 A, B) \quad and \quad E(A, B) \cong E(A, \Sigma^2 B).$$

Proof. It follows from Bott periodicity that Σ^2 is isomorphic to $\mathbb{C}$ in the E-theory category, and this proves the second part of the theorem. With the periodicity isomorphisms available, we obtain an inverse to the suspension map by simply suspending a second time.

Here then are the main theorems in the section:

Theorem 2.8. *Let B be a graded C^*-algebra and let I be an ideal in a separable C^*-algebra A. There is a functorial six-term exact sequence*

$$
\begin{array}{ccccc}
E(A/I, B) & \longrightarrow & E(A, B) & \longrightarrow & E(I, B) \\
\uparrow & & & & \downarrow \\
E(I, \Sigma B) & \longleftarrow & E(A, \Sigma B) & \longleftarrow & E(A/I, \Sigma B)
\end{array}
$$

Theorem 2.9. *Let A be a graded C^*-algebra and let J be an ideal in a separable C^*-algebra B. There is a functorial six-term exact sequence*

$$
\begin{array}{ccccc}
E(A, J) & \longrightarrow & E(A, B) & \longrightarrow & E(A, B/J) \\
\uparrow & & & & \downarrow \\
E(A, \Sigma B/J) & \longleftarrow & E(A, \Sigma B) & \longleftarrow & E(A, \Sigma J)
\end{array}
$$

For simplicity we shall discuss only the second of these two theorems (the proofs of the two theorems are similar, although the second is a little easier in some respects). For a full account of both see [27, Chapters 5 and 6].

The proof of Theorem 2.9 has two parts. The first is a construction borrowed from elementary homotopy theory, involving the following notion:

Definition 2.13. *Let $\pi\colon B \to C$ be a $*$-homomorphism of (graded) C^*-algebras. The mapping cone of π is the C^*-algebra*

$$C_\pi = \{\, b \oplus f \in B \oplus C[0,1] : \pi(b) = f(0) \quad and \quad f(1) = 0 \,\}.$$

Proposition 2.5. *Let $\pi\colon B \to C$ be a $*$-homomorphism. For every C^*-algebra A there is a long exact sequence of pointed sets*

$$\cdots \longrightarrow [A, \Sigma B] \longrightarrow [A, \Sigma C] \longrightarrow [A, C_\pi] \longrightarrow [A, B] \longrightarrow [A, C].$$

The proposition may be formulated for homotopy classes of asymptotic morphisms, as above, or for homotopy classes of ordinary $*$-homomorphisms (compare [59]). The proofs are the same in both cases. There are $*$-homomorphisms

$$\cdots \longrightarrow \Sigma B \longrightarrow \Sigma C \longrightarrow C_\pi \longrightarrow B \longrightarrow C,$$

which supply the maps in the proposition, and since the composition of any two successive $*$-homomorphisms in this sequence is null-homotopic, the composition of any two successive maps of the sequence in the proposition is trivial. Let us prove exactness at the $[\![A, B]\!]$ term. If the composition

$$A - \!\!\stackrel{\varphi}{-} \!\!\succ B \stackrel{\pi}{\longrightarrow} C$$

is null homotopic then a null homotopy gives an asymptotic morphism from $\Phi \colon A \dashrightarrow C[0, 1)$. The pair comprised of φ and Φ now determines an asymptotic morphism from A into C_π, as required. For more details see [27, Chapter 5].

Corollary 2.1. *Let $\pi \colon B \to C$ be a $*$-homomorphism. For every C^*-algebra A there is a functorial six-term exact sequence*

$$
\begin{array}{ccccc}
E(A, C_\pi) & \longrightarrow & E(A, B) & \longrightarrow & E(A, C) \\
\uparrow & & & & \downarrow \\
E(A, \Sigma C) & \longleftarrow & E(A, \Sigma B) & \longleftarrow & E(A, \Sigma C_\pi)
\end{array}
$$

This follows from Proposition 2.5 and Theorem 2.7. To prove Theorem 2.9 it remains to replace C_π with J in the above corollary, in the case where $\pi \colon B \to C$ is a surjection with kernel J. To this end, observe that there is an inclusion $j \mapsto j \oplus 0$ of J into C_π. Using the following construction one can show that this inclusion is an isomorphism in the E-theory category.

Theorem 2.10. ([27, Chapter 5].) *Let J be an ideal in a separable graded C^*-algebra A. There is a norm-continuous family $\{u_t\}_{t \in [1, \infty)}$ of degree-zero elements in J such that*

(a) $0 \leq u_t \leq 1$ *for all t,*
(b) $\lim_{t \to \infty} \| u_t j - j \| = 0$, *for all $j \in J$, and*
(c) $\lim_{t \to \infty} \| u_t a - a u_t \| = 0$, *for all $a \in A$.*

If $s \colon A/J \to A$ is any set-theoretic section of the quotient mapping then the formula

$$\varphi_t(f \otimes x) = f(u_t)s(x)$$

defines an asymptotic morphism from $\Sigma A/J$ into J. $\square$

Theorem 2.11. ([27, Proposition 5.14].) *Let J be an ideal in a separable, graded C^*-algebra A. The asymptotic morphism associated to the extension*

$$0 \longrightarrow \Sigma J \longrightarrow A[0,1) \longrightarrow C_\pi \longrightarrow 0$$

determines an element of $E(\Sigma C_\pi, \Sigma J)$ which is inverse to the element of $E(\Sigma J, \Sigma C_\pi)$ which is determined by the inclusion of J into C_π.   □

In view of Theorem 2.7 it now follows that $J \cong C_\pi$ in the E-theory category, and the proof of Theorem 2.9 is complete.

2.7 Equivariant Theory

We are now going to define an equivariant version of E-theory which will be particularly useful for computing the K-theory of group C^*-algebras. To keep matters as simple as possible we shall work here with countable and *discrete* groups, although it is possible to consider arbitrary second countable, locally compact groups.

The following definition provides the main idea behind the equivariant theory:

Definition 2.14. *Let G be a countable discrete group and let A and B be graded G-C^*-algebras (that is, graded C^*-algebras equipped with actions of G by grading-preserving $*$-automorphisms). An equivariant asymptotic morphism from A to B is an asymptotic morphism $\varphi \colon A \dashrightarrow B$ such that*

$$\varphi_t(g \cdot a) - g \cdot (\varphi_t(a)) \to 0, \quad as\ t \to \infty,$$

for all $a \in A$ and all $g \in G$.

Homotopy is defined just as in the non-equivariant case, and we set

$$[A, B]^G = \{ \text{Homotopy classes of asymptotic morphisms from } A \text{ to } B \}.$$

If B is a G-C^*-algebra then so is the asymptotic algebra $\mathfrak{A}(B)$, and an equivariant asymptotic morphism from A to B is the same thing, up to equivalence, as an equivariant $*$-homomorphism from A to $\mathfrak{A}(B)$.[7] Thanks to this observation it is a straightforward matter to define an equivariant version of the asymptotic category that we constructed in Section 2.2. The higher asymptotic algebras $\mathfrak{A}^n(B)$ are G-C^*-algebras; we define $[A, B]^G_n$ to be the set of n-homotopy classes of equivariant $*$-homomorphism from A to $\mathfrak{A}^n(B)$; and we define

$$[A, B]^G_\infty = \varinjlim [A, B]^G_n.$$

These are the morphism sets of a category, using the composition law described in Proposition 2.2, and this category may be 'amplified', as in Section 1.3. Finally, if A is separable (and assuming, as we shall throughout, that G is countable) then the canonical map gives an isomorphism

$$[A, B]^G \xrightarrow{\cong} [A, B]^G_\infty.$$

---

[7] This is one place where our assumption that G is discrete is helpful: if G is not discrete then the action of G on $\mathfrak{A}(B)$ is not necessarily continuous.

See [27] for details.

To define the equivariant E-theory groups it remains to introduce a stabilization operation which is appropriate to the equivariant context.

Definition 2.15. *Let G be a countable discrete group. The standard G-Hilbert space $\mathcal{H}_G$ is the infinite Hilbert space direct sum*

$$\mathcal{H}_G = \oplus_{n=0}^{\infty} \ell^2(G),$$

equipped with the regular representation of G on each summand and graded so the even numbered summands are even and the odd numbered summands are odd.

The standard G-Hilbert space has the following universal property:

Lemma 2.5. *If $\mathcal{H}$ is any separable graded G-Hilbert space[8] then the tensor product Hilbert space $\mathcal{H} \otimes \mathcal{H}_G$ is unitarily equivalent to $\mathcal{H}_G$ via a grading-preserving, G-equivariant unitary isomorphism of Hilbert spaces.*

Proof. Denote by $\mathcal{H}_0$ the Hilbert space $\mathcal{H}$ equipped with the trivial G-action. The formula $v \otimes [g] \mapsto g^{-1} \cdot v \otimes [g]$ defines a unitary isomorphism from $\mathcal{H} \otimes \ell^2(G)$ to $\mathcal{H}_0 \otimes \ell^2(G)$, and from it we obtain a unitary isomorphism

$$\mathcal{H} \widehat{\otimes} \mathcal{H}_G \xrightarrow[\cong]{} \mathcal{H}_0 \widehat{\otimes} \mathcal{H}_G.$$

Since $\mathcal{H}_0 \widehat{\otimes} \mathcal{H}_G$ is just a direct sum of copies of $\mathcal{H}_G$ it is clear that $\mathcal{H}_0 \widehat{\otimes} \mathcal{H}_G \cong \mathcal{H}_G$. Hence $\mathcal{H} \widehat{\otimes} \mathcal{H}_G \cong \mathcal{H}_G$, as required.

Definition 2.16. *Let G be a countable discrete group and let A and B be graded, separable G-C^*-algebras. Denote by $E_G(A, B)$ the set of homotopy classes of equivariant asymptotic morphisms from $\mathcal{S} \widehat{\otimes} A \widehat{\otimes} \mathcal{K}(\mathcal{H}_G)$ to $B \widehat{\otimes} \mathcal{K}(\mathcal{H}_G)$,*

$$E_G(A, B) = [\mathcal{S} \widehat{\otimes} A \widehat{\otimes} \mathcal{K}(\mathcal{H}_G), B \widehat{\otimes} \mathcal{K}(\mathcal{H}_G)]^G.$$

Remark 2.5. The virtue of working with the Hilbert space $\mathcal{H}_G$, as in the above definition, is that if $\mathcal{H}$ is *any* separable graded G-Hilbert space and if $\varphi \colon \mathcal{S} \widehat{\otimes} A \dashrightarrow B \widehat{\otimes} \mathcal{K}(\mathcal{H})$ is an equivariant asymptotic morphism then φ determines an element of $E_G(A, B)$. To see this, simply tensor φ by $\mathcal{K}(\mathcal{H}_G)$ and apply Lemma 2.5.

Remark 2.6. The construction described in the previous remark has a generalization which will be important in Lecture 4. Suppose that $\mathcal{H}$ is a separable, graded Hilbert space which is equipped with a *continuous family* of unitary G-actions, parametrized by $t \in [1, \infty)$. The continuity requirement here is pointwise strong continuity, so that if $g \in G$ and $k \in \mathcal{K}(\mathcal{H})$ then $g \cdot_t k$ is norm-continuous in t. Suppose now that A and B are G-C^*-algebras and that

---

[8] A *graded G-Hilbert space* is a graded Hilbert space equipped with unitary representations of G on its even and odd grading-degree summands.

$$\varphi \colon \mathcal{S} \widehat{\otimes} A \dashrightarrow B \widehat{\otimes} \mathcal{K}(\mathcal{H})$$

is an asymptotic morphism which is equivariant with respect to the given family of G-actions, in the sense that

$$\lim_{t \to \infty} \| \varphi_t(g \cdot x) - g \cdot_t (\varphi_t(x)) \| = 0,$$

for all $g \in G$ and $x \in \mathcal{S} \widehat{\otimes} A$. Then φ too determines an element of $E_G(A, B)$. Indeed, after we tensor with $\mathcal{K}(\mathcal{H}_G)$ and apply the procedure in the proof of Lemma 2.5 we obtain an asymptotic morphism into $B \widehat{\otimes} \mathcal{K}(\mathcal{H}_0 \widehat{\otimes} \mathcal{H}_G)$ which is equivariant in the usual sense for the single, fixed representation of G on $\mathcal{H}_0 \widehat{\otimes} \mathcal{H}_G$.

Remark 2.7. One final comment: it is essential that in Definition 2.16 we include a factor of $\mathcal{K}(\mathcal{H}_G)$ in both arguments. If we were to leave one out then we would obtain a quite different (and not very useful) object.

By comparing the definition of $E_G(A, B)$ to the construction of the equivariant, amplified asymptotic category we immediately obtain the following result:

Theorem 2.12. *The E_G-theory groups $E_G(A, B)$ are the morphism sets of an additive category whose objects are the separable graded G-C^*-algebras. There is a functor from the homotopy category of graded G-C^*-algebras and graded G-equivariant $*$-homomorphisms into the equivariant E-theory category which is the identity on objects.* ☐

The equivariant E theory category has a tensor product $\widehat{\otimes}_{max}$. Moreover there are six-term exact sequences of E-theory groups associated to short exact sequences of G-C^*-algebras. The precise statements and proofs are only minor modifications of what we saw in the non-equivariant case, and we shall omit them here. See [27].

2.8 Crossed Products and Descent

In order to apply equivariant E-theory to the problem of computing C^*-algebra K-theory one must first apply a descent operation which transfers computations in equivariant E-theory to computations in the nonequivariant theory. This involves the notion of crossed product C^*-algebra, and we begin with a rapid review of the basic definitions (see [53]) for more details).

Definition 2.17. *Let G be a discrete group and let A be a G-C^*-algebra. A covariant representation of A in a C^*-algebra B is a pair (φ, π) consisting of a $*$-homomorphism φ from A into a C^*-algebra B and a group homomorphism π from G into the unitary group of the multiplier algebra of B which are related by the formulas*

$$\pi(g)\varphi(a)\pi(g^{-1}) = \varphi(g \cdot a), \quad \text{for all } a \in A,\ g \in G.$$

Definition 2.18. *Let G be a discrete group and let A be a G-C^*-algebra. The linear space $C_c(G, A)$ of finitely-supported, A-valued functions on G is an involutive algebra with respect to the convolution multiplication and involution defined by*

$$f_1 \star f_2(g) = \sum_{h \in G} f_1(h) \, (h \cdot (f_2(h^{-1}g)))$$

$$f^*(g) = g \cdot (f(g^{-1})^*)$$

Observe that a covariant representation of A in a C^*-algebra B determines a $*$-homomorphism $\varphi \times \pi$ from $C_c(G, A)$ into B by the formula

$$(\varphi \times \pi)f = \sum_{g \in G} \varphi(f(g))\pi(g) \quad \text{for all } f \in C_c(G, A).$$

Definition 2.19. *The* full crossed product *C^*-algebra $C^*(G, A)$ is the completion of the $*$-algebra $C_c(G, A)$ in the smallest C^*-algebra norm which makes all the $*$-homomorphisms $\varphi \times \pi$ continuous.*

Example 2.4. Setting $A = \mathbb{C}$ we obtain the *full group C^*-algebra $C^*(G)$.*

If A is graded, and if G acts by grading - preserving automorphisms, then $C^*(G, A)$ has a natural grading too (the grading automorphism acts pointwise on functions in $C_c(G, A)$).

Remark 2.8. The C^*-algebra $C^*(G, A)$ contains a copy of A and the multiplier algebra of $C^*(G, A)$ contains a copy of G within its unitary group. Elements of $C_c(G, A)$ can be written as finite sums $\sum_{g \in G} a_g \cdot g$, where $a_g \in A$ and $a_g = 0$ for almost all g. It will usually be convenient to use this means of representing elements. For example the grading automorphism is

$$\sum_{g \in G} a_g \cdot g \mapsto \sum_{g \in G} \alpha(a_g) \cdot g.$$

The full crossed product is a functor from G-C^*-algebras to C^*-algebras which is (extending the terminology of Section 2.3 in the obvious way) both continuous and exact. As a result, there is a *descent functor* from the equivariant asymptotic category to the asymptotic category,

$$[\![A, B]\!]_{\infty}^G \longrightarrow [\![C^*(G, A), C^*(G, B)]\!]_{\infty}.$$

In order to obtain a corresponding functor in E-theory we need the following computation:

Lemma 2.6. *Let G be a discrete group, let B be a G-C^*-algebra and let $\mathcal{H}$ be a G-Hilbert space on which the group element $g \in G$ acts as the unitary operator $U_g \colon \mathcal{H} \to \mathcal{H}$. The formula*

$$\sum_{g \in G} (b_g \widehat{\otimes} k_g) \cdot g \mapsto \sum_{g \in G} (b_g \cdot g) \widehat{\otimes} k_g U_g$$

determines an isomorphism of C^*-algebras

$$C^*(G, B\widehat{\otimes}\mathcal{K}(\mathcal{H})) \xrightarrow[\cong]{} C^*(G, B)\widehat{\otimes}\mathcal{K}(\mathcal{H}).$$

Proof. The formula defines an algebraic $*$-isomorphism from $C_c(G, B\widehat{\odot}\mathcal{K}(\mathcal{H}))$ to $C_c(G, B)\widehat{\odot}\mathcal{K}(\mathcal{H})$. Examining the definitions of the norms for the max tensor product and full crossed product we see that the $*$-isomorphism extends to a $*$-isomorphism of C^*-algebras.

Combining the lemma with the descent functor between asymptotic categories we obtain the following result:

Theorem 2.13. *There is a descent functor from the equivariant E-theory category to the E-theory category which maps the G-C^*-algebra A to the full crossed product C^*-algebra $C^*(G, A)$, and which maps the E-theory class of a G-equivariant $*$-homomorphism $\varphi\colon A \to B$ to the E-theory class of the induced $*$-homomorphism from $C^*(G, A)$ to $C^*(G, B)$.* $\square$

Corollary 2.2. *Let G be a countable discrete group. Suppose that A and B are separable G-C^*-algebras and that A and B are isomorphic objects in the equivariant E-theory category. Then $K(C^*(G, A))$ is isomorphic to $K(C^*(G, B))$.* $\square$

2.9 Reduced Crossed Products

We also wish to apply equivariant E-theory to the computation of K-theory for *reduced* crossed products. Here the operation of descent works smoothly for a large class of groups, as the following discussion shows, but not so well for all groups, as we shall see in Lecture 6.[9]

In the following definition we shall use, in a very modest way, the notion of Hilbert module. See [45] for a treatment of this subject.

Definition 2.20. *Let A be a G-C^*-algebra and denote by $\ell^2(G, A)$ the Hilbert A-module comprised of functions $\xi\colon G \to A$ for which the series $\sum_g \xi(g)^*\xi(g)$ is norm-convergent in A. The regular representation of A is the covariant representation (φ, π) into the bounded, adjoinable operators on $\ell^2(G, A)$ given by the formulas*

$$(\varphi(a)\xi)(h) = (h^{-1} \cdot a)\xi(h), \qquad \xi \in \ell^2(G, A),$$

and

$$(\pi(g)\xi)(h) = \xi(g^{-1}h), \qquad \xi \in \ell^2(G, A).$$

The regular representation determines a $*$-homomorphism from the crossed product algebra $C^*(G, A)$ into the C^*-algebra of bounded, adjoinable operators on $\ell^2(G, A)$.

---

[9] It should be pointed out here that Kasparov's KK-theory has no such limitation in this respect. However it has other shortcomings. Indeed as we shall see in Lecture 6 there is no ideal bivariant K-theory for C^*-algebras.

Definition 2.21. *Let A be a G-C^*-algebra. The* reduced crossed product algebra *$C^*_\lambda(G, A)$ is the image of $C^*(G, A)$ in the regular representation.*

Example 2.5. Setting $A = \mathbb{C}$ we obtain the *reduced group C^*-algebra $C^*_\lambda(G)$.*

Like the full crossed product, the reduced crossed product is a functor from (graded) G-C^*-algebras to (graded) C^*-algebras. However unlike the full crossed product the reduced crossed product is not exact for every G (although inexact examples are hard to come by — see Lecture 6). This prompts us to make the following definition:

Definition 2.22. *A discrete group G is* exact *if the functor $A \mapsto C^*_\lambda(G, A)$ is exact in the sense of Definition 2.8.*

There is a very simple and beautiful characterization of exact groups, due to Kirchberg and Wassermann [43].

Proposition 2.6. *A discrete group G is exact if and only if its reduced group C^*-algebra $C^*_\lambda(G)$ is exact.*

Proof (Proof (sketch)). Exactness of $C^*_\lambda(G)$ is implied by exactness of G since in the case of trivial G-actions the reduced crossed product $C^*_\lambda(G, A)$ is the same thing as $A \otimes_{min} C^*_\lambda(G)$ (note that $C^*_\lambda(G)$ is trivially graded, so $\otimes_{min} = \widehat{\otimes}_{min}$ here). The reverse implication is argued as follows. If $C^*_\lambda(G)$ is exact then the sequence

$$0 \longrightarrow C^*(G, J) \otimes_{min} C^*_\lambda(G) \longrightarrow C^*(G, A) \otimes_{min} C^*_\lambda(G)$$
$$\longrightarrow C^*(G, A/J) \otimes_{min} C^*_\lambda(G) \longrightarrow 0$$

is exact. But for any G-C^*-algebra D there is a functorial embedding

$$C^*_\lambda(G, D) \longrightarrow C^*(G, D) \otimes_{min} C^*_\lambda(G)$$

defined by the formulas $g \mapsto g \otimes g$ and $d \mapsto d \otimes 1$, and moreover a functorial, continuous and linear left-inverse defined by $d \otimes 1 \mapsto d$, $g \otimes g \mapsto g$ and $g \otimes h \mapsto 0$ if $g \neq h$. It follows that the sequence

$$0 \longrightarrow C^*_\lambda(G, J) \longrightarrow C^*_\lambda(G, A) \longrightarrow C^*_\lambda(G, A/J) \longrightarrow 0$$

is a direct summand of the minimal tensor product exact sequence above, and is therefore exact itself. For more details see Section 5 of [43]

Exercise 2.4. If B is an exact C^*-algebra and if B_1 is a C^*-subalgebra of B then B then B_1 is also exact.

Thanks to the exercise and to Proposition 2.6 it is possible to show that many classes of groups are exact. For example all discrete subgroups of connected Lie groups are exact and all hyperbolic groups (these will be discussed in Lecture 5) are exact too. Every amenable group is exact since in this case the reduced and full crossed product functors are one and the same. For more information on exactness see for example [67]. We shall also return to the subject in Section 4.5.

By retracing the steps we took in the previous section we arrive at the following result:

Theorem 2.14. *Let G be an exact, countable, discrete group. There is a* descent func-tor *from the equivariant E-theory category to the E-theory category which maps a G-C*-algebra A to the reduced crossed product C*-algebra $C_\lambda^*(G, A)$, and which maps the class of a G-equivariant *-homomorphism $\varphi: A \to B$ to the class of the induced *-homomorphism from $C_\lambda^*(G, A)$ to $C_\lambda^*(G, B)$.*   □

Corollary 2.3. *Let G be an exact, countable, discrete group. Suppose that A and B are separable G-C*-algebras and that A and B are isomorphic objects in the equivariant E-theory category. Then $K(C_\lambda^*(G, A))$ is isomorphic to $K(C_\lambda^*(G, B))$.*
□

2.10 The Baum-Connes Conjecture

In this lecture we shall formulate the Baum-Connes conjecture and prove it in some simple cases, for example for finite groups and free abelian groups. We shall also sketch the proof of the conjecture for so-called 'proper' coefficient C^*-algebras. This result will play an important role in the next chapter. The proof for proper algebras is not difficult, but it is a little long-winded, and we shall refer the reader to the monograph [27] for the details.

We shall continue to work exclusively with *discrete* groups. Our formulation of the conjecture, which uses E-theory, is equivalent to the formulation in [7] which uses KK-theory. Indeed there is a natural transformation from KK to E which determines an isomorphism from the KK-theoretic 'left-hand side' of the Baum-Connes conjecture to its E-theoretic counterpart. The isomorphism can be proved either by a Mayer-Vietoris type of argument (see for example Lecture 5) or by di-rectly constructing an inverse. See also the discussion in Section 4.6 which in many cases reduces the conjecture to a statement in K-theory, independent of both E-theory and KK-theory.[10] Our treatment using E-theory is quite well suited to the theorems we shall formulate and prove in Lecture 4. However a major drawback of E-theory is that it is not well suited to dealing with inexact groups. In any case, the E-theoretic and KK-theoretic developments of the Baum-Connes theory are very similar, and having studied by himself the basics of KK-theory the reader could de-velop the Baum-Connes conjecture in KK-theory simply by replacing E with KK throughout this lecture.

---

[10] In fact the argument of Section 4.6 can be made to apply to any discrete group, but we shall not go into this here.

2.11 Proper G-Spaces

Let G be a countable discrete group. Throughout this lecture we shall be dealing with Hausdorff and paracompact topological spaces X equipped with actions of G by homeomorphisms.

Definition 2.23. *A G-space X is proper if for every $x \in X$ there is a G-invariant open subset $U \subseteq X$ containing x, a finite subgroup H of G, and a G-equivariant map from U to G/H.*

The definition says that locally the orbits of G in X look like G/H.

Example 2.6. If H is a finite subgroup of G then the discrete homogeneous space G/H is proper. Moreover if Y is any (Hausdorff and paracompact) space with an H-action then the *induced space* $X = G \times_H Y$ (the quotient of $G \times Y$ by the diagonal action of H, with H acting on G by right multiplication) is proper.

In fact every proper G-space is locally induced from a finite group action:

Lemma 2.7. *A G-space X is proper if and only if for every $x \in X$ there is a G-invariant open subset $U \subseteq X$ containing X, a finite subgroup H of G, an H-space Y, and a G-equivariant homeomorphism from U to $G \times_H Y$.* $\square$

Many proofs involving proper spaces proceed by reducing the case of a general proper G-space to the case of the local models $G \times_H W$, and hence to the case of finite group actions, using the lemma.

Lemma 2.8. *A locally compact G-space X is proper if and only if the map from $G \times X$ to $X \times X$ which takes (g, x) to (gx, x) is a proper map of locally compact spaces (meaning that the inverse image of every compact set is compact).* $\square$

Example 2.7. If G is a discrete subgroup of a Lie group L, and if K is a compact subgroup of L, then the quotient space L/K is a proper G-space.

2.12 Universal Proper G-Spaces

Definition 2.24. *A proper G-space X is universal if for every proper G-space Y there exists a G-equivariant continuous map $Y \to X$, and if moreover this map is unique up to G-equivariant homotopy.*

It is clear from the definition that any two universal proper G-spaces are G-equivariantly homotopy equivalent. For this reason let us introduce the notation eG for a universal proper G-space (with the understanding that different models for eG will agree up to equivariant homotopy).

Proposition 2.15 *Let G be a countable discrete group. There exists a universal proper G-space.* $\square$

Here is one simple construction (due to Kasparov and Skandalis [36]). Let X_1 be the space of (countably additive) measures on G with total mass 1 or less. This is a compact space in the topology of pointwise convergence. Let $X_{\frac{1}{2}}$ be the closed subspace of X_1 consisting of measures of total mass $\frac{1}{2}$ or less. The set-theoretic difference $X = X_1 \setminus X_{\frac{1}{2}}$ is a locally compact proper G-space which is universal.

In examples one can usually provide a much more concrete model. See [7] for examples (and see also Lectures 4 and 5 below). The following result, which we shall not prove, gives the general flavor of these constructions.

Proposition 2.7. *Let M be a complete and simply connected Riemannian manifold of nonpositive sectional curvature. If a discrete group G acts properly and isometrically on M then M is a universal G-space.* $\square$

Remark 2.9. The manifold here could be infinite-dimensional.

2.13 G-Compact Spaces

Definition 2.25. *A proper G-space X is G-compact if there is a compact subset $K \subseteq X$ whose translates under the G-action cover X.*

If X is a G-compact proper G-space then X is locally compact and the quotient X/G is compact.

Definition 2.26. *Let X be a G-compact proper G-space. A cutoff function for X is a continuous function $\theta \colon X \to [0,1]$ such that*

(a) $\operatorname{supp}(\theta)$ *is compact, and*
(b) $\sum_{y \in G} \theta^2(g \cdot x) = 1$, *for all $x \in X$.*

Observe that the sum in (b) is locally finite. Every G-compact proper G-space admits a cutoff function. Moreover any two cutoff functions are, in a sense, homotopic: if θ_0 and θ_1 are cutoff functions then the functions

$$\theta_t = \sqrt{t\theta_1^2 + (1-t)\theta_0^2}, \quad t \in [0,1]$$

are all cutoff functions.

Lemma 2.9. *Let θ be a cutoff function for the G-compact proper G-space X. The formula*

$$p(g)(x) = \theta(g^{-1}x)\,\theta(x).$$

defines a projection in $C_c(G, C_c(X))$, and hence in $C^*(G, X)$. The K-theory class of this projection is independent of the choice of cutoff function.

Remark 2.10. We are using here the streamlined notation $C^*(G, X)$ in place of $C^*(G, C_0(X))$.

Proof (Proof of the Lemma). A computation shows that p is a projection (note that the sum involved in the definition of p is in fact finite). If θ_0 and θ_1 are cutoff functions then associated to the homotopy of cutoff functions θ_t defined above there is a homotopy of projections p_t, and therefore θ_0 and θ_1 give rise to the same K-theory class, as required.

Definition 2.27. *We will call the unique K-theory class of projections associated to cutoff functions the unit class:*

$$[p] \in K(C^*(G, X)) \cong E(\mathbb{C}, C^*(G, X)).$$

Exercise 2.16 *(See [54, Thm 6.1].) Let X be a proper G-space. Show that the full and reduced crossed products $C^*(G, X)$ and $C^*_\lambda(G, X)$ are isomorphic.*

Hint: *One approach is to show that if $f \in C_c(G, C_c(X))$, and if $1 + f$ is invertible in $C^*_\lambda(G, X)$, then the inverse actually lies in $1 + C_c(G, C_c(X))$. It follows that $1 + f$ is invertible in $C^*(G, X)$ too, and therefore, the map $C^*(G, X) \to C_\lambda(G, X)$ is spectrum-preserving, and hence isometric.*

Remark 2.11. As a result of the exercise, we can obviously define a unit class in $K(C^*_\lambda(G, X))$ too.

2.14 The Assembly Map

In this section we shall further streamline our notation and write $E_G(X, D)$ in place of $E_G(C_0(X), D)$. Observe that $E_G(X, D)$ is covariantly functorial on the category of G-compact proper G-spaces X.

Definition 2.28. *Let G be a countable discrete group and let D be a separable G-C^*-algebra. The assembly map*

$$\mu \colon E_G(X, D) \to K(C^*(G, D))$$

is the composition

$$E_G(X, D) \xrightarrow{\text{descent}} E(C^*(G, X), C^*(G, D)) \xrightarrow{[p]} E(\mathbb{C}, C^*(G, D))$$

where the first map is the descent homomorphism of Section 2.8 and the second is composition with the unit class $[p] \in E(\mathbb{C}, C^*(G, X))$.

Definition 2.29. *Let G be a countable group and let D be a G-C^*-algebra. The topological K-theory of G with coefficients in a G-C^*-algebra D is defined by*

$$K^{top}(G, D) = \varinjlim_{\substack{X \subseteq eG \\ G\text{-inv, } G\text{-cpt}}} E_G(X, D),$$

where the limit is taken over the collection of G-invariant and G-compact subspaces $X \subseteq eG$, directed by inclusion.

To explain the limit, note that if $X \subseteq Y \subseteq eG$ are G-compact proper G-spaces then X is a closed subset of Y and restriction of functions defines a G-equivariant $*$-homomorphism from $C_0(Y)$ to $C_0(X)$. This induces a homomorphism from $E_G(X, D)$ to $E_G(Y, D)$.

If $X \subseteq Y \subseteq eG$ are G-compact proper G-spaces then under the restriction map from $E(\mathbb{C}, C^*(G, Y))$ to $E(\mathbb{C}, C^*(G, X))$ the unit class for Y maps to the unit class for X; consequently the assembly maps for the various G-compact subsets of eG are compatible and pass to the direct limit:

Definition 2.30. *The* (full) Baum-Connes assembly map *with coefficients in a separable G-C^*-algebra D is the map*

$$\mu : K^{top}(G, D) \to K(C^*(G, D))$$

which is obtained as the limit of the assembly maps of Definition 2.28 for G-compact subspaces $X \subset eG$.

Definition 2.31. *The* reduced Baum-Connes assembly map *with coefficients in a separable G-C^*-algebra D is the map*

$$\mu_\lambda : K^{top}(G, D) \to K(C^*_\lambda(G, D))$$

obtained by composing the full Baum-Connes assembly map μ with the map from $K(C^*(G, D))$ to $K(C^*_\lambda(G, D))$ induced from the quotient mapping from $C^*(G, D)$ onto $C^*_\lambda(G, D)$.

Remark 2.12. If G is exact and if X is a G-compact proper G-space then there is a reduced assembly map

$$\mu : E_G(C_0(X), D) \to K(C^*_\lambda(G, D)),$$

defined by means of a composition

$$E_G(C_0(X), D) \xrightarrow{\text{descent}} E(C^*_\lambda(G, X), C^*_\lambda(G, D)) \xrightarrow{[p]} E(\mathbb{C}, C^*_\lambda(G, D))$$

involving the reduced descent functor of Section 2.9. The Baum-Connes assembly map μ_λ may then be equivalently defined as a direct limit of such maps.

2.15 Baum-Connes Conjecture

The following is known as the *Baum-Connes Conjecture with coefficients* (the 'coefficients' being of course the auxiliary C^*-algebra D).

Conjecture 2.1. Let G be a countable discrete group. The Baum-Connes assembly map

$$\mu_\lambda : K^{top}(G, D) \to K(C^*_\lambda(G, D)).$$

is an isomorphism for every separable G-C^*-algebra D.

Not a great deal is known about this conjecture. We shall prove one of the main results (which covers, for example, amenable groups) in the next section. Unfortunately, thanks to some recent constructions of Gromov, the Baum-Connes conjecture with coefficients appears to be false, in general. See Lecture 6.

In the next conjecture, which is the official *Baum-Connes conjecture* for discrete groups, the coefficient algebra B is specialized to $D = \mathbb{C}$ and $D = C_0(0,1)$. We shall use the notations $K_*^{top}(G)$ and $K_*(C_\lambda^*(G))$ to denote topological and C^*-algebra K-theory in these two cases (this of course is customary usage in K-theory).

Conjecture 2.2. Let G be a countable discrete group. The Baum-Connes assembly map

$$\mu_\lambda : K_*^{top}(G) \to K_*(C_\lambda^*(G)).$$

is an isomorphism.

Somewhat more is known about this conjecture, thanks largely to the remarkable work of Lafforgue [44, 62]. For example, the conjecture is proved for all hyperbolic groups (we shall define these in Lecture 5). What is especially interesting is that, going beyond discrete groups, the Baum-Connes conjecture has now been proved for all reductive Lie and p-adic groups (this is part of what Lafforgue accomplished using his Banach algebra version of bivariant K-theory, although by invoking a good deal of representation theory many cases here had been confirmed prior to Lafforgue's work). Unfortunately we shall not have the time to discuss either Lafforgue's work or the topic of K-theory for non-discrete groups.

At the present time, the major open question seems to be whether or not the Baum-Connes conjecture (with or without coefficients, according to one's degree of optimism) is true for discrete subgroups of connected Lie groups. Even the case of uniform lattices in semisimple groups remains open.

Considerably more is known about the *injectivity* of the Baum-Connes assembly map, and fortunately this is all that is required in some of the key applications of the conjecture to geometry and topology. We shall say more about injectivity in Lecture 5.

Remark 2.13. We shall discuss in Lecture 6 the reason for working with $C^*(G, D)$ in place of $C_\lambda^*(G, D)$.

2.16 The Conjecture for Finite Groups

The reader can check for himself that the Baum-Connes conjecture is true (in fact it is a tautology) for the trivial, one-element group. Next come the finite groups. Here the conjecture is a theorem, and it is basically equivalent to a well-known result of Green and Julg which identifies equivariant K-theory and the K-theory of crossed product algebras in the case of finite groups. See [23, 35]. What follows is a brief account of this.

Theorem 2.17 (Green-Julg). *Let G be a finite group and let D be a G-C^*-algebra. The Baum-Connes assembly map*

$$\mu : K^{top}(G, D) \to K(C^*(G, D))$$

is an isomorphism for every G-C^*-algebra D.

Remark 2.14. If G is finite then $C^*(G, D) = C^*_\lambda(G, D)$ for every D.

If G is a finite group then eG can be taken to be the one point space. So the theorem provides an isomorphism

$$E_G(\mathbb{C}, D) \xrightarrow[\cong]{\mu} E(\mathbb{C}, C^*(G, D)).$$

The unit projection $p \in C^*(G)$ which is described in Lemma 2.9 is the function $p(g) = 1/|G|$, which is the central projection in $C^*(G)$ corresponding to the trivial representation of G (it acts as the orthogonal projection onto the G-fixed vectors in any unitary representation of G).

Theorem 2.17 is proved by defining an inverse to the assembly map μ. For this purpose we note that $C^*(G, D)$ may be identified with a fixed point algebra,

$$C^*(G, D) \xrightarrow[\cong]{} [D \otimes \mathrm{End}(\ell^2(G))]^G,$$

by mapping d to $\sum_{g \in G} g \cdot d \otimes p_g$ (where p_g is the projection onto the functions supported on $\{g\}$) and by mapping g to $1 \otimes \rho(g)$, where ρ is the right regular representation (the fixed point algebra is computed using the left regular representation). The displayed $*$-homomorphism can be thought of as an equivariant $*$-homomorphism from $C^*(G, D)$, equipped with the trivial action of G, into $D \otimes \mathrm{End}(\ell^2(G))$. It induces a homomorphism

$$E(\mathbb{C}, C^*(G, D)) \longrightarrow E_G(\mathbb{C}, D \otimes \mathrm{End}(\ell^2(G))).$$

But the left hand side here is $K(C^*(G, D))$ and the right hand side is $K^{top}(G, D)$, and it is not difficult to check that the above map inverts the assembly map μ, as required. For details see [27, Thm. 11.1].

2.17 Proper Algebras

Theorem 2.17 has an important extension to the realm of infinite groups, involving the following notion:

Definition 2.32. *A G-C^*-algebra B is* proper *if there exists a locally compact proper G-space X and an equivariant $*$-homomorphism φ from $C_0(X)$ into the grading-degree zero part of the center of the multiplier algebra of B such that $\varphi[C_0(X)] \cdot B$ is norm-dense in B.*

Remark 2.15. We shall say that B, as in the definition, is *proper over Z*. Throughout the lecture we shall deal with proper algebras which are *separable*.

The notion of proper algebra is due essentially to Kasparov [38], in whose work proper algebras appear in connection with RKK-theory, a useful elaboration of KK-theory. We shall not develop RKK here, or even its E-theoretic counterpart. While this limits the amount of machinery we must introduce, it will also make some of the arguments in this and later lectures a little clumsier than they need be.

Examples 2.18 *If G is finite every G-C^*-algebra is proper over the one point space. If Z is a proper G-space then $C_0(Z)$ is a proper G-C^*-algebra. If B is proper over Z then, for every G-C^*-algebra D, the tensor product $B\widehat{\otimes}D$ is also proper.*

Exercise 2.5. Prove that if B is proper then $C^*(G, B) = C^*_\lambda(G, B)$.

A guiding principle is that the action of a group on a proper algebra is more or less the same thing as the action of a *finite* group on a C^*-algebra. With this in mind the following theorem should not be surprising.

Theorem 2.19. [27, Theorem 13.1] *Let G be a countable discrete group and let B be a proper G-C^*-algebra. The Baum-Connes assembly map*

$$\mu : K^{top}(G, B) \to K(C^*(G, B))$$

is an isomorphism.

Remark 2.16. Thanks to Exercise 2.5, the assembly map μ_λ into $K(C^*_\lambda(G, B))$ is an isomorphism as well.

The proof of Theorem 2.19 is not difficult, but with the tools we have to hand it is rather long. So we shall just give a quick outline. The following computation is key not just to the proof of Theorem 2.19 but also to a number of results in Lecture 5.

Proposition 2.8. [27, Lemma 12.11] *Let H be a finite subgroup of a countable group G and let W be a locally compact space equipped with an action of H by homeomorphisms. If D is any G-C^*-algebra there is a natural isomorphism*

$$E_H(C_0(W), D) \cong E_G(C_0(G \times_H W), D),$$

where on the left hand side D is viewed as an H-C^*-algebra by restriction of the G-action.

Proof. The space W is included into $G \times_H W$ as the open set $\{e\} \times W$, and as a result there is an H-equivariant map from $C_0(W)$ into $C_0(G \times_H W)$. Composition with this map defines a 'restriction' homomorphism

$$E_G(C_0(G \times_H W), D) \xrightarrow{\text{Res}} E_H(C_0(W), D)$$

To construct an inverse, the important observation to make is that every H-equivariant asymptotic morphism from $C_0(W)$ into D extends uniquely to a G-equivariant asymptotic morphism from $C_0(G \times_H W)$ into $D \otimes \mathcal{K}(\ell^2(G/H))$. Decorating this construction with copies of $\mathcal{S}$ and $\mathcal{K}(\mathcal{H})$ we obtain an inverse map

$$E_H(C_0(W), D) \longrightarrow E_G(C_0(G \times_H W), D)$$

as required.

Proposition 2.8 has the following immediate application:

Lemma 2.10. *Let G be a countable group. If the assembly map*

$$\mu \colon K^{top}(G, B) \to K(C^*(G, B))$$

is an isomorphism for every G-C^*-algebra B which is proper over a G-compact space Z, then it is an isomorphism for every G-C^*-algebra.

Proof. Every proper algebra is a direct limit of G-C^*-algebras which are proper over G-compact spaces. Since K-theory commutes with direct limits (see Exercise 1.4), as does the crossed product functor, to prove the lemma it suffices to prove that the same is true for the functor $D \mapsto K^{top}(G, D)$. In view of the definition of $K^{top}(G, D)$ it suffices to prove that if Z is a G-compact proper G-simplicial complex then the functor $E_G(C_0(Z), D)$ commutes with direct limits. By a Mayer-Vietoris argument the proof of this reduces to the case where Z is a proper homogeneous space G/H. But here we have a sequence of isomorphisms

$$E_G(C_0(G/H), D) \simeq E_H(\mathbb{C}, D) \cong K(C^*(G, D)),$$

the first by Proposition 2.8 and the second by Theorem 2.17. Since K-theory commutes with direct limits the lemma is proved.

Lemma 2.11. *Let G be a countable group. If the assembly map*

$$\mu \colon K^{top}(G, B) \to K(C^*(G, B))$$

is an isomorphism for every G-C^*-algebra B which is proper over a proper homogeneous space $Z = G/H$ then it is an isomorphism for every G-C^*-algebra which is proper over a G-compact proper G-space.

Proof. This is another Mayer-Vietoris argument, this time in the B-variable. Observe that if B is proper over Z then to each G-invariant open set U in Z there corresponds an ideal $J = C_0(U) \cdot B$ of B. Using this, together with the long exact sequences in E-theory and the five lemma, an induction argument can be constructed on the number of G-invariant open sets needed to cover Z, each of which admits a G-map to a proper homogeneous space.

The proof of Theorem 2.19 therefore reduces to the case where B is proper over some proper homogeneous space G/H. Observe now that if B is proper over G/H then B is a direct sum of ideals corresponding to the points of G/H, and the ideal B_e corresponding to eH is an H-C^*-algebra. The proof is completed by developing a variant of the isomorphism in Proposition 2.8, and producing a commuting diagram

$$
\begin{array}{ccc}
K^{top}(G,B) & \xrightarrow{\ \mu\ } & K(C^*(G,B)) \\
{\scriptstyle\cong}\big\downarrow & & \big\downarrow{\scriptstyle\cong} \\
K^{top}(H,B_e) & \xrightarrow[\cong]{\ \mu\ } & K(C^*(H_e,B)).
\end{array}
$$

See [27, Chapter 12] for details.

2.18 Proper Algebras and the General Conjecture

The following simple theorem provides a strategy for attacking the Baum-Connes conjecture for general coefficient algebras. The theorem, or its extensions and relatives, is invoked in nearly all approaches to the Baum-Connes conjecture. As we shall see in Lecture 5 the theorem is particularly useful as a tool to prove results about the injectivity of the Baum-Connes map.

Theorem 2.20. *Let G be a countable discrete group. Suppose there exists a proper G-C^*-algebra B and morphisms $\beta \in E_G(\mathbb{C}, B)$ and $\alpha \in E_G(B, \mathbb{C})$ in the equivariant E-theory category such that*

$$
\alpha \circ \beta = 1 \in E_G(\mathbb{C}, \mathbb{C}).
$$

Then the Baum-Connes assembly map $\mu : K^{top}(G,D) \to K(C^*(G,D))$ is an isomorphism for every separable G-C^*-algebra D. If in addition G is exact then the reduced Baum-Connes assembly map

$$
\mu_\lambda : K^{top}(G,D) \to K(C^*_\lambda(G,D))
$$

is an isomorphism.

Proof. Let G be a countable discrete group, let D be a separable G-C^*-algebra and let α and β be as in the statement of the theorem. Consider the following diagram:

$$
\begin{array}{ccc}
K^{top}(G,\mathbb{C}\widehat{\otimes}D) & \xrightarrow{\ \mu\ } & K(C^*(G,\mathbb{C}\widehat{\otimes}D)) \\
{\scriptstyle\beta_*}\big\downarrow & & \big\downarrow{\scriptstyle\beta_*} \\
K^{top}(G,B\widehat{\otimes}D) & \xrightarrow[\cong]{\ \mu\ } & K(C^*(G,B\widehat{\otimes}D)) \\
{\scriptstyle\alpha_*}\big\downarrow & & \big\downarrow{\scriptstyle\alpha_*} \\
K^{top}(G,\mathbb{C}\widehat{\otimes}D) & \xrightarrow{\ \mu\ } & K(C^*(G,\mathbb{C}\widehat{\otimes}D)).
\end{array}
$$

The horizontal maps are the assembly maps; the vertical maps are induced from E-theory classes $\beta \otimes 1 \in E_G(\mathbb{C}\widehat{\otimes}D, B\widehat{\otimes}D)$ and $\alpha \otimes 1 \in E_G(B\widehat{\otimes}D, \mathbb{C}\widehat{\otimes}D)$. The diagram is commutative. Since the C^*-algebra B is proper, so is the tensor product $B\widehat{\otimes}D$ and therefore by the Theorem 2.19 the middle horizontal map is an isomorphism. By assumption, the compositions of the vertical maps on the left, and hence also on the right hand side are the identity. It follows that the top horizontal map is an isomorphism too. The statement concerning reduced crossed products is proved in exactly the same way.

2.19 Crossed Products by the Integers

In this section we shall apply the approach outlined in the previous section to just about the simplest example possible beyond finite groups: the free abelian group $G = \mathbb{Z}^n$. What follows will serve as a model for the more elaborate constructions in the next lecture. For this reason it might be worth the reader's while to study the present case quite carefully.

Let G act by translations on $\mathbb{R}^n$ in the usual way and then let G act on the graded C^*-algebra $\mathcal{C}(\mathbb{R}^n)$ that we introduced in Lecture 1 by $(g \cdot f)(v) = f(g \cdot v)$.

Exercise 2.6. With this action of the free abelian group $\mathbb{Z}^n$, the C^*-algebra $\mathcal{C}(\mathbb{R}^n)$ is proper.

We are going to produce a factorization

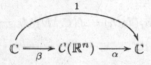

in $\mathbb{Z}^n$-equivariant E-theory. The elements α and β are very small modifications of the objects we defined in Lecture 1 while studying Bott Periodicity.

Definition 2.33. *Denote by* $\beta: S \to \mathcal{C}(\mathbb{R}^n)$ *the* *-homomorphism that was introduced in Definition 1.26, and for* $t \geq 1$ *denote by* $\beta_t: S \to \mathcal{C}(\mathbb{R}^n)$ *the* *-homomorphism* $\beta_t(f) = \beta(f_t)$, *where* $f_t(x) = f(t^{-1}x)$.

Thus $\beta_t(f) = f(t^{-1}C)$, where C is the Clifford operator introduced in Lecture 1.

Lemma 2.12. *The asymptotic morphism* $\beta: S \dashrightarrow \mathcal{C}(\mathbb{R}^n)$ *given by the above family of* *-homomorphisms* $\beta_t: S \to \mathcal{C}(\mathbb{R}^n)$ *is* $\mathbb{Z}^n$*-equivariant.*

Proof. We must show that if $f \in S$ and $g \in \mathbb{Z}^n$ then

$$\lim_{t \to \infty} \|f(t^{-1}C) - g(f(t^{-1}C))\| = 0.$$

Since the set of all $f \in S$ for which this holds (for all g) is a C^*-subalgebra of S it suffices to prove the limit formula for the generators $f = (x \pm i)^{-1}$ of S. For these we have

$$\|f(t^{-1}C) - g(f(t^{-1}C))\| = \|(t^{-1}C \pm i)^{-1} - (t^{-1}g(C) \pm i)^{-1}\|$$
$$\leq t^{-1}\|C - g(C)\|$$

by the resolvent identity. Since the Clifford algebra-valued function $C - g(C)$ is bounded on $\mathbb{R}^n$ the lemma is proved.

Definition 2.34. *Denote by* $\beta \in E_{\mathbb{Z}^n}(\mathbb{C}, \mathcal{C}(\mathbb{R}^n))$ *the class of the asymptotic morphism* $\beta: \mathcal{S} \dashrightarrow \mathcal{C}(\mathbb{R}^n)$.

Definition 2.35. *If* $g \in \mathbb{Z}^n$ *and* $v \in \mathbb{R}^n$, *and if* $s \in [0,1]$, *then denote by* $g \cdot_s v$ *the translation of* v *by* $sg \in \mathbb{R}^n$. *Denote by* $g \cdot_s f$ *the corresponding action of* $g \in \mathbb{Z}^n$ *on elements of the* C^*-*algebra* $\mathcal{C}(\mathbb{R}^n)$ *and also on operators on the Hilbert space* $\mathcal{H}(\mathbb{R}^n)$ *that was introduced in Definition 1.27.*

To define the class $\alpha \in E_{\mathbb{Z}^n}(\mathcal{C}(\mathbb{R}^n), \mathbb{C})$ that we require we shall use the asymptotic morphism $\alpha: \mathcal{S} \widehat{\otimes} \mathcal{C}(\mathbb{R}^n) \dashrightarrow \mathcal{K}(\mathcal{H}(\mathbb{R}^n))$ that we defined in Proposition 1.5, but we shall interpret it as an *equivariant* asymptotic morphism in the following way:

Lemma 2.13. *If* $f \widehat{\otimes} h \in \mathcal{S} \widehat{\otimes} \mathcal{C}(\mathbb{R}^n)$, $g \in \mathbb{Z}^n$, *and* $t \in [1, \infty)$ *then*

$$\lim_{t \to \infty} \|\alpha_t(f \widehat{\otimes} g \cdot h) - g \cdot_{t^{-1}} \alpha_t(f \widehat{\otimes} h)\| = 0.$$

Proof. The Dirac operator D is translation invariant, and so $g \cdot_{t^{-1}} f(t^{-1}D) = f(t^{-1}D)$ for all t. But $g \cdot_{t^{-1}} M_{h_t} = M_{(g \cdot h)_t}$ for all t. The lemma therefore follows from the formula

$$\alpha_t(f \widehat{\otimes} h) = f(t^{-1}D) M_{h_t}$$

for the asymptotic morphism α.

Definition 2.36. *Denote by* $\alpha \in E_{\mathbb{Z}^n}(\mathcal{C}(\mathbb{R}^n), \mathbb{C})$ *the E-theory class of the equivariant asymptotic morphism* $\alpha: \mathcal{S} \widehat{\otimes} \mathcal{C}(\mathbb{R}^n) \dashrightarrow \mathcal{K}(\mathcal{H}(\mathbb{R}^n))$, *where* $\mathcal{K}(\mathcal{H}(\mathbb{R}^n))$ *is equipped with the family of actions* $(g, k) \mapsto g \cdot_{t^{-1}} k$ *(compare Remark 2.6).*

Proposition 2.21 *Continuing with the notation above,* $\alpha \circ \beta = 1 \in E_{\mathbb{Z}^n}(\mathbb{C}, \mathbb{C})$.

Proof. Let $s \in [0,1]$ and denote by $\mathcal{C}_s(\mathbb{R}^n)$ the C^*-algebra $\mathcal{C}(\mathbb{R}^n)$, but with the scaled $\mathbb{Z}^n$-action $(g, h) \mapsto g \cdot_s h$. The algebras $\mathcal{C}_s(\mathbb{R}^n)$ form a continuous field of $\mathbb{Z}^n$-C^*-algebras over the unit interval (since the algebras are all the same this just means that the $\mathbb{Z}^n$-actions vary continuously). Denote by $\mathcal{C}_{[0,1]}(\mathbb{R}^n)$ the $\mathbb{Z}^N$-C^*-algebra of continuous sections of this field (namely the continuous functions from $[0,1]$ into $\mathcal{C}(\mathbb{R}^n)$, equipped with the $\mathbb{Z}^N$-action $(g \cdot h)(s) = g \cdot_s h(s)$). In a similar way, form the continuous field of $\mathbb{Z}^N$-C^*-algebras $\mathcal{K}_s(\mathcal{H}(\mathbb{R}^n))$ and denote by $\mathcal{K}_{[0,1]}(\mathcal{H}(\mathbb{R}^n))$ the $\mathbb{Z}^N$-C^*-algebra of continuous sections. With this notation, what we want to prove is that the composition

$$\mathbb{C} \xrightarrow{\ \beta\ } \mathcal{C}_1(\mathbb{R}^n) \xrightarrow{\ \alpha\ } \mathbb{C}$$

is the identity in equivariant E-theory.

The asymptotic morphism $\alpha \colon \mathcal{S} \widehat{\otimes} \mathcal{C}(\mathbb{R}^n) -- \rightarrow \mathcal{K}(\mathcal{H}(\mathbb{R}^n))$ induces an asymptotic morphism

$$\bar{\alpha} \colon \mathcal{S} \widehat{\otimes} \mathcal{C}_{[0,1]}(\mathbb{R}^n) -- \rightarrow \mathcal{K}_{[0,1]}(\mathcal{H}(\mathbb{R}^n)) \,,$$

and similarly the asymptotic morphism $\beta \colon \mathcal{S} -- \rightarrow \mathcal{C}(\mathbb{R}^n)$ determines an asymptotic morphism

$$\bar{\beta} \colon \mathcal{S} -- \rightarrow \mathcal{C}_{[0,1]}(\mathbb{R}^n)$$

by forming the tensor product of β with the identity on $C[0,1]$ and then composing with the inclusion $\mathcal{S} \subseteq \mathcal{S}[0,1]$ as constant functions. Consider then the diagram of equivariant E-theory morphisms

$$
\begin{array}{ccccc}
\mathbb{C} & \xrightarrow{\bar{\beta}} & \mathcal{C}_{[0,1]}(\mathbb{R}^n) & \xrightarrow{\bar{\alpha}} & C[0,1] \,, \\
= \downarrow & & \downarrow{\varepsilon_s} & & \downarrow{\varepsilon_s} \\
\mathbb{C} & \xrightarrow{\beta} & \mathcal{C}_s(\mathbb{R}^n) & \xrightarrow{\alpha} & \mathbb{C}
\end{array}
$$

where ε_s denotes the element induced from evaluation at $s \in [0,1]$. Observe that ε_s is an isomorphism in equivariant E-theory, for every s (indeed, ε_s, considered as a $*$-homomorphism, is an equivariant homotopy equivalence). Set $s = 0$. In this case the bottom composition is the identity element of $E_{\mathbb{Z}^n}(\mathbb{C}, \mathbb{C})$. This is because when $s = 0$ the action of $\mathbb{Z}^n$ on $\mathbb{R}^n$ is trivial and the asymptotic morphism $\beta \colon \mathcal{S} -- \rightarrow \mathcal{C}_0(\mathbb{R}^n)$ is homotopic to the (trivially equivariant) $*$-homomorphism $\beta \colon \mathcal{S} \rightarrow \mathcal{C}(\mathbb{R}^n)$ of Definition 1.26. So the required formula $\alpha \circ \beta = 1$ follows from Proposition 2.4. Since the bottom composition in the diagram is the identity it follows that the top composition is an isomorphism too.[11] Now set $s = 1$. Since, as we just showed, the top composition in the diagram is the identity, it follows that the bottom composition is the identity too. The proposition is proved.

3 Groups with the Haagerup Property

3.1 Affine Euclidean Spaces

Recall that we are using the term *Euclidean vector space* to refer to a real vector space equipped with a positive-definite inner product. In this lecture we shall be studying Euclidean spaces of possibly infinite dimension.

Definition 3.1. *An* affine Euclidean space *is a set E equipped with a simply-transitive action of the additive group underlying a Euclidean vector space V. An* affine subspace *of E is an orbit in E of a vector subspace of V. A subset X of E generates E if the smallest affine subspace of E which contains X is E itself.*

---

[11] It is the identity once $C[0,1]$ is identified with $\mathbb{C}$ via evaluation at any s, or equivalently once $C[0,1]$ is identified with $\mathbb{C}$ via the inclusion of $\mathbb{C}$ into $C[0,1]$ as constants.

Remark 3.1. Note that even if E is infinite-dimensional we are not assuming any completeness here (and moreover affine subspaces need not be closed).

Example 3.1. Every Euclidean vector space is of course an affine Euclidean space (over itself).

The following prescription makes E into a metric space.

Definition 3.2. *Let E be an affine Euclidean space over the Euclidean vector space V. If $e_1, e_2 \in E$, and if v is the unique vector in V such that $e_1 + v = e_2$, then we define the distance between e_1 and e_2 to be $d(e_1, e_2) = \|v\|$.*

Let Z be a subset of an affine Euclidean space E and let $b \colon Z \times Z \to \mathbb{R}$ be the square of the distance function: $b(z_1, z_2) = d^2(z_1, z_2)$. This function has the following properties:

(a) $b(z, z) = 0$, for all $z \in Z$,
(b) $b(z_1, z_2) = b(z_2, z_1)$, for all $(z_1, z_2) \in Z \times Z$, and
(c) for all n, all $z_1, \ldots, z_n \in Z$, and all $a_1, \ldots, a_n \in \mathbb{R}$ such that $\sum_{i=1}^n a_i = 0$,

$$\sum_{i,j=1}^n a_i b(z_i, z_j) a_j \leq 0.$$

(To prove the inequality, identify E with V and identify the sum with the quantity $-2\|\sum_{i=1}^n a_i z_i\|^2$.)

Proposition 3.1. *Let Z be a set and let $b \colon Z \times Z \to \mathbb{R}$ be a function with the above three properties. There is a map $\Phi \colon Z \to E$ of Z into an affine Euclidean space such that the image of f generates E and such that*

$$b(z_1, z_2) = d^2(\Phi(z_1), \Phi(z_2)),$$

for all $z_1, z_2 \in Z$. If $\Phi' \colon Z \to E'$ is another such map into another Euclidean space then there is a unique isometry $h \colon E \to E'$ such that $h(\Phi(z)) = \Phi'(z)$, for every $z \in Z$.

Proof. Denote by $\mathbb{R}_0[Z]$ the vector space of finitely supported, real-valued functions on Z which sum to zero:

$$\mathbb{R}_0[Z] = \{ f \in \mathbb{R}[Z] : \sum f(z) = 0 \}$$

If we equip $\mathbb{R}_0[Z]$ with the positive semidefinite form

$$\langle f_1, f_2 \rangle = -\frac{1}{2} \sum_{z_1, z_2 \in Z} f(z_1) b(z_1, z_2) f(z_2)$$

then the set of all $f \in \mathbb{R}_0[Z]$ for which $\langle f, f \rangle = 0$ is a vector subspace $\mathbb{R}_0^0[Z]$ of $\mathbb{R}_0[Z]$ (this is thanks to the Cauchy-Schwarz inequality) and the quotient

$$V = \mathbb{R}_0[Z]/\mathbb{R}_0^0[Z]$$

has the structure of a Euclidean vector space. Consider now the set of all finitely supported functions on Z which sum to 1. Let us say that two functions in this set are equivalent if their difference belongs to $\mathbb{R}_0^0[Z]$. The set of equivalence classes is then an affine Euclidean space E over V. If $\Phi\colon Z \to E$ is defined by $\Phi(z) = \delta_z$ then $d^2(\Phi(z_1), \Phi(z_2)) = b(z_1, z_2)$, as required. If $\Phi'\colon Z \to E'$ is another such map then the unique isometry h as in the statement of the lemma is given by the formula

$$h(f) = \sum f(z)\Phi'(z)$$

(note that in an affine space one can form linear combinations so long as the coefficients sum to 1).

Exercise 3.1. Justify the parenthetical assertion at the end of the proof. Prove that if h is an isometry of affine Euclidean spaces then

$$\sum a_i = 1 \quad \Rightarrow \quad h(\sum a_i e_i) = \sum a_i h(e_i).$$

This completes the uniqueness argument above.

Definition 3.3. *Let Z be a set. A function $b\colon Z \times Z \to \mathbb{R}$ is a* negative-type kernel *if b has the properties (a), (b) and (c) listed prior to Proposition 3.1.*

Thus, according to the proposition, maps into affine Euclidean spaces are classified, up to isometry, by negative-type kernels.

3.2 Isometric Group Actions

Let E be an affine Euclidean space and suppose that a group G acts on E by isometries. If e is any point of E then the function $g \mapsto g \cdot e$ maps G into E, and there is an associated negative-type function

$$b(g_1, g_2) = d^2(g_1 \cdot e, g_2 \cdot e).$$

Since G acts by isometries the function b is G-invariant, in the sense that

$$b(g_1, g_2) = b(gg_1, gg_2), \qquad \forall g, g_1, g_2 \in G,$$

and as a result it is determined by the one-variable function $b(g) = b(e, g)$, which is a negative-type function on G in the sense of the following definition.

Definition 3.4. *Let G be a group. A function $b\colon G \to \mathbb{R}$ is a* negative-type function *on G if it has the following three properties:*

(a) $b(e) = 0$,
(b) $b(g) = b(g^{-1})$, for all $g \in G$, and

(c) $\sum_{i,j=1}^{n} a_i b(g_i^{-1} g_j) a_j \leq 0$, for all n, all $g_1, \ldots, g_n \in X$, $s_i \in G$ and all $a_1, \ldots a_n \in \mathbb{R}$ such that $\sum_{i=1}^{n} a_i = 0$.

Proposition 3.2. *Let G be a set and let b be a negative-type function on G. There is an isometric action of G on an affine Euclidean space E and a point $e \in E$ such that the orbit of e generates E, and such that*

$$b(g) = d^2(e, g \cdot e),$$

for all $g \in G$.

Proof. Let E be the affine space associated to the kernel $b(g_1, g_2) = b(g_1^{-1} g_2)$, as in the statement of Proposition 3.1. There is therefore a map from G into E, which we shall write as $g \mapsto \bar{g}$, whose image generates E, and for which

$$b(g_1, g_2) = d^2(\bar{g}_1, \bar{g}_2).$$

Fix $h \in G$ and consider now the map $g \mapsto \overline{hg}$. Since

$$b(g_1, g_2) = d^2(\bar{g}_1, \bar{g}_2) = d^2(\overline{hg_1}, \overline{hg_2})$$

it follows from the uniqueness part of Proposition 3.1 that there is a (unique) isometry of E mapping $\bar{g}$ to $\overline{hg}$. The map which associates to $h \in G$ this isometry is the required action, and $e = \bar{e}$ is the required point in E. ∎

Remark 3.2. There is also a uniqueness assertion: if E' is a second affine Euclidean space equipped with an isometric G-action, and if $e' \in E$ is a point such that $b(g) = d(e', g \cdot e')^2$, for all $g \in G$, then there is a G-equivariant isometry $h \colon E \to E'$ such that $h(e) = e'$.

Remark 3.3. Proposition 3.2 is of course reminiscent of the GNS construction in C^*-algebra theory, which associates to each state of a C^*-algebra a Hilbert space representation and a unit vector in the representation space.

Exercise 3.2. Let E be an affine Euclidean space over the Euclidean vector space V. Suppose that a group G acts on E by isometries. Show that there is a linear representation π of G by orthogonal transformations on V such that

$$g \cdot (e + v) = g \cdot e + \pi(g)v,$$

for all $g \in G$, all $e \in E$, and all $v \in V$.

Exercise 3.3. According to the previous exercise, if V is viewed as an affine space over itself then for every isometric action of G on V there is a linear representation π of G by orthogonal transformations on V such that

$$g \cdot v = g \cdot 0 + \pi(g)v.$$

Show that for every $s \in [0, 1]$ the 'scaled' actions $g \cdot_s v = s(g \cdot 0) + \pi(g)(v)$ are also isometric actions of G on E.

3.3 The Haagerup Property

Definition 3.5. *Let G be a countable discrete group. An isometric action of G on an affine Euclidean space E is* metrically proper *if for some (and hence for every) point e of E,*

$$\lim_{g \to \infty} d(e, g \cdot e) = \infty.$$

In other words, an action is metrically proper if for every $R > 0$ there are only finitely many $g \in G$ such that $d(e, g \cdot e) \leq R$.

Definition 3.6. *A countable discrete group G has the* Haagerup property *if it admits a metrically proper isometric action on an affine Euclidean space.*

In view of Proposition 3.2, the Haagerup property may characterized as follows:

Proposition 3.3. *A group G has the Haagerup property if and only if there exists on G a proper, negative-type function $b \colon G \to \mathbb{R}$ (that is, a negative-type function for which the inverse image of each bounded set of real numbers is a finite subset of G).* □

Groups with the Haagerup property are also called (by Gromov [5]) *a-T-menable.* This terminology is justified by the following two results. The first is due to Bekka, Cherix and Valette [8].

Theorem 3.1. *Every countable amenable group has the Haagerup property.*

Proof. A function $\varphi \colon G \to \mathbb{C}$ is said to be *positive-definite* if $\varphi(e) = 1$,[12] if $\varphi(g) = \varphi(g^{-1})$, and if for all $g_1, \ldots, g_n \in G$, and all $\lambda_1, \ldots, \lambda_n \in \mathbb{C}$,

$$\sum_{i,j=1}^{n} \bar{\lambda}_i \varphi(g_i^{-1} g_j) \lambda_j \geq 0.$$

Observe that if φ is positive-definite then $1 - \operatorname{Re} \varphi$ is a negative-type function. Now one of the many characterizations of amenability is that G is amenable if and only if there exists a sequence $\{\varphi_n\}$ of finitely supported positive-definite functions on G which converges pointwise to the constant function 1. Given such a sequence we can find a subsequence such that the series $\sum_k (1 - \operatorname{Re} \varphi_{n_k})$ converges at every point of G. The limit is a proper, negative-type function.

The next result is essentially due to Delorme [18].

Theorem 3.2. *If G is a discrete group with Kazhdan's property T, and if G has in addition the Haagerup property, then G is finite.*

---

[12] This normalization is not always incorporated into the definition, but it is convenient here. We should also remark that the next condition $\varphi(g) = \varphi(g^{-1})$ is actually implied by the condition $\sum_{i,j=1}^{n} \bar{\lambda}_i \varphi(g_i^{-1} g_j) \lambda_j \geq 0$.

Proof. If G has property T then every isometric action of G on an affine Hilbert space has a fixed point (this is Delorme's theorem).[13] But if an isometric action has a fixed point it cannot be metrically proper, unless G is finite.

Remark 3.4. The reader is referred to [17] for a comprehensive introduction to the theory of property T groups. We shall also return to the subject in the last lecture.

Various classes of discrete groups are known to have the Haagerup property. Here is an incomplete list.

- Amenable groups (see above),
- Finitely generated free groups [30], or more generally, groups which act properly on locally finite trees.
- Coxeter groups [9],
- Discrete subgroups of $SO(n, 1)$ and $SU(n, 1)$ [56, 55],
- Thompson's groups [20, 51].

For more information about the Haagerup property consult [12].

3.4 The Baum-Connes Conjecture

The main objective of this lecture is to discuss the proof of the following theorem:

Theorem 3.3. *Let G be a countable discrete group with the Haagerup property. There exists a proper G-C^*-algebra B and E_G-theory elements $\alpha \in E_G(B, \mathbb{C})$ and $\beta \in E_G(\mathbb{C}, B)$ such that $\alpha \circ \beta = 1 \in E_G(\mathbb{C}, \mathbb{C})$.*

Thanks to the theory developed in the last lecture this has the following consequence:

Corollary 3.1. *Let G be a countable discrete group with the Haagerup property and let D be a G-C^*-algebra. The maximal Baum-Connes assembly map with coefficients in D is an isomorphism. Moreover if G is exact then the reduced Baum-Connes assembly map with coefficients in D is also an isomorphism*

Remark 3.5. The theorem and its corollary are also true for locally compact groups with the Haagerup property.

Remark 3.6. In fact the final conclusion is known to hold whether or not G is exact, but the proof involves supplementary arguments which we shall not develop here. In any case, perhaps the most striking application of the corollary is to amenable groups, and here of course the full and reduced assembly maps are one and the same (since the full and reduced crossed product C^*-algebras are one and the same).

In connection with the last remark it is perhaps worth noting that the following problem remains unsolved:

Problem 3.4 *Is every countable discrete group with the Haagerup property C^*-exact?*

---

[13] In fact the converse is true as well.

3.5 Proof of the Main Theorem, Part One

Let E bé an affine Euclidean space equipped with a metrically proper, isometric action of a countable group G. In this section we shall build from E a proper G-C^*-algebra $\mathcal{A}(E)$. In the next section we shall construct equivariant E-theory elements α and β, as in Theorem 3.3, and in Section 3.7 we shall prove that $\alpha \circ \beta = 1$.

Notation 3.5 *From here on we shall fix an affine Euclidean space E over a Euclidean vector space V. We shall be working extensively with finite-dimensional affine subspaces of E, and we shall denote these by E_a, E_b and so on. We shall denote by V_a the vector subspace of V corresponding to the finite-dimensional affine subspace E_a. If $E_a \subseteq E_b$ then we shall denote by V_{ba} the orthogonal complement of E_a in E_b. This is the orthogonal complement of V_a in V_b. Note that*

$$E_b = V_{ba} + E_a,$$

and that this is a direct sum decomposition in the sense that every point of E_b has a unique decomposition $e_b = v_{ba} + e_a$.

The following definition extends to affine spaces a definition we previously made for linear spaces. The change is only very minor.

Definition 3.7. *Let E_a be a finite-dimensional affine Euclidean subspace of E. Let $\mathcal{C}(E_a) = C_0(E_a, \mathrm{Cliff}(V_a))$.*

Here is the counterpart of Proposition 1.13:

Lemma 3.1. *Let $E_a \subseteq E_b$ be a nested pair of finite-dimensional subspaces of E. The correspondence $h \mapsto h_1 \widehat{\otimes} h_2$, where $h(v + e) = h_1(v)h_2(e)$ determines an isomorphism of graded C^*-algebras*

$$\mathcal{C}(E_b) \cong \mathcal{C}(V_{ba}) \widehat{\otimes} \mathcal{C}(E_a). \qquad \square$$

In Lecture 1 we made extensive use of the Clifford operator C. Recall that this was the function $C(v) = v$ from the Euclidean vector space V into the Clifford algebra $\mathrm{Cliff}(V)$. In the present context of affine spaces the Cliiford operator is not generally available since to define it we have to identify affine spaces with their underlying vector spaces, and we want to avoid doing this, at least for now. But we shall work with Clifford operators associated to various vector spaces which appear as orthogonal complements.

The following is a minor variation on Definition 1.26.

Definition 3.8. *Let V_a be a finite-dimensional linear subspace of V and denote by C_a the corresponding Clifford operator. Define a $*$-homomorphism*

$$\beta_a : \mathcal{S} \to \mathcal{S} \widehat{\otimes} \mathcal{C}(V_\dashv)$$

by the formula

$$\beta_a(f) = f(X \widehat{\otimes} 1 + 1 \widehat{\otimes} C_a).$$

Remark 3.7. The definition uses the language of unbounded multipliers. An alternative formulation, using the 'comultiplication' Δ, is that β_a is the composition

$$\mathcal{S} \xrightarrow{\;\Delta\;} \mathcal{S}\widehat{\otimes}\mathcal{S} \xrightarrow{\;1\widehat{\otimes}\beta\;} \mathcal{S}\widehat{\otimes}\mathcal{C}(\mathcal{V}_\dashv),$$

where $\beta\colon \mathcal{S} \to \mathcal{C}(\mathcal{V}_\dashv)$ is the $*$-homomorphism $\beta(f) = f(C_a)$ of Definition 1.26.

We are now going to construct a C^*-algebra $\mathcal{A}(E)$ as a direct limit of C^*-algebras $\mathcal{S}\widehat{\otimes}\mathcal{C}(\mathcal{E}_\dashv)$ associated to finite-dimensional affine subspaces E_a of E.

Definition 3.9. *Let $E_a \subseteq E_b$ be a nested pair of finite-dimensional affine subspaces of E. Define a $*$-homomorphism*

$$\beta_{b,a} : \mathcal{S}\widehat{\otimes}\mathcal{C}(\mathcal{E}_\dashv) \to \mathcal{S}\widehat{\otimes}\mathcal{C}(\mathcal{E}_\lfloor)$$

by using the identification $\mathcal{S}\widehat{\otimes}\mathcal{C}(\mathcal{E}_\lfloor) \cong \mathcal{S}\widehat{\otimes}\mathcal{C}(\mathcal{V}_{\lfloor\dashv})\widehat{\otimes}\mathcal{C}(\mathcal{E}_\dashv)$ and the formula

$$\mathcal{S}\widehat{\otimes}\mathcal{C}(\mathcal{E}_\dashv) \ni \{\widehat{\otimes}\langle \longmapsto \beta_{\lfloor\dashv}(\{)\widehat{\otimes}\langle \in \mathcal{S}\widehat{\otimes}\mathcal{C}(\mathcal{V}_{\lfloor\dashv})\widehat{\otimes}\mathcal{C}(\mathcal{E}_\dashv) \cong \mathcal{S}\widehat{\otimes}\mathcal{C}(\mathcal{E}_\lfloor),$$

where $\beta_{ba}\colon \mathcal{S} \to \mathcal{S}\widehat{\otimes}\mathcal{C}(\mathcal{V}_{\lfloor\dashv})$ is the $*$-homomorphism of Definition 3.8.

Lemma 3.6 *Let $E_a \subseteq E_b \subseteq E_c$ be finite-dimensional affine subspaces of E. We have $\beta_{c,b} \circ \beta_{b,a} = \beta_{c,a}$.*

Proof. Compute using the generators $u(x) = e^{-x^2}$ and $v(x) = xe^{-x^2}$ of $\mathcal{S}$.

As a result the graded C^*-algebras $\mathcal{S}\widehat{\otimes}\mathcal{C}(\mathcal{E}_\dashv)$, where E_a ranges over the finite-dimensional affine subspaces of E, form a directed system, as required, and we can make the following definition:

Definition 3.10. *Let E be an affine Euclidean space. The C^*-algebra of E, denoted $\mathcal{A}(E)$, is the direct limit C^*-algebra*

$$\mathcal{A}(E) = \varinjlim_{\substack{E_a \subseteq E \\ \textit{fin. dim.} \\ \textit{affine sbsp.}}} \mathcal{S}\widehat{\otimes}\mathcal{C}(\mathcal{E}_\dashv).$$

An action of G by isometries on E makes $\mathcal{A}(E)$ into a G-C^*-algebra. To see this, first define $*$-isomorphisms

$$g_{**}\colon \mathcal{C}(E_a) \to \mathcal{C}(gE_a)$$

by $(g_{**}f)(e) = g_*((f(g^{-1}e))$, where here $g_*\colon \mathrm{Cliff}(V_a) \to \mathrm{Cliff}(gV_a)$ is induced from the linear isometry of V associated to $g\colon E \to E$ (see Exercise 3.2).

Lemma 3.7 *The following diagram commutes:*

$$
\begin{array}{ccc}
\mathcal{S}\widehat{\otimes}\mathcal{C}(\mathcal{E}_{\dashv}) & \xrightarrow{\ \beta_{b,a}\ } & \mathcal{S}\widehat{\otimes}\mathcal{C}(\mathcal{E}_{|}) \\
{\scriptstyle 1\widehat{\otimes}g_{**}}\Big\downarrow & & \Big\downarrow{\scriptstyle 1\widehat{\otimes}g_{**}} \\
\mathcal{S}\widehat{\otimes}\mathcal{C}(\}\mathcal{E}_{\dashv}) & \xrightarrow[\ \beta_{gb,ga}\ ]{} & \mathcal{S}\widehat{\otimes}\mathcal{C}(\}\mathcal{E}_{|}).
\end{array}
$$

□

The lemma asserts that the maps g_{**} are compatible with the maps in the directed system which is used to define $\mathcal{A}(E)$. Consequently, we obtain a map g_{**} on the direct limit. In this way $\mathcal{A}(E)$ is made into a G-C^*-algebra, as required.

Theorem 3.8. *Let E be an affine Euclidean space equipped equipped with a metrically proper action of a countable discrete group G. Then the C^*-algebra $\mathcal{A}(E)$ is a proper G-C^*-algebra.*

Proof. Denote by $\mathcal{Z}(E_a)$ the grading-degree zero part of the center of the C^*-algebra $\mathcal{S}\widehat{\otimes}\mathcal{C}(\mathcal{E}_{\dashv})$. It is isomorphic to the algebra of continuous functions, vanishing at infinity, on the locally compact space $[0,\infty) \times E_a$. The linking map $\beta_{b,a}$ embeds $\mathcal{Z}(E_a)$ into $\mathcal{Z}(E_b)$, and so we can form the direct limit $\mathcal{Z}(E)$, which is a C^*-subalgebra of $\mathcal{A}(E)$, and is contained in the grading-degree zero part of the center of $\mathcal{A}(E)$ (in fact it is the entire degree zero part of the center). The C^*-subalgebra $\mathcal{Z}(E)$ has the property that $\mathcal{Z}(E) \cdot \mathcal{A}(E)$ is dense in $\mathcal{A}(E)$. The Gelfand spectrum of $\mathcal{Z}(E)$ is the locally compact space $Z = [0,\infty) \times \overline{E}$, where $\overline{E}$ is the metric space completion of E and Z is given the weakest topology for which the projection to $\overline{E}$ is weakly continuous[14] and the function $t^2 + d^2(e_0, e)$ is continuous, for some (hence any) fixed $e_0 \in E$. If G acts metrically properly on V then the induced action on the locally compact space Z is proper.

Remark 3.8. The above elegant argument is due to G. Skandalis.

3.6 Proof of the Main Theorem, Part Two

In this section we shall assume that E is a *countably* infinite-dimensional affine Euclidean space on which G acts by isometries (this simplifies one or two points of our presentation). For later purposes it will be important to work with actions which are not necessarily proper. Note however that if G has the Haagerup property then G will act properly and isometrically on some countably infinite-dimensional affine space E.

We are going to construct classes $\alpha \in E_G(\mathcal{A}(E), \mathbb{C})$ and $\beta \in E_G(\mathbb{C}, \mathcal{A}(E))$. We shall begin with the construction of β, and for this purpose we fix a point $e_0 \in E$.

---

[14] Observe that $\overline{E}$ is an affine space over the Hilbert space $\overline{V}$; by identifying $\overline{E}$ as an orbit of $\overline{V}$ we can transfer the weak topology of the Hilbert space $\overline{V}$ to $\overline{E}$.

This point is, by itself, an affine subspace of E, and there is therefore an inclusion $*$-homomorphism

$$\beta: \mathcal{S} \to \mathcal{A}(\mathcal{E}).$$

The image of β lies in all those subalgebras $\mathcal{S}\widehat{\otimes}\mathcal{C}(\mathcal{E}_\dashv)$ for which $e_0 \in E_a$, and considered as a map into $\mathcal{S}\widehat{\otimes}\mathcal{C}(\mathcal{E}_\dashv)$ the $*$-homomorphism β is given by the formula

$$\beta: f \mapsto f(C_{a,0}),$$

where $C_{a,0}: E_a \to \text{Cliff}(V_a)$ is defined by $C_{a,0}(e) = e - e_0 \in V_a$.

Lemma 3.2. *If $\beta: \mathcal{S} \dashrightarrow \mathcal{A}(\mathcal{E})$ is the asymptotic morphism defined by*

$$\beta_t(f) = \beta(f_t),$$

where $f_t(x) = f(t^{-1}x)$, then β is G-equivariant.

Proof. We must show that if e_0 and e_1 are two points in a finite-dimensional affine space E_a, then for every $f \in \mathcal{S}$,

$$\lim_{t\to\infty} \|f(t^{-1}C_{a,0}) - f(t^{-1}C_{a,1})\| = 0,$$

where $C_{a,0}$ is as above and similarly $C_{a,1}(e) = e - e_1$. It suffices to compute the limit for the functions $f(x) = (x \pm i)^{-1}$. For these one has

$$\|f(t^{-1}C_{a,0}) - f(t^{-1}C_{a,1})\| = t^{-1}\|C_{a,0} - C_{a,1}\| = t^{-1}d(e_0, e_1).$$

The proof is complete.

Definition 3.11. *The element $\beta \in E_G(\mathbb{C}, \mathcal{A}(E))$ is the E-theory class of the equivariant asymptotic morphism $\beta: \mathcal{S} \dashrightarrow \mathcal{A}(\mathcal{E})$ defined by $\beta_t: f \mapsto \beta(f_t)$.*

The definition of α is a bit more involved. It will be the E-theory class of an asymptotic morphism

$$\alpha: \mathcal{A}(E) \dashrightarrow \mathcal{K}(\mathcal{H}(E)),$$

and our first task is to associate a Hilbert space $\mathcal{H}(E)$ to the infinite-dimensional affine Euclidean space E. We begin by broadening Definition 1.27 to the context of affine spaces.

Definition 3.12. *Let E_a be a finite-dimensional affine subspace of E, with associated linear subspace V_a. The Hilbert space of E_a is the space of square integrable $\text{Cliff}(V_a)$-valued functions on E_a:*

$$\mathcal{H}(E_a) = L^2(E_a, \text{Cliff}(V_a)).$$

This is a graded Hilbert space, with grading inherited from that of $\text{Cliff}(V_a)$.

The following is the Hilbert space counterpart of Lemma 3.1.

Lemma 3.3. *Let* $E_a \subset E_b$ *be a nested pair of finite-dimensional subspaces of the affine space* E *and let* V_{ba} *be the orthogonal complement of* E_a *in* E_b. *The correspondence* $h \leftrightarrow h_1 \widehat{\otimes} h_2$, *where* $h(v + e) = h_1(v)h_2(e)$ *determines an isomorphism of graded Hilbert spaces* $\mathcal{H}(E_b) \cong \mathcal{H}(V_{ba}) \widehat{\otimes} \mathcal{H}(E_a)$. $\square$

Following the same path that we took in the last section, the next step is to assemble the spaces $\mathcal{H}(E_a)$ into a directed system.

Definition 3.13. *If* W *is a finite-dimensional Euclidean vector space* V *then the* basic vector $f_W \in \mathcal{H}(W)$ *is defined by*

$$f_W(w) = \pi^{-\frac{1}{4}\dim(W)} e^{-\frac{1}{2}\|w\|^2}.$$

Thus f_W *maps* $w \in W$ *to the multiple* $\pi^{-\frac{1}{4}\dim(W)} e^{-\frac{1}{2}\|w\|^2}$ *of the identity element in* $\mathrm{Cliff}(W)$.

Remark 3.9. The constant $\pi^{-\frac{1}{4}\dim(W)}$ is chosen so that $\|f_W\| = 1$.

Using the basic vectors $f_{ba} \in \mathcal{H}(V_{ba})$ we can organize the Hilbert spaces $\mathcal{H}(E_a)$ into a directed system as follows.

Definition 3.14. *If* $E_a \subseteq E_b$ *then define an isometry of graded Hilbert spaces* $V_{ba} \colon \mathcal{H}(E_a) \to \mathcal{H}(E_b)$ *by*

$$\mathcal{H}(E_a) \ni f \longmapsto f_{ba} \widehat{\otimes} f \in \mathcal{H}(V_{ba}) \widehat{\otimes} \mathcal{H}(E_a) \cong \mathcal{H}(E_b).$$

Lemma 3.9 *Let* $E_a \subseteq E_b \subseteq E_c$ *be finite-dimensional affine subspaces of* E. *Then* $V_{ca} = V_{cb}V_{ba}$. $\square$

We therefore obtain a directed system, as required, and we can make the following definition:

Definition 3.15. *Let* E *be an affine Euclidean space. The graded Hilbert space* $\mathcal{H}(E)$ *is the direct limit*

$$\mathcal{H}(E) = \varinjlim_{\substack{E_a \subseteq E \\ \text{fin. dim.} \\ \text{affine sbsp.}}} \mathcal{H}(E_a),$$

in the category of Hilbert spaces and graded isometric inclusions.

If G acts isometrically on E then $\mathcal{H}(E)$ is equipped with a unitary representation of G, just as $\mathcal{A}(E)$ is equipped with a G-action.

We are now almost ready to begin the definition of the asymptotic morphism $\alpha \colon \mathcal{A}(E) \dashrightarrow \mathcal{K}(\mathcal{H}(E))$. What we are going to do is construct a family of asymptotic morphisms,

$$\alpha^a \colon \mathcal{S} \widehat{\otimes} \mathcal{C}(\mathcal{E}_\dashv) \dashrightarrow \mathcal{K}(\mathcal{H}(\mathcal{E})),$$

one for each finite-dimensional subspace of E, and then prove that if $E_a \subseteq E_b$ then the diagram

$$
\begin{array}{ccc}
S \widehat{\otimes} \mathcal{C}(\mathcal{E}_\dashv) & \xrightarrow{\alpha^a} & \mathcal{K}(\mathcal{H}(E)) \\
\beta_{ba} \downarrow & & \downarrow = \\
S \widehat{\otimes} \mathcal{C}(\mathcal{E}_\llcorner) & \xrightarrow[\alpha^b]{} & \mathcal{K}(\mathcal{H}(E))
\end{array}
$$

is asymptotically commutative. Once we have done that we shall obtain a asymptotic morphism defined on the direct limit $\varinjlim S \widehat{\otimes} \mathcal{C}(\mathcal{E}_\dashv)$, as required.

To give the basic ideas we shall consider first a simpler 'toy model', as follows. Suppose for a moment that E is itself a *finite-dimensional* space. Fix a point in E; call it $0 \in E$; use it to identify E with its underlying linear space V; and use this identification to define scaling maps $e \mapsto t^{-1}e$ on E, for $t \geq 1$, with the common fiexed point $0 \in E$. If $h \in \mathcal{C}(E_a)$ and if $0 \in E_a$ then define $h_t \in \mathcal{C}(E_a)$ by the usual formula $h_t(e) = h(t^{-1}e)$.

Lemma 3.4. *Let E_a be an affine subspace of a finite-dimensional affine Euclidean space E. Denote by D_a the Dirac operator for E_a and denote by $B_{a\perp} = C_{a\perp} + D_{a\perp}$ the Clifford-plus-Dirac operator for $E_a^\perp$. The formula*

$$
\alpha_t^a: f \widehat{\otimes} h \mapsto f_t(B_{a\perp} \widehat{\otimes} 1 + 1 \widehat{\otimes} D_a)(1 \widehat{\otimes} M_{h_t})
$$

defines an asymptotic morphism

$$
\alpha^a: S \widehat{\otimes} \mathcal{C}(\mathcal{E}_\dashv) \dashrightarrow \mathcal{K}(\mathcal{H}(\mathcal{E})).
$$

Proof. The operator $B_{a\perp}$ is essentially self-adjoint and has compact resolvent (see Section 1.13). So we can define $*$-homomorphisms $\gamma_t: S \to \mathcal{K}(\mathcal{H}(\mathcal{E}_\dashv^\perp))$ by $\gamma_t(f) = f_t(B_{a\perp})$. Moreover we saw in Section 1.12 that the formula

$$
\alpha_t: f \widehat{\otimes} h \mapsto f_t(D_a) M_{h_t}
$$

defines an asymptotic morphism $\alpha: S \widehat{\otimes} \mathcal{C}(\mathcal{E}_\dashv) \dashrightarrow \mathcal{K}(\mathcal{H}(\mathcal{E}_\dashv))$. The formula for α^a in the statement of the lemma is nothing but the formula for the composition

$$
S \widehat{\otimes} \mathcal{C}(\mathcal{E}_\dashv) \xrightarrow{\Delta \widehat{\otimes} 1} S \widehat{\otimes} S \widehat{\otimes} \mathcal{C}(\mathcal{E}_\dashv) \xrightarrow{\gamma \widehat{\otimes} \alpha} \mathcal{K}(\mathcal{H}(E_a^\perp)) \widehat{\otimes} \mathcal{K}(\mathcal{H}(E_a)).
$$

So α^a is an asymptotic morphism, as required.

Lemma 3.5. *Let $E_a \subseteq E_b$ be a nested pair of affine subspaces of a finite-dimensional affine Euclidean space E. Denote by D_a and D_b the Dirac operators for E_a and E_b, and denote by*

$$
\alpha^a: S \widehat{\otimes} \mathcal{C}(\mathcal{E}_\dashv) \dashrightarrow \mathcal{K}(\mathcal{H}(\mathcal{E})) \quad \text{and} \quad \alpha^\llcorner: S \widehat{\otimes} \mathcal{C}(\mathcal{E}_\llcorner) \dashrightarrow \mathcal{K}(\mathcal{H}(\mathcal{E}))
$$

the asymptotic morphisms of Lemma 3.4. The diagram

$$\mathcal{S}\widehat{\otimes}\mathcal{C}(\mathcal{E}_{\dashv}) \xrightarrow{\;\alpha^a\;} \mathcal{K}(\mathcal{H}(E))$$

$$\beta_{ba}\Big\downarrow \qquad\qquad \Big\downarrow =$$

$$\mathcal{S}\widehat{\otimes}\mathcal{C}(\mathcal{E}_{\llcorner}) \xrightarrow[\;\alpha^b\;]{} \mathcal{K}(\mathcal{H}(E))$$

is asymptotically commutative.

Proof. We shall do a computation using the generators $u(x) = e^{-x^2}$ and $v(x) = xe^{-x^2}$ of $\mathcal{S}$. Denote by E_{ba} the orthogonal complement of E_a in E_b, so that

$$E = E_b^{\perp} \oplus E_{ba} \oplus E_a$$

and

$$\mathcal{H}(E) \cong \mathcal{H}(E_b^{\perp})\widehat{\otimes}\mathcal{H}(E_{ba})\widehat{\otimes}\mathcal{H}(E_a).$$

To do the computation we need to note that under the isomorphism of Hilbert spaces $\mathcal{H}(E_b) \cong \mathcal{H}(E_{ba})\widehat{\otimes}\mathcal{H}(E_a)$ the Dirac operator D_b corresponds to $D_{ba}\widehat{\otimes}1 + 1\widehat{\otimes}D_a$ (to be precise, the self-adjoint closures of these essentially self-adjoint operators correspond to one another). Similarly $B_{a\perp}$ corresponds to $B_{b\perp}\widehat{\otimes}1 + 1\widehat{\otimes}B_{ba}$ under the isomorphism $\mathcal{H}(E_a^{\perp}) \cong \mathcal{H}(E_b^{\perp})\widehat{\otimes}\mathcal{H}(E_{ba})$. Hence by making these identifications of Hilbert spaces we get

$$\exp(-t^{-2}D_b^2) = \exp(-t^{-2}D_{ba}^2)\widehat{\otimes}\exp(-t^{-2}D_a^2)$$

and

$$\exp(-t^{-2}B_{a\perp}^2) = \exp(-t^{-2}B_{b\perp}^2)\widehat{\otimes}\exp(-t^{-2}B_{ba}^2).$$

Now, applying α_t^a to the element $u\widehat{\otimes}h \in \mathcal{S}\widehat{\otimes}\mathcal{C}(\mathcal{E}_{\dashv})$ we get

$$\exp(-t^{-2}B_{b\perp}^2)\widehat{\otimes}\exp(-t^{-2}B_{ba}^2)\widehat{\otimes}\exp(-t^{-2}D_a^2)M_{h_t}$$

in $\mathcal{K}(\mathcal{H}(E_b^{\perp}))\widehat{\otimes}\mathcal{K}(\mathcal{H}(E_{ba}))\widehat{\otimes}\mathcal{K}(\mathcal{H}(E_a))$, while applying $\alpha_t^b \circ \beta_{ba}$ to $u\widehat{\otimes}h$ we get

$$\exp(-t^{-2}B_{b\perp}^2)\widehat{\otimes}\exp(-t^{-2}D_{ba}^2)\exp(-t^{-2}C_{ba}^2)\widehat{\otimes}\exp(-t^{-2}D_a^2)M_{h_t}.$$

But we saw in Section 1.13 that the two families of operators $\exp(-t^2B_{ba}^2)$ and $\exp(-t^{-2}D_{ba}^2)\exp(-t^{-2}C_{ba}^2)$ are asymptotic to one another, as $t \to \infty$. It follows that $\alpha_t^a(u\widehat{\otimes}h)$ is asymptotic to $\alpha_t^b(\beta_{ba}(u\widehat{\otimes}h))$, as required. The calculation for $v\widehat{\otimes}h$ is similar.

Turning to the infinite-dimensional case, it is clear that the major problem is to construct a suitable operator $B_{a\perp}$. We begin by assembling some preliminary facts. Suppose that we fix for a moment a finite-dimensional affine subspace E_a of E. Denote by $E_a^{\perp}$ its orthogonal complement in E. This is an infinite-dimensional subspace of V, but in particular it is a Euclidean space in its own right, and we can form the direct limit Hilbert space $\mathcal{H}(E_a^{\perp})$ as in Definition 3.15.

Lemma 3.6. *Let E_a be a finite-dimensional affine subspace of E and let $E_a^\perp$ be its orthogonal complement in E. The isomorphisms*

$$\mathcal{H}(E_b) \cong \mathcal{H}(V_{ba}) \widehat{\otimes} \mathcal{H}(E_a) \qquad (E_a \subseteq E_b)$$

of Lemma 3.3 combine to provide an isomorphism

$$\mathcal{H}(E) \cong \mathcal{H}(E_a^\perp) \widehat{\otimes} \mathcal{H}(E_a). \quad \square$$

Definition 3.16. *Let E_a be a finite-dimensional subspace of an affine Euclidean space E. The Schwartz space of E_a, denoted $\mathfrak{s}(E_a)$ is*

$$\mathfrak{s}(E_a) = \{\, \text{Schwartz-class } \mathrm{Cliff}(V_a)\text{-valued functions on } E_a \,\}.$$

The Schwartz space $\mathfrak{s}(E)$ is the algebraic direct limit of the Schwartz spaces $\mathfrak{s}(E_a)$:

$$\mathfrak{s}(E) = \varinjlim_{\substack{E_a \subseteq E \\ \text{fin. dim.} \\ \text{affine sbsp.}}} \mathfrak{s}(E_a),$$

using the inclusions $V_{ba} : \mathfrak{s}(E_a) \to \mathfrak{s}(E_b)$.

We now want to define a suitable operator $B_{a\perp}$ on $\mathcal{H}(E_a^\perp)$ with domain $\mathfrak{s}(E_a^\perp)$. A very interesting possibility is as follows. If $V \subseteq E_a^\perp$ is a finite-dimensional subspace then the operator $B_V = C_V + D_V$ acts on every Schwartz space $\mathfrak{s}(W)$, where $V \subseteq W$: just use the formula

$$(B_V f)(w) = \sum_1^n x_i e_i (f(w)) + \sum_1^n \widehat{e}_i \left(\frac{\partial f}{\partial x_i}(w) \right),$$

from Lecture 1, where $e_1, \ldots, e_n$ is an orthonormal basis for V and $x_1, \ldots, x_n$ are the dual coordinates on V, extended to coordinates on W by orthogonal projection. The actions on the Schwartz spaces $\mathfrak{s}(W)$ are compatible with the inclusions used to define the direct limit $\mathfrak{s}(E_a^\perp) = \varinjlim \mathfrak{s}(W)$, and we obtain an unbounded, essentially self-adjoint operator on $\mathcal{H}(E_a^\perp)$ with domain $\mathfrak{s}(E_a^\perp)$. Let us now make the following key observation:

Lemma 3.7. *Suppose that $E_a^\perp$ is decomposed as an algebraic direct sum of pairwise orthogonal, finite-dimensional subspaces,*

$$E_a^\perp = V_0 \oplus V_1 \oplus V_2 \oplus \cdots.$$

If $f \in \mathfrak{s}(E_a^\perp)$ then the sum

$$B_{a\perp} f = B_0 f + B_1 f + B_2 f + \cdots,$$

where $B_j = C_j + D_j$ is the Clifford-Dirac operator on V_j, has only finitely many nonzero terms. The operator defined by the sum is essentially self-adjoint on $\mathfrak{s}(E_a^\perp)$ and is independent of the direct sum decomposition of $E_a^\perp$ used in its construction.

Proof. Observe that

$$s(E_a^\perp) = \varinjlim_n s(V_0 \oplus \cdots \oplus V_n).$$

Therefore if $f \in s(E_a^\perp)$ then f belongs to some $s(V_0 \oplus \cdots \oplus V_n)$. Its image in $s(V_0 \oplus \cdots \oplus V_{n+k})$ under the linking map in the directed system is a function of the form

$$f_k(v_0 + \cdots v_{n+k}) = \text{constant} \cdot f(v_0 + \cdots + v_n) e^{-\frac{1}{2}\|v_{n+1}\|} \cdot \ldots \cdot e^{-\frac{1}{2}\|v_{n+k}\|^2}.$$

Since $e^{-\frac{1}{2}\|v_{n+k}\|^2}$ is in the kernel of B_{n+k} we see that $B_{n+k}f = 0$ for all $k \geq 1$. This proves the first part of the lemma. Essential self-adjointness follows from the existence of an eigenbasis for $B_{a\perp}$, which in turn follows immediately from the existence of eigenbases in the finite-dimensional case (see Corollary 1.1). The fact that $B_a^\perp$ is independent of the choice of direct sum decomposition follows from the formula

$$B_{a\perp} f = B_W f \qquad \text{if } f \in s(W) \subseteq s(E_a^\perp),$$

which in turn follows from the formula $B_{W_1} + B_{W_2} = B_{W_1 \oplus W_2}$ in finite dimensions.

Unfortunately the operator $B_{a\perp}$ above does *not* have compact resolvent. Indeed

$$B_{a\perp}^2 = B_0^2 + B_1^2 + B_2^2 + \cdots,$$

from which it follows that the eigenvalues for $B_{a\perp}^2$ are the sums

$$\lambda = \lambda_0 + \lambda_1 + \lambda_2 + \cdots,$$

where λ_j is an eigenvalue for B_j^2 and where almost all λ_j are zero. It therefore follows from Proposition 1.16 that while the eigenvalue 0 occurs with multiplicity 1, each positive integer is an eigenvalue of $B_{a\perp}^2$ of infinite multiplicity.

Because $B_{a\perp}$ fails to have compact resolvent we cannot immediately follow Lemma 3.4 to obtain our asymptotic morphisms α^a. Instead we first have to 'perturb' the operators $B_{a\perp}$ in a certain way.

Notation 3.10 *We are now going to* fix *an increasing sequence* $E_0 \subseteq E_1 \subseteq E_2 \subseteq \cdots$ *of finite-dimensional affine subspaces of E whose union is E. We shall denote by V_n the orthogonal complement of E_{n-1} in E_n (and write $V_0 = E_0$), so that there is an algebraic orthogonal direct sum decomposition*

$$E = V_0 \oplus V_1 \oplus V_2 \oplus V_3 \oplus \cdots.$$

Later on we shall want to arrange matters so that this decomposition is compatible with the action of G on E, but for now any decomposition will do.

Having chosen a direct sum decomposition as above, let us make the following definitions:

Definition 3.17. *Let E_a be a finite-dimensional affine subspace of E. An algebraic orthogonal direct sum decomposition*

$$E_a^\perp = W_0 \oplus W_1 \oplus W_2 \oplus \cdots$$

is standard *if it is of the form*

$$E_a^\perp = V_a \oplus V_n \oplus V_{n+1} \oplus \cdots,$$

for some finite-dimensional linear space V_a and some $n \geq 1$, where the spaces V_n are the members of the fixed decomposition of E given above.

Definition 3.18. *Let E_a be a finite-dimensional affine subspace of E. An algebraic orthogonal direct sum decomposition*

$$E_a^\perp = Z_0 \oplus Z_1 \oplus Z_2 \oplus \cdots$$

into finite-dimensional linear subspaces is acceptable *if there is a standard decomposition*

$$E_a^\perp = W_0 \oplus W_1 \oplus W_2 \oplus \cdots$$

such that

$$W_0 \oplus \cdots \oplus W_n \subseteq Z_n \oplus \cdots \oplus Z_n \subseteq W_0 \oplus \cdots \oplus W_{n+1}$$

for all sufficiently large n.

We are now going to define perturbed operators $B_{a^\perp,t}$ which depend on a choice of acceptable decomposition, as well as on a parameter $t \in [1, \infty)$.

Definition 3.19. *Let E_a be a finite-dimensional affine subspace of E and let*

$$E_a^\perp = Z_0 + Z_1 + Z_2 + \cdots$$

be an acceptable decomposition of $E_a^\perp$ as an orthogonal direct sum of finite-dimensional linear subspaces. For each $t \geq 1$ define an unbounded operator $B_{a^\perp,t}$ on $\mathcal{H}(E_a^\perp)$, with domain $\mathfrak{s}(E_a^\perp)$, by the formula

$$B_{a^\perp,t} = t_0 B_0 + t_1 B_1 + t_2 B_2 + \cdots$$

where $t_j = 1 + t^{-1}j$, where $B_b = C_n + D_n$, and where C_n and D_n are the Clifford and Dirac operators on the finite-dimensional spaces Z_n.

It follows from Lemma 3.7 that the infinite sum actually defines an operator with domain $\mathfrak{s}(E_a^\perp)$. The perturbed operators $B_{a^\perp,t}$ have the key compact resolvent property that we need:

Lemma 3.8. *Let E_a be a finite-dimensional affine subspace of E and let*

$$E_a^\perp = Z_0 \oplus Z_1 \oplus Z_2 \oplus \cdots$$

be an acceptable decomposition of $E_a^\perp$ as an orthogonal direct sum of finite-dimensional linear subspaces. The operator

$$B_{a^\perp,t} = t_0 B_0 + t_1 B_1 + t_2 B_2 + \cdots$$

is essentially self-adjoint and has compact resolvent.

Proof. The proof of self-adjointness follows the same argument as the proof in Lemma 3.7: one shows that there is an orthonormal eigenbasis for $B_{a^\perp,t}$ in $\mathfrak{s}(E_a^\perp)$. As for compactness of the resolvent, the formula

$$B_{a^\perp,t}^2 = t_0^2 B_0^2 + t_1^2 B_1^2 + t_2^2 B_2^2 +$$

implies that the eigenvalues of $B_{a^\perp,t}^2$ are the sums

$$\lambda = t_0^2 \lambda_0 + t_1^2 \lambda_1 + t_2^2 \lambda_2 + \cdots,$$

where λ_j is an eigenvalue for B_j^2 and where almost all λ_j are zero. Since the lowest positive eigenvalue for B_j is 1, and since $t_j \to \infty$ as $j \to \infty$ (for fixed t), it follows that for any R there are only finitely many eigenvalues for $B_{a^\perp,t}{}^2$ of size R or less. This proves that $B_{a^\perp,t}$ has compact resolvent, as required.

We can now define the asymptotic morphisms $\alpha^a \colon S \widehat{\otimes} C(\mathcal{E}_\dashv) \dashrightarrow \mathcal{K}(\mathcal{H}(\mathcal{E}))$ that we need. Fix a point 0 in E and use it to define scaling automorphisms $h \mapsto h_t$ on each $C(E_a)$ for which $0 \in E_a$.

Proposition 3.4. *Let E_a be a finite-dimensional affine subspace of E for which $0 \in E_a$ and let $B_{a^\perp,t}$ be the operator associated to some acceptable decomposition of $E_a^\perp$. The formula*

$$\alpha_t^a \colon f \widehat{\otimes} h \mapsto f_t(B_{a^\perp,t} \widehat{\otimes} 1 + 1 \widehat{\otimes} D_a)(1 \widehat{\otimes} M_{h_t})$$

defines an asymptotic morphism $\alpha^a \colon S \widehat{\otimes} C(\mathcal{E}_\dashv) \dashrightarrow \mathcal{K}(\mathcal{H}(\mathcal{E}_\dashv^\perp)) \widehat{\otimes} \mathcal{K}(\mathcal{H}(\mathcal{E}_\dashv))$, and hence, thanks to the isomorphism of Lemma 3.6, an asymptotic morphism

$$\alpha^a \colon S \widehat{\otimes} C(\mathcal{E}_\dashv) \dashrightarrow \mathcal{K}(\mathcal{H}(\mathcal{E})).$$

Proof. This is proved in exactly the same way as was Lemma 3.4.

It should be pointed out that operator $B_{a^\perp,t}$ *does* depend on the choice of accepable decomposition, and so our definition of α^a appears to depend on quite a bit of extraneous data. But the situation improves in the limit as $t \to \infty$. The basic calculation here is as follows:

Lemma 3.9. *Let E_a be a finite-dimensional affine subspace of E and denote by $B_t = B_{a^\perp, t}$ and $B_t' = B_{a^\perp, t}'$ be the operators associated to two acceptable decompositions of $E_a^\perp$. Then for every $f \in \mathcal{S}$,*

$$\lim_{t \to \infty} \| f(B_t) - f(B_t') \| = 0.$$

Proof. We shall prove the following special case: we shall show that if the summands in the acceptable decompositions are Z_n and Z_n', and if

$$Z_0 \oplus \cdots \oplus Z_n \subseteq Z_0' \oplus \cdots \oplus Z_n' \subseteq Z_0 \oplus \cdots \oplus Z_{n+1}$$

for *all* n, then $\lim_{t \to \infty} \| f(B_t) - f(B_t') \| = 0$. (For the general case, which is not really any harder, see [32] and [34].)

Denote by X_n the orthogonal complement of Z_n in Z_n', and by Y_n the orthogonal complement of Z_{n-1}' in Z_n (set $Y_0 = Z_0$). There is then a direct sum decomposition

$$E_a^\perp = Y_0 \oplus X_0 \oplus Y_1 \oplus X_1 \oplus \cdots,$$

with respect to which the operators $B_{a^\perp, t}$ and $B_{a^\perp, t}'$ can be written as infinite sums

$$B_t = t_0 B_{Y_0} + t_1 B_{X_0} + t_1 B_{Y_1} + t_2 B_{X_1} + \cdots$$

and

$$B_t' = t_0 B_{Y_0} + t_0 B_{X_0} + t_1 B_{Y_1} + t_1 B_{X_1} + \cdots.$$

Since $t_j - t_{j-1} = t^{-1}$ it follows that

$$B_t - B_t' = t^{-1} B_{X_0} + t^{-1} B_{X_1} + \cdots,$$

and therefore that

$$(B_t - B_t')^2 = t^{-2} B_{X_0}^2 + t^{-2} B_{X_1}^2 + \cdots.$$

In contrast,

$$B_t^2 = t_0^2 B_{Y_0}^2 + t_1^2 B_{X_0}^2 + t_1^2 B_{Y_1}^2 + t_2^2 B_{X_1}^2 + \cdots,$$

and since $t_j^2 \geq 1$ it follows that $\| (B_t - B_t') f \| \leq t^{-1} \| B_t f \|$ for every $f \in \mathfrak{s}(E_a^\perp)$. This implies that if $f(x) = (x \pm i)^{-1}$ then

$$\| f(B_t) - f(B_t') \| = \| (B_t' \pm i)^{-1} (B_t' - B_t)(B_t \pm i)^{-1} \| \leq \| (B_t' - B_t)(B_t \pm i)^{-1} \| \leq t^{-1}.$$

An approximation argument involving the Stone-Weierstrass theorem (which we have seen before) now finishes the proof.

For later purposes we note the following simple strengthening of Lemma 3.9. It is proved by following exactly the same argument.

Lemma 3.10. *With the hypotheses of the previous lemma, is $s \in [1, \infty)$*

$$\lim_{t \to \infty} \| f(sB_t) - f(sB_t') \| = 0,$$

uniformly in s. $\square$

It follows from Lemma 3.9 that our definition of the asymptotic morphism α^a is independent, up to asymptotic equivalence, of the choice of acceptable decomposition of $E_a^\perp$ (compare the proof of Lemma 3.4).[15]

Proposition 3.5. *The diagram*

$$
\begin{array}{ccc}
\mathcal{S}\hat{\otimes}\mathcal{C}(\mathcal{E}_{\dashv}) & \xrightarrow{\;\alpha^a\;} & \mathcal{K}(\mathcal{H}(E)) \\[1mm]
{\scriptstyle \beta_{ba}}\Big\downarrow & & \Big\downarrow{\scriptstyle =} \\[1mm]
\mathcal{S}\hat{\otimes}\mathcal{C}(\mathcal{E}_{\vert}) & \xrightarrow[\;\alpha^b\;]{} & \mathcal{K}(\mathcal{H}(E))
\end{array}
$$

is asymptotically commutative.

Proof. Using the computations we made in Section 1.13, as we did in the proof of Lemma 3.5, we see that the composition $\alpha^b \circ \beta_{ba}$ is asymptotic to the asymptotic morphism

$$f\hat{\otimes}h \mapsto f_t(B_t'\hat{\otimes}1 + 1\hat{\otimes}D_a)(1\hat{\otimes}M_{h_t}),$$

where, if α^b is computed using the acceptable decomposition

$$E_b^\perp = Z_0 \oplus Z_1 \oplus Z_2 \oplus \cdots,$$

then B_t' is the operator of Definition 3.19 associated to the decomposition

$$E_a^\perp = (E_{ba} \oplus Z_0) \oplus Z_1 \oplus Z_2 \oplus \cdots.$$

But this is an acceptable decomposition for $F_a^\perp$, and so $\alpha^b \circ \beta_{ba}$ is asymptotic to α^a, as required.

It follows that the asymptotic morphisms α^a combine to form a single asymptotic morphism

$$\alpha\colon \mathcal{A}(E) \dashrightarrow \mathcal{K}(\mathcal{H}(E)).$$

Our definition of the class $\alpha \in E_G(\mathfrak{A}(E), \mathbb{C})$ is therefore almost complete. It remains only to discuss the equivariance of α.

Suppose that the countable group G acts isometrically on E. Using the point $0 \in E$ that we chose prior to the proof of Proposition 3.4, indentify E with its underlying Euclidean vector space V, and thereby define a family of actions on E, parametrized by $s \in [0, 1]$ by

$$g \cdot_s e = s(g \cdot 0) + \pi(g)v, \qquad (0 + v = e)$$

(see Exercise 3.3). Thus the action $g \cdot_1 e$ is the original action, while $g \cdot_0 e$ has a global fixed point (namely $0 \in E$).

---

[15] It should be added however that α^a *does* depend on the choice of initial direct sum decomposition, as in 3.10.

Lemma 3.11. *There exists a direct sum decompostion*

$$E = V_0 \oplus V_1 \oplus V_2 \oplus + \cdots$$

as in 3.10 such that, if $E_n = V_0 \oplus \cdots \oplus V_n$, then for every $g \in G$ there is an $N \in \mathbb{N}$ for which

$$n > N \quad \Rightarrow \quad g \cdot E_n \subseteq E_{n+1}, \quad \text{for all } s \in [0,1]. \quad \square$$

Proposition 3.6. *If the direct sum decomposition*

$$E = V_0 \oplus V_1 \oplus V_2 \oplus + \cdots$$

is chosen as in Lemma 3.11 then the asymptotic morphism $\alpha \colon \mathcal{A}(E) \dashrightarrow \mathcal{K}(\mathcal{H}(E))$ is equivariant in the sense that

$$\lim_{t \to \infty} \|\alpha_t(g \cdot a) - g \cdot_{t^{-1}} (\alpha_t(x))\| = 0,$$

for all $a \in \mathcal{A}(E)$ and all $g \in G$.

Proof. Examining the definitions, we see that on $S \widehat{\otimes} \mathcal{C}(\mathcal{E}_\dashv)$ the asymptotic morphism $a \mapsto g^{-1} \cdot_{t^{-1}} (\alpha_t(g \cdot a))$ is given by exactly the same formula used to define α^a, except for the choice of acceptable direct sum decomposition of $E_a^\perp$. But we already noted that different choices of acceptable direct sum decomposition give rise to asymptotically equivalent asymptotic morphisms, so the proposition is proved.

Definition 3.20. *Denote by $\alpha \in E_G(\mathcal{A}(E), \mathbb{C})$ the class of the asymptotic morphism $\alpha \colon \mathcal{A}(E) \dashrightarrow \mathcal{K}(\mathcal{H}(E))$.*

3.7 Proof of the Main Theorem, Part Three

Here we show that $\alpha \circ \beta = 1 \in E_G(\mathbb{C}, \mathbb{C})$. The proof is almost exactly the same as the proof of Proposition 2.21 in the last lecture.

Lemma 3.12. *Suppose that the action of G on the affine Euclidean space E has a fixed point. Then the composition*

$$\mathbb{C} \xrightarrow{\ \beta\ } \mathcal{A}(E) \xrightarrow{\ \alpha\ } \mathbb{C}$$

in equivariant E-theory is the identity morphism on $\mathbb{C}$.

Proof. The proof has three parts. First, recall that in the definition of the asymptotic morphism β we began by fixing a point of E. It is clear from the proof of Lemma 3.2 that different choices of point give rise to asymptotically equivalent asymptotic morphisms, so we might as well choose a point which is fixed for the action of G on E. But having done so each *-homomorphism $\beta_t(f) = \beta(f_t)$ in the asymptotic morphism β is individually G-equivariant. It follows that the equivariant asymptotic morphism $\beta \colon S \dashrightarrow \mathcal{A}(\mathcal{E})$ is equivariantly homotopy equivalent to the equivariant ∗-homomorphism $\beta \colon S \to \mathcal{A}(\mathcal{E})$. Using this fact, it follows that we may compute the composition $\alpha \circ \beta$ in equivariant E-theory by computing the composition of the asymptotic morphism α with the ∗-homomorphism β. But the results in Section 1.13 show that this composition is asymptotic to $\gamma \colon S \to \mathcal{K}(\mathcal{H}(\mathcal{E}))$, where

$$\gamma_t(f) = f_t(B_t),$$

and B_t is the operator of Definition 3.19 associated to any acceptable decomposition of E. This in turn is homotopic to the asymptotic morphism $f \mapsto f(B_t)$. Finally this is homotopic to the asymptotic morphism defining $1 \in E_G(\mathbb{C}, \mathbb{C})$ by the homotopy

$$f \mapsto \begin{cases} f(sB_t) & s \in [1, \infty) \\ f(0)P & s = \infty, \end{cases}$$

where P is the projection onto the kernel of B_t (note that all the B_t have the same 1-dimensional, G-fixed kernel).

Theorem 3.11. *The composition $\alpha \circ \beta \in E_G(\mathbb{C}, \mathbb{C})$ is the identity.*

Proof. Let $s \in [0, 1]$ and denote by $\mathcal{A}_s(E)$ the C^*-algebra $\mathcal{A}(E)$, but with the scaled G-action $(g, h) \mapsto g \cdot_s h$. The algebras $\mathcal{A}_s(E)$ form a continuous field of G-C^*-algebras over the unit interval. Denote by $\mathcal{A}_{[0,1]}(E)$ the G-C^*-algebra of continuous sections of this field. In a similar way, form the continuous field of G-C^*-algebras $\mathcal{K}_s(\mathcal{H}(E))$ and denote by $\mathcal{K}_{[0,1]}(\mathcal{H}(E))$ the G-C^*-algebra of continuous sections. The asymptotic morphism $\alpha \colon \mathcal{A}(E) \dashrightarrow \mathcal{K}(\mathcal{H}(E))$ induces an asymptotic morphism

$$\bar{\alpha} \colon \mathcal{A}_{[0,1]}(E) \dashrightarrow \mathcal{K}_{[0,1]}(\mathcal{H}(E)),$$

and similarly the asymptotic morphism $\beta \colon S \dashrightarrow \mathcal{A}(E)$ determines an asymptotic morphism

$$\bar{\beta} \colon S \dashrightarrow \mathcal{A}_{[0,1]}(E).$$

From the diagram of equivariant E-theory morphisms

$$\begin{array}{ccccc} \mathbb{C} & \xrightarrow{\bar{\beta}} & \mathcal{A}_{[0,1]}(E) & \xrightarrow{\bar{\alpha}} & C[0,1] \,, \\ \Big\| & & \Big\downarrow{\varepsilon_s} & & \cong\Big\downarrow{\varepsilon_s} \\ \mathbb{C} & \xrightarrow{\beta} & \mathcal{A}_s(E) & \xrightarrow{\alpha} & \mathbb{C} \end{array}$$

where ε_s denotes the element induced from evaluation at $s \in [0, 1]$, we see that if the bottom composition is the identity for some $s \in [0, 1]$ then it is the identity for all $s \in [0, 1]$. But by Lemma 3.12 the composition is the identity when $s = 0$ since the action $(g, e) \mapsto g \cdot_0 e$ has a fixed point. It follows that the composition is the identity when $s = 1$, which is what we wanted to prove.

3.8 Generalization to Fields

We conclude this lecture by quickly sketching a simple extension of the main theorem to a situation involving fields of affine spaces over a compact parameter space. This generalization will be used in the next lecture to prove injectivity results about the Baum-Connes assembly map.

Definition 3.21. *Let Z be a set. Denote by $\mathrm{NT}(Z)$ the set of negative-type kernels $b\colon Z \times Z \to \mathbb{R}$. Equip $\mathrm{NT}(Z)$ with the topology of pointwise convergence, so that $b_\alpha \to b$ in $\mathrm{NT}(Z)$ if and only if $b_\alpha(z_1, z_2) \to b(z_1, z_2)$ for all $z_1, z_2 \in Z$.*

Suppose now that X is a compact Hausdorff space and that we are given a continuous map $x \mapsto b_x$ from X into $\mathrm{NT}(Z)$. For each $x \in X$ we can construct a Euclidean vector space V_x and an affine space E_x over V_x following the prescription laid out in the proof of Proposition 3.1 (thus for example V_x is a quotient of the space of finitely supported functions on Z which sum to 0, and E_x is a quotient of the space of finitely supported functions on Z which sum to 1). We obtain in this way some sort of 'continuous field' of affine Euclidean spaces over X (we shall not need to make this notion precise).

The C^*-algebras $\mathcal{A}(E_x)$ may be put together to form a *continuous field* of C^*-algebras over X. (See [19] for a proper discussion of continuous fields.) To do so we must specify which sections $x \mapsto f_x \in \mathcal{A}(E_x)$ are to be deemed continuous, and for this purpose we begin by deeming to be continuous certain families of isometries from $\mathbb{R}^n$ into the affine spaces E_x.

Definition 3.22. *Let U be an open subset of X and let $h_u\colon \mathbb{R}^n \to E_u$ be a family of isometries of $\mathbb{R}^n$ into the affine Euclidean spaces E_u ($u \in U$) defined above. We shall say that the family is continuous if there is a finite subset $F \subseteq Z$ and if there are functions $f_{j,u}\colon Z \to \mathbb{R}$ (where $j = 1, \ldots, n$, and $u \in U$) such that*

(a) *Each function $f_{j,u}$ sums to one, and is supported in F (and therefore each $f_{j,u}$ determines a point of E_u).*
(b) *For each $z \in Z$ and each j, the value $f_{j,u}(z)$ is a continuous function of $u \in U$.*
(c) *The isometry h_u maps the standard basis element e_j of $\mathbb{R}^n$ to the point of E_u determined by $f_{j,u}$.*

Definition 3.23. *Let us say that a function $x \mapsto f_x$ which assigns to each point $x \in X$ an element of the C^*-algebra $\mathcal{A}(E_x)$ is a continuous section if for every $x \in X$ and every $\varepsilon > 0$ there is an open set U containing x, a continuous family of isometries $h_u\colon \mathbb{R}^n \to E_u$ as above, and an element $f \in \mathcal{S}\widehat{\otimes}\mathcal{C}(\mathbb{R}^\backslash)$ such that*

$$\|h_{u,**}(f) - f_u\| < \varepsilon \quad \forall u \in U.$$

Here by $h_{u,**}$ we are using an abbreviated notation for the inclusion of $\mathcal{S}\widehat{\otimes}\mathcal{C}(\mathbb{R}^\backslash)$ into $\mathcal{A}(E_u)$ induced from the isometry $h_u\colon \mathbb{R}^n \to h_u[\mathbb{R}^n] \subseteq E_u$ by forming the composition

$$\mathcal{S}\widehat{\otimes}\mathcal{C}(\mathbb{R}^\backslash) \xrightarrow{1\otimes h_{u,**}} \mathcal{S}\widehat{\otimes}\mathcal{C}(\langle_\sqcap[\mathbb{R}^\backslash]) \xrightarrow{\subseteq} \mathcal{A}(E).$$

Lemma 3.13. *With the above definition of continuous section the collection of C^*-algebras $\{\mathcal{A}(E_x)\}_{x \in X}$ is given the structure of a continuous field of C^*-algebras over the space X.* □

Definition 3.24. *Denote by $\mathcal{A}(X, E)$ the C^*-algebra of continuous sections of the above continuous field.*

If a group G acts on the set Z then G acts on $\mathrm{NT}(Z)$ by the formula $(g \cdot b)(z_1, z_2) = b(g^{-1}z_1, g^{-1}z_2)$. In what follows we shall be solely interested in the case where $Z = G$ and the action is by left translation.

Definition 3.25. *Let X be a compact space equipped with an action of a countable discrete group G by homeomorphisms. An equivariant map $x \mapsto b_x$ from X into $\mathrm{NT}(G)$ is proper-valued if for every $S \geq 0$ there is a finite set $F \subseteq G$ such that*

$$b_x(g_1, g_2) \leq S \quad \Rightarrow \quad g_1^{-1}g_2 \in F.$$

The following is a generalization of Theorem 3.8.

Proposition 3.7. *If $b \colon X \to \mathrm{NT}(G)$ is a G-equivariant and proper-valued map then the G-C^*-algebra $\mathcal{A}(X, E)$ is proper.* □

By carrying out the constructions of the previous sections fiberwise we obtain the following result (which is basically due to Tu [18]).

Theorem 3.12. *Let G be a countable discrete group and let X be a compact metrizable G-space. Assume that there exists a proper-valued, equivariant map from X into $\mathrm{NT}(G)$. Then $\mathcal{A}(X, E)$ is a proper G-C^*-algebra and there are E theory classes*

$$\alpha \in E_G(\mathcal{A}(X, E), C(X)) \quad \text{and} \quad \beta \in E_G(C(X), \mathcal{A}(X, E))$$

for which the composition $\alpha \circ \beta$ is the identity in $E_G(C(X), C(X))$. □

By trivially adapting the simple argument used to prove Theorem 2.20 we obtain the following important consequence of the above:

Corollary 3.2. *Let G be a countable discrete group and let X be a compact metrizable G-space. If there exists a proper-valued, equivariant map from X into $\mathrm{NT}(G)$ then for every G-C^*-algebra D the Baum-Connes assembly map*

$$\mu : K^{\mathrm{top}}(G, D(X)) \to K(C^*(G, D(X)))$$

is an isomorphism. If G is exact then the same is true for the assembly map into $K(C^*_\lambda(G, D(X)))$. □

4 Injectivity Arguments

The purpose of this lecture is to prove that in various cases the Baum-Connes assembly map

$$K^{top}(G, D) \rightarrow K(C^*(G, D))$$

is injective. A great deal more is known about the injectivity of the assembly map than its surjectivity. In a number of cases, injectivity is implied by a *geometric* property of G, whereas surjectivity seems to require the understanding of more subtle issues in harmonic analysis.

In all but the last section we shall work with the full crossed product $C^*(G, D)$, but all the results have counterparts for $C^*_\lambda(G, D)$. If G is exact then arguments below applied in the reduced case; otherwise different arguments are needed.

4.1 Geometry of Groups

Let G be a discrete group which is generated by a finite set S. The *word-length* of an element $g \in G$ is the minimal length $\ell(g)$ of a string of elements from S and S^{-1} whose product is g. The (left-invariant) distance function on G associated to the length function ℓ is defined by

$$d(g_1, g_2) = \ell(g_1^{-1} g_2).$$

The word-length metric depends on the choice of generating set S. Nevertheless, the 'large-scale' geometric structure of G endowed with a word-length metric is independent of the generating set: the metrics associated to two finite generating sets S and T are related by inequalities

$$\frac{1}{C} \cdot d_S(g_1 g_2) - C \leq d_T(g_1, g_2) \leq C \cdot d_S(g_1, g_2) + C,$$

where the constant $C > 0$ depends on S and T but not of course on g_1 and g_2.

Definition 4.1. *Let Z be a set and let d and δ be two distance functions on Z. They are* coarsely equivalent *if for every $R > 0$ there exists a constant $S > 0$ such that*

$$d(z_1, z_2) < R \quad \Rightarrow \quad \delta(z_1, z_2) < S$$

and

$$\delta(z_1, z_2) < R \quad \Rightarrow \quad d_1(z_1, z_2) < S$$

Thus any two word-length metrics on a finitely generated group are coarsely equivalent. When we speak of 'geometric' properties of a finitely generated group we shall mean (in this lecture) properties shared by all metrics on G which are coarsely equivalent to a word-length metric. This geometry may often be visualized using Theorem 4.1 below.

Definition 4.2. *A curve in a metric space X is a continuous map from a closed interval into X. The length of a curve $\gamma\colon [a,b] \to X$ is the quantity*

$$\text{length}(\gamma) = \sup_{a=t_0<t_1<\cdots<t_n=b} \sum_{i=1}^{n} d(\gamma(t_i), \gamma(t_{i-1})).$$

A metric space X is a length space if for all $x_1, x_2 \in X$, $d(x_1, x_2)$ is the infimum of the lengths of curves joining x_1 and x_2.

Theorem 4.1. *Let G be a finitely generated discrete group acting properly and cocompactly by isometries on a length space X. Let x be a point of X which is fixed by no nontrivial element of G. Then the distance function*

$$\delta(g_1,g_2) = d(g_1 \cdot x, g_2 \cdot x)$$

on G is coarsely equivalent to the word-length metric on G. □

See [48, 64] for the original version of this theorem and [10] for an up to date treatment. In the context of the above theorem we shall say that the *space G is coarsely equivalent* to the space X (see [57, 58] for a development of the notion of coarse equivalence between metric spaces, of which our notion of coarse equivalence between two metrics on a single space is a special case).

Example 4.1. If G is the fundamental group of a closed Riemannian manifold M then G is coarsely equivalent to the universal covering space $\widetilde{M}$.

Example 4.2. Any finitely generated group is coarsely equivalent to its Cayley graph. For example free groups are coarsely equivalent to trees.

4.2 Hyperbolic Groups

Gromov's hyperbolic groups provide a good example of how geometric hypotheses on groups lead to theorems in C^*-algebra K-theory. In this section we shall sketch very briefly the rudiments of the theory of hyperbolic groups. Later on in the lecture we shall prove the injectivity of the Baum-Connes assembly map

$$\mu\colon K^{top}(G,D) \to K(C^*(G,D))$$

for hyperbolic groups.[16] The first injectivity result in this direction is due to Connes and Moscovici [15] who essentially proved rational injectivity of the assembly map in the case $D = \mathbb{C}$. Our arguments here will however be quite different.

Definition 4.3. *Let X be a metric space. A geodesic segment in X is a curve $\gamma\colon [a,b] \to X$ such that*

$$d(\gamma(s),\gamma(t)) = |s-t|$$

for all $a \leq s \leq t \leq b$.

---

[16] We shall see that every hyperbolic group is exact; hence the reduced assembly map μ_λ is injective too.

Observe that if γ is a geodesic segment from x_1 to x_2 then the length of γ is precisely $d(x_1, x_2)$.

Definition 4.4. *A geodesic metric space is a metric space in which each two points are joined by a geodesic segment.*

Definition 4.5. *A geodesic triangle in a metric space X is a triple of points of X, together with three geodesic segments in X connecting the points pairwise. A geodesic triangle is D-slim for some $D \geq 0$ if each point on each edge lies within a distance D of some point on one of the other two edges.*

Example 4.3. Geodesic triangles in trees are 0-slim. An equilateral triangle of side R in Euclidean space is $\frac{\sqrt{3}}{4} R$-slim.

Definition 4.6. *A geodesic metric space X is D-hyperbolic if every geodesic triangle Δ in X is D-slim and hyperbolic if it is D-hyperbolic for some $D \geq 0$.*

Thus trees are hyperbolic metric spaces but Euclidean spaces of dimension 2 or more are not.

Definition 4.6 is attributed by Gromov to Rips [24]. It is equivalent to a wide variety of other conditions, for which we refer to the original work of Gromov [24] or one of a number of later expositions, for example [22, 10]. (The reader is also referred to these sources for further information on everything else in this section.)

Definition 4.7. *A finitely generated discrete group G is* word-hyperbolic, *or just* hyperbolic, *if its Cayley graph is a hyperbolic metric space.*

This definition leaves open the possibility that the Cayley graph of G constructed with respect to one finite set of generators is hyperbolic while that constructed with respect to another is not. But the following theorem asserts that this is impossible:

Theorem 4.2. *If a finitely generated group G is hyperbolic for one finite generating set then it is hyperbolic for any other.* $\square$

Examples 4.3 *Every tree is a 0-hyperbolic space and a finitely generated free group is 0-hyperbolic. The Poincaré disk Δ is a hyperbolic metric space. If G is a proper and cocompact group of isometries of Δ then G is a hyperbolic group. In particular, the fundamental group of a Riemann surface of genus 2 or more is hyperbolic.*

If G is a finitely generated group and if $K \geq 0$ then the *Rips complex* $\mathrm{Rips}(G, K)$ is the simplicial complex with vertex set G, for which a $(p+1)$-tuple $(g_0, \ldots, g_p)$ is a p-simplex if and only if $d(g_i, g_j) \leq K$, for all i and j. In the case of hyperbolic groups the Rips complex provides a simple model for the universal space eG:

Theorem 4.4. [7, 47] *Let G be a hyperbolic group. If $K \gg 0$ then the Rips complex $\mathrm{Rips}(G, K)$ is a universal proper G-space.* $\square$

In the following sections we shall need one additional construction, as follows:

Definition 4.8. *A geodesic ray in a hyperbolic space X is a continuous function*

$$c : [0, \infty) \to X$$

such that the restriction of c to every closed interval $[0, l]$ is a geodesic segment. Two geodesic rays in X are equivalent *if*

$$\limsup_{t \to \infty} d(c_1(t), c_2(t)) < \infty.$$

The Gromov boundary *of a hyperbolic metric space X is the set of equivalence classes of geodesic rays in X. The* Gromov boundary ∂G *of a hyperbolic group G is the Gromov boundary of its Cayley graph.*

The Gromov boundary ∂G does not depend on the choice of generating set. It is equipped in the obvious way with an action of G. It may also be equipped with a compact metrizable topology, on which G acts by homeomorphisms. Moreover the disjoint union $\overline{G} = G \cup \partial G$ may be equipped with a compact metrizable topology in such a way that G acts by homeomorphisms, that G is an open dense subset of $\overline{G}$, and that a sequence of points $g_n \in G$ converges to a point $x \in \partial G$ iff $g_n \to \infty$ in G and there is a geodesic ray c representing x such that

$$\sup_n d(g_n, c) < \infty.$$

From our point of view, a key feature of $\overline{G} = G \cup \partial G$ is that the action of G on $\overline{G}$ is *amenable*. We shall discuss this notion in Section 4.5.

4.3 Injectivity Theorems

In this section we shall formulate several results which assert the injectivity of the Baum-Connes map μ under various hypotheses.

Our first injectivity result is a theorem which is essentially due to Kasparov (an improved version of it, which invokes his RKK-theory, underlies his approach to the Novikov conjecture).

Theorem 4.5. *Let G be a countable discrete group. Suppose there exists a proper G-C^*-algebra B and elements $\alpha \in E_G(B, \mathbb{C})$ and $\beta \in E_G(\mathbb{C}, B)$ such that for every finite subgroup H of G the composition $\gamma = \alpha \circ \beta \in E_G(\mathbb{C}, \mathbb{C})$ restricts to the identity in $E_H(\mathbb{C}, \mathbb{C})$. Then for every G-C^*-algebra D the Baum-Connes assembly map*

$$\mu \colon K^{top}(G, D) \to K(C^*(G, D))$$

is injective.

Proof. We begin by considering the same diagram we introduced in the proof of Theorem 2.20:

$$
\begin{array}{ccc}
K^{top}(G, \mathbb{C}\widehat{\otimes}D) & \xrightarrow{\ \mu\ } & K(C^*(G, \mathbb{C}\widehat{\otimes}D)) \\
\beta_* \downarrow & & \downarrow \beta_* \\
K^{top}(G, B\widehat{\otimes}D) & \xrightarrow[\cong]{\ \mu\ } & K(C^*(G, B\widehat{\otimes}D)) \\
\alpha_* \downarrow & & \downarrow \alpha_* \\
K^{top}(G, \mathbb{C}\widehat{\otimes}D) & \xrightarrow[\mu]{} & K(C^*(G, \mathbb{C}\widehat{\otimes}D)).
\end{array}
$$

The middle assembly map is an isomorphism since $B\widehat{\otimes}D$ is a proper G-C^*-algebra. We want to show that the top assembly map is injective, and for this it suffices to show that the top left-hand vertical map $\beta_*: K^{top}(G, D) \to K^{top}(G, D\widehat{\otimes}B)$ is injective. For this we shall show that the composition

$$
K^{top}(G, \mathbb{C}\widehat{\otimes}D) \xrightarrow{\ \beta_*\ } K^{top}(G, B\widehat{\otimes}D) \xrightarrow{\ \alpha_*\ } K^{top}(G, \mathbb{C}\widehat{\otimes}D)
$$

is an isomorphism. In view of the definition of K^{top} it suffices to show that if Z is a G-compact proper G-space then the map

$$
\gamma_* = \alpha_* \circ \beta_*: E_G(Z, D) \longrightarrow E_G(Z, D)
$$

is an isomorphism. The proof of this is an induction argument on the number n of G-invariant open sets U needed to cover Z, each of which admits a map to a proper homogeneous space G/H. If $n = 1$, so that Z itself admits such a map, then $Z = G \times_H W$, where W is a compact space equipped with an action of H. There is then a commuting diagram of restriction isomorphisms

$$
\begin{array}{ccc}
E_G(G \times_H W, D) & \xrightarrow{\ \gamma_*\ } & E_G(G \times_H W, D) \\
\mathrm{Res} \downarrow \cong & & \cong \downarrow \mathrm{Res} \\
E_H(W, D) & \xrightarrow{\ \gamma_*\ } & E_H(W, D),
\end{array}
$$

(see Proposition 2.8), and the bottom map is an isomorphism (in fact the identity) since $\gamma = 1$ in $E_H(\mathbb{C}, \mathbb{C})$. If $n > 1$ then choose a G-invariant open set $U \subseteq Z$ which admits a map to a proper homogeneous space, and for which the space $Z_1 = Z \setminus U$ may be covered by $n - 1$ G-invariant open sets, each admitting a map to a proper homogeneous space. By induction we may assume that the map γ_* is an isomorphism for Z_1. Applying the five lemma to the diagram

$$
\begin{array}{ccccccc}
\cdots \longrightarrow & E_G(Z_1, D) & \longrightarrow & E_G(Z, D) & \longrightarrow & E_G(U, D) & \longrightarrow \cdots \\
& \gamma_* \downarrow \cong & & \gamma_* \downarrow & & \gamma_* \downarrow \cong & \\
\cdots \longrightarrow & E_G(Z_1, D) & \longrightarrow & E_G(Z, D) & \longrightarrow & E_G(U, D) & \longrightarrow \cdots
\end{array}
$$

we conclude that γ_* is an isomorphism for Z too.

Remark 4.1. The proof actually shows that the assembly map is *split* injective.

The second result is taken from [10] and is as follows.

Theorem 4.6. *Let X be a compact, metrizable G-space and assume that X is H-equivariantly contractible, for every finite subgroup H of G. Let D be a separable G-C^*-algebra. If the Baum-Connes assembly map*

$$\mu : K^{top}(G, D(X)) \to K(C^*(G, D(X)))$$

is an isomorphism then the Baum-Connes assembly map

$$\mu : K^{top}(G, D) \to K(C^*(G, D))$$

is split injective.

Proof. The inclusion ι of D into $D(X)$ as constant functions gives rise to a commutative diagram

$$
\begin{array}{ccc}
K^{top}(G, D(X)) & \xrightarrow{\ \mu\ } & K(C^*(G, D(X))) \\
\ \ \uparrow{\scriptstyle \iota_*} & & \ \ \uparrow{\scriptstyle \iota_*} \\
K^{top}(G, D) & \xrightarrow[\ \mu\ ]{} & K(C^*(G, D)).
\end{array}
$$

We shall prove the theorem by showing that the left vertical map is an isomorphism. In view of the definition of K^{top} it suffices to show that if Z is any G-compact proper G-space then the map

$$\iota_* : E_G(C_0(Z), D) \to E_G(C_0(Z), D(X))$$

is an isomorphism. By a Mayer-Vietoris argument like the one we used in the proof of Theorem 4.6 it actually suffices to consider the case where Z admits a map to a proper homogeneous space G/H. In this case there is a compact space W equipped with an action of H such that $Z = G \times_H W$. Consider now the following commuting diagram of restriction isomorphisms:

$$
\begin{array}{ccc}
E_G(C_0(G \times_H W), D) & \xrightarrow{\ \iota_*\ } & E_G(C_0(G \times_H W), D(X)) \\
{\scriptstyle \mathrm{Res}}\downarrow{\scriptstyle \cong} & & {\scriptstyle \cong}\downarrow{\scriptstyle \mathrm{Res}} \\
E_H(C(W), D) & \xrightarrow[\ \iota_*\ ]{} & E_H(C(W), DX)).
\end{array}
$$

The bottom horizontal map is an isomorphism (since ι is a homotopy equivalence of H-C^*-algebras) and therefore the top horizontal map is an isomorphism too.

The last injectivity result is an analytic version of a result of Carlsson-Pedersen [11]. We will not discuss the proof, but refer the reader to the original paper of Higson [10, Thm. 1.2 & 5.2]. We include it only because it applies more or less directly to the case of hyperbolic groups.

Definition 4.9. *Let G be a discrete group, let X be a G-compact, proper G-space, and let $\overline{X}$ be a metrizable compactification of G to which the action of G on X extends to an action by homeomorphisms. The extended action is* small at infinity *if for every compact set $K \subseteq X$,*

$$\lim_{g \to \infty} \operatorname{diameter}(gK) = 0,$$

where the diameters are computed using a metric on $\overline{X}$.

Theorem 4.7. *Let G be a countable discrete group. Suppose there is a G-compact model for eG having a metrizable compactification $\overline{eG}$ satisfying*

(a) *the G action on eG extends continuously to $\overline{eG}$,*
(b) *the action of G on $\overline{eG}$ is small, and*
(c) *$\overline{eG}$ is H-equivariantly contractible, for every finite subgroup H of G.*

Then for every separable G-C^*-algebra D the Baum-Connes assembly map

$$\mu : K^{top}(G, D) \to K(C^*(G, D))$$

is injective.   □

4.4 Uniform Embeddings in Hilbert Space

We are now going to apply the second theorem of the previous section to prove injectivity of the Baum-Connes assembly map for a quite broad class of groups.

Definition 4.10. *Let X and Y be metric spaces. A* uniform embedding *of X into Y is a function $f : X \to Y$ with the following two properties:*

(a) *For every $R \geq 0$ there exists some $S \geq 0$ such that*

$$d(x_1, x_2) \leq R \quad \Rightarrow \quad d(f(x_1), f(x_2)) \leq S.$$

(b) *For every $S \geq 0$ there exists $R \geq 0$ such that*

$$d(x_1, x_2) \geq R \quad \Rightarrow \quad d(f(x_1), f(x_2)) \geq S.$$

Example 4.4. If f is a bi-Lipschitz homeomorphism from X onto its image in Y then f is a uniform embedding. But note however that if the metric space X is bounded then *any* function from X to Y is a uniform embedding (in particular, uniform embeddings need not be one-to-one).

Exercise 4.1. Let G a finitely generated group and let H be a finitely generated subgroup of G. If G and H are equipped with their word-length metrics then the inclusion $H \subseteq G$ is a uniform embedding.

Remark 4.2. In the context of groups, any function satisfying condition (a) of Definition 4.10 is in fact a Lipschitz function. But condition (b) is more delicate. For example it is easy to find examples for which the optimal inequality in (b) is something like

$$d(x_1, x_2) \geq e^S \quad \Rightarrow \quad d(f(x_1), f(x_2)) \geq S$$

If a finitely generated group G acts metrically-properly on an affine Euclidean space E, and if $e \in E$, then the map $g \mapsto g \cdot e$ is a uniform embedding of G into E. We are going to prove the following result, which partially extends the main theorem of the last lecture:

This theorem is due to Tu [18] and Yu [68] (in both cases, in a somewhat disguised form).

Theorem 4.8. *Let G be a countable discrete group. If G is uniformly embeddable in a Euclidean space then for every G-C^*-algebra D the Baum-Connes assembly map*

$$\mu : K^{top}(G, D) \to K(C^*(G, D))$$

is (split) injective.

The first step of the proof is to convert a uniform embedding, which is something purely metric in nature, into something more G-equivariant. For this purpose let us recall the following object from general topology:

Definition 4.11. *Let X be a discrete set. The* Stone-Cech compactification *of X is the set βX of all nonzero, finitely additive, $\{0, 1\}$-valued probability measures on the algebra of all subsets of X. We equip βX with the topology of pointwise convergence (with respect to which it is a compact Hausdorff space).*

Thus a point of βX is a function μ from the subsets of X into $\{0, 1\}$ which is additive on finite disjoint unions and which is not identically zero. A net μ_α converges to μ if and only if $\mu_\alpha(E)$ converges to $\mu(E)$, for every $E \subseteq X$.

Example 4.5. If x is a point of X then the measure μ_x, defined by the formula

$$\mu_x(E) = \begin{cases} 1 & \text{if } x \in E \\ 0 & \text{if } x \notin E \end{cases}$$

is a point of βX. In this way X is embedded into βX as a dense open subset.

Remark 4.3. The fact that the measures μ assume only the values 0 and 1 will matter little in what follows, and we could equally well work with arbitrary, finitely additive measures for which $\mu(X) = 1$.

If b is a *bounded* complex function on X, and if μ is a finitely additive measure on X, then we may form the integral

$$\int_X b(x) \, d\mu(x)$$

as follows. First, if b assumes only finitely many values (in other words if b is a *simple* function) then define

$$\int_X b(x)\,d\mu(x) = \sum_\lambda \lambda \cdot \mu\{x : b(x) = \lambda\}.$$

Second, if b is a general bounded function, write b as a uniform limit of simple functions and define the integral of b to be the limit of the integrals of the approximants.

Exercise 4.2. If b is a bounded function then the map $\mu \mapsto \int_X b(x)\,d\mu(x)$ is a continuous function from βX into $\mathbb{C}$.

Remark 4.4. The virtue of $\{0,1\}$-valued measures is that this integration process makes sense in very great generality — it is possible to integrate any function from X into *any* compact space.

Suppose now that G is a finitely generated discrete group. The compact space βG is equipped with a continuous action of G by the formula

$$(g \cdot \mu)(E) = \mu(Eg).$$

Let $f \colon G \to E$ be a uniform embedding into an affine Euclidean space and let $b \colon G \times G \to \mathbb{R}$ be the associated negative type kernel:

$$b(g_1, g_2) = d^2(f(g_1), f(g_2)).$$

According to part (a) of Definition 4.10 the function $g \mapsto b(gg_1, gg_2)$ is bounded, for every $g_1, g_2 \in G$. As a result, we may define negative type kernels b_μ, for $\mu \in \beta G$, by integration:

$$b_\mu(g_1, g_2) = \int_G b(gg_1, gg_2)d\mu(g).$$

Observe that $b_{g \cdot \mu}(g_1, g_2) = b_\mu(g^{-1}g_1, g^{-1}g_2)$, so that our integration construction defines an equivariant map from βG into the negative type kernels on G.

Lemma 4.1. *For every $S \geq 0$ there exists $R \geq 0$ so that if $d(g_1, g_2) \geq R$ then $b_\mu(g_1, g_2) \geq S$, for every $\mu \in \beta G$.*

Proof. This is a consequence of part (b) of Definition 4.10.

Since G is finitely generated, for every R there is a finite set F so that if $d(g_1, g_2) < R$ then $g_1^{-1}g_2 \in F$. The map $\mu \mapsto b_\mu$ is therefore *proper-valued* in the sense of Definition 3.25, and we have proved the following result:

Proposition 4.1. *If a finitely generated group G may be uniformly embedded into an affine Euclidean space then there is an equivariant, proper-valued continuous map from βG into the space $\mathrm{NT}(G)$ of negative-type kernels on G.* $\square$

It will now be clear that to prove Theorem 4.8 we mean to apply Theorems 3.12 and 4.6. To do so we must replace βG by a compact G-space which is smaller (second countable) and more connected (in fact contractible) than βG. This is done as follows.

Lemma 4.2. *Let G be a countable group, let X be a compact G-space and let $b: X \to \text{NT}(G)$ be a continuous and G-equivariant map from X into the negative type kernels on G. There is a metrizable compact G-space Y and a G-map from X to Y through which the map b factors.*

Proof. Take Y to be the Gelfand dual of the separable C^*-algebra of functions on X generated by the functions $x \mapsto b_{gx}(g_1, g_2)$, for all $g, g_1, g_2 \in G$.

Lemma 4.3. *Let G be a countable group, let Y be a compact metrizable G-space and let $b: Y \to \text{NT}(G)$ be a proper-valued, G-equivariant continuous map. There is a metrizable compact G-space Z which is H-equivariantly contractible, for every finite subgroup H of G, and a proper-valued, G-equivariant continuous map from Z into $\text{NT}(G)$.*

Proof. Let Z be the compact space of Borel probability measures on Y (we give Z the weak* topology it inherits as a subset of the dual of $C(Y)$; note that we are speaking now of *countably additive* measures defined on the Borel σ-algebra). If $\mu \in Z$ then define $b_\mu \in \text{NT}(G)$ by integration:

$$b_\mu(g_1, g_2) = \int_Y b_y(g_1, g_2) \, d\mu(y).$$

The map $\mu \mapsto b_\mu$ has the required properties.

Proof (Proof of Theorem 4.8). Proposition 4.1 and the lemmas above show that the hypotheses of Theorem 3.12 and Corollary 3.2 are met. Theorem 4.6 then implies injectivity of the assembly map, as required.

4.5 Amenable Actions

In this section we shall discuss a means of constructing uniform embeddings of groups into affine Hilbert spaces.

Definition 4.12. *Let G be a discrete group. Denote by $\text{prob}(G)$ the set of functions $f: G \to [0, 1]$ such that $\sum_{g \in G} f(g) = 1$. Equip $\text{prob}(G)$ with the topology of pointwise convergence, so that $f_\alpha \to f$ if and only if $f_\alpha(g) \to f(g)$ for every $g \in G$. Equip $\text{prob}(G)$ with an action of G by homeomorphisms via the formula $(g \cdot f)(h) = f(g^{-1}h)$.*

Definition 4.13. *Let G be a countable discrete group. An action of G by homeomorphisms on a compact Hausdorff space X is amenable if there is a sequence of continuous maps $f_n : X \to \text{prob}(G)$ such that for every $g \in G$*

$$\lim_{n \to \infty} \sup_{x \in X} \|f_n(g \cdot x) - g \cdot (f_n(x))\|_1 = 0.$$

Here, if k is a function on G, then we define $\|k\|_1 = \sum_{g \in G} |k(g)|$.

We are going to prove the following result:

Proposition 4.9 *If a finitely generated group G acts amenably on a compact space X then G is uniformly embeddable in a Hilbert space.*

Remark 4.5. The method below can easily be modified to show that if a countable group G (which is not necessarily finitely generated) acts amenably on some compact space X then there is an equivariant, proper-valued map from X into the negative-type kernels on G. The methods of the previous section then show that the Baum-Connes assembly map is injective for G.

Examples 4.10 *Every hyperbolic group acts amenably on its Gromov boundary. If G is a discrete subgroup of a connected Lie group H then G acts amenably some compact homogeneous space H/P. If G is a discrete group of finite asymptotic dimension then G acts amenably on the Stone-Cech compactification βG. See [33] for a discussion of all these cases (along with references to proofs).*

Definition 4.14. *Let Z be a set. A function $\varphi \colon Z \times Z \to \mathbb{C}$ is a positive-definite kernel on the set Z if $\varphi(z, z) = 1$ for all z, if $\varphi(z_1, z_2) = \overline{\varphi(z_2, z_1)}$, for all $z_1, z_2 \in Z$, and if*

$$\sum_{i,j=1}^{k} \overline{\lambda_i} \varphi(z_i, z_j) \lambda_j \geq 0$$

for all positive integers k, all $\lambda_1, \dots, \lambda_k \in \mathbb{C}$ and all $z_1, \dots, z_k \in Z$.

Remark 4.6. The normalization $\varphi(z, z) = 1$ is not always made, but it is useful here. As is the case with positive-definite functions on groups (which we discussed in Lecture 4), the condition $\varphi(z_1, z_2) = \overline{\varphi(z_2, z_1)}$ is actually implied by the last condition.

Comparing definitions, the following is immediate:

Lemma 4.4. *If φ is a positive-definite kernel on a set Z if $\operatorname{Re} \varphi$ denotes its real part, then $1 - \operatorname{Re} \varphi$ is a negative type kernel on Z.* $\square$

Proof (Proof of Proposition 4.9). Suppose that G acts amenably on a compact space X, and let $f_n \colon X \to \operatorname{prob}(G)$ be a sequence of functions as in Definition 4.13. After making suitable approximations to the f_n we may assume that for each n there is a finite set $F \subseteq G$ such that for every $x \in X$ the function $f_n(x) \in \operatorname{prob}(G)$ is supported in F. Now let $h_n(x, g) = f_n(x)(g)^{1/2}$. Then fix a point $x \in X$ and define functions $\varphi_n \colon G \times G \to \mathbb{C}$ by

$$\varphi_n(g_1, g_2) = \sum_{g \in G} h_n(g_1 x, g_1 g) h_n(g_2 x, g_2 g).$$

These are positive definite kernels on $G \times G$. For every finite subset $F \subseteq G$ and every $\varepsilon > 0$ there is some $N \in \mathbb{N}$ such that

$$n > N \quad \text{and} \quad g_1^{-1}g_2 \in F \quad \Rightarrow \quad |\varphi_n(g_1, g_2) - 1| < \varepsilon.$$

In addition, for every $n \in \mathbb{N}$ there exists a finite subset $F \subseteq G$ such that

$$g_1^{-1}g_2 \notin F \quad \Rightarrow \quad \varphi_n(g_1, g_2) = 0.$$

It follows that for a suitable subsequence the series $\sum_j (1 - \mathrm{Re}\,\varphi_{n_j})$ is pointwise convergent everywhere on $G \times G$. But each function $1 - \mathrm{Re}\,\varphi_{n_j}$ is a negative type kernel, and therefore so is the sum. The map into affine Euclidean space which is associated to the sum is a uniform embedding.

Remark 4.7. In fact it is possible to characterize the amenability of a group action in terms of positive definite kernels. See [2] for a clear and rapid presentation of the facts relevant here, and [3] for a comprehensive account of amenability. The existence of a sequence of positive definite kernels on G which have the two properties displayed in the proof of the lemma is equivalent to the amenability of the action of G on its Stone-Cech compactification βG. See [2] again, and see also Section 5.6 for more on this topic.

Remark 4.8. The theory of amenable actions is very closely connected to the theory of exact groups. To see why, suppose that G admits an amenable action on some compact space X. Then using the theory of positive-definite kernels it may be shown that

$$C^*(G, X) = C_\lambda^*(G, X)$$

and moreover that the crossed product C^*-algebra is *nuclear*. This means that the for any C^*-algebra D,

$$C^*(G, X) \otimes_{max} D = C^*(G, D) \otimes_{min} D.$$

See [2] for a discussion of these results. It follows of course that the crossed product C^*-algebra is *exact*, in the sense of Definition 2.10. But then it follows that $C_\lambda^*(G)$, which is a subalgebra of $C^*(G, X) = C_\lambda^*(G, X)$, is exact too. Therefore, by Proposition 2.6 the group G is exact. To summarize: *if G acts amenably on some compact space then G is exact.* In fact the converse to this is true too: see Section 5.6.

4.6 Poincaré Duality

We conclude this lecture with a few remarks concerning a 'dual' formulation of the Baum-Connes conjecture for certain groups. With an application to Lecture 6 in mind we shall formulate the following theorem in the context of reduced crossed products.

Theorem 4.11. *Let G be a countable exact group and let A be a separable proper G-C^*-algebra. Suppose that there is a class $\alpha \in E_G(A, \mathbb{C})$ with the property that for every finite subgroup H of G the restricted class $\alpha \mid_H \in E_H(A, \mathbb{C})$ is invertible. Then the Baum-Connes assembly map*

$$\mu_\lambda \colon K^{top}(G, D) \to K(C_\lambda^*(G, D))$$

is an isomorphism for a given separable G-C^*-algebra D if and only if the map

$$\alpha_* \colon K(C^*_\lambda(G, A \widehat{\otimes} D)) \to K(C^*_\lambda(G, D))$$

induced from α is an isomorphism.

Remark 4.9. In the proof we shall identify μ with α_*, so that μ will be for example injective if and only if α_* is injective. As usual, analogous statements may be proved for reduced crossed products, either in the same way if G is exact, or with some additional arguments otherwise.

Proof. Consider the diagram

$$
\begin{array}{ccc}
K^{top}(G, A \widehat{\otimes} D) & \xrightarrow{\;\mu_\lambda\;} & K(C^*_\lambda(G, A \widehat{\otimes} D)) \\
{\scriptstyle \alpha_*} \big\downarrow & & \big\downarrow {\scriptstyle \alpha_*} \\
K^{top}(G, D) & \xrightarrow{\;\mu_\lambda\;} & K(C^*_\lambda(G, D)),
\end{array}
$$

in which the horizontal arrows are the Baum-Connes assembly maps and the vertical arrows are induced by composition with α in E_G-theory and by composition with the element descended from α in nonequivariant E-theory. The diagram commutes. By Theorem 2.20 the top horizontal map is an isomorphism. Furthermore, an argument like the ones used in Section 4.3 shows that the left hand vertical map is an isomorphism. Therefore the bottom horizontal map is an isomorphism if and only if the right vertical map is an isomorphism, as required.

The theorem in effect reformulates the Baum-Connes conjecture entirely in the framework of K-theory (hence the term 'Poincaré duality', since we have replaced the K-homological functor K^{top} of G with K-theory). It has an important application to groups which act isometrically on Riemannian manifolds. We shall not go into details, but here is a rapid summary of the relevant facts. The Clifford algebra constructions we developed in Lecture 1 may be generalized to complete Riemannian manifolds M. We denote by $\mathcal{C}(M)$ the C^*-algebra of sections of the bundle of Clifford algebras $\text{Cliff}(T_x M)$ associated to the tangent spaces of M. There is a Dirac operator on M (an unbounded self-adjoint operator acting on the Hilbert space of L^2-sections of the Clifford algebra bundle on M), and it defines a class

$$\alpha \in E(\mathcal{C}(M), \mathbb{C})$$

in almost exactly the same way that we defined α for linear spaces. Moreover if a group G acts isometrically on M then the Dirac operator defines an equivariant class

$$\alpha \in E_G(\mathcal{C}(M), \mathbb{C}).$$

Now if M happens to be a universal proper G-space then the hypotheses of Theorem 4.11 are met:

Proposition 4.12 *Let* M *be a complete Riemannian manifold and suppose that a countable group* G *acts on* M *by isometries. Assume further that* M *is a universal proper* G-*space. The Dirac operator on* M *defines an equivariant* E-*theory class*

$$[D] \in E_G(\mathcal{C}(M), \mathbb{C}),$$

which, restricting from G *to and finite subgroup* $H \subseteq G$, *determines invertible elements*

$$[D]\,|_H \in E_H(\mathcal{C}(M), \mathbb{C}). \qquad \square$$

The proposition applies for example when G is a lattice in a semisimple group (take M to be the associated symmetric space), and in this case (which is perhaps the most important case of the Baum-Connes conjecture yet to be resolved) the conjecture reduces to a statement which can be formulated purely within K-theory.

5 Counterexamples

In this lecture we shall present a miscellany of examples and counterexamples. Together they show that the Baum-Connes conjecture is the weakest conjecture of its type which one can reasonably formulate. They also point to shortcomings in the machinery we have developed in these lectures. The counterexamples involve Kazhdan's property T and expander graphs.

5.1 Property T

Definition 5.1. *A discrete group* G *has property* T *if the trivial representation is an isolated point in the unitary dual of* G.

See the monograph [17] for an extensive discussion of property T. We shall use the following equivalent formulations of property T:

Theorem 5.1. *Let* G *be a discrete group. The following are equivalent:*

(a) G *has property* T.
(b) *Every isometric action of* G *on an affine Euclidean space has a fixed point.*
(c) *There is a central projection* $p \in C^*(G)$ *with the property that in any unitary representation of* G, *on a Hilbert space* $\mathcal{H}$ *the operator* p *acts as the orthogonal projection onto the* G-*fixed vectors in* $\mathcal{H}$. $\quad \square$

The projection $p \in C^*(G)$ will be called the *Kazhdan projection* for the property T group G.

Remark 5.1. If G is finite then the Kazhdan projection p is the sum

$$p = \frac{1}{|G|} \sum_{g \in G} g.$$

in the group algebra $\mathbb{C}[G] = C^*(G)$ (in the formula we are regarding G as a unitary subgroup of $C^*(G)$). If G is infinite then p is a very mysterious object. For example if we (mistakenly) regard p as an infinite formal series of group elements, $p = \sum a_g \cdot g$, then from the easily proved relation $g \cdot p = p$ we conclude that all the scalars a_g are equal, while from the fact that p acts as 1 in the trivial representation we conclude that the scalars a_g sum to 1. Thus we arrive at a formula for p like the one displayed above, where the sum is infinitely large and the normalizing constant $1/|G|$ is infinitely small.

It is quite difficult to exhibit infinite property T groups, but they do exist. For example Kazhdan proved that lattices in semisimple groups of real rank 2 or more have property T. It is also known that there are many hyperbolic groups with property T.

Lemma 5.1. *If G is an infinite property T group then the quotient mapping from $C^*(G)$ onto $C_\lambda^*(G)$ does not induce an isomorphism in K-theory.*

Proof. The central projection p generates a cyclic direct summand of $K(C^*(G))$ which is mapped to zero in $K(C_\lambda^*(G))$.

It follows immediately that if G is an infinite property T group then the Baum-Connes assembly maps into $K(C^*(G))$ and $K(C_\lambda^*(G))$ cannot both be isomorphisms. We shall not go into the matter in detail here but in fact it is the assembly map into $K(C^*(G))$ which is the problem. This can be seen quite easily in some examples. For instance it is not hard to show that if G has property T then associated to *each* irreducible, finite-dimensional and unitary representation of G is a distinct central projection in $C^*(G)$ (the Kazhdan projection is the one associated to the trivial representation). So if a property T group G has infinitely many irreducible, finite-dimensional and unitary representations (this will happen if G is an infinite linear group) then $K(C^*(G))$ will contain a free abelian subgroup of infinite rank, whereas $K^{top}(G)$ will very often be finitely generated.

Unfortunately the main method we have applied to prove cases of the Baum-Connes conjecture treats the full and reduced C^*-algebra more or less equally. Hence property T causes the method to fail:

Proposition 5.1. *If G is an exact, infinite property T group then G does not satisfy the hypotheses of Theorem 2.20.*

Proof. If G did satisfy the hypotheses then by Theorem 2.20 the quotient mapping $K(C^*(G))$ to $K(C_\lambda^*(G))$ would be an isomorphism.

5.2 Property T and Descent

Proposition 5.1 indicates that our basic strategy for proving the Baum-Connes conjecture for a group G, which involves proving an identity in equivariant, bivariant K-theory, will not work for infinite property T groups (at least if these groups are exact).

However one can ask whether it is possible to prove the conjecture for a given group G by proving an identity in bivariant K-theory for crossed product algebras. We noted in Lecture 5 that if eG is a complete manifold M then the Baum-Connes conjecture for G is equivalent to the assertion that the map

$$\alpha_* \colon K_*(C^*_\lambda(G, \mathcal{C}(M))) \to K_*(C^*_\lambda(G)),$$

induced from the Dirac operator class $\alpha \in E_G(\mathcal{C}(M), \mathbb{C})$, is an isomorphism. One might hope that in fact the descended class

$$\alpha \in E(C^*_\lambda(G, \mathcal{C}(M)), C^*_\lambda(G))$$

is an isomorphism. This is not (always) the case, as the following theorem of Skandalis [60] shows:

Theorem 5.2. *Let G be an infinite, hyperbolic property T group. Then $C^*_\lambda(G)$ is not equivalent in E-theory to any nuclear C^*-algebra.*

Recall from the last lecture that a C^*-algebra A is *nuclear* if $A \widehat{\otimes}_{min} D = A \widehat{\otimes}_{max} D$, for all D. Since the C^*-algebra $C^*_\lambda(G, \mathcal{C}(M))$ is easily proved to be nuclear we obtain the following result:

Corollary 5.1. *Let G be an infinite, hyperbolic, property T group and assume that G acts on a complete Riemannian manifold M by isometries. The Dirac operator class*

$$\alpha \in E(C^*_\lambda(G, \mathcal{C}(M)), C^*_\lambda(G))$$

is not invertible. □

Remark 5.2. The corollary applies to discrete, cocompact subgroups of the Lie groups $Sp(n, 1)$ (M is quaternionic hyperbolic space). See [17]. Despite this, it follows from the work of Lafforgue [44] that in this case α as above *does* induce an isomorphism on K-theory. This shows that E-theory is not a perfect weapon with which to attack the Baum-Connes conjecture.[17]

To prove Theorem 5.2 we shall use the following result.

Theorem 5.3. *Let G be a hyperbolic group and let ∂G be its Gromov boundary. There is a compact, metrizable topology on the disjoint union $\overline{G} = G \cup \partial G$ with the following properties:*

(a) *The set G is an open, discrete subset of $\overline{G}$.*
(b) *The left action of G on itself extends continuously to an amenable action of G on $\overline{G}$.*
(c) *The right action of G on itself extends continuously to an action on $\overline{G}$ which is trivial on $\overline{G}$.* □

---

[17] Exactly the same remarks apply here to KK-theory.

Remark 5.3. Item (c) is essentially the assertion that the natural action on the Gromov compactification is small at infinity, in the sense of Definition 4.9.

We shall also require a few simple representation-theoretic ideas.

Definition 5.2. *Let G be a discrete group. The* left regular, right regular *and* adjoint representations *of G on $\ell^2(G)$ are defined by the formulas*

$$(\lambda_g \xi)(h) = \xi(g^{-1}h)$$
$$(\rho_g \xi)(h) = \xi(hg)$$
$$(\alpha_g \xi)(h) = \xi(g^{-1}hg)$$

for all $g, h \in G$ and $\xi \in \ell^2(G)$. The biregular representation *of $G \times G$ on $\ell^2(G)$ is defined by the formula*

$$(\alpha_{g_1 g_2} \xi)(h) = \xi(g_1^{-1} h g_2)$$

for all $g_1, g_2, h \in G$ and $\xi \in \ell^2(G)$.

The left and right regular representations determine representations λ and ρ of $C_\lambda^*(G)$ in $\mathcal{B}(\ell^2(G))$. Since these representations commute with one another, together they determine a C^*-algebra representation

$$\beta \colon C_\lambda^*(G) \otimes_{max} C_\lambda^*(G) \longrightarrow \mathcal{B}(\ell^2(G)),$$

which is of course the biregular representation on $G \times G \subseteq C_\lambda^*(G) \otimes_{max} C_\lambda^*(G)$.

Definition 5.3. *Denote by J the kernel of the quotient homomorphism from $C_\lambda^*(G) \otimes_{max} C_\lambda^*(G)$ onto $C_\lambda^*(G) \otimes_{min} C_\lambda^*(G)$, so that there is an exact sequence*

$$0 \longrightarrow J \longrightarrow C_\lambda^*(G) \otimes_{max} C_\lambda^*(G) \longrightarrow C_\lambda^*(G) \otimes_{min} C_\lambda^*(G) \longrightarrow 0.$$

Lemma 5.2. *The C^*-algebra representation β maps the ideal J of $C_\lambda^*(G) \otimes_{max} C_\lambda^*(G)$ into the ideal $\mathcal{K}(\ell^2(G))$ of $\mathcal{B}(\ell^2(G))$.*

Proof. Denote by $\mathcal{Q}(\ell^2(G))$ the Calkin algebra for $\ell^2(G)$ — the quotient of the bounded operators by the ideal of compact operators. We are going to construct a $*$-homomorphism from $C_\lambda^*(G) \otimes_{min} C_\lambda^*(G)$ into $\mathcal{Q}(\ell^2(G))$ which makes the following diagram commute:

$$
\begin{array}{ccc}
C_\lambda^*(G) \otimes_{min} C_\lambda^*(G) & \dashrightarrow & \mathcal{Q}(\ell^2(G)) \\
\uparrow & & \uparrow \\
C_\lambda^*(G) \otimes_{max} C_\lambda^*(G) & \xrightarrow[\lambda \otimes \rho]{} & \mathcal{B}(\ell^2(G)).
\end{array}
$$

Here the vertical arrows are the quotient mappings. Commutativity of the diagram will prove the lemma.

We begin by constructing a *-homomorphism from $C(\partial G)$ into $\mathcal{Q}(\ell^2(G))$, as follows. If $f \in C(\partial G)$ then extend f to a continuous function on $\overline{G}$, restrict the extension to the open set $G \subseteq \overline{G}$, and then let the restriction act on $\ell^2(G)$ by pointwise multiplication. Two different extensions of $f \in C(\partial G)$ will determine two pointwise multiplication operators which differ by a compact operator. Hence our procedure defines a *-homomorphism $\varphi \colon C(\partial G) \to \mathcal{Q}(\ell^2(G))$, as required. Now let G act on $C(\partial G)$ via the (nontrivial) left action of G (see Theorem 5.3) and define a *-homomorphism

$$\varphi \colon C^*(G, \partial G) \to \mathcal{Q}(\ell^2(G))$$

by the formula

$$\varphi\Big(\sum_{g \in G} f_g \cdot g\Big) = \sum_{g \in G} \varphi(f_g)\lambda(g)$$

(we are using $\lambda(g)$ to denote both the unitary operator on $\ell^2(G)$ and its image in the Calkin algebra). Next, thanks to part (c) of Theorem 5.3 the *right* regular representation commutes with the algebra $\varphi[C(\partial G)] \subseteq \mathcal{Q}(\ell^2(G))$. We therefore obtain a *-homomorphism

$$C^*(G, \partial G) \otimes_{max} C^*_\lambda(G) \longrightarrow \mathcal{Q}(\ell^2(G)).$$

But since the action of G on ∂G is amenable the C^*-algebra $C^*(G, \partial G)$ is nuclear, so that the maximal tensor product above is the same as the minimal one. Moreover amenability also implies that $C^*(G, \partial G)$ agrees with $C^*_\lambda(G, \partial G)$. See Remark 4.8. It follows that the *-homomorphism displayed above is the same thing as a *-homomorphism

$$C^*_\lambda(G, \partial G) \otimes_{min} C^*_\lambda(G) \longrightarrow \mathcal{Q}(\ell^2(G)).$$

The lemma now follows by restricting this *-homomorphism to the subalgebra $C^*_\lambda(G) \otimes_{min} C^*_\lambda(G)$ of $C^*_\lambda(G, \partial G) \otimes_{min} C^*_\lambda(G)$.

Lemma 5.3. *The K-theory group $K(J)$ is nonzero.*

Proof. Let $\Delta \colon C^*(G) \to C^*_\lambda(G) \otimes_{max} C^*_\lambda(G)$ be the *-homomorphism $g \mapsto g \otimes g$. Let $p \in C^*(G)$ be the Kazhdan projection and let $q = \Delta(p)$. Then $q \in J$. To see this, observe that the composition

$$C^*(G) \overset{\Delta}{\longrightarrow} C^*_\lambda(G) \otimes_{max} C^*_\lambda(G) \longrightarrow C^*_\lambda(G) \otimes_{min} C^*_\lambda(G) \subseteq \mathcal{B}(\ell^2(G) \otimes \ell^2(G))$$

corresponds to the tensor product of two copies of the regular representation, and observe also that this representation has no nonzero G-fixed vectors. Hence, the image of the Kazhdan projection in $C^*_\lambda(G) \otimes_{min} C^*_\lambda(G)$ is zero. We shall now prove that $[q] \neq 0$ in $K(J)$. Note first that the representation

$$\beta \colon C^*_\lambda(G) \otimes_{max} C^*_\lambda(G) \longrightarrow \mathcal{B}(\ell^2(G))$$

maps q to a nonzero projection operator. Indeed the composition

$$C^*(G) \xrightarrow{\;\Delta\;} C^*_\lambda(G) \otimes_{max} C^*_\lambda(G) \xrightarrow{\;\beta\;} \mathcal{B}(\ell^2(G))$$

is the representation α of $C^*(G)$ associated to the adjoint representation of G, which *does* have nonzero G-fixed vectors, and β maps q to the orthogonal projection onto these fixed vectors. But by Lemma 5.2 the representation β maps J into the compact operators, and every nonzero projection in $\mathcal{K}(\ell^2(G))$ determines a nonzero K-theory class. Hence the map from $K(J)$ to $K(\mathcal{K}(\ell^2(G)))$ takes $[q]$ to a nonzero element, and therefore the class $[q] \in K(J)$ is itself nonzero.

Proof (Proof of Theorem 5.2). Let us suppose that there is a separable nuclear C^*-algebra A and an invertible E-theory element $\varphi \in E(C^*_\lambda(G), A)$. Since $C^*_\lambda(G)$ is an exact C^*-algebra there are invertible elements

$$\varphi \otimes_{max} 1 \in E(C^*_\lambda(G) \otimes_{max} C^*_\lambda(G), A \otimes_{max} C^*_\lambda(G))$$

and

$$\varphi \otimes_{min} 1 \in E(C^*_\lambda(G) \otimes_{min} C^*_\lambda(G), A \otimes_{min} C^*_\lambda(G)).$$

We therefore arrive at the following commuting diagram in the E-theory category:

$$
\begin{array}{ccc}
C^*_\lambda(G) \otimes_{max} C^*_\lambda(G) & \xrightarrow[\cong]{\varphi \otimes_{max} 1} & A \otimes_{max} C^*_\lambda(G)) \\
\downarrow & & \downarrow \\
C^*_\lambda(G) \otimes_{min} C^*_\lambda(G) & \xrightarrow[\varphi \otimes_{min} 1]{\cong} & A \otimes_{min} C^*_\lambda(G)
\end{array}
$$

But since A is nuclear the right hand vertical map is an isomorphism (even at the level of C^*-algebras). It follows that the left hand vertical map is an isomorphism in the E-theory category too. As a result, the K-theory map

$$K(C^*_\lambda(G) \otimes_{max} C^*_\lambda(G)) \longrightarrow K(C^*_\lambda(G) \otimes_{min} C^*_\lambda(G))$$

is an isomorphism of abelian groups. But thanks to the K-theory long exact sequence this contradicts Lemma 5.3.

5.3 Bivariant Theories

In the previous section we showed that it is not possible to prove the Baum-Connes conjecture for certain groups (for example uniform lattices in $Sp(n, 1)$) by working purely within E-theory (or for that matter within KK-theory). In this section we shall prove a theorem, also due to Skandalis [61], which points to another sort of weakness of bivariant K-theory. Recall that the bivariant theory we constructed — namely E-theory — has long exact sequences in both variables but that we could not equip it with a minimal tensor product operation (since the operation of minimal tensor product does not in general preserve exact sequences). Kasparov's KK-theory has minimal tensor products but the long exact sequences are only constructed under some hypothesis or other related to C^*-algebra nuclearity. One might ask whether or not there is an 'ideal' theory which has both desirable properties. The answer is no:

Theorem 5.4. *There is no bivariant K-theory functor on separable C^*-algebras which has both a minimal tensor product operation and long exact sequences in both variables.*

Remark 5.4. By the term 'bivariant K-theory functor' we mean a bifunctor which, like E-theory and KK-theory, is equipped with an associative product allowing us to create from it an additive category. The homotopy category of separable C^*-algebras should map to this category, and the ordinary one-variable K-theory functor should factor through it.

To prove Theorem 5.4 we shall need one additional computation from representation theory.

Lemma 5.4. [41] *Let G be a residually finite discrete group. The biregular representation β of $G \times G$ on $\ell^2(G)$ extends to a representation of the minimal tensor product $C^*(G) \otimes_{min} C^*(G)$.*

Proof. Let $\{G_n\}$ be a decreasing family of finite index normal subgroups of G for which the intersection $\cap G_n$ is the trivial one-element subgroup of G. If $x \in \mathbb{C}[G] \odot \mathbb{C}[G]$ then denote by x_n the corresponding 'quotient' element of $\mathbb{C}[G/G_n] \odot \mathbb{C}[G/G_n]$ and denote by β_n the biregular representation of $G/G_n \times G/G_n$ on $\ell^2(G/G_n)$. Thanks to the functoriality of $\otimes_{min}$ it is certainly the case that

$$\|x\|_{C^*(G) \otimes_{min} C^*(G)} \geq \sup_n \|x_n\|_{C^*(G/G_n) \otimes_{min} C^*(G/G_n)}.$$

In addition

$$\|x_n\|_{C^*(G/G_n) \otimes_{min} C^*(G/G_n)} \geq \|\beta_n(x_n)\|_{\mathcal{B}(\ell^2(G/G_n))}$$

(observe that since $C^*(G/G_n)$ is finite-dimensional the minimal tensor product here is equal to the maximal one). Now, it is easily checked that

$$\sup_n \|\beta_n(x_n)\|_{\mathcal{B}(\ell^2(G/G_n))} \geq \|\beta(x)\|_{\mathcal{B}(\ell^2(G))}.$$

Putting together all the inequalities we conclude that

$$\|x\|_{C^*(G) \otimes_{min} C^*(G)} \geq \|\beta(x)\|_{\mathcal{B}(\ell^2(G))},$$

as required.

Lemma 5.5. *Let I be the kernel of the quotient map π from $C^*(G)$ onto $C^*_\lambda(G)$, so that there is a short exact sequence*

$$0 \longrightarrow I \longrightarrow C^*(G) \overset{\pi}{\longrightarrow} C^*_\lambda(G) \longrightarrow 0.$$

If there is a bivariant theory $F(A, B)$ which has long exact sequences in both variables, and if C_π is the mapping cone of π, then the inclusion $I \subseteq C_\pi$ determines an invertible element of $F(I, C_\pi)$.

Proof. Consider the commuting diagram

$$
\begin{array}{ccccccccc}
0 & \longrightarrow & I & \longrightarrow & C^*(G) & \overset{\pi}{\longrightarrow} & C^*_\lambda(G) & \longrightarrow & 0 \\
& & \downarrow & & \downarrow & & \downarrow = & & \\
0 & \longrightarrow & C_\pi & \longrightarrow & Z_\pi & \longrightarrow & C^*_\lambda(G) & \longrightarrow & 0,
\end{array}
$$

where $Z_\pi = \{\, a \oplus f \in C^*(G) \oplus C^*_\lambda(G)[0,1] \, : \, \pi(a) = f(0)\,\}$. The inclusion of $C^*(G)$ into Z_π (as constant functions) is a homotopy equivalence, and therefore by applying F-theory to the diagram and then the five lemma we see that the inclusion $I \subseteq C_\pi$ induces isomorphisms

$$
F(A, I) \overset{\cong}{\longrightarrow} F(A, C_\pi) \quad \text{and} \quad F(C_\pi, B) \overset{\cong}{\longrightarrow} F(I, B)
$$

for every A and B. It follows that the inclusion determines an invertible element of $F(I, C_\pi)$ as required.

Proof (Proof of Theorem 5.4). If the bivariant 'F-theory' has a minimal tensor product then it follows from the lemma above that the inclusion

$$
I \otimes_{min} C^*(G) \subseteq C_\pi \otimes_{min} C^*(G)
$$

determines an invertible element in F-theory and therefore an isomorphism on K-theory groups. We shall prove the theorem by showing that the map on K-theory induced from the above inclusion fails to be surjective.

Consider the short exact sequence

$$
0 \longrightarrow L \longrightarrow C^*(G) \otimes_{min} C^*(G) \overset{\pi \otimes 1}{\longrightarrow} C^*_\lambda(G) \otimes_{min} C^*(G) \longrightarrow 0,
$$

where the ideal L is by definition the kernel of the quotient mapping $\pi \otimes 1$. The mapping cone of $\pi \otimes 1$ is (canonically isomorphic to) $C_\pi \otimes C^*(G)$, and therefore the inclusion $L \subseteq C_\pi \otimes_{min} C^*(G)$ induces an isomorphism in K-theory. Observe now that we have a sequence of inclusions

$$
I \otimes_{min} C^*(G) \subseteq L \subseteq C_\pi \otimes_{min} C^*(G).
$$

We wish to prove that the overall inclusion fails to be surjective in K-theory, and since the second inclusion is an *isomorphism* in K-theory it suffices to prove that the first inclusion fails to be surjective. From here the proof is more or less the same as the proof of Lemma 5.3, and we shall be very brief. There is a diagonal map

$$
\Delta \colon C^*(G) \to C^*(G) \otimes_{min} C^*(G)
$$

and we denote by $q \in C^*(G) \otimes_{min} C^*(G)$ the image under Δ of the Kazhdan projection. It is an element of the ideal L. According to Lemma 5.4 the biregular representation of $G \times G$ on $\ell^2(G)$ extends to $C^*(G) \otimes_{min} C^*(G)$. From the proof of Lemma 5.2 we obtain a commuting diagram

$$C^*_\lambda(G) \otimes_{min} C^*_\lambda(G) \longrightarrow \mathcal{Q}(\ell^2(G))$$

$$\uparrow \qquad\qquad\qquad\qquad\qquad \uparrow$$

$$C^*(G) \otimes_{min} C^*(G) \xrightarrow{\ \ \beta\ \ } \mathcal{B}(\ell^2(G)),$$

which shows that the C^*-algebra representation β maps the ideal L into the compact operators. Consider now the sequence of maps

$$I \otimes_{min} C^*(G) \xrightarrow{\ \subseteq\ } L \xrightarrow{\ \beta\ } \mathcal{K}(\ell^2(G)) \ .$$

The composition is zero. But the projection p is mapped to a nonzero element in $\mathcal{K}(\ell^2(G))$, and the K-theory class of $[q]$ is mapped to a nonzero element in the K-theory of $\mathcal{K}(\ell^2(G))$. This shows that the class $[q] \in K(L)$ is not the image of any K-theory class for $I \otimes_{min} C^*(G)$, and this completes the proof of the theorem.

5.4 Expander Graphs

The purpose of this section and the next is to present a counterexample to the Baum-Connes conjecture with coefficients, contingent on some assertions of Gromov.

Definition 5.4. *Let Γ be a finite graph (a finite, 1-dimensional simplicial complex) and let $V = V(\Gamma)$ be the set of vertices of Γ. The* Laplace operator $\Delta\colon \ell^2(V) \to \ell^2(V)$ *is the linear operator defined by the quadratic form*

$$\langle f, \Delta f\rangle = \sum_{d(v,v')=1} |f(v) - f(v')|^2$$

The sum is over all (unordered) pairs of adjacent vertices, or in other words over the edges of Γ. We shall denote by $\lambda_1(\Gamma)$ the first nonzero eigenvalue of Δ.

If the graph Γ is connected then the kernel of Δ consists precisely of the constant functions on V. In this case

$$\left.\begin{matrix} f \in \ell^2(V) \\ \sum_{x \in V} f(v) = 0 \end{matrix}\right\} \Longrightarrow \|f\|^2 \le \frac{1}{\lambda_1(\Gamma)} \langle \Delta f, f\rangle. \qquad (1)$$

Definition 5.5. *Let k be a positive integer and let $\varepsilon > 0$. A finite graph Γ is a (k, ε)-expander if it is connected, if no vertex of Γ is incident to more than k edges, and if $\lambda_1(\Gamma) \ge \varepsilon$.*

See [46] for an extensive discussion of the theory of expander graphs.

The following observation of Gromov shows that expander graphs give rise to examples of metric spaces which cannot be uniformly embedded in affine Euclidean spaces.

Proposition 5.2. *Let k be a positive integer, let $\varepsilon > 0$, and let $\{\Gamma_n\}_{n=1}^{\infty}$ be a sequence of (k, ε)-expander graphs for which $\lim_{n\to\infty} |V(\Gamma_n)| = \infty$. Let V be the disjoint union of the sets $V_n = V(\Gamma_n)$ and suppose that V is equipped with a distance function which restricts to the path-distance function on each V_n. Then the metric space V may not be embedded in an affine Euclidean space.*

Proof. Suppose that f is a uniform embedding into an affine Euclidean space E. We may assume that E is complete and separable, and we may then identify it isometrically with $\ell^2(\mathbb{N})$. By restricting f to each V_n, and by adjusting each f_n by a translation in $\ell^2(\mathbb{N})$ (that is, by adding suitable constant vector-valued functions to each f_n) we can arrange that each f_n is orthogonal to every constant function in the Hilbert space of functions from V_n to $\ell^2(\mathbb{N})$ (we just have to arrange that $\sum_{x \in X_n} f(x) = 0$). Now the Laplace operator can be defined on $\ell^2(\mathbb{N})$-valued functions just as it was on scalar functions, and the expander property (1) carries over to the vector-Laplacian (compute using coordinates in $\ell^2(\mathbb{N})$). However

$$\langle \Delta f_n, f_n \rangle = \sum_{d(v,v')=1} |f_n(v) - f_n(v')|^2$$
$$\leq \sum_{d(v,v')=1} 1$$
$$\leq \frac{k}{2}|V_n|.$$

It therefore follows from the expander property that

$$\sum_{v \in V_n} \|f_n(v)\|^2 = \|f_n\|^2 \leq \frac{1}{\varepsilon}\langle \Delta f_n, f_n \rangle \leq \frac{k}{2\varepsilon}|V_n|.$$

Thus for all n, and for at least half of the points $v \in V_n$, we have $\|f_n(v)\|^2 \leq \frac{k}{\varepsilon}$. This contradicts the definition of uniform embedding since among this half there must be points v_n and v'_n with $\lim_{n\to\infty} d(v_n, v'_n) = \infty$.

In a recent paper [26], M. Gromov has announced the existence of finitely generated groups which do not uniformly embed into Hilbert space. Complete details of the construction have not yet appeared, but the idea is to construct within the Cayley graph of a group a sequence of images of expander graphs. Let us make this a little more precise, as follows.

Definition 5.6. *Let us say that a finitely generated discrete group G is a Gromov group if for some positive integer k and some $\varepsilon > 0$ there is a sequence of (k, ε)-expander graphs Γ_n and a sequence of maps $\varphi_n : V(\Gamma_n) \to G$ such that :*

(a) There is a constant R, such that if v and v' are adjacent vertices in some graph Γ_n then $d(\varphi_n(v), \varphi_n(v')) \leq R$.

(b) $\lim_{n\to\infty} \left(\max\{ \frac{|\varphi_n^{-1}[g]|}{|V(\Gamma_n)|} : g \in G \} \right) = 0.$

Remark 5.5. The second condition implies that $\lim_{n\to\infty} |V(\Gamma_n)| = \infty$.

It appears that Gromov's ideas prove that Gromov groups, as above, exist. In any case, we shall explore below some of the properties of Gromov groups. We conclude this section with a simple extension of Proposition 5.2, the proof of which is left to the reader.

Proposition 5.5 *If G is a Gromov group then G cannot be uniformly embedded in an affine Euclidean space.* □

5.5 The Baum-Connes Conjecture with Coefficients

We shall prove that, contingent on the existence of a Gromov group as in the last section, there exists a separable, commutative C^*-algebra D, and an action of a countable group G on D, for which the Baum-Connes map

$$\mu_\lambda \colon K^{top}(G, D) \to K(C^*_\lambda(G, D))$$

fails to be an isomorphism.

Lemma 5.6. *Let G be a countable group and let J be an ideal in a G-C^*-algebra A. If the Baum-Connes assembly map μ_λ is an isomorphism for G, with coefficients all the separable C^*-subalgebras A and A/J, then the K-theory sequence*

$$K(C^*_\lambda(G, J)) \longrightarrow K(C^*_\lambda(G, A)) \longrightarrow K(C^*_\lambda(G, A/J))$$

is exact in the middle.

Proof. Since exactness of the sequence is preserved by direct limits it suffices to consider the case in which A itself is separable. The proof then follows from a chase around the diagram of assembly maps

$$
\begin{array}{ccc}
K^{top}(G, A) & \longrightarrow & K^{top}(G, A/J) \\
\downarrow & & \downarrow \\
K(C^*(G, J)) \longrightarrow K(C^*(G, A)) & \longrightarrow & K(C^*(G, A/J)) \\
\downarrow & \downarrow & \downarrow \\
K(C^*_\lambda(G, J)) \longrightarrow K(C^*_\lambda(G, A)) & \longrightarrow & K(C^*_\lambda(G, A/J))
\end{array}
$$

and the fact that the middle row is exact in the middle.

We shall prove that if G is a Gromov group then for a suitable A and J the conclusion of the lemma fails.

Definition 5.7. *Let A be the C^*-algebra of bounded complex-valued functions on $G \times \mathbb{N}$ for which the restriction to each subset $G \times \{n\}$ is a c_0-function. Denote by J the ideal in A consisting of c_0-functions on $G \times \mathbb{N}$.*

Thus $A \cong \ell^\infty(\mathbb{N}, c_0(G))$ and $J \cong c_0(\mathbb{N}, c_0(G))$.

Now let G act on A be the *right* translation action of G on $G \times \mathbb{N}$.

Lemma 5.7. *The (right regular) covariant representation of A on $\ell^2(G \times \mathbb{N})$ determines a faithful representation of the reduced crossed product algebra $C^*_\lambda(G, A)$ as operators on $\ell^2(G)$.* $\square$

From here on we shall assume that G is a Gromov group. For simplicity we shall now assume that the maps $\varphi_n \colon V_n \to G$ which appear in Definition 5.6 are *injective*. For the general case see [11].

Let V be the disjoint union of the V_n. Let us map V_n via φ_n to the nth copy of G in $G \times \mathbb{N}$, and thereby embed V into $G \times \mathbb{N}$. We can now identify $\ell^2(V)$ with a closed subspace of $\ell^2(G \times \mathbb{N})$.

Definition 5.8. *Denote by $\Delta \colon \ell^2(G \times \mathbb{N}) \to \ell^2(G \times \mathbb{N})$ the direct sum of the Laplace operators on each $\ell^2(V_n) \subseteq \ell^2(V)$ with the identity on the orthogonal complement of $\ell^2(V) \subseteq \ell^2(G \times \mathbb{N})$.*

Lemma 5.8. *The operator $\Delta - I$ belongs to $C^*_\lambda(G, A) \subseteq \mathcal{B}(\ell^2(G \times \mathbb{N}))$ (it is in fact in the algebraic crossed product).*

Proof. First, some notation. Let us continue to identify the vertex set $V_n = V(\Gamma_n)$, via φ_n, with a subset of G. We shall write $[g, g'] \in E(\Gamma_n)$ if the group elements g and g' correspond to vertices in V_n which are adjacent in the graph Γ_n. Finally if g corresponds to a vertex of Γ_n we shall write $k_n(g)$ for its valence, minus 1.

The Hilbert space $\ell^2(G \times \mathbb{N})$ has canonical basis elements f_{gn} and in this basis the formula for Δ is

$$\begin{cases} (\Delta - I) \colon f_{gn} \mapsto k(g) f_{gn} - \displaystyle\sum_{[g,g'] \in E(\Gamma_n)} f_{g'n} & \text{if } g \in V_n \\[2ex] (\Delta - I) \colon f_{gn} \mapsto 0 & \text{if } g \notin V_n. \end{cases}$$

We can therefore write $\Delta - I$ as a finite sum

$$\Delta - I = e \cdot a_e + \sum_{h \neq e} h \cdot a_h,$$

where the coefficient functions $a_g \in A$ are defined by

$$a_e(g, n) = \begin{cases} k_n(g) & \text{if } g \in V_n \\ 0 & \text{if } g \notin V_n \end{cases}$$

and, for $h \neq e$,

$$a_h(g, n) = \begin{cases} -1 & \text{if } [gh^{-1}, g] \in E(\Gamma_n) \\ 0 & \text{if } [gh^{-1}, g] \in E(\Gamma_n). \end{cases}$$

(The sum is finite thanks to the first item in Definition 5.6.)

Since the graphs Γ_n are (k, ε)-expanders the point 0 is isolated in the spectrum of Δ, and therefore we can make the following definition:

Definition 5.9. *Let G be a Gromov group and assume that the maps $\varphi_n \colon V_n \to G$ are injective. Denote by $E \in C_\lambda^*(G, A)$ orthogonal projection onto the kernel of Δ.*

The operator E is the orthogonal projection onto the ℓ^2-functions on $G \times \mathbb{N}$ which are constant on each V_n and zero on the complement of V.

Lemma 5.9. *The class of E in $K(C_\lambda^*(G, A))$ is not in the image of the map $K(C_\lambda^*(G, J)) \to K(C_\lambda^*(G, A))$.*

Proof. Let $A_n = c_0(G \times \{n\})$, which is a quotient of A, and denote by

$$\pi_n : C_\lambda^*(G, A) \to C_\lambda^*(G, A_n)$$

the quotient mapping. We get maps

$$\pi_{n*} : K(C_\lambda^*(G, A)) \to K(C_\lambda^*(G, A_n)) = \mathbb{Z}.$$

Since $\pi_n(E)$ is a rank one projection, we find $\pi_n([E]) = 1$, for all n. Therefore the K-theory class of p in $K(C_\lambda^*(G, A))$ does not come from $K(C_\lambda^*(G, J))$ (which maps to the direct sum $\oplus_{n \in \mathcal{N}} \mathbb{Z}$ under $\oplus \pi_n$).

Lemma 5.10. *The image of E in $C_\lambda^*(G, A/J)$ is zero.*

To prove the lemma we shall need some means of determining when elements in reduced crossed product algebras $C_\lambda^*(G, D)$ are zero. For this purpose, recall that the C^*-algebra $C_\lambda^*(G, D)$ is faithfully represented as operators on the Hilbert D-module $\ell^2(G, D)$.

Exercise 5.1. If P_g denotes the orthogonal projection onto the functions in $\ell^2(G, D)$ supported on $\{g\}$, and if $T \in C_\lambda^*(G, D)$, then $P_g T P_e$ is an operator from functions supported on $\{e\}$ to functions supported on $\{g\}$. If all the elements $P_g T P_e$ are equal to 0 then $T = 0$.

Exercise 5.2. The operator $P_g T P_e$ can be identified with an element $T_g \in D$ via the formula
$$(P_g T P_e \xi)(g) = T_g \cdot \xi(e), \qquad \forall \xi \in \ell^2(G, D).$$
If T is a finite sum $T = \sum d_g \cdot g$ in the algebraic crossed product (where $d_g \in D$) then $T_g = d_g$. If $\varphi \colon D \to D'$ is a G-equivariant $*$-homomorphism and if Φ is the induced map on crossed products then $\Phi(T)_g = \varphi(T_g)$.

By checking on finite sums we see that if an operator $T \in C_\lambda^*(G, A)$ has matrix coefficients $T_{gn,g'n'}$ for the canonical basis of $\ell^2(G, \mathbb{N})$ then the functions $T_g \in A$ associated to T are defined by

$$T_g(h, n) = T_{hgn, hn}.$$

Proof (Proof of Lemma 5.10). The projection $E \colon \ell^2(G \times \mathbb{N}) \to \ell^2(G \times \mathbb{N})$ is comprised of the sequence projections E_n onto the constant functions in $\ell^2(V_n)$. The matrix coefficients of E are therefore described by the formula

$$\begin{cases} E \colon f_{gn} \mapsto \sum_{g' \in V_n} \frac{1}{|V_n|} f_{g'n} & \text{if } g \in V_n \\ E \colon f_{gn} \mapsto 0 & \text{if } g \notin V_n. \end{cases}$$

As a result, the functions $E_g \in A$ associated to the projection E, as in the exercises, are given by the formula

$$E_g(h, n) = \begin{cases} \frac{1}{|V_n|} & \text{if } hg, h \in V_n \\ 0 & \text{if } hg \notin V_n \text{ or } h \notin V_n. \end{cases}$$

This shows that $E_g \in J$, for all $g \in G$. It follows that the elements $E_n \in A/J$ associated to the image of E in $C^*_\lambda(G, A/J)$ are 0, and so the projection E is itself 0 in $C^*_\lambda(G, A/J)$.

The two lemmas show that the K-theory sequence

$$K(C^*_\lambda(G, J)) \longrightarrow K(C^*_\lambda(G, A)) \longrightarrow K(C^*_\lambda(G, A/J))$$

fails to be exact in the middle. Hence:

Theorem 5.6. *Let G be a Gromov group. There is a separable, commutative G-C^*-algebra D for which the Baum-Connes assembly map*

$$\mu_\lambda \colon K^{top}(G, D) \to K(C^*_\lambda(G, D))$$

fails to be an isomorphism. □

5.6 Inexact Groups

The following result (see [28, 29, 15]) shows that Gromov groups fail to be exact.

Theorem 5.7. *If a finitely generated discrete group G is exact then G embeds uniformly in a Hilbert space.*

To prove the theorem we shall use a difficult characterization of separable exact C^*-algebras, due to Kirchberg [42] (see also [66] for an exposition). It involves the following notion:

Definition 5.10. *Let A and B be unital C^*-algebras. A unital linear map $\Phi \colon A \to B$ of C^*-algebras is completely positive if for all $k \in \mathbb{N}$ the linear map $\Phi_k \colon M_k(A) \to M_k(B)$ defined by applying Φ entrywise to a matrix of elements of A is positive (meaning it maps positive matrices to positive matrices).*

Theorem 5.8. *A separable C^*-algebra A is exact if and only if every injective $*$-homomorphism $A \to \mathcal{B}(\mathcal{H})$ can be approximated in the point norm topology by a sequence of unital completely positive maps, each of which factors, via unital, completely positive maps, through a matrix algebra.* $\square$

Kirchberg's theorem has the following consequence:

Corollary 5.2. *If G is a countable exact group then there exists a sequence of completely positive maps $\Phi_n \colon C^*_\lambda(G) \to \mathcal{B}(\ell^2(G))$ which converge pointwise in norm to the natural inclusion of $C^*_\lambda(G)$ into $\mathcal{B}(\ell^2(G))$ and which have the property that for every $n \in \mathbb{N}$ the operator valued function $g \mapsto \Phi_n(g)$ is supported on a finite subset of G.*

Proof. By Theorem 5.8 there exists a sequence of unital completely positive maps which converge pointwise in norm to the natural inclusion of $C^*_\lambda(G)$ into $\mathcal{B}(\ell^2(G))$, and which individually factor through matrix algebras. Let us write these maps as compositions

$$C^*_\lambda(G) \xrightarrow{\Theta_n} M_{k_n}(\mathbb{C}) \xrightarrow{\Psi_n} \mathcal{B}(\ell^2(G)) . \tag{2}$$

Now, a linear map $\Theta \colon C^*_\lambda(G) \to M_k(\mathbb{C})$ is completely positive if and only if the linear map $\theta \colon M_k(C^*_\lambda(G)) \to \mathbb{C}$ defined by the formula

$$\theta([f_{ij}]) = \frac{1}{k} \sum_{i,j=1}^{k} \Theta(f_{ij})_{ij}$$

is a state. Moreover the correspondence $\Theta \leftrightarrow \theta$ is a bijection between completely positive maps and states. In addition, if $h_1, \dots, h_k$ are finitely supported functions on G which determine a unit vector in the k-fold direct sum $\ell^2(G) \oplus \cdots \oplus \ell^2(G)$, then the vector state

$$\theta([f_{ij}]) = \sum_{i,j=1}^{k} \langle h_i, \lambda(f_{ij})h_j \rangle$$

on $M_k(C^*_\lambda(G))$ corresponds to a completely positive map Θ which is finitely supported, as a function on G, as in the statement of the lemma. But the convex hull of the vector states associated to a faithful representation of a C^*-algebra is always weak*-dense in the set of all states (this is a version of the Hahn-Banach theorem). It follows that the set of those completely positive maps from $C^*_\lambda(G)$ into $M_k(\mathbb{C})$ which are finitely supported as functions on G is dense, in the topology of pointwise norm-convergence, in the set of all completely positive maps from $C^*_\lambda(G)$ into $M_k(\mathbb{C})$. By approximating the maps Θ_n in the compositions (2) we obtain completely positive maps from $C^*_\lambda(G)$ into $\mathcal{B}(\ell^2(G))$ with the required properties.

Proof (Proof of Theorem 5.7). According to Corollary 5.2 there exists a sequence of unital completely positive maps $\Phi_n \colon C^*_\lambda(G) \to \mathcal{B}(\ell^2(G))$ which converge pointwise in norm to the natural inclusion of $C^*_\lambda(G)$ into $\mathcal{B}(\ell^2(G))$ and which are individually finitely supported as functions on G. Define a sequence of functions

$$\varphi_n \colon G \times G \to \mathbb{C}$$

by

$$\varphi_n(g_1, g_2) = \langle [g_1^{-1}], \Phi_n(g_1^{-1} g_2)[g_2^{-1}] \rangle.$$

The functions φ_n are *positive-definite kernels* on the set G, in the sense of Definition 4.14. (To prove the inequality $\sum \overline{\lambda_i} \varphi_n(g_i, g_j) \lambda_j \geq 0$ write the sum as a matrix product

$$\begin{bmatrix} g_1 \cdots g_k \end{bmatrix} \begin{bmatrix} \overline{\lambda_1} \Phi_n(g_1^{-1} g_1) \lambda_1 & \cdots & \overline{\lambda_1} \Phi_n(g_1^{-1} g_k) \lambda_k \\ \vdots & \ddots & \vdots \\ \overline{\lambda_k} \Phi_n(g_k^{-1} g_1) \lambda_1 & \cdots & \overline{\lambda_k} \Phi_n(g_k^{-1} g_k) \lambda_k \end{bmatrix} \begin{bmatrix} g_1 \\ \vdots \\ g_k \end{bmatrix}$$

and apply the definition of complete positivity.) The functions φ_n converge pointwise to 1, and moreover for every finite subset $F \subseteq G$ and every $\varepsilon > 0$ there is some $N \in \mathbb{N}$ such that

$$n > N \quad \text{and} \quad g_1^{-1} g_2 \in F \quad \Rightarrow \quad |\varphi_n(g_1, g_2) - 1| < \varepsilon.$$

In addition, for every $n \in \mathbb{N}$ there exists a finite subset $F \subset G$ such that

$$g_1^{-1} g_2 \notin F \quad \Rightarrow \quad \varphi_n(g_1, g_2) = 0.$$

It follows that for a suitable subsequence the series $\sum_j (1 - \varphi_{n_j})$ is pointwise convergent everywhere on $G \times G$. But each function $1 - \varphi_{n_j}$ is a negative type kernel, and therefore so is the sum. The map into affine Euclidean space which is associated to the sum is a uniform embedding.

Remark 5.6. This proof is obviously very similar to that of Proposition 4.9. In fact, according to Remark 4.7 the above argument shows that if a countable group G is exact then G acts amenably on its Stone-Cech compactification βG [28, 29, 15]. As a result: *if a countable group G is exact then the Baum-Connes assembly map*

$$\mu_\lambda \colon K^{top}(G, D) \to K(C_\lambda^*(G, D))$$

is injective, for every D.

References

1. John Frank Adams. *Infinite loop spaces.* Princeton University Press, Princeton, N.J., 1978.
2. C. Anantharaman-Delaroche. Amenability and exactness for dynamical systems and their C^*-algebras. Preprint, 2000.
3. C. Anantharaman-Delaroche and J. Renault. *Amenable groupoids.* L'Enseignement Mathématique, Geneva, 2000. With a foreword by Georges Skandalis and Appendix B by E. Germain.

4. M. F. Atiyah. *K-Theory*. Benjamin Press, New York, 1967.
5. M. F. Atiyah. Bott periodicity and the index of elliptic operators. *Quart. J. Math. Oxford Ser. (2)*, 19:113–140, 1968.
6. Michael Atiyah and Raoul Bott. On the periodicity theorem for complex vector bundles. *Acta Math.*, 112:229–247, 1964.
7. P. Baum, A. Connes, and N. Higson. Classifying space for proper actions and K-theory of group C^*-algebras. *Contemporary Mathematics*, 167:241–291, 1994.
8. M. E. Bekka, P. A. Cherix, and A. Valette. Proper affine isometric actions of amenable groups. In Ferry et al. [21], pages 1–4.
9. M. Bożejko, T. Januszkiewicz, and R. Spatzier. Infinite Coxeter groups do not have Kazhdan's property. *J. Operator Theory*, 19:63–67, 1988.
10. M. Bridson and A. Haefliger. *Metric Spaces of Non-Positive Curvature*, volume 319 of *Grundlehren der Mathematischen Wissenschaften*. Springer Verlag, 1999.
11. G. Carlsson and E. Pedersen. Controlled algebra and the Novikov conjectures for K- and L-theory. *Topology*, 34:731–758, 1995.
12. Pierre-Alain Cherix, Michael Cowling, Paul Jolissaint, Pierre Julg, and Alain Valette. *Groups with the Haagerup property*. Birkhäuser Verlag, Basel, 2001. Gromov's a-T-menability.
13. A. Connes and N. Higson. Almost homomorphisms and $\hat{K}K$-theory. unpublished manuscript, http://math.psu.edu/higson/Papers/CH.dvi, 1989.
14. A. Connes and N. Higson. Déformations, morphismes asymptotiques et K-théorie bivariante. *C. R. Acad. Sci. Paris, Série I*, 311:101–106, 1990.
15. A. Connes and H. Moscovici. Cyclic cohomology, the Novikov conjecture, and hyperbolic groups. *Topology*, 29:345–388, 1990.
16. H. L. Cycon, R. G. Froese, W. Kirsch, and B. Simon. *Schrödinger operators with application to quantum mechanics and global geometry*. Springer-Verlag, Berlin, study edition, 1987.
17. P. de la Harpe and A. Valette. La propriété (T) de Kazhdan pour les groupes localement compacts. *Astérisque*, 175:1–158, 1989.
18. Patrick Delorme. 1-cohomologie des représentations unitaires des groupes de Lie semisimples et résolubles. Produits tensoriels continus de représentations. *Bull. Soc. Math. France*, 105(3):281–336, 1977.
19. J. Dixmier. *C*-algebras*. North Holland, Amsterdam, 1970.
20. D. Farley. *Finiteness and CAT(0) Properties of Diagram Groups*. PhD thesis, Binghamton Univ., 2000.
21. S. Ferry, A. Ranicki, and J. Rosenberg, editors. *Novikov Conjectures, Index Theorems and Rigidity*. Number 226, 227 in London Mathematical Society Lecture Notes. Cambridge University Press, 1995.
22. E. Ghys and P. de la Harpe. *Sur les Groups Hyperboliques d'aprés Mikhael Gromov*, volume 83 of *Progress in Mathematics*. Birkhäuser, Boston, 1990.
23. P. Green. Equivariant K-theory and crossed product C^*-algebras. In R. Kadison, editor, *Operator Algebras and Applications*, volume 38 of *Proceedings of Symposia in Pure Mathematics*, pages 337–338, Providence, RI, 1982. American Mathematical Society.
24. M. Gromov. Hyperbolic groups. In S. Gersten, editor, *Essays in Group Theory*, volume 8 of *MSRI Publ.*, pages 75–263. Springer Verlag, 1987.
25. M. Gromov. *Asymptotic Invariants of Infinite Groups*, pages 1–295. Number 182 in London Mathematical Society Lecture Notes. Cambridge University Press, 1993.
26. Misha Gromov. Spaces and questions. *Geom. Funct. Anal.*, (Special Volume, Part I):118–161, 2000. GAFA 2000 (Tel Aviv, 1999).

27. E. Guentner, N. Higson, and J. Trout. *Equivariant E-Theory for C*-Algebras*, volume 148 of *Memoirs of the AMS*. American Mathematical Society, 2000.

28. E. Guentner and J. Kaminker. Exactness and the Novikov conjecture. To appear in Topology, 1999.

29. E. Guentner and J. Kaminker. Addendum to "Exactness and the Novikov conjecture". To appear in Topology, 2000.

30. U. Haagerup. An example of a non-nuclear C^*-algebra with the metric approximation property. *Invent. Math.*, 50:279–293, 1979.

31. N. Higson. Bivariant K-theory and the Novikov conjecture. *Geom. Funct. Anal.*, 10:563–581, 2000.

32. N. Higson, G. Kasparov, and J. Trout. A Bott periodicity theorem for infinite dimensional Euclidean space. *Advances in Mathematics*, 135:1–40, 1998.

33. N. Higson and J. Roe. Amenable actions and the Novikov conjecture. Preprint, 1998.

34. Nigel Higson and Gennadi Kasparov. E-theory and KK-theory for groups which act properly and isometrically on Hilbert space. *Invent. Math.*, 144(1):23–74, 2001.

35. P. Julg. k-théorie équivariante et produits croisés. *Comptes Rendus Acad. Sci. Paris*, 292:629–632, 1981.

36. G. Kasparov and G. Skandalis. Groups acting on "bolic" spaces and the Novikov conjecture. Preprint, 2000.

37. G. G. Kasparov. The operator K-functor and extensions of C^*-algebras. *Math. USSR Izvestija*, 16(3):513–572, 1981.

38. G. G. Kasparov. Equivariant KK-theory and the Novikov conjecture. *Invent. Math.*, 91:147–201, 1988.

39. G. G. Kasparov. *K-Theory, Group C*-Algebras and Higher Signatures (Conspectus)*, pages 101–146. Volume 1 of Ferry et al. [21], 1995. First circulated 1981.

40. E. Kirchberg. On non-semisplit extensions, tensor products and exactness of group C^*-algebras. *Invent. Math.*, 112:449–489, 1993.

41. Eberhard Kirchberg. The Fubini theorem for exact C^*-algebras. *J. Operator Theory*, 10(1):3–8, 1983.

42. Eberhard Kirchberg. On subalgebras of the CAR-algebra. *J. Funct. Anal.*, 129(1):35–63, 1995.

43. Eberhard Kirchberg and Simon Wassermann. Exact groups and continuous bundles of C^*-algebras. *Math. Ann.*, 315(2):169–203, 1999.

44. V. Lafforgue. K-théorie bivariante pour les algèbres de banach et conjecture de baum-connes. *Inventiones Mathematicae*, 2002.

45. E. C. Lance. *Hilbert C*-modules*. Cambridge University Press, Cambridge, 1995. A toolkit for operator algebraists.

46. A. Lubutzky. *Discrete Groups, Expanding Graphs and Invariant Measures*, volume 125 of *Progress in Mathematics*. Birkhäuser, Boston, 1994.

47. David Meintrup and Thomas Schick. A model for the universal space for proper actions of a hyperbolic group. *New York J. Math.*, 8:1–7 (electronic), 2002.

48. J. Milnor. A note on curvature and the fundamental group. *J. Differential Geom.*, 2:1–7, 1968.

49. John Milnor. *Introduction to algebraic K-theory*. Princeton University Press, Princeton, N.J., 1971. Annals of Mathematics Studies, No. 72.

50. V. Lafforgue N. Higson and G. Skandalis. Counterexamples to the Baum-Connes conjecture. Preprint, 2001.

51. G. Niblo and L. Reeves. Groups acting on CAT(0) cube complexes. *Geometry and Topology*, 1:1–7, 1997.

52. N. Ozawa. Amenable actions and exactness for discrete groups. Preprint OA/0002185, 2000.

53. G. K. Pedersen. *C*-algebras and their Automorphism Groups*, volume 14 of *London Mathematical Society Monographs*. Academic Press, London, 1979.

54. N. C. Pillips. *Equivariant K-Theory for Proper Actions*, volume 178 of *Pitman Research Notes in Math*. Longmann Scientific and Technical, Essex, England, 1989.

55. G. Robertson. Addendum to "crofton formulae and geodesic distance in hyperbolic spaces". *J. of Lie Theory*, 8:441, 1998.

56. G. Robertson. Crofton formulae and geodesic distance in hyperbolic spaces. *J. of Lie Theory*, 8:163–172, 1998.

57. J. Roe. *Coarse Cohomology and Index Theory on Complete Riemannian Manifolds*, volume 104 of *Memoirs of the AMS*. American Mathematical Society, 1993.

58. J. Roe. *Index Theory, Coarse Geometry and Topology of Manifolds*. Number 90 in CBMS Regional Conference Series in Math. American Mathematical Society, 1996.

59. J. Rosenberg. The role of K-theory in non-commutative algebraic topology. *Contemporary Mathematics*, 10:155–182, 1982.

60. G. Skandalis. Une notion de nuclèaritè en K-thèorie. *K-Theory*, 1:549–573, 1988.

61. G. Skandalis. Le bifoncteur de Kasparov n'est pas exact. *Comptes Rendus Acad. Sci. Paris, Sèrie I*, 313:939–941, 1991.

62. G. Skandalis. Progrès récents sur la conjecture de Baum-Connes. Contribution de Vincent Lafforgue. *Séminaire Bourbaki*, 1999.

63. G. Skandalis, J. L. Tu, and G. Yu. Coarse Baum-Connes conjecture and groupoids. Preprint, 2000.

64. A. Svarc. Volume invariants of coverings. *Dokl. Akad. Nauk. SSSR*, 105:32–34, 1955.

65. J. L. Tu. La conjecture de Baum-Connes pour les feuilletages moyennables. *K-Theory*, 17:215–264, 1999.

66. S. Wassermann. *Exact C*-Algebras and Related Topics*, volume 19 of *Lecture Note Series*. Seoul National University, Seoul, 1994.

67. Simon Wassermann. *C*-exact groups. In *C*-algebras (Münster, 1999)*, pages 243–249. Springer, Berlin, 2000.

68. Guoliang Yu. The coarse Baum-Connes conjecture for spaces which admit a uniform embedding into Hilbert space. *Invent. Math.*, 139(1):201–240, 2000.

Geometric and Analytic Properties of Groups

Erik Guentner[1] and Jerome Kaminker[2],[*]

[1] Department of Mathematics, University of Hawaii,
Manoa, 2565 The Mall, Honolulu, HI 86802
erik@math.hawaii.edu

[2] Department of Mathematical Sciences, IUPUI, Indianapolis, IN 46202-3216
kaminker@math.iupui.edu

Abstract We will present a survey of the connections between the harmonic analysis of a discrete group and the asymptotic properties of the group considered as a metric space.

1 Introduction

Let Γ be a finitely presented group. By a geometric property of Γ we mean a property invariant under coarse equivalence. This entails treating Γ as a metric space with metric obtained from a length function. On the other hand, an analytic property of Γ is one which depends on the harmonic analysis, or unitary representation theory, of the group. Specifically, we will consider properties which are expressible in terms of, $C_r^*(\Gamma)$, the reduced C^*-algebra of Γ. It is an interesting development that there are some unexpected connections between these notions. They have been discovered in work surrounding the Baum-Connes conjecture which relates the K-homology of the classifying space of a group to the K-theory of its reduced C^*-algebra.

We will concentrate on the property of exactness, which is defined as an analytic property, but, somewhat surprisingly, turns out to be a geometric property as well. Further, it has the virtue of being related to the Baum-Connes and Novikov conjectures. Some of the results described here are essentially folk theorems and will be stated without attribution.

Some geometric properties of groups which one may consider are

1) uniform embeddability of Γ in a Hilbert space,
2) Property A of Yu,
3) amenability.

Some analogous analytic properties of $C_r^*(\Gamma)$ are

---

[*] The first author was supported in part by NSF Grant DMS-0071402. The second author was supported in part by NSF Grant DMS-0071435

a) exactness,
b) approximation properties for $C^*_r(\Gamma)$,
c) nuclearity.

There are several relations between these concepts and one of our goals in this survey is to indicate what is known and what questions remain.

2 Coarse Equivalence, Quasi-Isometries and Uniform Embeddings

The notion of uniform embeddability is best formulated in the coarse category, c.f. [16]. Recall that a coarse map between metric spaces X and Y is a metrically proper function $f : X \to Y$ which has the property that, for every $R > 0$, there exists an $S > 0$ such that

$$d_X(x, y) \leq R \Rightarrow d_Y(f(x), f(y)) \leq S.$$

A coarse equivalence is a coarse map $f : X \to Y$ for which there is a coarse map $g : Y \to X$ and a $K > 0$ such that $d_Y(f \circ g(y), y) < K$ and $d_X(g \circ f(x), x) < K$, for all $x \in X$ and $y \in Y$.

Definition 2.1. *A metric space X is uniformly embeddable in a metric space Y if X is coarsly equivalent to a subspace of Y.*

The coarse map implementing the equivalence is referred to as a uniform embedding. By choosing the range, Y, carefully one imposes conditions on X. For example, one of the most useful situations is when Y is chosen to be a Hilbert space. This implies that the Coarse Baum-Connes conjecture holds for X. One may also consider letting Y be an L^p-space. While some results are known in these cases about when uniform embeddings are possible, [4], there do not exist strong geometric implications as in the Hilbert space case.

Proposition 2.1. *Let $f : X \to Y$ be a coarse map. Then the following statements are equivalent.*

i) *f is a uniform embedding,*
ii) *f is a coarse equivalence between X and $f(X) \subseteq Y$,*
iii) *for every $R > 0$ there is an $S > 0$ such that*

$$d_X(x, y) \geq S \Rightarrow d_Y(f(x), f(y)) \geq R,$$

iv) *there exists functions $\rho_\pm : \mathbb{R}_+ \to \mathbb{R}_+$, both increasing to $+\infty$, such that*

$$\rho_-(d_X(x, y)) < d_Y(f(x), f(y)) < \rho_+(d_X(x, y)).$$

for all $x, y \in X$.

We will mainly be interested in (infinite) finitely generated groups, Γ, viewed as metric spaces. There are many metrics one might put on a group, but we shall restrict to those obtained from a length function—that is, a function $l : \Gamma \to \Gamma$ satisfying that $l(\gamma) \geq 0$, $l(\gamma\gamma') \leq l(\gamma) + l(\gamma')$, and $l(e) = 0$. The associated left invariant metric is given by $d(\gamma, \gamma') = l(\gamma^{-1}\gamma')$. One possible choice is the word length function associated to a symmetric set of generators, $\{s_1, \ldots, s_n\}$, where $l_w(t) = k$ if $t = s_1 \cdots s_k$ and k is the fewest generators needed. Choosing different sets of generators will provide different word length metrics, but they will all be coarsely equivalent.

Geometric properties of groups are those invariant under coarse equivalence. It is more customary to use the relation of quasi-isometry, but for the metric spaces obtained from finitely presented groups the notions are equivalent. This is a consequence of the following notions which we formulate in the discrete case only.

Definition 2.2. *A discrete metric space is geodesic if for any two points x and y there is a finite sequence of points $x = x_1, \ldots, x_n = y$ with $d(x, y) = \Sigma_1^{n-1} d(x_i, x_{i+1})$. A discrete metric space is quasi-geodesic if there exists $\lambda > 0$ and $\delta > 0$ so that for any x and y there exists points $x = x_1, \ldots, x_n = y$ with $d(x_i, x_{i+1}) \leq \delta$ and $d(x, y) \geq \lambda \Sigma_1^{n-1} d(x_i, x_{i+1})$.*

Recall that a uniform embedding is a *quasi-isometry* if the functions $\rho_{\pm}(r)$ can be taken to be of the form $\rho_+(r) = Cr + D$ and $\rho_-(r) = \dfrac{1}{C}r - D$ for some $C \geq 1$ and $D \geq 0$.

Proposition 2.2. *Let $f : X \to Y$ be a coarse equivalence of metric spaces. If X is quasi-geodesic then one can take $\rho_+(r) = Cr + D$. If in addition Y is quasi-geodesic then one may also take $\rho_-(r) = \dfrac{1}{C}r - D$.*

Thus, a coarse equivalence of quasi-geodesic metric spaces is a quasi-isometry. Since a finitely generated group with its word length metric is a geodesic space, one obtains that such groups are coarsely equivalent if and only if they are quasi-isometric. Indeed, one can carry this one step further.

Proposition 2.3. *Let $f : X \to Y$ be a uniform embedding. Assume that X is quasi-geodesic. Then f is a quasi-isometry if and only if $f(X) \subseteq Y$ is also a quasi-geodesic space with its induced metric from Y.*

Note that this is not always the case. If Γ is a finitely generated group then the inclusion of any subgroup, $i : \Gamma' \to \Gamma$ is a uniform embedding. However, the subgroup $\mathbb{Z} \subseteq \Gamma$, where Γ is the discrete 3-dimensional Heisenberg group, is not quasi-geodesic, c.f. [5].

If one now specialize to the case when Y is a separable Hilbert space, $\mathcal{H}$, then there are several useful implications. We shall discuss them from an analytic point of view in the next section.

3 Exact Groups

In this section we will introduce the relevant analytic properties of discrete groups. They are expressed in terms of the reduced C^*-algebra of Γ, $C_r^*(\Gamma)$. Recall that this is the C^*-algebra generated by the elements of Γ acting as unitary operators via left translation on $l^2(\Gamma)$, $s \cdot \xi(t) = \xi(s^{-1}t)$, so $C_r^*(\Gamma) \subseteq \mathcal{B}(l^2(\Gamma))$.

If $\mathcal{A}$ and $\mathcal{B}$ are C^*-algebras then, in general, there are many norms that can be put on the algebraic tensor product. The resulting normed algebra may then be completed to obtain an C^*-algebra. There is largest and a smallest norm and the associated completed tensor products are denoted by $\otimes_{min}$ and $\otimes_{max}$ norms. Since $\otimes_{max}$ has better functorial properties then $\otimes_{min}$, it is convenient to have the following condition satisfied.

Definition 3.1. *A C^*-algebra $\mathcal{A}$ is nuclear if $\mathcal{B} \otimes_{max} \mathcal{A} \cong \mathcal{B} \otimes_{min} \mathcal{A}$ for all C^*-algebras $\mathcal{B}$.*

Now, because of the functorial properties of $\otimes_{max}$ one has the following.

Proposition 3.1. *Let $0 \to \mathcal{I} \to \mathcal{B} \to \mathcal{B}/\mathcal{I} \to 0$ be an exact sequence of C^*-algebras. Then $0 \to \mathcal{I} \otimes_{max} \mathcal{A} \to \mathcal{B} \otimes_{max} \mathcal{A} \to \mathcal{B}/\mathcal{I} \otimes_{max} \mathcal{A} \to 0$ is exact for any C^*-algebra $\mathcal{A}$.*

Thus, if $\mathcal{A}$ is nuclear, one also has $0 \to \mathcal{I} \otimes_{min} \mathcal{A} \to \mathcal{B} \otimes_{min} \mathcal{A} \to \mathcal{B}/\mathcal{I} \otimes_{min} \mathcal{A} \to 0$ exact. This remains true if $\mathcal{A}$ is only a subalgebra of a nuclear algebra, although $\mathcal{A}$ may not be nuclear itself. This leads to the notion of an exact C^*-algebra.

Definition 3.2. *A C^*-algebra $\mathcal{A}$ is exact if $0 \to \mathcal{I} \to \mathcal{B} \to \mathcal{B}/\mathcal{I} \to 0$ an exact sequence implies that $0 \to \mathcal{I} \otimes_{min} \mathcal{A} \to \mathcal{B} \otimes_{min} \mathcal{A} \to \mathcal{B}/\mathcal{I} \otimes_{min} \mathcal{A} \to 0$ is as well.*

Much is known about exact C^*-algebras, a significant amount of this due to E. Kirchberg and S. Wassermann, [13, 14]. For example, one has the following result of Kirchberg.

Theorem 3.1. *A separable C^*-algebra $\mathcal{A}$ is exact if and only if it is a subquotient of a nuclear C^*-algebra.*

It is not terribly difficult to obtain examples of non-exact C^*-algebras. For example, $C^*(\mathbb{F}_2)$, the *full* C^*-algebra of the free group on two generators is not exact, [19]. On the other hand, we will see in the next paragraph that $C_r^*(\mathbb{F}_2)$ is exact. This leads to the main definition of this section.

Definition 3.3. *A group Γ is exact if $C_r^*(\Gamma)$ is an exact C^*-algebra.*

Since $C_r^*(\Gamma)$ being nuclear is equivalent to Γ being amenable, one sees that exactness is a generalization of amenability. While finding nonexact C^*-algebras is a manageable task, it is much more difficult to find nonexact groups. In fact, for a while it was thought that all groups might be exact. This hope was shattered by the discovery by M. Gromov of examples of nonexact groups, [6] .

The main task of this section is to explain how exactness can be viewed as a geometric property. This will lead to the examples of Gromov. For this it is convenient to bring in the notion of topological groupoid. Associated to the discrete group Γ is the locally compact groupoid $\beta\Gamma \rtimes \Gamma$, where $\beta\Gamma$ is the Stone-Čech compactification of Γ. The left action by translation of Γ on itself extends to an action on $\beta\Gamma$ and $\beta\Gamma \rtimes \Gamma$ is the transformation group groupoid. It is r-discrete and has a Haar system obtained from counting measure. There is a notion of amenability for such groupoids which is equivalent to the reduced C^*-algebra of the groupoid, $C(\beta\Gamma) \rtimes \Gamma$, being nuclear. Since $\beta\Gamma$ is compact, the algebra $C_r^*(\Gamma)$ is contained in $C(\beta\Gamma) \rtimes \Gamma$ and hence if it is nuclear, then Γ is exact. This applies to the group $\mathbb{F}_2$, since $\mathbb{F}_2$ acts amenably on its Gromov boundary, $\partial\mathbb{F}_2$, so that the crossed product $C(\partial\mathbb{F}_2) \rtimes_r \mathbb{F}_2$ is nuclear. It turns out that this criterion actually gives a necessary and sufficient condition for exactness. To state this, let Δ_R denote the strip $\{(s,t) \mid d(s,t) < R\}$.

Proposition 3.2 ([9, 15, 1]). *Let Γ be a finitely generated group. Then the following are equivalent.*

i) *Γ is exact,*
ii) *$\beta\Gamma \rtimes \Gamma$ is an amenable groupoid,*
iii) *there is a sequence of positive type functions $\mu_n : \Gamma \times \Gamma \to \mathbb{R}$ such that*
   a) *each μ_n is supported in a strip, Δ_R, for some R,*
   b) *$\mu_n \to 1$ uniformly on strips.*
iv) *Γ has Property A of Yu.*

Since Property A is known to be a quasi-isometry invariant, [20], we obtain,

Proposition 3.3. *Exactness is a geometric property of a finitely generated group.*

To contrast this with the notions in Section 2 one notes the following.

Proposition 3.4 ([9]). *Let Γ be a finitely generated group. Then the following are equivalent.*

i) *Γ is uniformly embeddable in a Hilbert space,*
ii) *$\beta\Gamma \rtimes \Gamma$ is an a-T-menable groupoid,*
iii) *there is a sequence of positive type functions $\mu_n : \Gamma \times \Gamma \to \mathbb{R}$ such that*
   a) *each μ_n is in $C_0(\Gamma \times \Gamma, \Delta)$, the set of functions which go to zero away from the diagonal,*
   b) *$\mu_n \to 1$ uniformly on strips.*

The property of a-T-menability for groupoids in (ii) was introduced by Tu as a generalization of the corresponding notion of Gromov for groups. It is also called the Haagerup Property, and we will use the terms interchangeably. Note that the only difference between 3.6 (iii) and 3.8 (iii) is the condition on the support of the $\mu_n's$, the support being contained in a strip in 3.6. The following result is now clear.

Proposition 3.5 ([9]). *Let Γ be a finitely generated group. If Γ is exact then it is uniformly embeddable in a Hilbert space.*

Since the definitions look so similar one may ask whether the converse holds. The answer is not known, but there is a numerical invariant of finitely generated groups, $R(\Gamma)$, which takes values in $[0, 1]$ and provides information on this question, [8]. If

$R(\Gamma) > 0$ then Γ is uniformly embeddable in a Hilbert space. If $R(\Gamma) > 1/2$ then Γ is exact. It is a geometric invariant and can be computed in several cases. Using it one obtains the following result.

Proposition 3.6. *Let Γ be a finitely generated group which is uniformly embeddable in a Hilbert space via $f : \Gamma \to \mathcal{H}$. If $f(\Gamma)$, with the metric induced as a subspace of $\mathcal{H}$, is quasi-geodesic then Γ is exact.*

It is worth noting that there is an equivariant version of the entire development outlined above. In that case one may essentially replace *exact* by *amenable* and the groupoid $\beta\Gamma \rtimes \Gamma$ by the group Γ itself. If one also requires that the maps involved are all equivariant, then uniform embeddability of Γ in a Hilbert space is replaced by the group Γ, (rather than the groupoid), having the Haagerup Property. The equivariant version of the invariant, $R_{equiv}(\Gamma)$, now satisfies that $R_{equiv}(\Gamma) > 0$ implies that Γ has the Haagerup property and $R_{equiv}(\Gamma) > 1/2$ implies that Γ is amenable.

If one considers the class of all finitely generated groups which are exact or uniformly embeddable in a Hilbert space, then there are several useful closure properties. Thus, exact groups are closed under extensions, taking direct limits, and free products with amalgamation. Groups uniformly embeddable in a Hilbert space have similar closure properties, except that they are not known to be closed under extensions. This is a curious situation and hopefully the relation between the two properties will be clarified soon, [8, 7].

4 Exactness and the Baum-Connes and Novikov Conjectures

The Baum-Connes conjecture with coefficients, for a group Γ, asserts that the index map

$$\mu_r : K_*^{top}(\Gamma; \mathcal{A}) \to K_*(\mathcal{A} \rtimes_r \Gamma) \tag{1}$$

is an isomorphism for any separable $\Gamma - C^*$-algebra $\mathcal{A}$.

A major breakthrough was obtained by the work of N. Higson and G. Kasparov, [12], where they proved that this holds for any group with the Haagerup Property. When $\mathcal{A} = \mathbb{C}$ the injectivity of the map μ_r implies that the Novikov conjecture on homotopy invariance of higher signatures holds for Γ. A second breakthrough was obtained by G. Yu, [20], (and also J. L. Tu, [18]) when it was proved that the Coarse Baum-Connes conjecture holds for any group that is uniformly embeddable in a Hilbert space. Recall that this states that the map

$$\mu_{coarse} : K_*^{coarse}(\Gamma) \to K_*(C^*(|\Gamma|)), \tag{2}$$

is an isomorphism. The group on the right is the K-theory of the Roe algebra of the group Γ viewed as a metric space. The Principle of Descent applies when Γ has a classifying space which is a finite complex. From this it follows that if μ_{coarse} in (2) is an isomorphism, then μ_r in (1) is an injection, so that the Novikov conjecture holds. Thus one obtains,

Proposition 4.1. *If the group Γ is exact then the Novikov conjecture holds for Γ.*

In fact, uniform embeddability of Γ in a Hilbert space is sufficient as was established in [17] based on an argument of Higson, [10]. It is interesting that two of the major classes of groups which act amenably on a compact space, and hence are exact, satisfy the Baum-Connes conjecture. They are the class of amenable groups, which act amenably on a point, and hyperbolic groups, which act amenably on the Gromov boundary. This suggests the question of whether the property of exactness of Γ can play a role in proving the surjectivity of the Baum-Connes index map, μ_r. Evidence for this will appear in the next section where we will see that the same phenomenon which obstructs the exactness of Gromov's group also allows one to exhibit a group for which the Baum-Connes conjecture with certain specific coefficients fails.

While the Coarse Baum-Connes conjecture holding for a group is a geometric property, the same is not known for the Baum-Connes conjecture itself. However, usually it is the case, based on ad-hoc arguments, if a group Γ satisfies the Novikov conjecture and Γ' is quasi-isometric to it, then Γ' does so as well. However, there does not seem to be a method of proof which extends to groups quasi-isometric to a given one.

5 Gromov Groups and Expanders

We first recall the definition of an expanding sequence of graphs and establish some notation. If X is a finite graph, then we will let $|A|$ denote the number of vertices in a subset $A \subseteq X$. The valence of a graph, $val(X)$, is the maximum number of edges coming out of a vertex

Definition 5.1. *A family of finite graphs, $\{X_n\}$, is an expanding sequence of graphs if the following properties hold.*

i)   *there is a $K > 0$ such that $val(X_n) \leq K$ for all n,*
ii) *there is a constant $C > 0$ so that if $A \subseteq X_n$ has $|A| > \frac{1}{2}$ then $|\partial A|/|A| > C$, where $\partial A = \{v \in X_n \mid d(v, A) \leq 1\}$,*
iii) *$|X_n| \to \infty$*

One can make a metric space out of the disjoint union $\amalg X_n$ in such a way that each X_n has its usual metric and $d(X_n, X_m) \to \infty$ as $m, n \to \infty$. Then one has the following basic property due to Gromov.

Proposition 5.1. *The space $\amalg X_n$ cannot be uniformly embedded in a Hilbert space.*

Indeed, one has the following stronger result due to N. Higson for a special class of expanding graphs.

Proposition 5.2. *Let $\{X_n\}$ be the Cayley graphs for the finite quotients of a residually finite, Property T, group with a fixed set of generators. The metric space $\amalg X_n$ satisfies*

i)   *the uniform Roe algebra, $UC^*(\amalg X_n)$, is not exact,*

ii) the Coarse Baum-Connes map,

$$\mu_{coarse} : K_*^{coarse}(\amalg X_n) \to K_*(UC^*(\amalg X_n))$$

is not surjective.

The main case of interest for the Baum-Connes conjecture is that of a group, and more specifically, because of its connection with the Novikov conjecture, finitely generated groups. To relate the notion of expanders to geometric properties of groups Gromov considered the condition that the Cayley graph of a group contained an expanding sequence of graphs. In order to weaken the notion of containment, Higson, Lafforgue and Skandalis, [11], made use of the following notion which is due to Gromov, [6].

Definition 5.2. *A metric space Y weakly contains an expanding sequence of graphs, $\{X_n\}$ if there are maps $f_n : X_n \to Y$ satisfying*

i)   each f_n is Lipschitz with the same constant for all n,

* $\lim_{n\to\infty} \sup_{x\in X_n} \dfrac{|f_n^{-1}(x)|}{|X_n|} = 0.$*

Thus, there would be a Lipschitz map of $\amalg X_n$ into Y which becomes closer to being one-one as n increases. This condition is sufficient to show that Y cannot be uniformly embedded in a Hilbert space. If it is possible to take $Y = \Gamma$, a finitely generated group , then one can obtain a counterexample, for Γ, to the Baum-Connes conjecture with coefficients. In fact, Gromov has shown, using probabilistic methods, that there exists a group which weakly contains an expanding sequence of graphs. If one takes $\mathcal{A} = l^\infty(\mathbb{N}, c_0(\Gamma))$ for the algebra, then the Baum-Connes index map,

$$\mu_r : K_*^{top}(\Gamma, \mathcal{A}) \to K_*(\mathcal{A} \rtimes \Gamma)$$

is not surjective, [11] .

6 Final Remarks

One aspect of the subject of exact groups that we have not discussed is the use of approximation properties for $C_r^*(\Gamma)$. In fact, the most useful characterizations of nuclearity and exactness are in terms of approximation properties involving completely positive maps. These are noncommutative analogs of summability methods, such as Cesaro summability for continuous functions in $C(S^1) = C^*(\mathbb{Z})$. These topics are currently under study and it will be interesting to see how directly they can be linked to more geometric properties of groups.

Two conditions on a group which imply that the Novikov conjecture hold have been introduced by A. Connes, H. Moscovici, and M. Gromov, [2, 3]. They are that all the cohomology of a group is Lipschitz, and that the K-theory of the classifying space of a group is exhausted by almost flat elements. It is not known how these ideas are related to exactness, but based on recent arguments of Dranishnikov, [4], it is possible that a connection can be established.

Alain Connes has suggested that understanding exactness of $C_r^*(\Gamma)$ in a more homological sense could lead to an adjustment in the formulation of the Baum-Connes conjecture to take into account the existence of nonexact groups.

Finally, it would be very useful to know to what extent expanding graphs capture the obstruction to exactness. Is it possible that a group is not exact if and only if it weakly contains an expander? Is a group uniformly embeddable in a Hilbert space if and only if it does not weakly contain an expander? Is the conjecture made by Dranishnikov in [4] that a group is exact if and only if it uniformly embeds in l^1 true?

References

1. C. Anantharaman-Delaroche, *Amenability and exactness for dynamical systems and their C^*-algebras*, pre-print, 2001.
2. A. Connes, M. Gromov, and H. Moscovici, *Group cohomology with Lipschitz control and higher signatures*, Geom. Funct. Anal. **3** (1993), no. 1, 1–78. MR 93m:19011
3. Alain Connes, Mikhaïl Gromov, and Henri Moscovici, *Conjecture de Novikov et fibrés presque plats*, C. R. Acad. Sci. Paris Sér. I Math. **310** (1990), no. 5, 273–277. MR 91e:57041
4. A.N. Dranishnikov, *Anti-čech approximation in coarse geometry*, IHES preprint, 2002.
5. M. Gromov, *Asymptotic invariants of infinite groups*, London Mathematical Society Lecture Notes, no. 182, pp. 1–295, Cambridge University Press, 1993.
6. M. Gromov, *Random walk in random groups*, IHES preprint, 2002.
7. E. Guentner and M. Dadarlat, *Uniform embeddability of free products and extensions of discrete groups*, preprint, 2002.
8. E. Guentner and J. Kaminker, *Exactness and uniform embeddability of discrete groups*, Preprint.
9. Erik Guentner and Jerome Kaminker, *Exactness and the Novikov conjecture*, Topology **41** (2002), no. 2, 411–418. MR 1 876 896
10. N. Higson, *Bivariant K-theory and the Novikov conjecture*, Geom. Funct. Anal. **10** (2000), no. 3, 563–581. MR 2001k:19009
11. N. Higson, V. Lafforgue, and G. Skandalis, *Counterexamples to the Baum-Connes conjecture*, Geom. Funct. Anal. **12** (2002), no. 2, 330–354. MR 1 911 663
12. Nigel Higson and Gennadi Kasparov, *E-theory and KK-theory for groups which act properly and isometrically on Hilbert space*, Invent. Math. **144** (2001), no. 1, 23–74. MR 2002k:19005
13. E. Kirchberg, *On non-semisplit extensions, tensor products and exactness of group C^*-algebras*, Invent. Math. **112** (1993), 449–489.
14. Eberhard Kirchberg and Simon Wassermann, *Permanence proerties of C^*-exact groups*, Doc. Math. **4** (1999), 513–558 (electronic). MR 2001i:46089
15. N. Ozawa, *Amenable actions and exactness for disrete groups*, C. R. Acad. Sci. Paris Sér. I Math. **330** (2000), no. 8, 691–695.
16. John Roe, *Index theory, coarse geometry, and topology of manifolds*, Published for the Conference Board of the Mathematical Sciences, Washington, DC, 1996. MR 97h:58155

17. G. Skandalis, J. L. Tu, and G. Yu, *Coarse Baum-Connes conjecture and groupoids*, Preprint, 2000.
18. J. L. Tu, *La conjecture de Baum-Connes pur les feuilletages moyennables*, K-Theory **17** (1999), 215–264.
19. S. Wassermann, *Exact C^*-algebras and related topics*, Lecture Note Series, vol. 19, Seoul National University, Seoul, 1994.
20. G. Yu, *The coarse Baum-Connes conjecture for spaces which admit a uniform embedding into Hilbert space*, Invent. Math. **139** (2000), no. 1, 201–240.

More Lectures on Algebraic Quantum Field Theory

J. E. Roberts*

Dipartimento di Matematica, Università di Roma "Tor Vergata"
I-00133 Roma, Italy
roberts@mat.uniroma2.it

Abstract In which the basic concepts of algebraic quantum field theory are developed, the main achievements outlined and the last part devoted to a detailed account of superselection sectors of field theories over curved spacetime.

1 Introduction

Algebraic quantum field theory is over forty years old now and, although it cannot be denied its place as part of non–commutative geometry, it remains the odd one out in this volume. This partly reflects the exigencies of physics and partly a sociological phenomenon in that, with few exceptions [29], [57], [58], we have preferred to stick to our tried methods rather than pioneer the use of the new methods of non–commutative geometry. This means that algebraic quantum field theory has continued to be non–commutative measure theory enriched with a local structure: the basic object is a net of von Neumann algebras. Many of those who pick up this volume in the search of enlightment will therefore have had no previous contact with algebraic quantum field theory and my contribution has this in mind. The first service I can perform for such readers is to list textbooks and other introductory material: [1], [5], [6], [12], [53], [73].

This contribution follows its own pedagogical ideas. It is divided into three parts of very different character. The first and longest part explains in a leisurely fashion some of the scientific cultural background which shaped the subject. It is all well known to anyone who has been working in the subject for some time. With the passage of time, newcomers perhaps do not even realize that they should absorb these ideas. My intention was to keep the first part from being too technical thus allowing it to be read through and understood quickly. The physical terminology might give difficulties to those with a background just in mathematics. Alas, the reader of this first part might conclude that algebraic quantum field theory has remained at the level of conceptual ideas, axioms and minor results.

* Research supported by MURST, CNR–GNAFA and INDAM–GNAMPA

For this reason there is a second part devoted to outlining some of the principal achievements of the subject with some, less prominent, comments on lack of achievement. I approached this with some trepidation since there must have been over 500 papers written on the subject and I have a detailed knowledge of just a small fraction, essentially those I have been involved in. My selection is biased towards what I know better hence towards my friends, towards sector theory, the leitmotif of this contribution, and towards the past, when fewer papers were written. I hope these words of explanation will mollify those whose important contributions are passed over.

The third part aims to give a innovative version of sector theory with the necessary details. It needs to be studied rather than read through quickly. The hope is that those who have been sufficiently motivated by the first two parts to work through the third will have an instrument at their disposal allowing them to make original contributions to the subject. This version has its origins in [50] and is motivated by the wish to develop sector theory for quantum field theory on curved spacetime. Further work, particularly on classes of models, is needed before being able to claim that the basic problems have now been solved.

2 Algebraic Quantum Field Theory

Algebraic quantum field theory is an approach to quantum field theory using operator algebras. I would distinguish three different aspects. The first is using operator algebras to formulate some basic physical laws, the laws of local relativistic quantum physics, designed to provide a framework for discussing elementary particle physics. The second aspect is using operator algebras to analyze structural features of elementary particle physics on the basis of these physical laws. Here we can make a rough division of the material into three categories. First there are the spectral properties, that is the properties related to the energy-momentum spectrum of the theory, such as covariance, positivity of the energy and the concepts of vacuum and particle. Then there is the local structure: local algebras, local commutativity, locality of interactions, causality, finite propagation speed and the Reeh-Schlieder Theorem. Finally, there is the study of superselection structure. This allows us to understand the origin of three of the characteristic aspects of elementary particle physics - the composition of charges, the classification of particle statistics and the charge conjugation which underlies the basic symmetry between particles and their antiparticles.

The third aspect is to uncover the structural possibilities inherent to quantum field theory again on the basis of these laws. Apart from leading to a better understanding of quantum field theory one hopes, of course, that this will lead to new insights into elementary particle physics. To illustrate what I mean, I would mention the question, probably ill posed, of what is a gauge theory. Little progress has been made in this direction but I would say the right approach is not to try to adapt the detailed notions of classical gauge theory, a subject which has never been of physical interest in its own right and is hence an unreliable guide. Instead, one should work within algebraic quantum field theory, concentrating on those aspects with direct physical significance such as Gauß' Theorem and the fact that electrical charges cannot be localized. They should appear as structural elements of a quantum field theory and could be used to distinguish one theory from another.

Those meeting these ideas for the first time should, I feel, be asking three questions.

Why is the theory formulated in terms of local observables and not in terms of fields in the conventional sense such as the electromagnetic field and what is the relation between the two concepts?

Why does one take an abstract "axiomatic" approach looking for laws of local relativistic quantum physics rather than looking at a specific concrete model such as quantum electrodynamics?

Why does one use abstruse mathematical objects like C^*-algebras and von Neumann algebras rather than working with operators and vectors in Hilbert space as in elementary quantum mechanics?

3 Quantum Fields and Local Observables

I come now to my first question: why is the theory formulated in terms of local observables and not in terms of fields in the conventional sense such as the electromagnetic field?

The first point can be understood when I explain what is meant by a quantum field because the concept is not as simple as one might a priori imagine. If you want to keep the ideas as concrete as possible, you might like to think in terms of the electromagnetic field. Now, in classical field theory, a field is a physical quantity depending on spacetime. A pure state is described by a field configuration and the simplest observables are the field strengths at points in spacetime.

In quantum field theory a pure state can no longer be described in terms of a field configuration any more than a pure state in quantum mechanics can be described in terms of the orbit of a particle. Furthermore, a quantum field is too singular an object for the field strength at a point to be an observable. This makes good physical sense, first, because if we think, for example, of measuring the electromagnetic field the best we can do is to measure some average field strength in the neighbourhood of a point whose details depend on the exact apparatus being used and are known only in rough outline. Actually, one should, I think, consider singular field configurations in classical physics, too, but at least you can explain the concept of field in terms of smooth field configurations. Secondly, it is against the spirit of the Heisenberg uncertainty relations to be able to achieve absolute spatial resolution. The better the spatial resolution, the higher the momentum transfer to the system has to be. Indeed, elementary particle physicists need to go to ever higher energies to probe the structure of matter on an ever smaller scale. For a discussion of the interpretation of measurements of field strengths I refer you to [9] and for a discussion of the intrinsic limitations of spatial resolutions on scales of the order of the Planck length [33].

What we need is a formalism which takes account of these unpleasant but physically comprehensible facts of life. If we insist on having operators on Hilbert space, we shall need some sort of average field strengths, $\int \varphi(x) f(x) dx$, and a formalism to deal with such objects. The standard formalism, generally referred to as the Wightman formalism, uses functions called test functions to model the averaging process and defines fields as unbounded operator-valued distributions. Let me suppose, for simplicity that I am dealing with a single scalar field, then one supposes that there is a dense linear subspace D in a Hilbert space $\mathcal{H}$ and for each smooth function f with compact support, $f \in \mathcal{D}(\mathbb{R}^4)$, a linear operator $\varphi(f)$ mapping D into D such that $\varphi(f)$ depends linearly on f, and $f \mapsto (\Phi, \varphi(f)\Psi)$ is continuous for each $\Phi, \Psi \in D$. This means that $f \mapsto (\Phi, \varphi(f)\Psi)$ is a distribution and an alternative way of saying this would be to demand that there is a sequence $g_n \in \mathcal{D}(\mathbb{R}^\triangle)$ depending on Φ, Ψ such that

$$(\Phi, \varphi(f)\Psi) = \lim \int g_n(x) f(x) dx.$$

$\varphi(f)$ is to be thought of informally as

$$\varphi(f) = \int \varphi(x) f(x) dx.$$

Finally, if φ is to be an Hermitian field, $\varphi = \varphi^*$, we must have

$$(\Phi, \varphi(f)\Psi) = (\varphi(\bar{f})\Phi, \Psi), \quad \Phi, \Psi \in D.$$

Actually, if you look at textbooks on Wightman field theory or axiomatic field theory as it is often called, you will find that a different space of test functions denoted $\mathcal{S}(\mathbb{R}^\triangle)$ is used and this yields what are called tempered distributions. For the development of axiomatic field theory it is indeed important that the class of test functions can be extended from $\mathcal{D}(\mathbb{R}^\triangle)$ to $\mathcal{S}(\mathbb{R}^\triangle)$ but this is not our aim and I shall stick to the space $\mathcal{D}(\mathbb{R}^\triangle)$ as it is closer to the spirit of a localized measurement.

There is no difficulty in modifying the above formalism to deal with several fields - the same domain D has to be used for each field - or with fields with several components. In the latter case one uses test functions with the same number of components.

Note, that there is a distinct shift of emphasis when passing from the physical idea of an average field strength to the mathematical idea of a distribution. I used to motivate the Wightman formalism in this way without being aware of this shift. It came home to me whilst working on a joint paper [33] with Doplicher and Fredenhagen on quantum spacetime when it became necessary to define a field on quantum spacetime. The right way to generalize was to say that in a field average $f(x)dx$ is a probability measure on Minkowski space so that when spacetime becomes non-commutative (and this is all the word quantum means here) $f(x)dx$ should be replaced by the non-commutative analogue of a probability measure, i.e. by a normalized positive linear functional, or state, on the C^*-algebra describing the quantum spacetime.

Thus the concept of quantum field has lost something of the simplicity of the classical concept of field. Furthermore, the resulting mathematical theory of unbounded operator-valued distributions, although it served well for the initial development of axiomatic field theory, does not function well from a mathematical point of view. There are too few results and too many pathologies. Furthermore, it does not fit in well with Hilbert space. The reason for this is that in order to perform simple algebraic operations with facility we have defined our fields on a common dense invariant domain D. However, from the theory of unbounded operators on Hilbert space we know that the correct domain will usually depend on the operator and that we shall want to ensure that the operator in question is closed or perhaps self-adjoint. Another illustration of the same type of phenomenon is that the adjoint $\varphi(f)^*$ of a smeared field operator must be expected to have a domain larger than D. In fact, our definition of an Hermitian field reads $\varphi(\bar{f}) \subset \varphi(f)^*$. Thus the algebra generated by the fields is not a *–algebra under the usual Hilbert space *–operation. Instead we have to define a new *–operation † setting

$$X^\dagger := X^* \upharpoonright D,$$

where X here denotes some polynomial in the fields.

Another disadvantage of formulating quantum field theory in terms of fields becomes apparent when you pose the question of when do two different sets of fields describe the same physical theory. I see no plausible answer to this question which can be given within the framework of axiomatic field theory. Furthermore, the question is not merely academic. There are examples known where apparently quite unrelated sets of fields in fact define the same physical theory.

There is yet another fact which must be taken into consideration. Not all quantum fields have a direct physical interpretation. The electromagnetic field, where the field strengths can be measured is the exception rather than the rule. For example, the strength of the electron field cannot be measured. In algebraic field theory it is regarded as a principle that the whole theory should be expressible in terms of observable quantities. This has proved to be a valuable principle which should at the very least be incorporated into axiomatic field theory in some form or other.

In fact, to get round these problems and arrive at algebraic quantum field theory there are two quite distinct steps to be taken. The first step is to replace the unbounded field operators by algebras of bounded operators. The elements of these algebras are to be thought of as being some bounded function of the unbounded fields. For example, if $f = \bar{f}$ we might think of using $e^{i\varphi(f)}$ in place of $\varphi(f)$. If $\varphi(f)$ were self-adjoint, then $e^{i\varphi(f)}$ would be unitary and hence bounded. This is more than can be expected but there are many cases where $\varphi(f)$ turns out to be essentially self-adjoint and in these cases $e^{i\varphi(f)^*}$ would be a suitable unitary operator. Notice that we are being drawn into the kind of technicality which is largely absent when working within algebraic quantum field theory. The passage from axiomatic field theory to algebraic quantum field theory is fraught with this kind of technicality which I would like to evade as far as possible. For this reason, I present one possible scheme which gives the general idea and works well enough in practice in most examples.

Let $\mathfrak{P}(\mathcal{O})$ denote the *-algebra of polynomials in the basic fields where the test functions are chosen to have support within the spacetime region $\mathcal{O}$. We will specify the class of regions $\mathcal{O}$ later on, and for the moment you might like to think of $\mathcal{O}$ as being any bounded open set. We consider the weak commutant of $\mathfrak{P}(\mathcal{O})$, $\mathfrak{P}(\mathcal{O})^w$, this is the set of $S \in \mathcal{B}(\mathcal{H})$ such that $(X^\dagger\Phi, S\Psi) = (S^*\Phi, X\Psi)$, $X \in \mathfrak{P}(\mathcal{O})$, $\Phi, \Psi \in D$. Note that this definition has been carefully chosen so that the problem of whether S maps D into D or not does not arise. A more naive definition of the commutant at once raises this problem. Obviously, if $S \in \mathfrak{P}(\mathcal{O})^w$ then $S^* \in \mathfrak{P}(\mathcal{O})^w$ and if $\mathcal{Y}$ is any such self-adjoint set of bounded operators then $\mathcal{Y}'$ is a *–algebra and in fact a von Neumann algebra since $\mathcal{Y}' = \mathcal{Y}'''$ or equivalently since $\mathcal{Y}'$ is closed in the weak operator topology. In any case if we now set

$$\mathfrak{F}(\mathcal{O}) := \mathfrak{P}(\mathcal{O})^{w'},$$

we get a net $\mathcal{O} \mapsto \mathfrak{F}(\mathcal{O})$ of von Neumann algebras over regions in spacetime. The term net here simply means that

$$\mathfrak{F}(\mathcal{O}_1) \subset \mathfrak{F}(\mathcal{O}_2) \quad \text{whenever} \quad \mathcal{O}_1 \subset \mathcal{O}_2.$$

We have now replaced one type of mathematical object, namely unbounded operator–valued tempered distributions which give a theory which neither functions smoothly nor fits in well with the structure of Hilbert space with another type of object, a net of von Neumann algebras. Von Neumann algebras do have a beautiful, rich and highly developed theory which is intimately related to Hilbert space. The net structure turns out to be very important and this is just as well because it is all that remains of the original idea of a field as being a physical quantity depending on a point in spacetime.

Let me make two remarks which may help to relate our construction to the idea of looking for bounded functions of the fields. First, any bounded function of the fields should commute with all operators which commute with the fields themselves. Secondly, it is a theorem that if you start with a self-adjoint operator, form the commutant and then form the commutant again then you end up with the von Neumann algebra of bounded Borel functions of the original operator.

We have also gone some way towards answering the question of when two sets of fields define the same physical theory. We now have a mathematically plausible answer, namely when the corresponding nets of von Neumann algebras are isomorphic. The role of the fields may therefore be though of as providing one possible set of 'generators' of the associated net of von Neumann algebras.

However, our mathematically plausible answer is not correct from the physical point of view, precisely because we have not yet taken account of the fact that not all field strengths are observable, i.e. can really be measured. In fact, what characterizes a physical theory is not the field net $\mathcal{O} \mapsto \mathfrak{F}(\mathcal{O})$ but a smaller net $\mathcal{O} \mapsto \mathfrak{A}(\mathcal{O})$ called the observable net or net of local observables. To construct it, we should start just from the observable fields and proceed as above although there is one point to watch. When we have a generating set of unobservable fields then certain combinations of these fields will be observable and these combinations must be included. For example,

$$j^\mu(x) = \bar{\psi}(x)\gamma^\mu\psi(x)$$

might be observable when $\psi(x)$ is not. In practice, it is very often possible to define $\mathcal{O} \mapsto \mathfrak{A}(\mathcal{O})$ starting from $\mathcal{O} \mapsto \mathfrak{F}(\mathcal{O})$. In such cases, there is a group G called the gauge group which acts as automorphisms of $\mathfrak{F}$ in such a way that $\mathfrak{A}$ is the fixed-point net:

$$\mathfrak{A}(\mathcal{O}) = \{F \in \mathfrak{F}(\mathcal{O}) : g(F) = F, g \in G\}.$$

This is one aspect of the principle of gauge invariance. From what I have said, the physically correct answer to our question can now be guessed: two field theories are to be regarded as equivalent when the associated nets of observables are isomorphic. One may want to modify this answer slightly to take account of further structure but this is just an embellishment.

4 Quantum Field Theory

I come now to my second question: why take an abstract 'axiomatic' approach looking for laws of local relativistic quantum physics rather than looking at a specific physically relevant concrete model such as quantum electrodynamics? The first point to make is that one should in principle be doing both. It is usually simpler to take an abstract approach as general structural features are recognized more quickly. Although it is, in principle, easier to prove the same general result in the particular case of a specific model, in practice, it is harder because irrelevant details and even notation can obscure the essential points. On the other hand, the abstract approach is never able to make predictions specific to one particular model such as predicting the mass of a particle in a model instead you can only predict general phenomena pertaining to a whole class of models. Furthermore, if you want a counter–example to some particular conjecture then you will be forced to look at some particular model. Unfortunately, this usually involves quite a lot of work. In any case, you need to begin by specifying the basic net of observables and this can only mean giving a set of generators. These generators are usually, but not always, a set of Wightman fields. Probably, the reason why most quantum field theorists do not make use of nets of algebras is that they spend most of their time working with specific models. So, why don't we look at that specific physically relevant model.

Well, the basic fact of life about quantum field theory is that it is extremely difficult to exhibit relativistic quantum fields with physically realistic interactions. There is, I should add, no difficulty in giving an explicit description of relativistic quantum fields which do not interact and the simplest such free field systems are associated with particles which do not interact. There is a lot that can be learned from a careful study of such systems and every beginner has sooner or later to get to grips with the free field.

Naturally, we have in some way to learn to deal with interactions in quantum field theory and various strategies have been adopted with some modest measure of success. The oldest strategy is to use renormalized perturbation theory. The basic idea is to start from a free field theory described in Fock space and perturb it by introducing an interaction. A physically realistic interaction should be relativistically invariant and local. The simplest such interaction term that one could envisage would be a self–interaction term of the form

$$\int \lambda \varphi(x)^4 dx$$

for a scalar field φ. The interaction is local because it involves the strength of the field at the point x and integrating over spacetime should guarantee relativistic invariance. However, whether we think in terms of sesquilinear forms or distributions, there is no obvious way of giving a meaning to $\varphi(x)^4$. This problem leads to what are called ultraviolet divergences but the integration over all spacetime is also not well defined and this leads to infrared divergences. What gets more to the root of the matter is to realize that a realistic interacting theory does not live in the same representation Fock space as the free field theory we are trying to perturb. To get any further, it becomes necessary to modify the interaction by introducing ultraviolet and infrared cutoffs and then to remove these cutoffs in a complicated limiting procedure in which counter terms are introduced and the original Fock space is exchanged for a new Hilbert space. This procedure is called renormalization. Of course the interacting theory is not really constructed this way. First, one only works to a certain finite order in perturbation theory. Secondly, the perturbation theory would anyway not converge and thirdly, instead of computing, say, the vacuum expectation value of certain observables one computes certain S–matrix elements. Thus one does not really try to get at the interacting theory but only at the resulting scattering theory which is what is of interest to elementary particle physics. The applications of relativistic quantum field theory are almost exclusively to elementary particle physics and, as one would expect naively, the methods of renormalized perturbation theory work best when the coupling constants are small. In particular, they can be applied to quantum electrodynamics to give results of fantastic accuracy and to the Salam-Weinberg theory, the theory unifying weak and electromagnetic interactions. However, they are of little use when we come to a theory of strong interactions such as quantum chromodynamics.

A second attempt to do better is constructive quantum field theory which goes right to the heart of the matter by trying to construct interacting field theories. It originated in the late sixties and, as a result of hard work by a number of mathematical physicists, made good progress during the first decade. In this period a whole class of interacting theories were constructed in two spacetime dimensions and two theories φ_3^4 and Y_3 in three spacetime dimensions. This demonstrated that it was possible to construct interacting relativistic theories using the ideas of renormalization theory, a fact which in the absence of models had been questioned. Unfortunately, the ultraviolet singularities of all theories in four spacetime dimensions are more severe and it has not proved possible to extend the methods to cover these cases. In retrospective there are two poignant comments on the efforts to construct φ_4^4. First, elementary particle physicists would not have cared about it one way or the other; by the end of the seventies only gauge theories were considered interesting. Secondly, it seems that φ_4^4 is trivial in the sense that attempts to construct the theory seem to lead to a free field theory. There is I think no proof, indeed it is going to be difficult to rule out the possibility that a different way of constructing the theory leads to a different result. More worrying is the fact that the balance of opinion is that quantum electrodynamics despite the successes of renormalized perturbation theory shares the same fate. One explains this result away by saying that a theory of electromagnetic interactions on its own is not consistent.

Constructive quantum field theory is not dead. Attention was turned to gauge theories and quite a different approach was adopted. Instead of introducing cutoffs, the theory is put on a lattice and this gets rid of the ultraviolet singularities. In the end one wants to take a limit as the lattice spacing goes to zero and obtain a theory on the usual spacetime continuum. But I won't attempt to give any futher details.

So the short answer to my question is that that specific physically relevant model just does not exist so that we come back to algebraic quantum field theory and the subject of this course. The reaction of this branch of quantum field theory to the absence of realistic interacting models is to claim that one can nevertheless learn something about the properties of such models. The idea is to set up and analyse general laws of local relativistic quantum physics. This yields physical insight and the goal of physics is not merely to be able to predict what happens but also to understand nature.

I hope that this discussion has given you some feeling for the diversity of quantum field theory in particular for how the same problems can be tackled in very different ways. A lot of work has gone into each of these different areas, each has had its successes and each is in some way threatened by its failures. Although I have not described all activities within quantum field theory, you would be correct in concluding that there is no approach to treating local relativistic quantum field theories which does not have some obvious weaknesses.

5 Spacetime and Its Symmetries

Among the most important symmetries of a physical system are those linked to spacetime such as spacetime translations, rotations, Galilei transformations and Lorentz transformations. In these lectures which are oriented towards applications to elementary particle physics we must take account of the special theory of relativity. On the other hand, since we only deal with microscopic systems, the effects of gravity are unimportant, at least down to distances of the order of the Planck length, so that we can ignore the general theory of relativity. Our spacetime for the moment will be Minkowski space, the spacetime of special relativity.

The basic difference between Minkowski space and the spacetime of Galilean physics lies in the causal structure. According to Einstein's special theory of relativity, no signal can propagate faster than the speed of light. If you fix one point x, the remaining points fall into five distinct classes: the points which can be influenced by x form what is called the *forward light cone* at x. Points on the boundary of this cone distinct from x can be reached by a light signal emanating from x and such points are termed *positive lightlike* with respect to x. Points in the interior of this cone are said to be positive timelike with respect to x. Points which can influence x form the *backward light cone* at x, points of the boundary distinct from x are said to be *negative lightlike*, points of the interior negative timelike with respect to x. The remaining points can neither influence x nor be influenced by x and are said to be spacelike with respect to x. This situation should be contrasted with the degenerate situation in 'Galilean' relativity, where we have only the future, the past and the present.

To describe these sets mathematically, we pick an origin and an inertial frame of reference making $\mathbb{M} \simeq \mathbb{R}^4$ and represent a point $x \in \mathbb{M}$ by coordinates $x^0 = ct$, x^1, x^2, x^3. We define a scalar product by

$$x \cdot y = x^0 y^0 - x^1 y^1 - x^2 y^2 - x^3 y^3 := x^0 y^0 - \mathbf{x} \cdot \mathbf{y}.$$

The forward light cone at the origin $\bar{V}_+$ is now given by

$$\bar{V}_+ = \{x \in \mathbb{M} : x \cdot x \geq 0,\, x^0 \geq 0\}$$

and its interior is denoted by V_+,

$$V_+ = \{x \in \mathbb{M} : x \cdot x > 0,\, x^0 > 0\}.$$

The backward light cone is $\bar{V}_- = -\bar{V}_+$, its interior $V_- = -V^+$.

The simplest bounded regions in Minkowski space are the *double cones* and we will use the open double cones to index the nets of von Neumann algebras. An open double cone is constructed as follows: pick two points v_+ and v_- in Minkowski space with v_+ positive timelike with respect to v_- then the double cone $\mathcal{O}$ with vertices v_+ and v_- is the intersection of the open backward light cone in v_+ with the open forward light cone in v_-. The set of open double cones will be denoted by $\mathcal{K}$. By restricting ourselves to double cones we avoid irrelevant geometric problems whilst still retaining sufficient information to relate our basic dynamical variables to spacetime. In fact, the set of open double cones forms a base for the topology on Minkowski space and we shall see when we discuss the property of additivity that we expect to have a natural way of extending the net to all open bounded sets.

The invariance group of Minkowski space is the Poincaré group $\mathcal{P}_+^\uparrow$. An element $L \in P_+^\uparrow$ is, in these coordinates, specified by a pair (a, Λ), where a is the image of the origin under L and Λ is a linear transformation on $\mathbb{R}^4$, thus $Lx = \Lambda x + a$, where

$$\Lambda x \cdot \Lambda y = x \cdot y.$$

I will further impose the conditions $\mathrm{Det}\Lambda > 0$, $e_0 \cdot \Lambda e_0 > 0$, where $e_0^0 = 1$, $e_0^i = 0$, $i = 1, 2, 3$. These conditions imply that reflections are ruled out and the direction of time is preserved. This is the meaning of the symbols $\uparrow$, $+$. The group law can be computed from the expression for Lx and gives

$$(a, \Lambda)(a', \Lambda') = (a + \Lambda a', \Lambda \Lambda').$$

The identity in the group is $(0, I)$ and the inverse is given by

$$(a, \Lambda)^{-1} = (-\Lambda^{-1}a, L^{-1}).$$

The subgroup $\{(a, I)\}$ is the subgroup T of translations and $\{(0, \Lambda)\}$ is the subgroup $\mathcal{L}_+^\uparrow$ of Lorentz transformations. Note that, as a subgroup of $\mathcal{P}_+^\uparrow$, $\mathcal{L}_+^\uparrow$ depends on the choice of origin, whereas T does not. The subgroup of elements $R \in \mathcal{L}_+^\uparrow$ such that $Re_0 = e_0$, the stability subgroup of e_0, must also map

$$\{x : e_0 \cdot x = 0\} = \mathbb{R}^3 \subset \mathbb{R}^4$$

into itself. This is just the subgroup of rotations in space, isomorphic to $SO(3)$.

After a suitable Poincaré transformation, any double cone has the form

$$\{x : |\mathbf{x}| \leq a - |x^0|\}.$$

Here $a > 0$ is the *radius* of the double cone and the ball which is the intersection of the hyperplane $x^0 = 0$ with the double cone is referred to as the *base* of the double cone.

We can now formulate as a physical principle for applications to elementary particle physics that Poincaré transformations are symmetries of our physical system. This is referred to as Poincaré covariance. I will give a precise formulation of this principle in the context of algebraic quantum field theory shortly. However, there is one complication I would like to mention now. You will know that fields fall into two classes, Bose fields and Fermi fields, and, as the names imply, this is intimately related to the distinction between Bose and Fermi statistics. It turns out that Fermi fields change sign under a rotation through 2π whereas Bose fields do not change sign. This is referred to as the connection between spin and statistics, a connection which can be well understood in the framework of quantum field theory. This means that it is, in general, not the Poincaré group which acts on the field nets but rather the covering group $\tilde{\mathcal{P}}_+^\uparrow$ of the Poincaré group.

The covering group can be realized as the set of pairs (a, A), where a is an element of the translation group and A is an element of $SL(2, \mathbb{C})$, the group of complex-valued 2×2-matrices with determinant 1. To understand the relationship of this group to the Poincaré group, we represent the points of Minkowski space, $x \in \mathbb{M}$, as self-adjoint 2×2-matrices:

$$x \mapsto \tilde{x} := \sum_\mu x^\mu \sigma^\mu,$$

where σ^μ, $\mu = 0, 1, 2, 3$ are the Pauli matrices, σ^0 being the unit matrix as usual. An $A \in SL(2, \mathbb{C})$ determines a Lorentz transformation $\Lambda(A)$ through the equation:

$$\tilde{x}' = A\tilde{x}A^*, \quad x' := \Lambda(A)x.$$

The group law in $\tilde{\mathcal{P}}_+^\uparrow$ is given by

$$(a, A)(a', A') = (a + \Lambda(A)a', AA')$$

and $(a, A) \mapsto (a, \Lambda(A))$ is a surjective homomorphism $\tilde{\mathcal{P}}_+^\uparrow \to \mathcal{P}_+^\uparrow$, the covering homomorphism, whose kernel consists of the unit and $(0, -I)$. We shall also use the symbol L to denote an element (a, A) of the covering group and we talk about Poincaré covariance even if we are dealing with a representation of the covering group.

Although these lectures are oriented towards elementary particle physics, I have decided to present first those parts of the theory which involve neither the Poincaré group nor even the translation group. All that is involved of spacetime is therefore the causal structure of Minkowski space. This has the advantage of collecting together those ideas which should make sense in more general curved spacetimes. In recent years algebraic quantum field theory has in fact made important contributions to quantum field theory on curved spacetimes. It should also be possible, albeit with more difficulty, to adapt these ideas to quantum field theory over quantum spacetime. However, if one tries to envisage incorporating quantum gravity where the causal structure of spacetime becomes state-dependent then it is not clear whether the ideas of algebraic quantum field theory are sufficiently flexible to provide a working formalism.

6 Local Observables

Let me now begin the development of algebraic quantum field theory by explaining some of the basic 'laws' of the subject so that we can then proceed to the real task of making deductions on the basis of these laws. Of course these basic laws will take on a very different form from the laws of physics we grew up on. This is only to be expected because our basic formalism is also very different. In fact, we have already taken a first important step toward algebraic quantum field theory by claiming effectively that it is not really the fields $\varphi(f)$ that are essential but rather the local structure, the fact that we have a net $\mathcal{O} \mapsto \mathfrak{F}(\mathcal{O})$ of field algebras. This bold claim has stood the test of time well.

An equally important step towards algebraic quantum field theory is to reinterpret this local structure in terms of measurements. One adds to quantum theory a new physical principle based on the observation that all physical measurements are carried out within spacetime and that, whatever else they establish, they always establish, within limits inherent in any measurement, when and where something has happened. This principle that measurements are localized in spacetime was first formulated by R. Haag [52]. It is formalized as follows: to each open double cone $\mathcal{O}$ in Minkowski space one associates a C^*–algebra $\mathfrak{A}(\mathcal{O})$ to be understood as the C^*–algebra generated by the observables which can, in principle, be measured within $\mathcal{O}$. We obviously have

$$\mathfrak{A}(\mathcal{O}_1) \subset \mathfrak{A}(\mathcal{O}_2), \quad \text{if} \quad \mathcal{O}_1 \subset \mathcal{O}_2;$$

more observations can be performed within a larger set. Our basic object is now not a single C^*–algebra but a correspondence $\mathcal{O} \mapsto \mathfrak{A}(\mathcal{O})$, i.e. a net $\mathfrak{A}$ of C^*–algebras over double cones in Minkowski space.

Since the set of open double cones is directed under inclusion, the union of all $\mathfrak{A}(\mathcal{O})$ is a *–algebra. Since C^*–algebras have a unique C^*–norm so does this *–algebra but it is not a C^*–algebra since it is not complete. Nevertheless, we can still define a state ω to be a positive normalized linear functional and we have $|\omega(A)| \leq \|A\|$ since each A lies in some C^*–algebra $\mathfrak{A}(\mathcal{O})$. Hence if we complete our *–algebra to form a C^*–algebra each state will extend uniquely to a state of the C^*–algebra. This C^*–algebra is called the C^*–algebra of local observables and is denoted by the same symbol $\mathfrak{A}$ as the net itself. This C^*–algebra therefore has the same set of states as the net itself and we can be sure that its self-adjoint elements, although not necessarily strictly localized can at least be treated in an idealized sense as observables. Although we may consider, for certain purposes, just the C^*–algebra of local observables it is the net structure of $\mathfrak{A}$ which is the new important element of structure.

Note that this way of formalizing the localization of measurements takes due account of the fact that measurements are never precisely localized at points in spacetime. It does not introduce anything resembling a spacetime position operator and treats space and time on an equal footing. It represents a fundamental advance in our understanding of the relation between quantum theory and spacetime.

This said, it must be admitted that measurements on microscopic systems are in practice well localized only on a macroscopic scale. The typical measuring apparatus of elementary particle physics is the particle counter which responds to fluctuations in field strengths near its sensitive area. It is an enormous extrapolation to envisualize observables localized in double cones with a radius of the order of $10^{-15}m$. Nevertheless the quantum field theory models used in the theoretical discussion of elementary particles provide candidates for the observable net where $\mathfrak{A}(\mathcal{O})$ has a rich structure for arbitrarly small double cones at least to the extent that the mathematical structure of such models is clear. In this rather indirect sense, the localization of observables is as well justified as field theory itself.

Although the interpretation of $\mathfrak{A}(\mathcal{O})$ as the C^*–algebra generated by the observables that can be measured in $\mathcal{O}$ lies at the ideological basis of algebraic quantum field theory and helps us to comes to terms with the net $\mathfrak{A}$, it does not in practice play an important role in applications. Applications to scattering theory are to a certain degree an exception to this rule but even there the interpretation is not needed for small sets $\mathcal{O}$. There are other conceivable interpretations, for example, in terms of localized operations that play a role in certain applications, as we shall see.

Essential for applications is the idea that the net $\mathfrak{A}$ completely determines the quantum field theoretical model under investigation. All relevant physical concepts are expressed directly or indirectly in terms of $\mathfrak{A}$. Let me illustrate the idea with a simple example. A physical state of the model determines and is determined by a state of $\mathfrak{A}$, i.e. two physical states agree if and only if the corresponding states on $\mathfrak{A}$ agree. This is not an empty statement because the net $\mathfrak{A}$ does not usually coincide with the field net $\mathfrak{F}$. Instead $\mathfrak{A}$ is merely a subnet of $\mathfrak{F}$ so that two distinct states of $\mathfrak{F}$ can determine the same physical state namely when their restrictions to $\mathfrak{A}$ coincide. We know that $\mathfrak{F}$ must be considered as distinct from $\mathfrak{A}$ as the basic fields need not be observable. Indeed, most of the fundamental fields used to describe the interactions of elementary particle physics are not observable, that is we have no way of measuring the corresponding field strengths.

The advent of quantum mechanics has taught us that not all observables can be measured 'simultaneously' and that this is linked mathematically to the fact that the observables do not all commute with one another. Furthermore, we understand the physical origin of this behaviour in the interaction of the system with the measuring apparatus. The state of the system is changed through the measuring procedure. Since measurements are localized in spacetime, the disturbance of the system will also have its origin in a localized region of spacetime. On the other hand Einstein's theory of relativity tells us that no signal can propagate faster than the speed of light. Hence this disturbance will only affect measurements made within the causal shadow of the localization region of the first measurement. So that measurements made within regions which are spacelike separated cannot interfere with each other and the corresponding observables should commute. This is the principle of locality.

To formalize this we introduce the notion of spacelike complement. If $\mathcal{O}$ is a subset of Minkowski space, then its spacelike complement $\mathcal{O}'$ is defined to be the set of points spacelike to every point of $\mathcal{O}$:

$$\mathcal{O}' := \{x : (x - y) \cdot (x - y) < 0, \quad y \in \mathcal{O}\}.$$

The condition for two sets $\mathcal{O}_1$ and $\mathcal{O}_2$ to be spacelike separated is that

$$\mathcal{O}_1 \subset \mathcal{O}_2'.$$

Hence locality can be formulated as follows:

$$A_1 A_2 = A_2 A_1, \quad A_1 \in \mathfrak{A}(\mathcal{O}_1), \quad A_2 \in \mathfrak{A}(\mathcal{O}_2), \quad \mathcal{O}_1 \subset \mathcal{O}_2'.$$

It is this principle of locality which allows us to draw many conclusions of physical interest from the net structure of $\mathfrak{A}$. In particular, it is the fundamental step towards an understanding of particle statistics. Although we can still talk of a net structure in a non-relativistic theory, there is no upper limit to the signal velocity in such theories. To see any vestiges of the principle of locality, it is necessary to localize observables in space at a fixed instant of time. In many cases the instantaneous propagation of effects even means that the time–translations do not act as automorphisms of the algebra of local observables. Although this is a consequence of the unphysical aspects of a non–relativistic theory, it must be borne in mind before trying to apply these ideas to non–relativistic models.

If we work with the field net then the principle of locality cannot be invoked. As I have already stressed, fields are, in general, not observables and there is no reason why spacelike separated elements of the field net should commute. Indeed, there is no a priori reason why there should be any simple commutation relations for spacelike separated elements of the field net. In practice, however, most field nets satisfy a rather simple variant of the principle of locality: they have Bose-Fermi commutation relations. This structure can be formalized as follows: there is a local automorphism α_k of the field net $\mathfrak{F}$, i.e.

$$\alpha_k(\mathfrak{F}(\mathcal{O})) = \mathfrak{F}(\mathcal{O}), \quad \mathcal{O} \in \mathcal{K},$$

whose square is the identity automorphism ι. We define

$$F_\pm := \frac{1}{2}(F \pm \alpha_k(F)),$$

F_+ and F_- are the Bose and Fermi parts of F respectively and $\alpha_k(F_\pm) = \pm F_\pm$. Then if $F \in \mathfrak{F}(\mathcal{O}_1)$ and $F' \in \mathfrak{F}(\mathcal{O}_2)$ with $\mathcal{O}_1 \subset \mathcal{O}_2'$, we have the following commutation relations:

$$F_+ F'_+ = F'_+ F_+; \quad F_- F'_+ = F'_+ F_-; \quad F_- F'_- = -F'_- F_-.$$

Thus the Fermi parts anticommute at spacelike separations rather than commuting. It must however be stressed that the status of these commutation relations is quite different to those pertaining to observables. In the latter case, they are the consequence of fundamental physical principles. In the former case, we have no a priori understanding. In fact, the relations are not universal and it is a challenge to understand why they nevertheless play such an important role. Algebraic quantum field theory has succeeded in providing a satisfactory explanation.

7 Additivity

There are properties of the net $\mathfrak{A}$ other than locality which seem to express laws of physics in the sense that they hold for any physically reasonable model. One of these properties is additivity. This expresses the fact that the net structure is entirely determined by what happens on arbitrarily small sets.

Let us first consider this property in a more general setting. Let $\mathfrak{N}$ be a net defined over a set S of open subsets of a topological space directed under inclusion so that $\mathfrak{N}(\mathcal{O}_1) \subset \mathfrak{N}(\mathcal{O}_2)$ whenever $\mathcal{O}_1 \subset \mathcal{O}_2$. The net $\mathfrak{N}$ is said to be additive if whenever we cover $\mathcal{O} \in S$ by subsets $\mathcal{O}_i \in S$, $\mathfrak{N}(\mathcal{O})$ is generated by the $\mathfrak{N}(\mathcal{O}_i)$, i.e. is the smallest subobject containing each $\mathfrak{N}(\mathcal{O}_i)$. We express additivity by writing

$$\mathfrak{N}(\mathcal{O}) = \bigvee_i \mathfrak{N}(\mathcal{O}_i) \quad \text{if} \quad \mathcal{O} = \cup_i \mathcal{O}_i.$$

I have deliberately left open the nature of $\mathfrak{N}(\mathcal{O})$, i.e. the category where $\mathfrak{N}$ takes values, for good reasons, as is clear from the following examples where we take our directed set to be the set $\mathcal{K}$ of double cones in Minkowski space. Let $\mathcal{D}(\mathcal{O})$ denote the set of smooth functions on Minkowski space with compact support contained in $\mathcal{O}$, then $\mathcal{D}$ is additive as a net of vector spaces (and even as a net of Abelian groups). In fact, if $\mathcal{O}_i$, $i \in I$ is a covering of $\mathcal{O}$ then using a partition of the identity subordinate to the covering any $f \in \mathcal{D}(\mathcal{O})$ may be written as a sum

$$f = \sum_{i \in I} f_i, \quad f_i \in \mathcal{D}(\mathcal{O}_i),$$

where only a finite number of terms in the sum are different from zero.

This is the source of additivity in the Wightman formalism since a polynomial net $\mathfrak{P}$ of fields is now obviously additive as a net of algebras. However, if you try to deduce the additivity of the net of fields or observables from this result, you run into the usual technical problems, which again, typically, can be overcome in concrete models. In fact, the situation for free fields is extremely simple. In the Bose case, the nets are generated by Weyl operators. More precisely, we have an additive net $\mathfrak{N}$ of vector spaces, which could be taken to be $\mathcal{D}$ in the case of a single scalar field, together with a (degenerate) symplectic form σ and unitary operators $W(f)$, $f \in \mathfrak{N}(\mathcal{O})$, such that

$$W(f)W(f') = e^{\frac{1}{2}i\sigma(f,f')}W(f + f').$$

If we now define $\mathfrak{F}(\mathcal{O})$ to be the algebra, or C^*–algebra or von Neumann algebra generated by the $W(f)$, $f \in \mathfrak{N}(\mathcal{O})$ then we shall always get an additive net.

The case of free Fermi fields is even simpler because the field operators $\psi(f)$ are bounded operators so that we may as well use the Hilbert space $\mathcal{H}$ in place of the Wightman domain D. In any case, defining $\mathfrak{F}(\mathcal{O})$ to be generated by the $\psi(f)$, $f \in \mathfrak{N}(\mathcal{O})$, will always give an additive net whenever $\mathfrak{N}$ is additive. I should add that we again have typical algebraic structure, the (self-dual) CAR-algebras, but this is irrelevant as far as the additivity goes.

Let me now point out an advantage of having additive nets by referring to the problem of extending a net so as to be defined on a larger directed set of open sets.

Let me call a subset of $\mathfrak{N}$ localized if it is contained in $\mathfrak{N}(\mathcal{O})$ for some fixed $\mathcal{O} \in \mathcal{S}$. Any such localized subset generates a subobject of $\mathfrak{N}(\mathcal{O})$, independent of the choice of $\mathcal{O}$, in the range categories under consideration. Hence if $\mathcal{S} \subset \mathcal{S}_1$ and each element of $\mathcal{S}_1$ is contained in an element of $\mathcal{S}$, i.e $\mathcal{S}$ is cofinal in $\mathcal{S}_1$, there is a natural way of extending $\mathfrak{N}$ to $\mathcal{S}_1$. If $\mathcal{O}_1 \in \mathcal{S}_1$, $\cup\{\mathfrak{N}(\mathcal{O}) : \mathcal{O} \subset \mathcal{O}_1\}$ is localized and we define $\mathfrak{N}(\mathcal{O}_1)$ to be the associated subobject.

We define a morphism of nets, $\alpha : \mathfrak{N} \to \mathfrak{N}'$, in the obvious manner, namely as a set $\alpha_{\mathcal{O}}$ of morphisms of $\mathfrak{N}(\mathcal{O}) \to \mathfrak{N}'(\mathcal{O})$ compatible with the inclusions. If α is a morphism of nets over $\mathcal{S}$ then there is a unique extension $\hat{\alpha}$ to a morphism of the canonical extensions to nets over $\mathcal{S}_1$ for the range categories under considerations (pullbacks exist). Of course if we are given a net over $\mathcal{S}_1$ then we get a net over $\mathcal{S}$ by restriction and the uniqueness of $\hat{\alpha}$ implies that the extension functor is a left adjoint of the restriction functor.

Now let us suppose that S forms a base for the topology and consider additive nets. Then the canonical extension of an additive net is the unique additive extension. Consequently, the functors of extension and restriction of additive nets are inverses of one another and the category of additive nets over S and over S_1 are isomorphic. Thus we could, if we wish, replace S by the set of open sets contained in an element of S or indeed by any cofinal base for the topology.

Returning to the situation of immediate physical interest, we have good reasons for defining nets just over the set K of open double cones: it keeps the basic notion as simple as possible and it helps to avoid irrelevant geometric problems. However, there are equally good reasons for wanting to extend the net to a wider class of open sets. Because in some theories there are interesting properties involving open sets which, unlike double cones, have interesting topological properties. As an example, consider the flux of the magnetic field through a surface. This observable depends only on the curve bounding the surface since it can be expressed as the integral of the vector potential along that curve. It will thus be associated with the algebra $\mathfrak{A}(\mathcal{O})$, where $\mathcal{O}$ is any open set containing a surface bounded by the curve. But if you take $\mathcal{O}$ to be an open torus containing the curve then the observable will not be associated with $\mathfrak{A}(\mathcal{O})$ since the vector potential is not an observable. This behaviour can be verified in the algebra of observables of the free electromagnetic field.

With an additive net, we are at liberty to adopt both points of view simultaneously. Indeed, without some sort of regularity property such as additivity there would not be much point in studying nets over bounded open sets.

There is one less trivial point I would like to clarify in this section. Suppose $\mathfrak{F}$ is an additive field net acted on by a compact group G of net automorphisms and set

$$\mathfrak{A}(\mathcal{O}) = \{F \in \mathfrak{F}(\mathcal{O}) : g(F) = F, \ g \in G\}$$

then is the net $\mathfrak{A}$ additive? This is not an academic question because we define the observable net of free fields in this way, interpreting G as a gauge group. A superficial attempt to settle the question would probably lead one to guess that the response is likely to be negative. In fact, the response is positive with one small finesse but depends on knowing more about the action of G on $\mathfrak{F}$. We need some definitions.

7.1 Definition A Hilbert space H in a C^*–algebra $\mathcal{A}$ with unit is a norm-closed linear subspace such that $\psi^*\psi \in \mathbb{C}I$ for $\psi \in H$. It is then a Hilbert space with a scalar product given by $\psi^*\psi' = (\psi, \psi')I$.

If the Hilbert space is finite-dimensional and ψ_i, $i = 1, 2, \ldots, d$ is an orthonormal basis then $\sigma_H(A) = \sum_i \psi_i A \psi_i^*$ is independent of the choice of basis and defines an endomorphism of $\mathcal{A}$, $\sigma_H(I)$ is called the support of H. Any (bounded) linear map of H into $\mathcal{A}$ is represented by left multiplication with a unique $X \in \mathcal{A}$ satisfying $X = X\sigma_H(I)$. In a von Neumann algebra we may even take H to be infinite dimensional.

If H is a Hilbert space in $\mathcal{A}$ and α is an automorphism of $\mathcal{A}$ then $\alpha(H)$ is another Hilbert space and $\psi \mapsto \alpha(\psi)$, $\psi \in H$, is a unitary mapping from H to $\alpha(H)$. It follows that when $\mathcal{A}$ is acted on by a group G of automorphisms and H is G–invariant, i.e. $g(H) = H$, $g \in G$, then we have a unitary representation of G on H.

The endomorphism σ_H restricts to give an endomorphism of the fixed-point algebra. A unitary mapping between finite–dimensional G–invariant Hilbert spaces H and H' commuting with the action of G is given by left multiplication with an isometry V in the fixed point algebra with initial projection $\sigma_H(I)$ and final projection $\sigma_{H'}(I)$.

7.2 Definition An automorphism group G of a C^*–algebra with unit $\mathcal{A}$ is said to have *full Hilbert spectrum* if every equivalence class of continuous irreducible representation is realized on a G–invariant Hilbert space in $\mathcal{A}$.

We shall need the following consequence of a full Hilbert spectrum.

7.3 Lemma Let G be a compact group of automorphisms acting continuously on a C^*–algebra with unit $\mathcal{A}$ or σ–continuously on a von Neumann algebra $\mathcal{M}$. Then if G has full Hilbert spectrum finite sums $\sum_k A_k \psi_k$, where the A_k are in the fixed point algebra and the ψ_k are in a finite–dimensional G–invariant Hilbert space, form a *–algebra dense in $\mathcal{A}$ or σ–dense in $\mathcal{M}$.

Proof By elementary harmonic analysis on G, the set of elements of $\mathcal{A}$ transforming irreducibly under G is total in $\mathcal{A}$. Now an irreducible subspace of $\mathcal{A}$ is equivalent as a G–module to some G–invariant Hilbert space H as G has full Hilbert spectrum. The corresponding intertwining operator is given by left multiplication with a unique $A \in \mathcal{A}$ such that $A\sigma_H(I) = A$. The intertwining property implies that A is actually in the fixed point algebra. Hence an element of our irreducible subspace is of the form $A\psi$ with $\psi \in H$. Thus finite sums $\sum_k A_k \psi_k$ as above are obviously dense in $\mathcal{A}$. In fact they correspond precisely to the set of elements contained in a finite dimensional G–invariant subspace of $\mathcal{A}$, i.e the set of G–finite elements. But the set of G–finite elements is obviously a *–algebra. The proof in the case of a von Neumann algebra just involves replacing the norm topology by the σ–topology.

We can now formulate the result on additivity.

7.4 Theorem Let $\mathfrak{F}$ be an additive net over bounded open sets and G a compact group of net automorphisms. Suppose that G has full Hilbert spectrum as an automorphism group of $\mathfrak{F}(\mathcal{O})$, for each (non-empty) bounded open set $\mathcal{O}$, then the net $\mathfrak{A}$ defined by

$$\mathfrak{A}(\mathcal{O}) := \{ A \in \mathfrak{F}(\mathcal{O}) : g(A) = A, \, g \in G \},$$

if $\mathcal{O}$ is connected and

$$\mathfrak{A}(\mathcal{O}) := \bigvee_i \mathfrak{A}(\mathcal{O}_i),$$

if $\mathcal{O}_i$ are the connected components of $\mathcal{O}$, is an additive net.

The proof rests on the following lemma.

7.5 Lemma Let $\mathcal{O}, \mathcal{O}_i$ be non-empty, bounded, connected open sets, $i \in I$ with $\mathcal{O} = \cup_i \mathcal{O}_i$. Given $i_0, i' \in I$ and an irreducible G–invariant Hilbert space H in $\mathfrak{F}(\mathcal{O}_{i'})$, let $\mathcal{M} := \bigvee_i \mathfrak{A}(\mathcal{O}_i)$ then

a) there exist $i_1, i_2, \ldots, i_n = i' \in I$ with $\mathcal{O}_{i_{m-1}} \cap \mathcal{O}_{i_m} \neq \emptyset$, $m = 1, 2, \ldots, n$.
b) there is a partial isometry $V \in \mathcal{M}$ with $V^*V = \sigma_H(I)$ and $VH \in \mathfrak{F}(\mathcal{O}_{i_0})$
c) $\sigma_H(\mathcal{M}) \subset \mathcal{M}$.

Proof Let $J \subset I$ denote the subset of elements of I which can be joined by a finite chain to i_0 as in a). Then $\cup_{i \in J} \mathcal{O}_i \cap \cup_{i \notin J} \mathcal{O}_i = \emptyset$. Since $\mathcal{O}$ is connected and each $\mathcal{O}_i$ is non–empty and open, $J = I$ proving a). If $\mathcal{O}_{i_0} \cap \mathcal{O}_{i'} \neq \emptyset$ then there is a G–invariant Hilbert space in $\mathfrak{F}(\mathcal{O}_{i_0} \cap \mathcal{O}_{i'})$ equivalent to H as a G–module. Hence there is a partial isometry $V \in \mathfrak{A}(\mathcal{O}_{i'})$ with $V^*V = \sigma_H(I)$ and $VH \in \mathfrak{F}(\mathcal{O}_{i_0})$. b) now follows from a). Obviously, $\sigma_H(\mathfrak{A}(\mathcal{O}_{i'})) \subset \mathfrak{A}(\mathcal{O}_{i'})$. So by b), $\sigma_H(\mathfrak{A}(\mathcal{O}_{i_0})) \subset V^*\mathfrak{A}(\mathcal{O}_{i_0})V \subset \mathcal{M}$ proving c).

Proof of Theorem 7.4 It suffices to show that if $\mathcal{O}, \mathcal{O}_i, i \in I$, are as in Lemma 7.5 then $\mathfrak{A}(\mathcal{O}) = \mathcal{M}$. Now finite sums $\sum_k A_k \psi_k$ where $A_k \in \mathcal{M}$ and ψ_k are elements of irreducible G–invariant Hilbert spaces in $\mathfrak{F}(\mathcal{O}_{i_0})$ form a *-subalgebra of $\mathfrak{F}(\mathcal{O})$ by c) of Lemma 5. This *–subalgebra contains all irreducible G–invariant Hilbert spaces in all $\mathfrak{F}(\mathcal{O}_i)$ by b) of Lemma 5. It is hence dense in $\mathfrak{F}(\mathcal{O})$ since $\mathfrak{F}(\mathcal{O}) = \bigvee_{i \in I} \mathfrak{F}(\mathcal{O}_i)$. Thus averaging these sums over G, a σ–continuous operation, yields the required result.

This still leaves open the question of why gauge groups acting on field nets have full Hilbert spectrum but this involves the action of the translation group and positivity of the energy and must be postponed.

In conclusion, I remark that there are analogies with sheaf theory. Of course, a net is not even a presheaf because the mappings go the other way round: it is a precosheaf. More significantly, we have insisted on our mappings being inclusions so that although additivity is the analogue of the sheaf property, additive nets are not cosheaves.

8 Local Normality

I now recall my third question: why does one use abstruse mathematical objects like C^*–algebras and von Neumann algebras rather than working with operators and vectors in Hilbert space as in elementary quantum mechanics?

In the algebraic approach to quantum field theory we are no longer bound to work on a fixed Hilbert space. We are free to look at any representation π we wish. Let me explain why this freedom is necessary. Associated with a representation is the set of vector states and the set of normal states of the representation. These are the states of the form

$$\omega_\Psi(A) = (\Psi, \pi(A)\Psi), \quad A \in A,$$

and

$$\omega_\rho(A) = \text{Tr}(\rho\pi(A)), \quad A \in A,$$

respectively, where Ψ is a unit vector and ρ is a density matrix of the representation space of π. Of course, every vector state is a normal state. Now if we were to start from the vacuum Hilbert space, as is often the case, then we would discover that the equilibrium states of the system are not normal states of the vacuum representation. Thus our Hilbert space would be of no use to us if we wanted to study these physically relevant states.

Of course, we could conceive of using the universal representation of the C^*-algebra of local observables. This is the direct sum over the GNS representations of the states of our algebra. In this representation every state is a vector state by construction. The trouble is that only a tiny fraction of the states of our system have any physical relevance and one of the basic problems in treating systems with an infinite number of degrees of freedom is to single out and classify the states or representations of physical relevance. Although substantial progress has been made on this point, there is no general solution even if we restrict our attention to the states of interest to elementary particle physics and rule out states of interest to statistical mechanics. Thus our universal Hilbert space has far too many normal states and we have no way, in general, of specifying a Hilbert space representation giving us just the states we want and nothing more. However, there is one aspect of the problem which has turned out to be rather simple.

Suppose we begin with any one state ω of particular physical significance such as a vacuum state, ground state or equilibrium state. We can then perform the GNS construction and look at the normal states of the representation π_ω. For technical reasons, one usually prefers to drop the normalization condition and consider the set of normal positive linear functionals of π_ω. This is the *folium* $f(\omega)$ associated with ω and the states in this folium should all be physically relevant. There is another larger set of positive linear forms associated with ω which I would like to call the *local folium* of ω, $lf(\omega)$. $\omega' \in lf(\omega)$ if given $\mathcal{O} \in \mathcal{K}$, there is a $\omega_\mathcal{O} \in f(\omega)$ such that $\omega'(A) = \omega_\mathcal{O}(A)$, $A \in \mathfrak{A}(\mathcal{O})$. The local folium is determined by the sets $f(\omega) \upharpoonright \mathfrak{A}(\mathcal{O}) = lf(\omega) \upharpoonright \mathfrak{A}(\mathcal{O})$, $\mathcal{O} \in \mathcal{K}$.

It would seem that the local folium is independent of the state ω of particular physical significance that we started with and that any physically relevant state belongs to the local folium. Thus determining the local folium using a state of particular physical significance is a first step towards characterizing the physically relevant states. Furthermore, despite the fact that $\mathfrak{A}(\mathcal{O})$ has an infinite number of degrees of freedom, we are locally in the same happy situation as in elementary quantum mechanics in that, in practice, we do not need to worry about which representation to choose.

Let us accept this as a fact of life and try to adjust our formalism so as to be able to exploit this additional information. We can do this by introducing von Neumann algebras. If ω is a state of particular physical significance then we have an associated net of von Neumann algebras $\mathcal{O} \mapsto \pi_\omega(\mathfrak{A}(\mathcal{O}))^-$ formed by taking weak operator closures in the representation π_ω. If ω' is another state of particular physical significance then we have a unique isomorphism $\eta_\mathcal{O}$ of $\pi_\omega(\mathfrak{A}(\mathcal{O}))^-$ onto $\pi_{\omega'}(\mathfrak{A}(\mathcal{O}))^-$ such that $\eta_\mathcal{O}\pi_\omega(A) = \pi_{\omega'}(A)$ for $A \in \mathfrak{A}(\mathcal{O})$ since π_ω and $\pi_{\omega'}$ have by hypothesis the same normal states. The uniqueness of $\eta_\mathcal{O}$ shows that we get in this way a net isomorphism η. Thus we have a unique net of von Neumann algebras and any physically relevant state extends uniquely to a locally normal state of this new net, i.e to a state which is normal on each local von Neumann algebra.

What one now does is to replace the original net by the net $\mathcal{O} \mapsto \pi_\omega(\mathfrak{A}(\mathcal{O}))^-$ of von Neumann algebras and restricts one's attention to locally normal states since any physically relevant state is assumed to belong to the local folium. Any self-adjoint element A of $\pi_\omega(\mathfrak{A}(\mathcal{O}))^-$ may be regarded as being, in an idealized sense, an observable localized in $\mathcal{O}$ since we may answer all questions related to a 'measurement' of A in any physically relevant state. Whenever possible, I will simplify notation and concepts, supposing that $\mathfrak{A}$ denotes a net of von Neumann algebras and considering locally normal states only. A point in favour of taking $\mathfrak{A}$ to be a net of von Neumann algebras is that it is, in practice, usually clear in a model how the net is to be defined whereas it is not usually so clear what criterion should be adopted to single out a C^*–subnet.

A representation π of a von Neumann algebra $\mathcal{M}$ is said to be *normal* if the mapping $\pi : \mathcal{M} \to \mathcal{B}(\mathcal{H}_\pi)$ is σ–continuous. The GNS representation corresponding to a normal state is normal. A representation π of a net $\mathfrak{A}$ of von Neumann algebras is said to be *locally normal* if the restriction of π to each $\mathfrak{A}(\mathcal{O})$ is normal. The GNS representation corresponding to a locally normal state is itself locally normal. Again when we have taken the observable net $\mathfrak{A}$ to be a net of von Neumann algebras, it would not be consistent even to consider using representations or states which are not locally normal.

There is one further interesting remark to be made about the physically relevant states. Suppose each such state determines the local folium and hence has a faithful locally normal GNS representation like the states of particular physical significance. Then the local von Neumann algebras must be factors. Even this strengthened version of local normality seems to hold in practice. It is the principle of *local definiteness* of Haag, Narnhofer and Stein [55]. The problem of determining the local folium becomes more acute in curved spacetime where states of particular physical significance are rather rare and interesting ideas have emerged. In due course they will modify the way we look at quantum field theory in Minkowski space.

9 Inclusions of von Neumann Algebras

In the algebraic approach to quantum field theory, the field theory aspect manifests itself through the net structure which thereby acquires a fundamental importance. The basic building block of the nets in question is an inclusion of von Neumann algebras. For this reason, we give here some relevant information on inclusions of von Neumann algebras. Whilst this material proves very useful when handling observable nets or field nets, it is obvious that the structural information of real importance to quantum field theory is not encoded in a single inclusion of von Neumann algebras. From a mathematical point of view, I should emphasize that the material presented here is motivated by quantum field theory and therefore has a rather different flavour than the theory of subfactors initiated by Jones.

Let me begin with some definitions.

9.1 Definition An inclusion of von Neumann algebras $\mathcal{L} \subset \mathcal{N}$ is said to have *Property S* if, given central projections $E \in \mathcal{L}$ and $F \in \mathcal{N}$, $EF = 0$ implies $E = 0$ or $F = 0$. The inclusion is said to have *Property B* if every non-zero projection in $\mathcal{L}$ is equivalent to I in $\mathcal{N}$. The inclusion is said to be *split* if there is an intermediate type I factor $\mathcal{M}$, i.e. if $\mathcal{L} \subset \mathcal{M} \subset \mathcal{N}$.

Borchers [11] discovered that the analogue of Property B applied to nets of observables is a consequence of positivity of the energy. Schlieder [75] remarked that Property S follows from Property B. The split property for nets of observables is an old conjecture of Borchers. The first proof that it held in a concrete example is due to Buchholz who proved the result for the free scalar field [17]. Much work has been done since then in particular by Doplicher and Longo who introduced the important notion of a standard split inclusion [37], to be discussed later.

If we suppose that our inclusion is concretely represented on a Hilbert space, then Property S is equivalent to saying that given projections $E \in \mathcal{L}$ and $F \in \mathcal{N}'$ then $EF = 0$ implies that either $E = 0$ or $F = 0$. The property was first looked at in this way and the equivalence follows by passing to the corresponding central supports. It was also shown by Roos [74] that the property is equivalent to saying that the *-subalgebra generated by $\mathcal{L}$ and $\mathcal{N}'$ is naturally isomorphic to the algebraic tensor product $\mathcal{L} \otimes \mathcal{N}'$ of $\mathcal{L}$ and $\mathcal{N}'$. To compare Property S with Property B we simply observe that Property S is obviously equivalent to saying that every non–zero projection in $\mathcal{L}$ has central support I in $\mathcal{N}$. It is therefore a consequence of Property B.

A factor of type I is a von Neumann algebra $\mathcal{M}$ with trivial centre and containing a minimal projection. It is therefore evident that a split inclusion has Property S but not necessarily Property B. A projection E in a von Neumann algebra is minimal if $E \neq 0$ and any smaller projection is either 0 or E itself. Thus the von Neumann algebra $E\mathcal{M}E$ has no non-trivial projections so that $E\mathcal{M}E = \mathbb{C}E$ and $E\mathcal{M}E = \omega(M)E$ where ω is a normal state of $\mathcal{M}$ with $\omega(E) = 1$. Before proceeding, let me recall that if F is the largest projection in $\mathcal{M}$ on which a normal positive linear functional ω vanishes then $\omega(FM) = \omega(MF) = 0$, $M \in \mathcal{M}$. Thus if $E := I - F$, the *support* of ω,

$$\omega(M) = \omega(EME), \quad M \in \mathcal{M}.$$

In our case the minimal projection E is the support of the associated state and hence by minimality of any state ω' dominated by ω. Thus

$$\omega'(M) = \omega'(EME) = \omega(M)\omega'(E) = \omega(M), \quad M \in \mathcal{M},$$

so that ω is a pure state. The corresponding GNS representation is irreducible and normal but also faithful so that $\mathcal{M}$ is isomorphic to some $\mathcal{B}(\mathcal{H})$. On the other hand any $\mathcal{B}(\mathcal{H})$ is a factor with minimal projections, the projections onto the 1–dimensional subspaces of $\mathcal{H}$.

Now suppose we have a type I factor $\mathcal{M}$ on a Hilbert space $\mathcal{H}$, let E be a minimal projection of $\mathcal{M}$ and pick a unit vector $\Phi \in \mathcal{H}$, with $E\Phi = \Phi$ then

$$(\Phi, MM'\Phi) = (\Phi, EMEM'\Phi) = (\Phi, M\Phi)(\Phi, M'\Phi).$$

Now we obviously have $E = [\mathcal{M}'\Phi]$ and setting $E' := [\mathcal{M}\Phi] \in \mathcal{M}'$ and noting that, since $\mathcal{M}$ and $\mathcal{M}'$ commute, the closed linear space spanned by vectors of the form

$MM'\Phi$ with $M \in \mathcal{M}$ and $M' \in \mathcal{M}'$ contains $\mathcal{M} \vee \mathcal{M}'\Phi = \mathcal{B}(\mathcal{H})\Phi = \mathcal{H}$, we see that the map

$$M\Phi \otimes M'\Phi \mapsto MM'\Phi, \quad M \in \mathcal{M}, \quad M' \in \mathcal{M}',$$

extends to a unitary operator identifying $E'\mathcal{H} \otimes E\mathcal{H}$ with $\mathcal{H}$. Since Φ induces a pure normal state on $\mathcal{M}$ and is cyclic for $\mathcal{M}$ on $E'\mathcal{H}$, $\mathcal{M}$ is just $\mathcal{B}(E'\mathcal{H}) \otimes I$ and, taking commutants $\mathcal{M}'$ is just $I \otimes \mathcal{B}(E\mathcal{H})$. E' is of course just a minimal projection of $\mathcal{M}'$.

Thus in any concrete representation of our inclusion on a Hilbert space $\mathcal{H}$, the Hilbert space will split as a tensor product $\mathcal{H} = \mathcal{J} \otimes \mathcal{K}$ of Hilbert spaces in such a way that

$$\mathcal{L} \subset \mathcal{B}(\mathcal{J}) \otimes I, \quad \mathcal{N}' \subset I \otimes \mathcal{B}(\mathcal{K}).$$

Recall that the tensor product of von Neumann algebras $\mathcal{M}_i$ on Hilbert spaces $\mathcal{H}_i$, $i = 1, 2$, is the von Neumann algebra $\mathcal{M}_1 \bar{\otimes} \mathcal{M}_2$ on $\mathcal{H}_1 \otimes \mathcal{H}_2$ generated by $\mathcal{M}_1 \otimes I$ and $I \otimes \mathcal{M}_2$. It does not depend on the choice of the Hilbert spaces. Thus the split property implies that the von Neumann algebra generated by $\mathcal{L}$ and $\mathcal{N}'$ is a tensor product.

Before going on to discuss standard split inclusions, I will have to talk about cyclic and separating vectors. Let me first explain when a von Neumann algebra $\mathcal{M}$ has a separating vector. Such a vector defines a faithful state ω but if E_i, $i \in I$, is a partition of the unit in $\mathcal{M}$, i.e. a set of mutually orthogonal projections with sum 1 then

$$\sum_{i \in I} \omega(E_i) = 1$$

so that I must be countable. Thus $\mathcal{M}$ must be a σ-finite or countably decomposable von Neumann algebra. Of course, any von Neumann algebra on a separable Hilbert space is σ-finite. To see that any such algebra does have a faithful normal state, we proceed as follows: pick a maximal set of non-zero normal positive linear functionals ω_i, $i \in I$, with mutually orthogonal supports E_i and $\sum_{i \in I} \omega_i(E_i) < +\infty$. Then

$$\omega(M) := \sum_i \omega_i(M), \quad M \in \mathcal{M}$$

defines a normal positive linear functional with support $E := \sum_i E_i$. We claim that ω is faithful. If this is not the case then $F := I - E \neq 0$. Let ω' be a normal linear functional with $\omega'(F) \neq 0$ then $M \mapsto \omega'(FMF)$ is a non-zero normal positive linear functional with support $\leq F$ contradicting the maximality of the set ω_i, $i \in I$. In other words, we have proved the following lemma.

9.2 Lemma A von Neumann algebra admits a faithful normal state if and only if it is σ-finite.

The faithful state in question is given by a density matrix but to get a separating vector we need to know that it is a vector state. Now there is a simple condition which guarantees that every normal state is a vector state.

9.3 Lemma Let $\mathcal{M}$ be a von Neumann algebra on a Hilbert space $\mathcal{H}$ whose commutant $\mathcal{M}'$ contains a Hilbert space H of countable infinite dimensions and support 1 then every normal state of $\mathcal{M}$ is a vector state.

Proof Let ω be a normal state of $\mathcal{M}$ then diagonalizing a density matrix defining ω yields vectors $\Phi_n \in \mathcal{H}$, $n \in \mathbb{N}$, such that

$$\omega(M) = \sum_{n \in \mathbb{N}} (\Phi_n, M\Phi_n).$$

Let ψ_n, $n \in \mathbb{N}$, be an orthonormal basis of H then $\sum_{n \in \mathbb{N}} \psi_n \Phi_n$ converges to a vector Φ and

$$(\Phi, M\Phi) = \sum_{m,n} (\psi_m \Phi_m, M\psi_n \Phi_n) = \omega(M).$$

We incorporate the above considerations into a definition.

9.4 Lemma The following conditions are equivalent and define a *properly infinite* von Neumann algebra $\mathcal{M}$.

a) $\mathcal{M}$ contains a Hilbert space of countably infinite dimension and support I.

b) $\mathcal{M}$ contains a countably infinite family of mutually orthogonal projections summing up to I and each equivalent to I,

c) $\mathcal{M} \simeq \mathcal{B}(H) \otimes \mathcal{M}$, where H is a Hilbert space of countably infinite dimensions.

Proof Let ψ_n be an orthonormal basis of the Hilbert space H in question, then the $\psi_n \psi_n^*$ are mutually orthogonal projections summing up to I and each equivalent to I. Conversely, if we are given such projections E_n and ψ_n is an isometry with $\psi_n \psi_n^* = E_n$, then $\psi_m^* \psi_n = \psi_m^* E_m E_n \psi_n = \delta_{mn} I$. Hence the ψ_n are the orthonormal basis of a Hilbert space with support I in $\mathcal{M}$ and we have shown that $a)$ and $b)$ are equivalent. Given H, the map $\psi \otimes \Phi \mapsto \psi\Phi$, $\psi \in H$, $\Phi \in \mathcal{H}$, the underlying Hilbert space of $\mathcal{M}$, defines a unitary from $H \otimes \mathcal{H}$ to $\mathcal{H}$ and hence an isomorphism of $\mathcal{B}(H)\bar{\otimes}\mathcal{B}(\mathcal{H})$ onto $\mathcal{B}(\mathcal{H})$. the image of $\mathcal{B}(H) \otimes I$ is (H, H) and of $I \otimes \mathcal{M}$ is $\sigma_H(\mathcal{M})$. The von Neumann algebra generated by (H, H) and $\sigma_H(\mathcal{M})$ is $\mathcal{M}$ since

$$M = \sum_{m,n} \psi_m \psi_n^* \sigma_H(\psi_m^* M\psi_n), \quad M \in \mathcal{M}.$$

Thus $a)$ implies $c)$. To show that $c)$ implies $a)$, it suffices to show that $\mathcal{B}(H)$ contains a Hilbert space of countably infinite dimension and support I. Such a Hilbert space can be constructed as follows: let e_n and e_{mn}, $m, n \in \mathbb{N}$, be two orthonormal bases of H and define $\psi_m e_n := e_{mn}$. A simple computation shows that the ψ_m are the orthonormal basis of a Hilbert space of support I in $\mathcal{B}(H)$.

Returning to inclusions, we make a further definition.

9.5 Definition An inclusion $\mathcal{L} \subset \mathcal{N}$ is said to have *infinite multiplicity* if the relative commutant $\mathcal{L}' \cap \mathcal{N}$ is properly infinite. It is said to be *σ-finite* if $\mathcal{N}$, and hence $\mathcal{L}$, are σ-finite von Neumann algebras.

To illustrate the relevance of these notions, we remark that if the inclusion is σ–finite and has infinite multiplicity, then Property S implies Property B since infinite projections are then equivalent to their central supports.

We are often interested in having joint cyclic and separating vectors for a whole set of von Neumann algebras and this is not as difficult as it might seem thanks to the work of Dixmier and Maréchal [31]. The reason is that if there are any cyclic vectors, there are very many cyclic vectors. In fact, the proofs of the Reeh–Schlieder Theorem showed that not just the vacuum vector was cyclic for the local algebras but any vector analytic for the energy so that we automatically have a dense set of cyclic vectors. Even more is true and this quite in general. First, remark that if Ω is cyclic for $\mathcal{M}$ and T is an invertible element of $\mathcal{M}$, then $T\Omega$ is evidently cyclic for $\mathcal{M}$. Dixmier and Maréchal show that every element of $\mathcal{M}$ is the strong limit of a sequence of invertible elements so this simple observation already implies that the set of cyclic vectors is either empty or dense. Secondly, this set is even a G_δ in our Hilbert space, i.e. it is a countable intersection of open sets, because if Ω is a cyclic vector, the set of cyclic vectors is just

$$\cap_{n\in\mathbb{N}} \cup_{M\in\mathcal{M}} \{\Phi \in \mathcal{H} : \|M\Phi - \Omega\| < \frac{1}{n}\}.$$

The final step in the analysis is to use the basic property of G_δ's: in a Baire space such as $\mathcal{H}$, the intersection of a countable number of dense G_δ's is again a dense G_δ. Thus we can sum up the analysis in the following result.

9.6 Proposition On a separable Hilbert space a countable set of properly infinite von Neumann algebras each with properly infinite commutant has a dense G_δ of joint cyclic and separating vectors.

10 Standard Split Inclusions

I would like to begin this section by discussing a problem arising frequently in algebraic quantum field theory and elsewhere, too. You are given a representation π of a C^*–algebra $\mathcal{A}$ on a Hilbert space $\mathcal{H}$ and a topological group G acting as automorphisms of $\mathcal{A}$ and want to define a continuous unitary representation V of G on H such that

$$V(g)\pi(A) = \pi(g(A))V(g), \quad g \in G.$$

The pair $\{\pi, V\}$ is then called a covariant representation. I will not attempt to touch on all aspects of this problem here. Some necessary conditions are clear: $\ker\pi$ must be G–invariant, the action should extend by continuity, say in the σ–topology, to the von Neumann algebra on $\mathcal{H}$ generated by $\pi(\mathcal{A})$ and, since we require V to be continuous and the induced action on the trace class operators is continuous, the original action must be continuous in the weak topology induced by the normal states of π. Thus without loss of generality, we may consider a von Neumann algebra $\mathcal{M}$ carrying a σ–continuous action of G.

Now even when one knows that the action of G on $\mathcal{M}$ is unitarily implemented there remains the problem of making a continuous coherent choice of that unitary so as to get a continuous unitary representation of G. This is a problem in group cohomology. However, there are two cases where a canonical choice of that unitary can be made thus yielding the continuous unitary representation directly. The first case is where the representation is the GNS representation of a G–invariant state. If Ω denotes the corresponding cyclic vector then $V(g)$ is determined by

$$V(g)M\Omega = g(M)\Omega, \quad M \in \mathcal{M}, \quad g \in G.$$

The second case is when the defining representation of $\mathcal{M}$ is the *standard representation*, a representation determined up to unitary equivalence. As we shall deal exclusively with σ–finite von Neumann algebras, we may define it to be the GNS representation of a faithful normal state. Its Hilbert space is denoted $L^2(\mathcal{M})$ or $L^2(\mathcal{M}, \Omega)$ if the cyclic vector needs to be specified.

I must now sketch the basic ideas of modular theory which revolutionized the theory of von Neumann algebras in the late sixties and had important applications to physics in particular to the concept of equilibruium state.

The theory starts from the map $M\Omega \mapsto M^*\Omega$, a densely defined antilinear involution whose closure is denoted by S. Being closed, S has a polar decomposition, which following the established notation, I write

$$S = J\Delta^{\frac{1}{2}}, \quad \Delta = S^*S.$$

J is obviously an antiunitary involution and the positive operator Δ is called the modular operator. The essential and non-trivial property of J and Δ is expressed by the following two equations:

$$J\mathcal{M}J = \mathcal{M}'; \quad \Delta^{it}\mathcal{M}\Delta^{-it} = \mathcal{M}, \quad t \in \mathbb{R}.$$

The unitary operators Δ^{it} therefore define automorphisms σ_t of $\mathcal{M}$ known as the *modular automorphisms*. The importance of these concepts for statistical mechanics lies in the fact that if ω_Ω is an equilibrium state and $\mathcal{M} = \mathfrak{A}^-$ then the restriction of $\sigma_{-\beta^{-1}t}$ to $\mathfrak{A}$ is just the time evolution. Here β^{-1} is, as usual, just kT.

Here, we are interested in quite a different aspect. The cyclic and separating vector defines for us a cone $\mathcal{P}_\Omega^\natural$ in $\Lambda^2(\mathcal{M})$:

$$\mathcal{P}_\Omega^\natural = \{JMJM\Omega : M \in \mathcal{M}\}^- = \{\Delta^{\frac{1}{4}}M\Omega : M \in \mathcal{M}_+\}^-.$$

It has the following fundamental property: every normal state of $\mathcal{M}$ is of the form ω_Φ for a *unique* vector $\Phi \in \mathcal{P}_\otimes^\natural$. Furthermore, since the vectors in $\mathcal{P}_\otimes^\natural$ are J–invariant, Φ is cyclic for $\mathcal{M}$ if and only if it is separating for $\mathcal{M}$, i.e. if and only if ω_Φ is a faithful state. The norm topology on the positive normal linear functionals corresponds to the Hilbert space topology on $\mathcal{P}_\otimes^\natural$. The cone also provides us with a canonical unitary implementation of the automorphism group $\mathrm{Aut}\mathcal{M}$, in the following sense: if $\alpha \in \mathrm{Aut}\mathcal{M}$ there is a unique unitary operator $V(\alpha)$ such that

$$V(\alpha)M = \alpha(M)V(\alpha), \quad M \in \mathcal{M}; \quad V(\alpha)\mathcal{P}^\natural_\Omega \subset \mathcal{P}^\natural_\Omega.$$

The uniqueness is clear here, because the action of $V(\alpha)$ on $\mathcal{P}^\natural_\Omega$ is determined by the action of α on normal states and $\mathcal{P}^\natural_\Omega$ spans the Hilbert space. An immediate consequence of the uniqueness is that V is a unitary representation of $\mathrm{Aut}\mathcal{M}$, i.e. $V(\alpha)V(\alpha') = V(\alpha\alpha')$. V becomes a continuous unitary representation if we give $\mathrm{Aut}\mathcal{M}$ the topology of pointwise norm convergence on the predual. Indeed, this topology is just the topology induced by the strong operator topology on the canonical unitary implementation. Note, that if $\omega_\Omega \circ \alpha = \omega_\Omega$, then $V(\alpha)\Omega = \Omega$ so that $V(\alpha)$ is just the unitary provided by the GNS–construction starting with the invariant state ω_Ω.

We can also deal in the same way with isomorphisms between von Neumann algebras in standard representations fixing a cone for each algebra involved. Anyway, we now have two important cases where automorphism groups can be unitarily implemented. As we know that our local algebras have joint cyclic and separating vectors we are in good shape. However, in the context of an inclusion $\mathcal{L} \subset \mathcal{N}$, we can ask for our implementing operators $V(g)$ to lie in $\mathcal{N}$. As our automorphisms have little tendency to be inner, there is a price to pay: we have to be content with implementing the automorphism just on $\mathcal{L}$. We shall later see the relevance of this question in the physical context.

We now need to examine in more detail a special class of split inclusions, the *standard split inclusions* introduced by Doplicher and Longo [37]. They define a triple $\Lambda = \{\mathcal{L}, \mathcal{N}, \Omega\}$ to be a standard split inclusion when $\mathcal{L}$ and $\mathcal{N}$ are von Neumann algebras on a Hilbert space $\mathcal{H}_\Lambda$ with an intermediate type I factor $\mathcal{M}$, $\mathcal{L} \subset \mathcal{M} \subset \mathcal{N}$ and $\Omega \in \mathcal{H}_\Lambda$ is a joint cyclic and separating vector for $\mathcal{L}$, $\mathcal{N}$ and $\mathcal{L}' \cap \mathcal{N}$. Note at once that if Λ is a standard split inclusion then so is $\Lambda' := \{\mathcal{N}', \mathcal{L}', \Omega\}$.

We now examine some of the consequences of the split property.

10.1 Lemma If $\Lambda = (\mathcal{L}, \mathcal{N}, \Omega)$ is a standard split inclusion then $\mathcal{H}_\Lambda$, the underlying Hilbert space, is separable.

Proof Let $\omega = \omega_\Lambda$ denote the state on $\mathcal{N}$ induced by Ω. ω is faithful since Ω separates $\mathcal{N}$. Hence, if $\mathcal{M}$ is a type I factor between $\mathcal{L}$ and $\mathcal{N}$, the restriction of ω to $\mathcal{M}$ will be a faithful normal state so that $\mathcal{M}$ is σ–finite by Lemma 9.2. Hence $\mathcal{M}$ is $\mathcal{B}(\mathcal{K})$ for some separable Hilbert space $\mathcal{K}$. Similarly regarding $\mathcal{M}'$ as $\mathcal{B}(\mathcal{K}')$, $\mathcal{K}'$ must also be separable. But we have seen in Sec. 8 that $\mathcal{H}$ and $\mathcal{K} \otimes \mathcal{K}'$ are isomorphic so that $\mathcal{H}$ is separable.

The next result is due to Doplicher and Longo [37] and came as a surprise. In a standard split inclusion $\Lambda = \{\mathcal{L}, \mathcal{N}, \Omega\}$ there is, by definition, a type I factor between $\mathcal{L}$ and $\mathcal{N}$ however there is even a canonical choice, $\mathcal{M}_\Lambda$, transforming correctly under isomorphisms of standard split inclusions.

If $\mathcal{M}$ is any intermediate type I factor, the vector Ω defines a faithful normal state of $\mathcal{M}$ and $\mathcal{M}'$ yielding a corresponding faithful normal product state on $\mathcal{M} \vee \mathcal{M}' = \mathcal{B}(\mathcal{H}_\Lambda)$. The restriction of this state to $\mathcal{L} \vee \mathcal{N}'$ is also a faithful normal product state $\tilde{\omega}$:

$$\tilde{\omega}(LN') = \tilde{\omega}(L)\tilde{\omega}(N'), \quad L \in \mathcal{L}, \quad N' \in \mathcal{N}'.$$

But $\mathcal{L} \vee \mathcal{N}'$ has a cyclic and separating vector, namely Ω, so that $\tilde{\omega}$ is induced by a cyclic vector. However, since I want to use this vector to construct $\mathcal{M}_\Lambda$ a canonical choice has to be made so we use modular theory picking the unique cyclic and separating vector $\Omega_\Lambda \in \mathcal{P}_\otimes^\natural$ inducing $\tilde{\omega}$, where the cone is defined relative to the von Neumann algebra $\mathcal{L} \vee \mathcal{N}'$. Note that $\Omega_\Lambda = \Omega_{\Lambda'}$. We can now realize $\mathcal{H}_\Lambda$ as a tensor product $\mathcal{H}_\Lambda \otimes_\Lambda \mathcal{H}_\Lambda$ in a canonical way by requiring

$$L\Omega \otimes_\Lambda N'\Omega := LN'\Omega_\Lambda, \quad L \in \mathcal{L}, \quad N' \in \mathcal{N}'.$$

The associated flip permuting the two factors in the tensor product will be denoted by θ_Λ:

$$\theta_\Lambda(\Phi \otimes_\Lambda \Psi) = \Psi \otimes_\Lambda \Phi.$$

There is a canonical endomorphism ψ_Λ called the *universal localizing map*, for reasons which will become apparent later,

$$\psi_\Lambda(X) := X \otimes_\Lambda I$$

and a canonical type I factor $\mathcal{M}_\Lambda := \psi_\Lambda(\mathcal{B}(\mathcal{H}_\Lambda))$. Now

$$L\Omega \otimes_\Lambda N'\Omega = LN'\Omega_\Lambda = N'L\Omega_{\Lambda'} = N'\Omega \otimes_{\Lambda'} L\Omega.$$

Hence $\Phi \otimes_\Lambda \Psi = \Psi \otimes_{\Lambda'} \Phi$, $\Phi, \Psi \in \mathcal{H}_\Lambda$. It follows that $\theta_{\Lambda'} = \theta_\Lambda$ and $X \otimes_\Lambda I = I \otimes_{\Lambda'} X$. Hence $\psi_{\Lambda'} = \mathrm{Ad}\theta_\Lambda\psi_\Lambda$ and

$$\mathcal{M}_{\Lambda'} = \theta_\Lambda \mathcal{M}_\Lambda \theta_\Lambda^{-1}.$$

We also have canonical Hilbert spaces $H_\Lambda := (\iota, \psi_{\Lambda'})$ and $H_{\Lambda'} := (\iota, \psi_\Lambda)$, where ι denotes the identity automorphism. The elements of H_Λ are the mappings of the form $\Phi \mapsto \Phi \otimes_\Lambda \Psi$ and of $H_{\Lambda'}$ the mappings of the form $\Phi \mapsto \Phi \otimes_{\Lambda'} \Psi = \Psi \otimes_\Lambda \Phi$ and we obviously have $(H_\Lambda, H_\Lambda) = \mathcal{M}_\Lambda$ and $(H_{\Lambda'}, H_{\Lambda'}) = \mathcal{M}_{\Lambda'}$. This description makes it clear that the canonical localizing map ψ_Λ is the canonical inner endomorphism $\sigma_{\Lambda'}$ defined by the Hilbert space $H_{\Lambda'}$.

Now letting σ_Λ denote the endomorphism induced by H_Λ, $\sigma_\Lambda(\mathcal{N}) = (H_\Lambda, H_\Lambda)' \cap \mathcal{N}$ so that the inclusion $\mathcal{M}_\Lambda \subset \mathcal{N}$ and a fortiori the inclusion $\mathcal{L} \subset \mathcal{N}$ have infinite multiplicity.

10.2 Corollary Any standard split inclusion has Property B.

Proof $\mathcal{L} \subset \mathcal{N}$ is split so it has Property S. However, it has infinite multiplicity and is σ–finite so it has Property B.

We now consider the group of automorphisms $\mathrm{Aut}\Lambda$ of a standard split inclusion $\Lambda = \{\mathcal{L}, \mathcal{N}, \Omega\}$. By this we mean the set of automorphisms α of $\mathcal{N}$ such that

$\alpha(\mathcal{L}) = \mathcal{L}$ and $(\Omega, \alpha(N)\Omega) = (\Omega, N\Omega)$, $N \in \mathcal{N}$. We give this group the topology of pointwise norm convergence on the predual of $\mathcal{N}$. As we know, this coincides with the strong operator topology on its canonical implementation, given by

$$V(\alpha)N\Omega = \alpha(N)\Omega, \quad N \in \mathcal{N},$$

as ω_Ω is an invariant state. Since Ω is cyclic for $\mathcal{L}$ this is at the same time the canonical implementation for $\alpha \upharpoonright \mathcal{L}$. Now $\alpha \in \text{Aut}\Lambda$ implies $\alpha(\mathcal{M}_\Lambda) = \mathcal{M}_\Lambda$ since $\mathcal{M}_\Lambda$ is the canonical type I factor. Hence our topology is also the topology of the pointwise norm convergence on the predual of $\mathcal{M}_\Lambda$. Doplicher and Longo [37] proved the following important result which is one of several indications that groups of net automorphisms leaving the vacuum state invariant, such as the group of unbroken gauge automorphisms or the group of internal symmetries, are compact metrizable.

10.3 Theorem If $\Lambda = \{\mathcal{L}, \mathcal{N}, \Omega\}$ is a standard split inclusion then $\text{Aut}(\Lambda)$ is compact and metrizable.

Proof Consider the faithful state $\omega_\Omega \upharpoonright \mathcal{M}_\Lambda$ and let $\rho \in \mathcal{M}_\Lambda$ be the corresponding density matrix and consider the group of unitaries of $\mathcal{M}_\Lambda$ commuting with ρ. Since $\omega_\Omega \upharpoonright \mathcal{M}_\Lambda$ is faithful, the eigenspaces of ρ are finite–dimensional so our group is just a product of a countable number of unitary groups of full matrix algebras and is hence compact metrizable. The image under the continuous map $U \mapsto \text{ad}U \in \text{Aut}(\mathcal{M}_\Lambda)$ is the group of automorphisms of $\mathcal{M}_\Lambda$ leaving $\omega_\Omega \upharpoonright \mathcal{M}_\Lambda$ invariant. This image is therefore compact metrizable so its canonical implementation on $\mathcal{H}_\Lambda$ is compact metrizable in the strong operator topology. The subgroup of the $V(\alpha)$ inducing automorphisms of both $\mathcal{L}$ and $\mathcal{N}$ is a closed subgroup and is hence compact and metrizable, too.

10.4 Proposition The action α of $\text{Aut}\Lambda$ on $\mathcal{N}$ has full Hilbert spectrum.

Proof Since Ω is invariant for V and cyclic for $\mathcal{N}$, the spectrum of V is the same as that of the action of α on $\mathcal{N}$. However, the spectrum of V coincides with the spectrum of α on H_Λ and this is included in the Hilbert spectrum. Since the action α is by definition faithful, we have full Hilbert spectrum.

We now come back to the question of implementing automorphisms on $\mathcal{L}$ using representations in $\mathcal{N}$. This can now be done efficiently using the universal localizing map: let $g \mapsto V(g)$ be a continuous unitary representation of a topological group on $\mathcal{H}_\Lambda$ then $g \mapsto \psi_\Lambda(V(g)) := V_\Lambda(g)$ is another such representation unitarily equivalent to a multiple of the original representation and taking values in the unitary group of $\mathcal{M}_\Lambda$. Writing $g(T) := V(g)TV(g)^*$, we see that if A and $g(A)$ are in $\mathcal{L}$ then

$$V_\Lambda(g)A = \psi_\Lambda(V(g)A) = \psi_\Lambda(g(A)V(g)) = g(A)V_\Lambda(g) \quad g \in G.$$

Thus V_Λ still implements that part of the action involving just $\mathcal{L}$.

11 Some Properties of Nets

After these two sections on inclusions, we return to the physical setting adapting the definitions to nets and making comments on the physical interpretation where appropriate.

You will recall that when I dicussed local normality, I said that the local observable algebras seem to be factors, in practice. I also provided a tentative physical interpretation, the principle of local definiteness which meant that each relevant state had a faithful locally normal GNS representation.

It was a standard open problem in the early days of algebraic quantum field theory to prove that a local algebra is a factor of type III. Now a factor of type III on a separable Hilbert space is a factor where each non–zero projection is equivalent to I. Borchers proved using positivity of the energy that the local algebras have a rather weaker property, namely that each non–zero projection in a local algebra is equivalent to I in a slightly larger algebra. To be precise, let us make a definition.

11.1 Definition A net $\mathfrak{A}$ of von Neumann algebras over $\mathcal{K}$ has the *Borchers Property* if given $\mathcal{O} \in \mathcal{K}$ there is an $\mathcal{O}_1 \subset \mathcal{O}$ such that any projection $E \neq 0 \in \mathfrak{A}(\mathcal{O}_1)$ satisfies $E \sim I$ in $\mathfrak{A}(\mathcal{O})$.

We have therefore adapted the definition of Property B from inclusions to nets. In the same way, we may adapt Property S and the split property.

As a consequence of the above discovery of Borchers, it no longer seemed vital to establish whether the local algebras are actually factors: the Borchers property sufficed for the most important applications. The problem is hence still open, although it is known in some generality that the algebra of a wedge region is a factor of type III_1.

The Borchers Property can also be expressed in terms of Hilbert spaces in von Neumann algebras if one wishes. It is equivalent to saying that every partition of the unit in $\mathfrak{A}(\mathcal{O}_1)$ results from an orthonormal basis of a Hilbert space of support I in $\mathfrak{A}(\mathcal{O})$. When we come to use the Borchers Property in the theory of superselection sectors then we shall really need the property not for the observable net but rather for the *dual net* $\mathfrak{A}^d$, where

$$\mathfrak{A}^d(\mathcal{O}) := \mathfrak{A}(\mathcal{O}')'$$

and the commutant is understood in the vacuum representation. The Borchers Property for the dual net has a simple interpretation in terms of the observable net. It means that given $\mathcal{O}_1^- \subset \mathcal{O}$ any non–trivial subrepresentation of the vacuum representation of $\mathfrak{A}(\mathcal{O}_1')$ becomes equivalent to the vacuum representation in restriction to $\mathfrak{A}(\mathcal{O}')$. Borchers original proof works well for the dual net, which is very much like the observable net, and, in practice, often even satisfies local commutativity. This will be discussed in the next section and turns out to be vital for superselection theory.

Haag and Kastler [54] were the first to stress the need to use abstract C^*–algebras in quantum field theory and in the same paper they raised the question of the statistical independence of the algebras corresponding to spacelike separated regions. To explain what is involved here, suppose you restrict a state ω on the observable net $\mathfrak{A}$ to a local algebra $\mathfrak{A}(\mathcal{O})$ then you get a state $\omega \restriction \mathcal{O}$ of $\mathfrak{A}(\mathcal{O})$. Conversely, if you begin with a state ω' of $\mathfrak{A}(\mathcal{O})$ then there is a state ω of $\mathfrak{A}$ such that $\omega' = \omega \restriction \mathcal{O}$. Any state on a C^*–subalgebra can be extended to a state on the whole algebra. Now suppose you have two local algebras $\mathfrak{A}(\mathcal{O}_1)$ and $\mathfrak{A}(\mathcal{O}_2)$ and you are given states ω_i

of $\mathfrak{A}(\mathcal{O}_i)$ then: is there a state ω of $\mathfrak{A}$ with $\omega \upharpoonright \mathcal{O}_i = \omega_i$, $i = 1, 2$? If this is the case then $\mathfrak{A}(\mathcal{O}_1)$ and $\mathfrak{A}(\mathcal{O}_2)$ are said to be *statistically independent*.

Well, obviously this will not be the case in general because if $\mathcal{O}_1 \subset \mathcal{O}_2$ then it would be necessary, and indeed sufficient, for the restriction of ω_2 to $\mathfrak{A}(\mathcal{O}_1)$ to coincide with ω_1. This is a trivial example but again on physical grounds there would be trouble if we can send a signal from $\mathcal{O}_1$ to $\mathcal{O}_2$ because if we have the information that an unknown state ω looks like ω_1 in $\mathcal{O}_1$ then the information we can gather in $\mathcal{O}_2$ must be consistent with the known situation in $\mathcal{O}_1$. In other words there will be restrictions on ω_2. Thus, we really only have a chance of getting statistical independence if $\mathcal{O}_1 \subset \mathcal{O}'_2$ and to avoid problems with the boundary we should perhaps suppose $\mathcal{O}_1^- \subset \mathcal{O}'_2$.

The following result of Roos [74] shows what statistical independence involves mathematically.

11.2 Theorem Let $\mathcal{A}$ and $\mathcal{B}$ be commuting C^*-subalgebras containing the unit of a C^*-algebra $\mathcal{C}$. Then the following conditions are equivalent.

a) $\mathcal{A}$ and $\mathcal{B}$ are statistically independent,
b) given $A \in \mathcal{A}$ and $B \in \mathcal{B}$ with $AB = 0$ then either $A = 0$ or $B = 0$,
c) the *-subalgebra generated by $\mathcal{A}$ and $\mathcal{B}$ is naturally isomorphic to the *algebraic tensor product* $\mathcal{A} \otimes \mathcal{B}$ of $\mathcal{A}$ and $\mathcal{B}$.

I will not give the proof since there are no physical ideas involved.

In our discussion of inclusions of von Neumann algebras, we saw that Property S follows from Property B. We therefore conclude from Theorem 11.2 that the algebras $\mathfrak{A}(\mathcal{O}_1)$ and $\mathfrak{A}(\mathcal{O}_2)$ with $\mathcal{O}_1^- \subset \mathcal{O}'_2$ are statistically independent.

Now that we know that the algebras of observables corresponding to suitably spacelike separated regions are indeed statistically independent, we should ask about the physical significance of statistical independence. I see none. The problem is again one of physical relevance of states. Even if we start with states of physical relevance over each of the two algebras we do not know that there is a physically relevant extension. As we do not have a general solution to the problem of determining which states are physically relevant, it might seem that there is not much point in pursuing the matter further. But this is too pessimistic.

If we work in a physically relevant representation such as the vacuum representation then, as I explained, every normal state of the representation should be regarded as being physically relevant. Furthermore, every physically relevant state should be normal in restriction to the local algebras. Hence if we consider two double cones $\mathcal{O}_1$ and $\mathcal{O}_2$ with $\mathcal{O}_1^- \subset \mathcal{O}'_2$ our questions amounts to the following: given normal states of $\mathfrak{A}(\mathcal{O}_1)$ and $\mathfrak{A}(\mathcal{O}_2)$ is there a *normal* state of the representation extending both of them? It seems that an even stronger property, the *split property* already discussed in the context of inclusions of von Neumann algebras, holds in physically reasonable theories, i.e. given any pair of double cones as above there is a factor of type I $\mathcal{M}$ with

$$\mathfrak{A}(\mathcal{O}_1) \subset \mathcal{M} \subset \mathfrak{A}(\mathcal{O}_2).$$

Borchers conjectured early on that the split property ought to be valid but it was not until 1974 that there was a proof by Buchholz [17] that it did at least hold in the simplest of models, the free massive scalar field. By now it is known to hold in most but not all models. Furthermore, the models where it is known to be violated such as most generalized free field models are known to be deficient from a physical point of view. Basically speaking they have too many local degrees of freedom.

I should perhaps add that properties which are not universally valid tend to be more interesting. A good example of this is the property of duality that we shall come to shortly.

We have seen that statistical independence for normal states does not suffice to motivate the split property. A physical interpretation of the split property was proposed by Buchholz, Doplicher and Longo [22]. We ask the following question: is it possible to prepare a normal state of $\mathfrak{A}(\mathcal{O}_1)$ by making a measurement in a slightly larger region $\mathcal{O}_2$? It is not clear how best to formalize the act of preparing a state but fortunately a simple version of preparation suffices for our purposes. If we are in a state ω and measure a projection E with $\omega(E) \neq 0$ accepting the value 1 for the next measurement and rejecting the value 0 then we have prepared the state ω_E for the subsequent measurement, where

$$\omega_E(A) := \omega(E)^{-1}\omega(EAE), \quad A \in \mathfrak{A}.$$

Now suppose the split property holds and take E to be a minimal projection in the type I factor $\mathcal{M}$. Then we see that ω_E coincides on $\mathcal{M}$ with the pure state defined by E, independently of the choice of ω with $\omega(E) \neq 0$. Buchholz, Doplicher and Longo call a projection in a local algebra which prepares a particular state on some smaller local algebra a *local filter* and they show that the existence of a local filter for each $\mathfrak{A}(\mathcal{O})$ is equivalent to the split property.

The split property has many applications. It was first systematically investigated and exploited by Doplicher and Longo [37] in connection with the question of the existence of local currents associated with the charges arising in the theory of superselection sectors. In this context, it is the notion of standard split inclusion which is important. As we have seen this notion involves not just an inclusion of von Neumann algebras but also a state. For this reason I ought to talk about the Reeh–Schlieder property first.

One of the earliest important results in algebraic quantum field theory was the discovery, known as the Reeh–Schlieder Theorem, that, in the vacuum representation, the vacuum vector Ω is cyclic for each local algebra $\mathfrak{F}(\mathcal{O})$, i.e. $\mathfrak{F}(\mathcal{O})\Omega$ is dense in $\mathcal{H}_0$, the vacuum Hilbert space. More accurately, the Reeh–Schlieder Theorem was a result in the Wightman formalism asserting that $\mathfrak{P}(\mathcal{O})\Omega$ is dense in $\mathcal{H}_0$ which was quickly adapted to the algebraic framework. The proof depends on additivity and positivity of the energy. Now a vector is cyclic for a von Neumann algebra if and only if it is separating for the commutant. Hence, in the case of an observable net, local commutativity implies that Ω is cyclic and separating for each $\mathfrak{A}(\mathcal{O})$. In fact, it is not difficult to extend this result to the local field algebras $\mathfrak{F}(\mathcal{O})$.

Now for the physical implications where for simplicity I suppose that we are dealing with the observable net $\mathfrak{A}$ which has been identified with its vacuum representation. I recall that the square of the variance of an observable A in a state ω is given by $\omega(A^2) - \omega(A)^2$ and is a measure of how much the observed values of A in the state ω fluctuate around their mean value $\omega(A)$. The variance vanishes in a vector state ω_Φ if and only if Φ is an eigenvector for A. Now suppose that $A \in \mathfrak{A}(\mathcal{O})$ and that $A\Omega = \lambda\Omega$ then $(A - \lambda I)\Omega = 0$ and, since Ω is separating, $A = \lambda I$. Hence all non–trivial local observables exhibit fluctuations in the vacuum state. In particular, there is no possibility of constructing an ideal particle counter; there will always be some probability of the counter responding in the vacuum state.

The Reeh–Schlieder Theorem also stopped one from thinking of a state of the form $\omega_{A\Omega}$ with $A \in \mathfrak{A}(\mathcal{O})$ as being 'localized' in $\mathcal{O}$. This cannot be a good idea if Ω is cyclic for $\mathfrak{A}(\mathcal{O})$. A state ω is said to be *strictly localized* in $\mathcal{O}$ if

$$\omega(A) = \omega_0(A), \quad A \in \mathfrak{A}(\mathcal{O}').$$

A state of the form $\omega_{A\Omega}$ with $A \in \mathfrak{A}(\mathcal{O})$ should be regarded as being approximately localized in $\mathcal{O}$ and better and better localized the smaller $\|A\|$ is. In fact, it is not difficult to show using the Reeh–Schlieder property that if $\|A\|$ takes on its minimum value 1 then $A^*A = I$, i.e. A is an isometry. In this case $\omega_{A\Omega}$ is strictly localized in $\mathcal{O}$.

As a consequence of the Reeh–Schlieder Theorem we know that whenever we have the split property, we actually have standard split inclusions. In the context of inclusions of von Neumann algebras, we have seen how these inclusions allow one to implement symmetries locally using the universal localizing map.

But let me, at this point, explain why Doplicher and Longo [32], [36] were interested in implementing symmetries locally. Superselection theory, which we will come to shortly, is based on the idea of creating charge. The resulting charge operators parameterize the superselection sectors and belong to the weak closure of the observable net on the vacuum Hilbert space of the field net. They are global quantities. However, we expect to be able to make local charge measurements and we have ideas derived from models on how this come about. There would seem to be a conserved current j^μ associated with a conserved charge. This is a Wightman field so there are the usual technical difficulties but formally you get a local operator related to the measurement of charge by taking $j^0(g_\delta f_R)$, where f_R and g_δ are smooth positive functions with compact support on space and time respectively, $f_R(\mathbf{x}) = 1$ for $|\mathbf{x}| \le R$, $\int g_\delta(t)dt = 1$ and $g_\delta(t) = 0$ for $|t| \ge \delta$. We expect these to be related to the global charge Q through

$$QF\Omega = \lim_{R\to\infty} [j^0(g_\delta f_R), F]\Omega,$$

where Ω is the vacuum vector. The unitary operator $e^{ij^0(g_\delta f_R)}$ should then induce the same action as e^{iQ} on operators contained in a double cone centred on the origin of radius $< R$. Therefore, we have a unitary operator, localized in one double cone which induces a gauge automorphism on some smaller double cone. This behaviour can be verified for the free charged scalar field. Thus the motivation of Doplicher and Longo was to prove the existence of local conserved currents. What they found was slightly different: the universal localizing map ψ_Λ of the standard split inclusion $\{\mathfrak{F}(\mathcal{O}_1), \mathfrak{F}(\mathcal{O}_2), \Omega\}, \mathcal{O}_1^- \subset \mathcal{O}_2$, in a field net applied to the representation $g \mapsto V(g)$ of the gauge group yields a *true* representation $g \mapsto \psi_\Lambda(V(g)) := V_\Lambda(g) \in \mathfrak{F}(\mathcal{O}_2)$ of the group of gauge automorphisms such that, for $F \in \mathfrak{F}(\mathcal{O}_1)$,

$$V_\Lambda(g)F = \psi_\Lambda(V(g)F) = \psi_\Lambda(g(F)V(g)) = g(F)V_\Lambda(g) \quad g \in G,$$

whereas the exponentials of the smeared currents would not have. It is not clear whether one should continue to search for the local conserved currents or whether the natural object in algebraic quantum field theory is what one gets with the universal localizing map.

Another application is to spacetime symmetries, supposing our net to be Poincaré covariant to illustrate the ideas involved. We then have a continuous unitary representation U of the covering group of the Poincaré group such that

$$U(L)FU(L)^* = \alpha_L(F), \quad F \in \mathfrak{F}.$$

Thus $U(L)\mathfrak{F}(\mathcal{O})U(L)^* = \mathfrak{F}(L\mathcal{O})$. We again apply the universal localizing map ψ_Λ setting

$$U_\Lambda(L) = \psi_\Lambda(U(L))$$

and U_Λ is a continuous unitary representation of the covering group of the Poincaré group in $\mathcal{M}_\Lambda \subset \mathfrak{F}(\mathcal{O}_2)$ and the analogous computation now yields

$$U_\Lambda(L)FU_\Lambda(L)^* = \alpha_L(F), \quad F, \alpha_L(F) \in \mathfrak{F}(\mathcal{O}_1).$$

Note that, as U satisfies the spectrum condition, U_Λ does, too.

12 Duality

I come now to a very important and interesting property known as duality. This is usually formulated as follows: the net $\mathfrak{A}$ of observables is said to satisfy duality for a set $\mathcal{O}$ in the representation π if

$$\pi(\mathfrak{A}(\mathcal{O}))^- = \pi(\mathfrak{A}(\mathcal{O}'))' \cap \pi(\mathfrak{A})^-.$$

Since there are always unbounded sets involved here, there is a minor but completely superfluous problem of defining $\mathfrak{A}(\mathcal{O})$ when $\mathcal{O}$ is unbounded. We have already introduced a definition to deal with the spacelike complement. I think the above notation does not really do justice to the notion of duality which only involves von Neumann algebras. It should be seen in terms of a net $\mathfrak{R}_{\hat{\pi}}$ of von Neumann subalgebras of $\mathcal{B}(\mathcal{H}_\pi)$. It can be defined over arbitrary subsets S of Minkowski space by setting

$$\mathfrak{R}_\pi(\mathcal{O}) := \pi(\mathfrak{A}(\mathcal{O})), \quad \mathcal{O} \in \mathcal{K}$$

and requiring $\mathfrak{R}_\pi(S)$ to be the von Neumann algebra generated by the $\mathfrak{R}_\pi(\mathcal{O})$ with $\mathcal{O} \in \mathcal{K}$ and $\mathcal{O} \subset S$. As this von Neumann algebra only depends on the interior of S we may work entirely with open sets using intS' in place of the spacelike complement. If $\mathfrak{A}$ is additive then so is $\mathfrak{R}_\pi$. In any case, duality for a set $\mathcal{O}$ in the representation π can now be written

$$\mathfrak{R}_\pi(\mathcal{O}) = \mathfrak{R}_\pi(\mathcal{O}')' \cap \mathfrak{R}_\pi(\mathbb{M}) =: \mathfrak{R}_\pi(\mathcal{O}')^c.$$

If duality holds for $\mathcal{O}$ then it also holds for $\mathcal{O}''$. If the ^{c} is an involution in $\mathfrak{R}_\pi(\mathbb{M})$ as is usually the case in practice then duality for $\mathcal{O}$ also implies duality for $\mathcal{O}'$.

Let me attempt a physical interpretation: if we consider just the normal states of the representation π then the self-adjoint elements of $\mathfrak{R}_\pi(\mathbb{M})$ are those operators which can be considered in an idealized sense as bounded observables. If such an element is in $\mathfrak{R}_\pi(\mathcal{O}')^c$ then it commutes with all observables that can be measured in $\mathcal{O}'$ and is therefore a candidate for an observable which can be measured in $\mathcal{O}$. Our equation asserts that every such element can, in an idealized sense, be measured in $\mathcal{O}$ because its expectation values in the normal states of π can be approximated arbitrarily well by observables localized in $\mathcal{O}$.

By far the most important case of duality is when $\mathcal{O} \in \mathcal{K}$ and π is the vacuum representation. In fact, when we talk about an observable net satisfying duality we mean simply that duality is satisfied for every $\mathcal{O} \in \mathcal{K}$ in the vacuum representation π.

I have already indicated that duality is particularly interesting because the property is not generally true. The breakdown of duality for double cones in the vacuum sector is indicative of the spontaneous breakdown of gauge symmetry. When duality holds for double cones in the vacuum sector it does not necessarily hold for other regions. It does not hold for disconnected sets if there are localized charges in the theory. It will not hold for sets which are not simply connected in the presence of a magnetic field. Finally, if it holds for the vacuum sector, it will not necessarily hold for other representations. For example, duality breaks down for a sector with parastatistics.

Duality is important primarily for its role in the analysis of superselection structure. Actually this analysis only requires a weaker property known as *essential duality*. Here we need the dual net $\mathfrak{A}^d$ introduced in the last section:

$$\mathfrak{A}^d(\mathcal{O}) = \cap_{\mathcal{O}_1 \subset \mathcal{O}'} \mathfrak{A}(\mathcal{O}_1)' = \mathfrak{R}_{\pi_0}(\mathcal{O}')'.$$

Let me comment here on this concept of dual net. If we call two nets $\mathfrak{A}$ and $\mathfrak{B}$ over $\mathcal{K}$ *relatively local* if

$$\mathfrak{A}(\mathcal{O}_1) \subset \mathfrak{B}(\mathcal{O}_2)' \quad \text{whenever } \mathcal{O}_1 \subset \mathcal{O}_2'.$$

then $\mathfrak{A}^d$ is the maximal net which is local relative to $\mathfrak{A}$. Thus $\mathfrak{A} \subset \mathfrak{B}$ implies $\mathfrak{B}^d \subset \mathfrak{A}^d$. Since $\mathfrak{A}^{dd}$ is the largest net local relative to $\mathfrak{A}^d$, $\mathfrak{A} \subset \mathfrak{A}^{dd}$. Thus passing to the dual net has the same formal properties as taking a commutant. In particular,

$\mathfrak{A}^d = \mathfrak{A}^{ddd}$. A net $\mathfrak{A}$ is local if $\mathfrak{A} \subset \mathfrak{A}^d$ and, consequently, $\mathfrak{A}^{dd}$ is local whenever $\mathfrak{A}$ is local. Applying the definition, we get the following expression for the second dual net:

$$\mathfrak{A}^{dd}(\mathcal{O}) = \cap_{\mathcal{O} \subset \mathcal{O}_1'} \mathfrak{R}(\mathcal{O}_1').$$

The double dual net plays a role in the theory of superselection sectors in that it has the same sectors as the original net.

Now one way of expressing duality for double cones in the vacuum sector is to write

$$\mathfrak{A} = \mathfrak{A}^d.$$

Essential duality may be expressed in three equivalent ways, we may say that the dual net $\mathfrak{A}^d$ is local, or that it satisfies duality, or we may write

$$\mathfrak{A}^d = \mathfrak{A}^{dd}.$$

Essential duality is a consequence of duality for wedge regions. A wedge region is a region of the form

$$W = L W_R,$$

where L denotes a Poincaré transformation and W_R, the right wedge, is defined by

$$W_R = \{x \in \mathbb{M} : x^1 > |x^0|\}.$$

The interior of its spacelike complement is the left wedge, $W_L = -W_R$. Duality for the right wedge in the vacuum sector therefore means that

$$\mathfrak{R}_{\pi_0}(W_R) = \mathfrak{R}_{\pi_0}(W_L)^c.$$

One basic geometric fact is involved in showing that wedge duality implies essential duality. Given two spacelike separated double cones there is a wedge containing the first and spacelike to the second. We can now proceed to the proof.

12.1 Lemma If an observable net represented irreducibly on a Hilbert space satisfies wedge duality, it satisfies essential duality.

Proof The basic geometric fact implies

$$\cap\{\mathfrak{R}_\pi(W) : \mathcal{O} \subset W\} \subset \cap\{\pi(\mathfrak{A}(\mathcal{O}_1))' : \mathcal{O}_1 \subset \mathcal{O}'\} = \pi(\mathfrak{A})^d(\mathcal{O}).$$

On the other hand, we actually have equality here if wedge duality holds since

$$\pi(\mathfrak{A})^d(\mathcal{O}) \subset \cap\{\mathfrak{R}_\pi(W')' : \mathcal{O} \subset W\}.$$

But the basic geometric fact now also implies that the dual net is local so that essential duality holds.

13 Intertwiners

The theory of superselection sectors was innovative both from the physical and the mathematical point of view. In these lectures, I would like to separate these two aspects as clearly as possible. We need to have a clear view of the mathematical machinery. It should by now be regarded basically as something to use rather than to develop despite the fact that certain future applications will lead us to develop the machinery further. Thus the sole purpose of this section is to introduce a formalism enabling one to work more efficiently with representations and endomorphisms and their intertwiners. Given two representations π and π' we write $T \in (\pi, \pi')$ to denote an intertwining operator, i.e. T is a bounded linear operator from $\mathcal{H}_\pi$ to $\mathcal{H}_{\pi'}$ such that

$$T\pi(A) = \pi'(A)T, \quad A \in \mathcal{A}.$$

One checks at once that $T^* \in (\pi', \pi)$ and that if $T' \in (\pi', \pi'')$ then $T' \circ T \in (\pi, \pi'')$. Intertwining operators express the relations that exist between representations. For example, π and π' are unitarily equivalent if (π, π') contains a unitary, π is unitarily equivalent to a subrepresentation of π' if (π, π') contains an isometry and π is unitarily equivalent to the direct sum of π_1 and π_2 if there are isometries $W_i \in (\pi_i, \pi)$, $i = 1, 2$ with $W_1 \circ W_1^* + W_2 \circ W_2^* = 1_\pi$. Nonetheless, the representations together with their intertwiners should be thought of as being aspects of a single entity, a C^*–category. The representations are the objects, the intertwiners are the arrows and there is a C^*–norm on the arrows which is just the operator norm.

We get another example of a C^*–category if we look at the unital endomorphisms of a C^*–algebra $\mathcal{A}$ and their intertwiners, where the set (ρ, ρ') of intertwiners between ρ and ρ' is defined by

$$(\rho, \rho') = \{R \in \mathcal{A} : R\rho(A) = \rho'(A)R, \ A \in \mathcal{A}\}.$$

(ρ, ρ') is a linear space. Again if $R \in (\rho, \rho')$ then $R^* \in (\rho', \rho)$. If, in addition, $R' \in (\rho', \rho'')$ then we write $R' \circ R$ to denote the product $R'R$ considered as an element of (ρ, ρ''). Furthermore, we write 1_ρ to denote the unit I considered as an element of (ρ, ρ).

However, $\mathrm{End}\mathcal{A}$ is more than just a C^*–category. Not only can we compose endomorphisms but we have a related product of intertwiners. If $R_1 \in (\rho_1, \rho_1')$ and $R_2 \in (\rho_2, \rho_2')$ then we write $R_1 \otimes R_2$ to denote $R_1\rho_1(R_2) = \rho_1'(R_2)R_1$ considered as an element of $(\rho_1\rho_2, \rho_1'\rho_2')$. Note that $\otimes$ unlike $\circ$ is defined for any pair of intertwiners. $\otimes$ is bilinear and associative and an elementary calculation shows that

$$(R_1 \otimes R_2)^* = R_1^* \otimes R_2^*,$$

$$R_1' \otimes R_2' \circ R_1 \otimes R_2 = (R_1' \circ R_1) \otimes (R_2' \circ R_2),$$

where the last equation, the interchange law, is valid whenever the right hand side is defined. It also illustrates our conventions with regard to the use of brackets: $\otimes$ is to be evaluated before $\circ$ unless the contrary is indicated by brackets. The following special case of the interchange law is often useful:

$$R_1 \otimes 1_{\rho_2'} \circ 1_{\rho_1} \otimes R_2 = R_1 \otimes R_2 = 1_{\rho_1'} \otimes R_2 \circ R_1 \otimes 1_{\rho_2}.$$

If ι denotes the identity automorphism then 1_ι acts as a unit for $\otimes$:

$$1_\iota \otimes R = R \otimes 1_\iota = R.$$

It should be noted that $\otimes$ has the algebraic properties of a tensor product and the algebraic rules we have defined turn our C^*-category into a (strict) *tensor* C^*-category. Note, too, that if we consider just automorphisms of $\mathcal{A}$, we get a tensor C^*-subcategory Aut$\mathcal{A}$.

A further relevant example of a tensor C^*-category is the category $\mathfrak{U}(G)$ of finite dimensional continuous unitary representations of a compact group G. The arrows in the category are again the intertwiners and the product structure is the usual tensor product of representations and intertwiners. The unit is the trivial representation on the complex numbers.

14 States of Relevance

Let us return to the problem of classifying in some way the states of physical relevance. This is a difficult problem and we must be content with some modest steps. First, we have states of particular physical significance such as vacuum states, ground states and equilibrium states. We have agreed that any normal state of the corresponding GNS representation should be regarded as being physically relevant. If these were all the physically relevant states we would have essentially solved our problem. However, this is not the case and even if we restrict ourselves to states of relevance to elementary particle physics, these are not necessarily normal states of the vacuum representation because they might correspond to non-zero baryon or lepton number.

Hence a first step to the classification problem would be to concentrate on the states of relevance to elementary particle physics. Without wishing to give a precise technical definition, the characteristic property of such states should be that they become like the vacuum state at large spacelike separations. This is the theoretical counterpart to pumping out to obtain a high vacua and shielding from the effects of cosmic rays with large quantities of concrete.

Let us concentrate, therefore, on finding and classifying the states of physical relevance associated with a given vacuum state ω_0. This is what the theory of superselection sectors is about. We are still dealing with a large set of states - the normal states of the vacuum sector alone are parametrized by the density matrices on $\mathcal{H}_0$ - so that we should try to define equivalence classes. We should also take account of the fact that forming convex combinations is a rather trivial operation. This can be done by first considering just pure states. Now the pure states of $\mathfrak{A}$ are precisely the vector states of the irreducible representations. Thus there is a very natural equivalence relation staring one in the face. Two pure states are said to be equivalent if they are vector states of the same irreducible representation. The equivalence classes are called sectors. Since there are many sectors of no physical relevance, I reserve the term superselection sector for sectors of relevance to elementary particle physics. Coherent superposition is only possible within the same sector. In view of these remarks, it is natural to shift the emphasis from states of physical relevance to representations of physical relevance.

For completeness, I should remark that if we want to go beyond elementary particle physics and include statistical physics with its equilibrium states then it is not possible to begin with irreducible representations and pure states. Instead you need factorial states, i.e. states where the GNS representation π has the property that

$$\pi(A)' \cap \pi(A)'' = \mathbb{C}I.$$

These states still have the property of describing one particular physical situation rather than some sort of mixture. In the case of equilibrium states these would be the pure phases.

15 Charges in Particle Physics

The theory of superselection sectors has its origins in elementary particle physics where the basic experiment is a scattering experiment. In such an experiment, two particles are made to collide and the outgoing particles are detected with an appropriate particle detector. One is not only interested in the momenta of the particles involved but also in the charges carried by the particles. Here the word 'charge' is being used in the generic sense and refers not only to electric charge but also the baryon number and the three lepton numbers. These would seem to be the only charges in nature and their importance in analyzing scattering experiments is that for each of the charges, the total charge is an invariant of the scattering process. Thus if you know the incoming partcles, there are already strong constraints on the nature of the outgoing particles.

Other types of charges have also been considered and I limit myself to one characteristic example, isospin (the total isospin rather than a component of the isospin). Here one imagines a world where there are only strong interactions and the isospin is a conserved quantity. As there are no electromagnetic interactions, a neutron cannot be distinguished from a proton and we have a single particle, the nucleon. What makes this example interesting is the more complicated law for adding isospin.

The goal of the theory of superselection sectors was to understand the nature and properties of charge. Charge was conceived of as labelling the equivalence classes of certain representations of the observable net. The term sector refers to the equivalence class of an irreducible representation. What is expected is that charges can be added, to every charge there is a conjugate charge reflecting the particle–antiparticle symmetry. Every sector should have an associated statistics describing the symmetry of the multiparticle wavefunctions of particles carrying the charge in question. The results of the theory will be summarized in the next section. They showed what could be achieved with algebraic quantum field theory and a few simple hypotheses.

In particular, we see that a charge like isospin is the generic case. The nucleon is a para–Fermion of order 2. The results are also well illustrated by the specific (free field) models at our disposal.

16 The Selection Criterion I

In this section, we wish to formulate a precise mathematical problem. It is here that physics enters and it is above all here that your critical attention is needed. There will be new developments in superselection theory in the future and the crucial step in finding new applications is to pose the right problem and find the right structural assumptions. The mathematical machinery then runs on its own and gives you the results.

The term superselection sector refers to the equivalence class of an irreducible representation and our physical intuition is that these classes correspond to the presence of different charges. One expects that these charges can be added and that every charge has a conjugate charge corresponding to the particle–antiparticle symmetry. To confirm this intuition, we need a criterion singling out just the right class of representations. Unfortunately, no such criterion is known. Instead, there are natural criteria singling out subclasses of representations where the structure of the corresponding charges can be analyzed.

The first and simplest such criterion [35] singles out localized charges: a representation π should satisfy

$$\pi \upharpoonright \mathcal{O}' \simeq \pi_0 \upharpoonright \mathcal{O}', \quad \mathcal{O} \in \mathcal{K}.$$

In other words, for each double cone $\mathcal{O}$ there is a unitary U with

$$U\pi(A) = \pi_0(A)U. \quad A \in \mathfrak{A}(\mathcal{O}_1), \quad \mathcal{O}_1 \subset \mathcal{O}'.$$

If we think of the states of π as differing from those of π_0 by the presence of a charge. The selection criterion says that this charge can be localized in any $\mathcal{O} \in \mathcal{K}$ in the sense that there is a normal state of π coinciding with the vacuum state on $\mathfrak{A}(\mathcal{O}')$.

Assuming duality in the vacuum representation π_0 together with the Borchers property, the analysis of [35] showed that the selected representations were closed under subrepresentations and countable direct sums and, together with their intertwiners, formed a tensor C^*–category admitting a natural symmetry (in more than 2 spacetime dimensions). This symmetry implies that every sector has an associated statistics describing the symmetry of the multiparticle wavefunctions of particles carrying the charge in question. For an irreducible representation, the permutation symmetry is characterized by a parameter, λ, the *statistics parameter*,

$$\lambda \in \{0\} \cup \{\pm \frac{1}{d} : d \in \mathbb{N}\}.$$

The possible statistics are thus classified by a sign, the *statistical phase*, distinguish-ing para–Bose from para–Fermi statistics and a dimension d, the *statistical dimen-sion*, giving the order of the parastatistics. Ordinary Bose and Fermi statistics corre-spond to $d = 1$ [35]. The case of infinite statistics, $d = \infty$, was later ruled out by a physically natural additional hypothesis [41]. In two spacetime dimensions, where the braid group replaces the permutation group, the statistics phase can be a complex number.

Every representation with finite statistics has a conjugate representation, unique up to unitary equivalence. The conjugate of an irreducible representation π is deter-mined up to equivalence as the unique irreducible represention $\bar{\pi}$ such that $\pi \otimes \bar{\pi}$ contains π_0 as a subrepresentation.

Prior to this, Borchers [10] had investigated representations satisfying the stron-ger condition

$$\pi \upharpoonright \mathfrak{A}(\mathcal{O})^c \simeq \pi_0 \upharpoonright \mathfrak{A}(\mathcal{O})^c, \quad \mathcal{O} \in \mathcal{K},$$

in the erroneous belief that this followed from positivity of the energy. Here $\mathfrak{A}(\mathcal{O})^c := \mathfrak{A}(\mathcal{O})' \cap \mathfrak{A}$ is the relative commutant. He showed that these represen-tations have ordinary Bose or Fermi statistics $d = 1$, and consequently ruled out parastatistics. Haag's lack of belief in this claim motivated [35].

The above selection criterion is too restricitive to cover all situations of interest in elementary particle physics. For example, it obviously does not cover the case of electric charge because by Gauß' Theorem, we can measure a localized electric charge by measuring the total flux of the electric field through the base of any suffi-ciently large double cone. On the other hand, the selection criterion is natural and can be easily analyzed and turns out to correspond, in a sense that can be made precise, to the sectors obtained from the vacuum by a net of fields obeying normal Bose–Fermi commutation relations.

It remained to Buchholz and Fredenhagen [26] to derive a selection criterion from hypotheses on the energy–momentum spectrum. But their hypotheses were stronger: the theory should have a mass gap and there should be massive 1–particle state in the sector in question. They concluded that

$$\pi \upharpoonright \mathfrak{A}(\mathcal{C})^c \simeq \pi_0 \upharpoonright \mathfrak{A}(\mathcal{C})^c,$$

for all spacelike cones $\mathcal{C}$, where a spacelike cone is a translate of $\{\lambda \mathcal{O} : \lambda > 0\}$ and $\mathcal{O}$ is a double cone with vertices satisfying $x \cdot x = -1$. This condition implies the more geometric condition

$$\pi \upharpoonright \mathcal{C}' \simeq \pi_0 \upharpoonright \mathcal{C}',$$

but nothing seems to be known about whether these last two criteria are equivalent. This result is very important because it says that, under the above conditions, the worst that can happen is localization in spacelike cones. Otherwise, Buchholz and Fredenhagen derived the analogues of the results of [35] with one important excep-tion. Spacetime dimension ≥ 4 is needed to be sure of getting a permutation symme-try. In dimension 3, as in dimension 2 for strictly localized charges, in general, one just gets a braiding.

Attempts to classify braid statistics have led to some interesting results [44], [62], [69]. They apply to the case where the statistics operator $\varepsilon(\rho, \rho)$ has just two or three eigenvalues. No general result was in sight when these investigations were abandoned. The basic problem would seem to be the complicated structure of the set of representations of the braid group on an infinite number of strings $\mathbb{B}_\infty$ as compared with the case of the permutation group $\mathbb{P}_\infty$. These remain the basic papers on algebraic field theory in low spacetime dimension. Some confident newcomer might be able to improve on these results.

Despite these rather complete results on the structure of sectors singled out by these selection criteria, one important question remained open. The tensor C^*–category of transportable localized endomorphisms with finite statistics and their intertwiners seemed to have all the properties of a tensor C^*–category of finite dimensional unitary representations of a compact group. But was it always such a category? It seemed to be the case in practice. This proved to be a problem of duality for compact groups and a positive response was finally given in [38]. Interestingly enough, a very similar result for the tensor category of finite dimensional representations of an algebraic group was proved shortly afterwards, motivated by algebraic geometry [30]. The result of [38] meant that, in the absence of infinite statistics, charges and their associated structure can be alternatively described by the equivalence classes of finite–dimensional unitary repesentations of a compact group, the gauge group, with a $\mathbb{Z}_2$–grading corresponding to the sign of the statistics [39]. Furthermore, the observable net is contained as a subset of a canonical field net with Bose–Fermi commutation relations. The irreducible vacuum representation of this canonical field net, when restricted to the observable net, decomposes into a direct sum of irreducible representations where each sector appears with a multiplicity equal to its statistical dimension [39].

Comparatively little work has been devoted to showing the absence of superselection sectors for a certain class of models, or more generally, to showing that a given class of superselection sectors in fact exhausts all sectors [40], [73], [65], [60].

17 Charges of Electromagnetic Type

The selection criteria discussed in the last section are too restrictive to cover all situations of interest in elementary particle physics. For example, the first criterion obviously does not cover the case of electric charge because, by Gauß' Theorem, we can measure a localized electric charge by measuring the total flux of the electric field through the base of any sufficiently large double cone. However, even the second criterion is violated: when the electric charge is non–zero there is a non–zero asymptotic Coulomb field specific to the unitary equivalence class of the representation π under consideration. If the asymptotic Coulomb field is non–zero for some direction within $\mathcal{C}'$, π_0 and π will not be equivalent on restriction to $\mathcal{C}'$.

The electric charge, therefore, presents a challenge to sector theory. It was, after all, one of the initial examples of a superselected quantity [82]. Sadly, algebraic quantum field has so far failed here, despite some interesting ideas [20], [46], [23], [24], [61]. One reason for this lack of success is that quantum electrodynamics does not exist outside of perturbation theory so that we cannot look to see what selection criterion is satisfied by representations of non–zero electric charge. Nor are there any other physically relevant models that might give us a clue as to the right selection criterion. But the difficulties are not confined to divining the selection criterion. The electric charge is the parameter of interest in quantum electrodynamics and takes on positive and negative integral values. It should label the equivalence classes of irreducible objects in a suitable tensor C^*–category but it does not parametrize the equivalence classes of physically relevant irreducible representations because to each electric charge there are a myriad of inequivalent representations differing in the infrared clouds or, put another way, in their asymptotic Coulomb fields. To arrive at the electric charge, one either has to take infrared equivalence classes or to make an appropriate choice of one representation for each value of the electric charge. However, I am currently optimistic that new methods will yield a good qualitative understanding of charges of electromagnetic type.

18 Solitonic Sectors

It follows from the selection criteria that states of a sector tend to the vacuum state for measurements tending spacelike to infinity. In a two dimensional spacetime one must really distinguish between tending spacelike to infinity to the left and spacelike to infinity to the right. If one uses the left spacelike complement in the selection criterion, we may find that the states of a sector, although tending to the vacuum state for measurements tending spacelike to infinity to the left, tend to a different vacuum state for measurements tending spacelike to infinity to the right. Such sectors are called solitonic. The genesis of such sectors in algebraic quantum field theory from cohomology was pointed out in [72] and non–trivial examples were analyzed in [45]. In fact, even if one restricts oneself to non–solitonic sectors of the observable net, the induced representation of the field net can prove to be solitonic.

In [42], Fredenhagen pointed out that endomorphisms could not be used to describe solitonic sectors. They should be regarded as mapping from, say, the vacuum on the left to that on the right and he proposed a law of composition of sectors taking this into account and making use of the translation covariance of the representations involved. A more detailed account can be found in [43].

A role in the analysis is played by nets of von Neumann algebras over wedges in a vacuum representation. There are two such nets, one over right wedges and one over left wedges, here referred to for brevity as the right net and left net, respectively.

The natural selection criterion is to single out representations equivalent to a vacuum representation, its left vacuum, on the spacelike complement of right wedges and to a possibily different vacuum, its right vacuum, on the spacelike complement of left wedges. When duality for wedges holds, the representation gives rise to a right morphism from the right net in the right vacuum representation to the right net in its left vacuum representation and a left morphism from the left net in its left vacuum representation to a left net in its right vacuum representation.

Thus the composition of right morphisms is not always defined. Vacuum representations, right morphisms and their intertwiners form a $2 - C^*$–category, rather than a tensor C^*–category [76]. If we use left morphisms, instead, the resulting $2 - C^*$–category arises by dualizing with respect to 1–arrows.

In fact, the formalism under discussion is quite general and in no way restricted to two spacetime dimensions. The sets of left and right wedges can be replaced by suitable sets of unbounded regions such as the set of translates of a spacelike cone. The resulting selection criterion can then be analyzed using the cohomological techniques of subsequent sections. The formalism provides a natural approach to charges that can be localized only in unbounded regions and might prove suitable for solitons in higher dimensions.

Müger [66] analyzed sectors of the observable net that lift to solitonic sectors of the field net in terms soliton automorphisms of the field net that is automorphisms that are one gauge automorphism on the left spacelike complement of a double cone and another on the right.

19 Scattering Theory

The aim of scattering theory is to bring to light the particle aspects and in particular to construct the many–particle states out of the 1–particle states. Scattering theory played an important part in the genesis of algebraic quantum field theory. In particular the need to treat the case of composite particles where the basic fields did not create 1–particle states from the vacuum [51] paved the way to the new theory. As the relevant limiting procedures were thus identified in the Wightman formalism before the advent of algebraic quantum field theory, little had to be changed to adapt scattering theory to the new environment. However to underpin the new philosophy that the local observables sufficed for a complete description, Araki and Haag [4] showed that collision cross section could, in principle, be expressed just in terms of the local observables.

A variant on the usual formalism for scattering theory, developed in [35], was motivated by the need to do scattering theory when the 1–particle states are not in the vacuum sector. This variant is so simple that it retains its interest even though we now know that there is a canonical field net allowing a conventional scattering theory. Otherwise Buchholz made an important technical advance by showing how to do scattering theory in the presence of zero mass particles [18], [19]. However the concept of particle in physics is subtle and has many facets. Wigner's idea of linking the concept of stable particle in the sense of scattering theory to irreducible subrepresentations of the physically relevant unitary representation of the covering group of the Poincaré group was only one important clarifying step. The notion of particle weight [28], [68], to be thought of as improper momentum eigenstates of a particle, also covers the case of infraparticles such as the electron, where the inevitable presence of soft photons means that there are no 1–electron states with a mass equal to the mass of the electron.

An interesting and physically well motivated potential application of superselection theory is to study charge and particle structure in the scaling limit [21]. In this limit colour should emerge as a charge akin to isospin and quarks should be para–Fermions of order three.

20 Modular Theory

With the advent of modular theory, I tried to see whether the theory would have implications for quantum field theory. The local von Neumann algebras do have the vacuum as a common cyclic and separating vector. Hence each such algebra has a canonical modular operator and a canonical modular conjugation. However there were, in general, no obvious relations between these modular data for different regions. I rashly concluded that modular data would have little relevance to algebraic quantum field theory. I was soon to be proved wrong. The work of Bisognano and Wichmann [7], [8] came as a surprise and related the modular operator of the von Neumann algebra of a wedge region to the generator of the 1–parameter group of boosts leaving the wedge invariant and its modular conjugation to the reflection in the line forming the edge of the wedge. I had been looking at the modular operators for the von Neumann algebra of a double cone where the geometric interpretation of the modular operator and conjugation fails except for zero–mass theories [56].

Bisognano and Wichmann also showed that duality held for wedge regions whenever there are Wightman fields generating the net of observables. This led Wichmann to define the von Neumann algebra of a double cone $\mathcal{O}$ as the intersection of the von Neumann algebra of all wedges containing $\mathcal{O}$. This coincides with the usual definition whenever duality holds for double cones otherwise it yields the dual net. More recently, Mund [67] has derived the Bisognano–Wichmann Theorem and the CPT Theorem for theories admitting a complete interpretation in terms of massive particles.

The applications of modular theory to algebraic quantum field theory remain somewhat mysterious. The geometric interpretation of the modular data for a wedge seems very particular to a wedge and the idea of an extended region like a wedge playing a significant role contrasts with the philosophy that the net, being in practice additive, is determined by the von Neumann algebras of small double cones just as Wichmann's proposal to start with the von Neumann algebras of wedges leaves additivity as a mystery.

This proposal of Wichmann has led to a new approach to constructing the net of von Neumann algebras for free field models based on ideas of Schroer and even to a net for a free massless particle with continuous helicity [16].

Although the original result of Bisognano and Wichmann used the existence of Wightman fields generating the net of von Neumann algebras, their theorem in time got turned into a principle. Thus Guido and Longo [48] derive a PCT and a Spin and Statistics Theorem where the principal assumption *modular covariance* was that the modular operator associated with the von Neumann algebra of a wedge W and the vacuum state induced on the net the automorphisms implementing the one parameter group of Lorentz boosts leaving the wedge invariant,

$$\Delta_W^{it} \mathfrak{A}(\mathcal{O}) \Delta_W^{-it} = \mathfrak{A}(\Lambda_W(t)\mathcal{O}), \quad \mathcal{O} \in \mathcal{K}.$$

A PCT Theorem in four spacetime dimensions asserts the existence of an antiunitary operator Θ with

$$\Theta \mathfrak{A}(\mathcal{O}) \Theta^{-1} = \mathfrak{A}(-\mathcal{O}), \quad \mathcal{O} \in \mathcal{K}$$

and mapping each sector with finite statistics onto the conjugate sector.

An even more radical proposal was *geometric modular action* [25], postulating the geometric action of the modular conjugation associated with the von Neumann algebras of wedgelike regions and designed as a selection criterion for interesting states of a quantum field theory in curved spacetime. The relation with the principle of modular covariance is discussed in [25].

21 Conformal Field Theory

In general, conformal field theory can be understood as a field theory where the conformal group replaces the Poincaré group as a covariance group. Here it will be understood in a rather more restrictive sense where we have a net $I \to \mathfrak{A}(I)$ of von Neumann algebras defined on open intervals of the circle whose complements have non–empty interiors. The covariance group is here $PSL(2, \mathbb{R})$, the group of conformal transformations on the complex plane, preserving the orientation and leaving the unit circle globally invariant. Of course, in a covariant theory, we have a continuous unitary representation U of its covering group with

$$U(g)\mathfrak{A}(I)U(g)^* = \mathfrak{A}(gI),$$

for each element g of the covering group and each interval I. Positivity of the energy here means positivity of the generator of the rotation subgroup.

The mathematical analogy with algebraic quantum field theory in Minkow–ski space is clear, the relation with physics less so. However, if we are willing to start from a two dimensional Minkowski space and a theory that is chirally invariant then the theory splits into a tensor product of a part moving to the left and a part moving to the right. These may be regarded as quantum field theories defined on the two lines making up the light cone in two dimensions. Compactifying the lines then gives us conformal theories defined on the circles.

In the context of conformal field theory, Guido and Longo [49] have proved the analogues of many of the results in Minkowski space and others, notably the equivalence of local and global intertwiners: if we take endomorphisms ρ, σ of $\mathfrak{A}$ with finite index localized in an interval I then

$$(\rho_I, \sigma_I) = (\rho, \sigma),$$

where ρ_I and σ_I denote the restrictions of ρ and σ to $\mathfrak{A}(I)$. They also showed that a conjugate for an endomorphism of finite index localized in I could be obtained by conjugating with the modular conjugation associated with $\mathfrak{A}(\tilde{I})$, where I and $\tilde{I}$ are open intervals with disjoint closures. Their goal however was to prove an analogue of the spin and statistics theorem, namely that

$$\kappa_\rho = s_\rho,$$

where κ_ρ is the statistics phase of the sector and s_ρ is the univalence labelling the projective representation of $PSL(2, \mathbb{R})$ associated with the sector. Related earlier work can be found in [44] and in the context of a field theory with localization in spacelike cones in a three dimensional Minkowski space in [47].

Kawahigashi, Longo and Müger [60] look at the inclusions associated with a disjoint union of intervals in a conformal field theory. They show that if the index of the inclusion is finite then it coincides with the global index of the sectors and that the associated tensor category is modular.

Another important recent result is that of Kawahigashi and Longo [59] who classify the diffeomorphism covariant local nets of von Neumann algebras on the circle with central charge less than one.

Conformal field theory also provides the most interesting examples where the sector structure has been explicitly computed. I refer here to the loop group models intensively studied by A. Wassermann [81] and others [78].

22 Curved Spacetime

The remainder of this contribution illustrates current research in algebraic quantum field theory by developing sector theory for quantum field theory on a globally hyperbolic spacetime in detail.

In principle, there can be different motives for passing to curved spacetime. One might, for example, see this as a first step towards quantum gravity, but as this first step is so minute and the second never taken, little good can result. A second motive might be to look for new phenomena within the wider framework. To date, relatively few new phenomenon have been discovered. There is the Casimir effect and above all the Hawking effect, the flagship of quantum field theory on curved spacetime. Aspects of the Hawking effect have been treated in algebraic quantum field theory [63], [77]. It is conceivable that curved spacetime yields new mechanisms for generating sectors. None have been found, but the subject is still in its infancy.

The motive here is to check our understanding of the underlying physics. Many of our basic physical laws such as locality and duality and even the Reeh–Schlieder property should remain valid on curved spacetime and have similar consequences. Others such as Poincaré or even translation covariance, positivity of the energy or the concept of a vacuum obviously refer to the special features of Minkowski space and need to be modified or even abandoned. However, apart from the vacuum, these play only a secondary role in sector theory, and, as we shall see, we do not even need a concept of vacuum in the strict Minkowskian sense. Hence, we expect that appropriate analogues of the basic results in Minkowski space can be found, at least for spacetimes sufficiently similar in nature to Minkowski space. Potential geometric obstacles arise in other classes of spacetime such as those with compact Cauchy surfaces. The results presented here indicate that these obstacles can be overcome.

An innovative mathematical aspect in the following is the use of general partially ordered sets with a causal disjointness relation, an approach first publically advocated in [6] and put to good use in [50]. In fact, the points of spacetime never enter into the theory but only some suitable set of subsets of spacetime ordered under inclusion and endowed with the causal disjointness relation derived from the metric. The motivation here was not additional generality but the wish to put into evidence the geometric properties actually needed in sector theory and to separate this relatively simple task from the potentially more difficult task of deducing them from the geometric properties of spacetime. But the innovation bore other fruit as it proved convenient, or even essential, to use other partially ordered sets and causal disjointness relations in the course of the reasoning.

The results presented here grew out of Sec. 3 of [50], where in addition there are two sections devoted to the connection between spin and statistics in curved spacetime. They would form interesting complementary reading because spacetime geometry enters in a more essential way, these results being valid only in special classes of spacetime.

The present discussion includes some interesting new results indicating how to treat reducible reference representations, developing in more detail what happens when the index set changes, in particular, when passing to a Cauchy surface and showing how to define the left inverses needed to classify the statistics. This goes some way to extending the results on sector structure in [50] to the case of globally hyperbolic spacetimes with compact Cauchy surfaces.

Whilst the methods of perturbation theory do not fall within the scope of these lecture notes, the reader may be served by my drawing his attention to recent advances extending perturbation theory to curved spacetime and inspired by algebraic quantum field theory [13], [14], [15].

23 Partially Ordered Sets

As I have already made clear, a characteristic feature of the following discussion is that the index set $\mathcal{K}$ of our net will be treated as an abstract partially ordered set, equipped with a relation $\perp$ of causal disjointness. We will do a large part of superselection theory in this context, exhibiting the relevant conditions on $\mathcal{K}$, $\perp$. A relation $\perp$ of causal disjointness on $\mathcal{K}$ is supposed to satisfy

a) $\mathcal{O}_1 \perp \mathcal{O}_2 \Rightarrow \mathcal{O}_2 \perp \mathcal{O}_1$.

b) $\mathcal{O}_1 \subset \mathcal{O}_2$ and $\mathcal{O}_2 \perp \mathcal{O}_3 \Rightarrow \mathcal{O}_1 \perp \mathcal{O}_3$.

c) Given $\mathcal{O}_1 \in \mathcal{K}$, there exists an $\mathcal{O}_2 \in \mathcal{K}$ such that $\mathcal{O}_1 \perp \mathcal{O}_2$.

We write $\mathcal{O}^\perp := \{\mathcal{O}_1 \in \mathcal{K} : \mathcal{O}_1 \perp \mathcal{O}\}$.

Of course, this structure is extremely general and may have nothing to do with sets in a curved spacetime. For example, we could take $\mathcal{K}$ to be the set of projections in a unital C^*–algebra with the usual order relation on projections and orthogonality as the relation of causal disjointness. We could also take $\mathcal{K}$ to be the set of closed real subspaces of a complex Hilbert space, ordered under inclusion. Here there are two relations of causal disjointness that play a role in the theory of free fields. For Bose fields $M_1 \perp M_2$ means that M_1 and iM_2 are orthogonal. Whilst for Fermi fields, it means that M_1 and M_2 are orthogonal. These two examples are of interest in that each has a canonically associated net of von Neumann algebras [2] and the nets of von Neumann algebras for free fields are just pullbacks for the appropriate morphisms of partially ordered sets with a causal disjointness relation [3].

Here, however, are some characteristic examples that have played a role in algebraic quantum field theory and serve here to illustrate various steps in the development of the theory of superselection sectors.

a) $\mathcal{K}$ is the set of open double cones in Minkowski space ordered under inclusion. A double cone in Minkowski space is a set of the form $a_+ + V^- \cap a_- + V^+$ defined by vertices a_+ and a_- with $a_+ - a_-$ positive timelike, where $V^\pm$ denotes the future and past open light cone. $\mathcal{O}_1 \perp \mathcal{O}_2$ is traditionally written $\mathcal{O}_1 \subset \mathcal{O}_2'$, where, to be precise, $\mathcal{O}_2'$ denotes the interior of the spacelike complement of $\mathcal{O}_2$.

b) $\mathcal{W}$ is the set of open wedges in Minkowski space ordered under inclusion. A wedge is a Poincaré transform of $\{x : x^1 > x^0\}$. $W_1 \perp W_2$ is again $W_1 \subset W_2'$. This is the partially ordered set behind the Bisognano–Wichmann Theorem[8], cf. Sec. 19.

c) $\mathcal{C}$ is the set of proper open intervals of the circle ordered under inclusion. $\perp$ is the relation of disjointness. This is the partially ordered set of chiral field theory, cf. Sec. 20.

d) Consider a spacelike hyperboloid in Minkowski space $\{n : n.n = -1\}$ with the causal structure inherited from Minkowski space. Let $\mathcal{D}$ denote the partially ordered set of open double cones on the hyperboloid defined by analogy with a) above. For more details see the appendix of [39]. $D_1 \perp D_2$ is again $D_1 \subset D_2'$. The spacelike hyperboloid is used to represent spacelike infinity as becomes clear from the next example.

e) Let D be a double cone on the above spacelike hyperboloid, then the spacelike cone S in Minkowski space with apex a and base D is the set of points

$$S := \cup_{\lambda>0}\{a + \lambda n : n \in D\}.$$

S is the set of open spacelike cones in Minkowski space ordered under inclusion and $\perp$ is derived as in a) and b) from the notion of causal complement. The results of [26] make this a natural partially ordered set to consider for theories with only massive particles.

f) Take $\mathcal{K}$ to be the set of regular diamonds in a globally hyperbolic spacetime ordered under inclusion. A regular diamond is the causal closure of a regular open subset of a Cauchy surface and $\perp$ is the relation of causal disjointness in the globally hyperbolic spacetime. For more details, see [50].

Although f) is the only example referring to curved spacetime, analogues of wedge regions also seem to play an important role in curved spacetime. They appear in the discussion of one version of the spin and statistics theorem in [50]. They play a basic role in algebraic holography [70] and are used as the basic class of regions when defining a quantum field theory in algebraic terms [25]. The same abstract partially ordered set with causal disjointness relation can appear as a set of wedges in different curved spacetimes and be use to transplant a field theory from one curved spacetime to the next.

The theoretical reason for working in terms of $\mathcal{K}, \perp$ is that the geometry of curved spacetime does not play a direct role in quantum field theory. It only enters through the chosen set $\mathcal{K}$ of open subsets ordered under inclusion and equipped with the natural causal relations derived from the metric. Using $\mathcal{K}, \perp$ puts clearly in evidence the most important part of the structure really used in field theory.

The task of deriving the principal results of superselection theory for field theories on a globally hyperbolic spacetime convinced me of the utility of writing proofs in terms of partially ordered sets. Had the original proofs in superselection theory been phrased in terms of $\mathcal{K}, \perp$, the task in curved spacetime would have reduced to verifying the necessary properties of $\mathcal{K}, \perp$. As it was, the old proofs had to be looked over to make sure that no specific properties of double cones were being used.

But the real benefits only emerge in the course of fulfilling this task: as $\mathcal{K}, \perp$ has no fixed meaning, it became natural to change its meaning as a means of overcoming the obstacles that presented themselves in the course of this work.

The continuing interest in superselection theory for quantum field theory on the circle provided further motivation. Of course, when causal disjointness is replaced by disjointness as in c), many other examples can be generated by replacing the circle by some topological space and the open intervals by a base of open sets.

You will notice that in the examples the elements of $\mathcal{K}$ are open subsets of some topological space of some simple geometric nature. In a), c) and e), we have a base for the topology. In b) and d), only a subbase. When we pass to partially ordered sets, we should not forget their origins but should treat them in a way analogous to topological spaces.

We first consider their cohomology. Cohomology has applications to many branches of mathematics; there is a cohomology of groups, of Lie algebras, of algebras and of topological spaces to name only the better known examples. It is not easy for the beginner to grasp the underlying common features and I do not have the time to give much more than a brief introduction and then pass on to the specific definitions we shall need. Underlying cohomology is what is called the simplicial category Δ^+ which can be realized in various ways. The simplest way is to take the objects of Δ^+ to be the finite ordinals, $n = \{0, 1, ..., n-1\}$ and to take the arrows to be the monotone mappings. All these monotone mappings are compositions of two particular simple types of mapping; the injective monotone mappings from one ordinal to the succeeding ordinal denoted $d_i^n : n \rightarrow n + 1$, where the superscript is usually omitted and the subscript tells you which element is omitted in the image; and the surjective monotone mappings from one ordinal to the preceding one denoted $s_i^n : n + 1 \rightarrow n$, where the subscript tells you which element is repeated in the image. The following identities allow you to compute effectively

$$d_i d_j = d_{j+1} d_i, \; i \leq j : \qquad s_j s_i = s_i s_{j+1}, \; i \leq j :$$

$$s_j d_i = d_i s_{j-1}, \; i < j : \qquad s_j d_j = s_j d_{j+1} = 1; \qquad s_j d_i = d_{i-1} s_j, \; i > j + 1.$$

Actually, each monotone map can be factorized uniquely as the composition of a surjective monotone map and an injective monotone map.

We also have a geometric interpretation of the simplicial category: the ordinal $n + 1$ is replaced by Δ_n, the standard n simplex;

$$\Delta_n := \{(\lambda^0, \lambda^1, \ldots, \lambda^n) : \lambda^0 \geq 0, \lambda^1 \geq 0, \ldots, \lambda^n \geq 0, \sum_i \lambda^i = 1\},$$

and $\lambda^0, \lambda^1, \ldots, \lambda^n$ are the barycentric coordinates. Δ_0 is of course a point and Δ_1 an interval, etc. The affine mappings $f_* : \Delta_n \rightarrow \Delta_m$ are derived from monotone mappings on the indices

$$f_*(\lambda^0, \lambda^1, \ldots, \lambda^n) = (\mu^0, \mu^1, \ldots, \mu^m), \; \mu^j = \sum_{f i = j} \lambda^i.$$

To arrive at the singular cohomology of topological spaces one proceeds as follows. Let M be a topological space and let $\Sigma_n(M)$ denote the set of continuous linear mappings from Δ_n to M. An element of $\Sigma_n(M)$ is called a singular n–simplex. Thus a 0–simplex is just a point of M and a 1–simplex is a path in M. Given f as above, there is a corresponding mapping $f^* : \Sigma_m(M) \rightarrow \Sigma_n(M)$ defined by $f^*(c) = c \circ f_*$ and $(fg)^* = g^* f^*$. This gives us a contravariant functor to the category of sets, i.e. a simplicial set. One writes ∂_i for d_i^* and σ_i for s_i^* and we obviously have

$$\partial_j \partial_i = \partial_i \partial_{j+1}, \; i \leq j; \qquad \sigma_i \sigma_j = \sigma_{j+1} \sigma_i, \; i \leq j;$$

$$\partial_i \sigma_j = \sigma_{j-1} \partial_i, \; i < j; \qquad \partial_j \sigma_j = \partial_{j+1} \sigma_j = 1; \qquad \partial_i \sigma_j = \sigma_j \partial_{i-1}, \; i > j + 1.$$

Note that ∂_i has a simple geometric interpretation; it associates to an n–simplex the $n - 1$–simplex which is its i–th face. σ_i associates to an n–simplex the degenerate $n + 1$–simplex got by repeating the i–th vertex.

We can proceed analogously in the case of partially ordered sets, regarding the standard n–simplex as the partially ordered set of its subsimplices ordered under inclusion. In this interpretation, the arrow $f : n+1 \rightarrow m+1$ is given by $f_* : \Delta_n \rightarrow \Delta_m$, where $f_*(S) := f(S)$, for any subsimplex S of Δ_n. Again $f_* g_* = (fg)_*$. An n–simplex of a partially ordered set $\mathcal{P}$ is an order–preserving map $\Delta_n \rightarrow \mathcal{P}$, where Δ_n is considered as a partially ordered set. $\Sigma_n(\mathcal{P})$ or just Σ_n will denote the partially ordered set of n–simplices of $\mathcal{P}$ with the pointwise ordering. Given $f : n+1 \to m+1$ as above, we define $f^* : \Sigma_m(\mathcal{P}) \to \Sigma_n(\mathcal{P})$, by $f^*(c)(S) = c(f(S))$. Then $g^* f^* = (fg)^*$ so we again have a contravariant functor from Δ^+ to the category of sets.

Thus a 0–simplex a of the partially ordered set $\mathcal{P}$ is just an element of $\mathcal{P}$ and a 1–simplex b consists of two 0–simplices denoted $\partial_0 b$ and $\partial_1 b$ contained in a third element $|b|$ of $\mathcal{P}$ called the *support* of b.

23.1 Lemma *Let $\mathcal{P}$ be a directed partially ordered set then $\Sigma_*(\mathcal{P})$ admits a contracting homotopy.*

Proof. We must define $h : \Sigma_n(\mathcal{P}) \to \Sigma_{n+1}(\mathcal{P})$ with $\partial_0 h = id$, $\partial_i h = h \partial_{i-1}$, $i = 1, 2, \ldots, n+1$. We do this by induction and, to simplify notation, let us write, for example, $c(012)$ for $c(\{0, 1, 2\})$. We pick an arbitrary element a_0 of $\mathcal{P}$ as base point and define ha for $a \in \Sigma_0$ by setting $ha(0) = a_0$, $ha(1) = a(0)$ and picking $ha(01)$ arbitrarily, subject to $a_0, a(0) \leq ha(01)$. Now suppose h has been defined on $\Sigma_p(\mathcal{P})$ for $p < n$ so as to satisfy the required relations. Then we have only to define $hc(01 \ldots n+1)$ because we already know that the faces of hc must be c, $h\partial_0 c, h\partial_1 c, \ldots, h\partial_n c$ and these faces are compatible with each other by virtue of the assumed relations between h and the ∂_i. Hence we may pick for $hc(01 \ldots n+1)$ an element of $\mathcal{P} \geq c(01 \ldots n), h\partial_0 c(01 \ldots n), \ldots, h\partial_n c(01 \ldots n)$. This is possible since $\mathcal{P}$ is directed. Hence we have extended h to $\Sigma_n(\mathcal{P})$ in such a way that the required relations are satisfied. This completes the proof.

A partially ordered set $\mathcal{P}$ is *connected* if given $a, a' \in \Sigma_0(\mathcal{P})$, there is a path from a' to a in $\mathcal{P}$, i.e. if there exist $b_0, b_1, \ldots, b_n \in \Sigma_1(\mathcal{P})$ with $\partial_0 b_0 = a$, $\partial_1 b_n = a'$ and $\partial_0 b_i = \partial_1 b_{i-1}$, $i = 1, 2, \ldots, n$. Obviously, if $\mathcal{P}$ is not connected, it is a disjoint union of its connected components. These notions are related to topological notions in the following way.

23.2 Lemma *Let $\mathcal{P}$ be a base for the topology of a space M and ordered under inclusion and suppose the elements of $\mathcal{P}$ are open, (non-empty) and path–connected. Then an open subset X of M is path–connected if and only if $\mathcal{P}_X := \{\mathcal{O} \in \mathcal{P} : \mathcal{O} \subset X\}$ is connected.*

Proof. Any two points of X are contained in elements of $\mathcal{P}_X$ so if this is connected and each of its elements are path–connected the two points can be joined by a path in X. Conversely, given $\mathcal{O}_0, \mathcal{O}_1 \in \mathcal{P}_X$, there is a path in X beginning in $\mathcal{O}_1$ and ending in $\mathcal{O}_0$, if X is pathwise connected. Since $\mathcal{P}$ is a base for the topology, it is easy to construct a path in $\mathcal{P}_X$ joining $\mathcal{O}_1$ and $\mathcal{O}_0$.

A subset $\mathcal{S}$ of $\mathcal{P}$ of the form $\mathcal{P}_X$ has the property that $\mathcal{O} \in \mathcal{S}$ and $\mathcal{O}_1 \subset \mathcal{O}$ implies $\mathcal{O}_1 \in \mathcal{S}$. Such subsets are referred to as *sieves*. If $\mathcal{P}$ is a base for the topology of M then a sieve $\mathcal{S}$ is a base for the topology of the open subset $X_{\mathcal{S}} := \cup\{\mathcal{O} : \mathcal{O} \in \mathcal{S}\}$. The connected components of a partially ordered set are sieves, the intersection of sieves is again a sieve.

23.3 Corollary *Under the hypotheses of Lemma 23.2, the connected components of $\mathcal{P}$ are of the form $\mathcal{P}_X$, where X runs over the path–connected components of M.*

Of the characteristic examples a)-e) all are connected except b), d) and e) in space dimension $s = 1$. The following result will prove useful in calculating the connected components of partially ordered sets.

23.4 Lemma *Let $i \mapsto \mathcal{P}_i$ be an order–preserving map from a partially ordered set I to the set of sieves of a partially ordered set $\mathcal{P}$ ordered under inclusion. Suppose that $\mathcal{P} = \cup_{i \in I} \mathcal{P}_i$. Let $\mathcal{C} \subset \mathcal{P}$ and set $\mathcal{C}_i := \mathcal{C} \cap \mathcal{P}_i$ then $\mathcal{C}$ is a union of components of $\mathcal{P}$ or empty if and only if $\mathcal{C}_i$ is a union of components of $\mathcal{P}_i$ or empty for each $i \in I$. If I is connected and $\mathcal{C}_i$ is either empty or a component of $\mathcal{P}_i$, $i \in I$, then $\mathcal{C}$ is either empty or a component of $\mathcal{P}$.*

Proof. If $\mathcal{C}$ is a union of components or empty and $b \in \Sigma_1(\mathcal{P}_i)$ with $\partial_1 b \in \mathcal{C}_i$ then $b \in \Sigma_1(\mathcal{P})$ so $\partial_0 b \in \mathcal{C} \cap \mathcal{P}_i = \mathcal{C}_i$ and $\mathcal{C}_i$ is a union of components or empty. Conversely, if each $\mathcal{C}_i$ is a union of components or empty and $b \in \Sigma_1(\mathcal{P})$ with $\partial_1 b \in \mathcal{C}$, then $|b| \in \mathcal{P}$ for some i. But $\mathcal{P}_i$ is a sieve so $b \in \Sigma_1(\mathcal{P}_i)$ and $\partial_1 b \in \mathcal{C} \cap \mathcal{P}_i$. Since $\mathcal{C}_i$ is a union of components, $\partial_0 b \in \mathcal{C}_i \subset \mathcal{C}$ so $\mathcal{C}$ is a union of components. Now $\mathcal{C}$ is a component or empty, if any given pair $a \in \mathcal{C}_i$ and $a' \in \mathcal{C}_{i'}$ can be joined by a path in $\mathcal{C}$. But I being connected, we may as well suppose i and i' have an upper bound $j \in I$. If $\mathcal{C}_j$ is a component, a and a' can even be joined by a path in $\mathcal{C}_j$, completing the proof of the lemma.

We let $\mathrm{Open}(M)$ denote the set of open subsets of M ordered under inclusion and $\mathrm{Sieve}(\mathcal{P})$ the set of sieves of $\mathcal{P}$, then defining for a open set X of M, $\mu(X)$ to be the set of $\mathcal{O} \in \mathcal{P}$ contained in X, μ is an injective order-preserving map from $\mathrm{Open}(M)$ to $\mathrm{Sieve}(\mathcal{P})$. If we define $\nu(\mathcal{S}) := X_{\mathcal{S}}$, then ν is order-preserving and a left inverse for μ.

We may regard $\mathrm{Sieve}(\mathcal{P})$ as a substitute for $\mathrm{Open}(M)$ even in the absence of a topological space M. Let $\mathcal{P}$ and $\mathcal{Q}$ be partially ordered sets and $f : \mathcal{P} \to \mathcal{Q}$ be an order preserving mapping. If $\mathcal{S}$ is a sieve of $\mathcal{P}$ then its inverse image

$$f^{-1}(\mathcal{S}) := \{\mathcal{O} \in \mathcal{P} : f(\mathcal{O}) \in \mathcal{S}\},$$

is a sieve of $\mathcal{P}$. If $\mathcal{S}_1 \subset \mathcal{S}_2$, then $f^{-1}(\mathcal{S}_1) \subset f^{-1}(\mathcal{S}_2)$ and f^{-1}: $\mathrm{Sieve}(\mathcal{Q}) \to \mathrm{Sieve}(\mathcal{P})$ is an order-preserving map.

If $\mathcal{K}$ is a partially ordered set equipped with a relation $\perp$ of causal disjointness and $\mathcal{S}$ is a sieve of $\mathcal{K}$ then we define the causal complement of $\mathcal{S}$ to be the sieve

$$\{\mathcal{O} \in \mathcal{K} : \mathcal{O} \perp \mathcal{O}_1, \mathcal{O}_1 \in \mathcal{S}\}.$$

If $\mathcal{K}$ is a base of open sets of a topological space M and the relation $\perp$ on $\mathcal{K}$ is induced by a relation $\perp$ on $\mathrm{Open}(M)$ satisfying a) and b) above and which is local in the sense that if $X \in \mathrm{Open}(M)$ and $X \subset \cup_i \mathcal{O}_i$, then $\mathcal{O}_i \perp \mathcal{O}$ for all i implies $X \perp \mathcal{O}$. This condition is obviously satisfied by the relation of causal disjointness on a globally hyperbolic spacetime. It implies that $\mu(X^{\perp}) = \mu(X)^{\perp}$. We also have $\nu(\mathcal{S})^{\perp} = \nu(\mathcal{S}^{\perp})$ for any sieve $\mathcal{S}$ in $\mathcal{K}$, where the causal complement $\mathcal{S}^{\perp}$ of the sieve $\mathcal{S}$ is the intersection of the sieves $\mathcal{O}^{\perp}$ for $\mathcal{O} \in \mathcal{S}$. A set X or a sieve $\mathcal{S}$ is said to be causally closed if $X = X^{\perp\perp}$ or $\mathcal{S} = \mathcal{S}^{\perp\perp}$, respectively.

23.5 Lemma *When restricted to causally closed open sets and sieves, the maps μ and ν are inverses of one another.*

Proof. If S is a sieve and $X := \nu(S)$, then $\mu(X)^\perp = S^\perp$. If S is causally closed, then $X^{\perp\perp} = \nu(S^{\perp\perp}) = \nu(S)$ and X is causally closed. On the other hand, if X is causally closed and we set $S := \mu(X)$, then $S^{\perp\perp} = \mu(X^{\perp\perp}) = \mu(X)$ and S is causally closed. It remains to show that $S = \mu\nu(S)$ if S is causally closed. But, in this case,

$$S \subset \mu\nu(S) \subset \mu\nu(S)^{\perp\perp} = S^{\perp\perp} = S,$$

completing the proof.

The remainder of this section may be omitted on first reading. It is devoted to computing the number of connected components of various partially ordered sets. It will be seen in Sec. 28 how this bears on the problem of determining the number of symmetries or braidings involved in sector theory.

In the course of these investigations, we need to introduce various partially ordered sets derived from $\mathcal{K}, \perp$. We have already defined $\mathcal{O}^\perp$, $\mathcal{O} \in \mathcal{K}$, and, of our characteristic examples a)–e), $\mathcal{O}^\perp, W^\perp, C^\perp, D^\perp$ and $S^\perp$ are connected except for $\mathcal{O}^\perp$ when $s = 1$ which has two components. $C^\perp$ and $W^\perp$ have maximal elements. Next there is the graph $\mathcal{G}^\perp$ of the relation $\perp$:

$$\mathcal{G}^\perp := \{(\mathcal{O}_1, \mathcal{O}_2) : \mathcal{O}_1 \perp \mathcal{O}_2\}$$

with the order inherited from $\mathcal{K} \times \mathcal{K}$ with the product ordering. Then we have already considered $\Sigma_n = \Sigma_n(\mathcal{K})$ as a partially ordered set and we let $\Sigma_1^\perp$ denote the set of 1–simplices b with $\partial_1 b \perp \partial_0 b$ with the induced ordering. We have an obvious order–preserving mapping $\Sigma_1^\perp \rightarrow \mathcal{G}^\perp$.

In practice, $\mathcal{G}^\perp$ has as many components as does $\mathcal{O}^\perp$ and here is a result in this direction.

23.6 Lemma *Suppose $\mathcal{O}_c^\perp$ and $\mathcal{O}_{\bar{c}}^\perp$ are components of $\mathcal{O}^\perp$, $\mathcal{O} \in \mathcal{K}$, such that $\mathcal{O}_1 \in \mathcal{O}_{2,c}^\perp$ if and only if $\mathcal{O}_2 \in \mathcal{O}_{1,\bar{c}}^\perp$ and such that $\mathcal{O}_{1,c}^\perp \cap \mathcal{O}_{2,c}^\perp \neq \emptyset$ for $\mathcal{O}_1, \mathcal{O}_2 \in \mathcal{K}$, then $\mathcal{G}_c^\perp := \{\mathcal{O}_1 \times \mathcal{O}_2 : \mathcal{O}_2 \in \mathcal{O}_{1,c}^\perp\}$ is connected.*

Proof. If $\mathcal{O}_3 \in \mathcal{O}_{1,c}^\perp$ and $\mathcal{O}_4 \in \mathcal{O}_{2,c}^\perp$, then we must show that there is a path in $\mathcal{G}_c^\perp$ from $\mathcal{O}_1 \times \mathcal{O}_3$ to $\mathcal{O}_2 \times \mathcal{O}_4$. Pick $\mathcal{O}_5 \in \mathcal{O}_{1,c}^\perp \cap \mathcal{O}_{2,c}^\perp$ then since $\mathcal{O}_{1,c}^\perp$ is connected, there is a path in $\mathcal{G}_c^\perp$ from $\mathcal{O}_1 \times \mathcal{O}_3$ to $\mathcal{O}_1 \times \mathcal{O}_5$. But $\mathcal{O}_1, \mathcal{O}_2 \in \mathcal{O}_{5,\bar{c}}^\perp$ which is connected so there is a path in $\mathcal{G}_c^\perp$ from $\mathcal{O}_1 \times \mathcal{O}_5$ to $\mathcal{O}_2 \times \mathcal{O}_5$ and, since $\mathcal{O}_{2,c}^\perp$ is connected, from $\mathcal{O}_2 \times \mathcal{O}_5$ to $\mathcal{O}_2 \times \mathcal{O}_4$, completing the proof.

Note that the lemma can, in particular, be applied when $\mathcal{O}^\perp$ is connected.

We also need the local versions of the sets introduced above.

$$\Sigma_1^\perp(\mathcal{O}) := \{b \in \Sigma_1^\perp : |b| \subset \mathcal{O}\},$$

$$\mathcal{G}^\perp(\mathcal{O}) := \{(\mathcal{O}_1, \mathcal{O}_0) : \mathcal{O}_1, \mathcal{O}_0 \subset \mathcal{O}, \mathcal{O}_1 \perp \mathcal{O}_0\}.$$

There is an obvious order–preserving injection $i : \mathcal{G}^\perp(\mathcal{O}) \rightarrow \Sigma_1^\perp(\mathcal{O})$. We simply consider $\mathcal{O}_i$ as $\partial_i b$ and $\mathcal{O}$ as $|b|$. Conversely, we have an order–preserving surjection

$s : \Sigma_1^{\perp}(\mathcal{O}) \to \mathcal{G}^{\perp}(\mathcal{O})$ mapping b to $(\partial_1 b, \partial_0 b)$. b lies in the same component of $\Sigma^{\perp}(\mathcal{O})$ as $i \circ s(b)$. Hence if $s(b)$ and $s(b')$ lie in the same component so do b and b' and we have computed the components of $\Sigma_1^{\perp}(\mathcal{O})$ in terms of those of $\mathcal{G}^{\perp}(\mathcal{O})$

There is a strategy for computing the components of $\Sigma_1^{\perp}$ in terms of those of $\Sigma_1^{\perp}(\mathcal{O})$. It is to look for coherent choices of components for the $\Sigma_1^{\perp}(\mathcal{O})$, i.e. we want a component $\Sigma_{1,c}^{\perp}(\mathcal{O})$ for each $\mathcal{O}$ such that

$$\Sigma_{1,c}^{\perp}(\mathcal{O}_1) = \Sigma_{1,c}^{\perp}(\mathcal{O}_2) \cap \Sigma_1^{\perp}(\mathcal{O}_1), \quad \mathcal{O}_1 \subset \mathcal{O}_2.$$

23.7 Lemma *Given a coherent choice of components $\mathcal{O} \mapsto \Sigma_{1,c}^{\perp}(\mathcal{O})$, then $\Sigma_{1,c}^{\perp} :=$ $\{b \in \Sigma_1 : b \in \Sigma_{1,c}^{\perp}(|b|)\}$ is a component of $\Sigma_1^{\perp}$.*

Proof. $\mathcal{K}$ being connected, the result will follow from Lemma 23.4 once we show that

$$\Sigma_{1,c}^{\perp}(\mathcal{O}) = \Sigma_{1,c}^{\perp} \cap \Sigma_1^{\perp}(\mathcal{O}).$$

But if $b \in \Sigma_{1,c}^{\perp}(\mathcal{O})$, $|b| \subset \mathcal{O}$ and since we have a coherent choice of components, $b \in \Sigma_{1,c}^{\perp}(|b|)$ giving an inclusion. The reverse inclusion is trivial, completing the proof.

Now when $\mathcal{K}$ denotes the set of regular diamonds in a globally hyperbolic spacetime with dimension > 2, then $\Sigma_1^{\perp}(\mathcal{O})$ has a single component so that $\Sigma_1^{\perp}$ is connected by Lemma 23.7. It remains to consider the case of a globally hyperbolic spacetime of dimension two. We know that each $\Sigma_1^{\perp}(\mathcal{O})$ now has two components and that one passes from one component to the other by reversing the orientation of the 1–simplices. We need a way of specifying a coherent choice of components. If the Cauchy surfaces are non–compact, then $\mathcal{G}^{\perp}$ also has two components and one passes from one component to the other by interchanging the two double cones. Hence mapping b to $(\partial_1 b, \partial_0 b)$ must map the two components of $\Sigma_1^{\perp}(\mathcal{O})$ into different components of $\mathcal{G}^{\perp}$. Denoting the two components of $\mathcal{G}^{\perp}$ by $\mathcal{G}_{\ell}^{\perp}$ and $\mathcal{G}_r^{\perp}$, the inverse images under the above map give us a coherent choice of components. Lemma 23.7 then shows us that $\Sigma_1^{\perp}$ has precisely two components and that one passes from one component to the other by reversing the orientation of 1–simplices.

On the other hand, in a globally hyperbolic spacetime (M, g) of dimension two with compact Cauchy surfaces, we know from the discussion in Sec. 3.1 of [50] that $\mathcal{G}^{\perp}$ is connected. However, $\Sigma_1^{\perp}$ continues to have two components and we need a different procedure for making a coherent choice of local components. To this end, we pick a nowhere vanishing timelike vector field and restricting this to a regular diamond $\mathcal{O}$, we have, by the discussion following Lemma 2.2 of [50], a coherent way of distinguishing the left and right components of the spacelike complement of a point in the regular diamond and hence left and right components of $\mathcal{G}^{\perp}(\mathcal{O})$ and $\Sigma_1^{\perp}(\mathcal{O})$. Thus by Lemma 23.7, $\Sigma_1^{\perp}$ has two connected components and one passes from one component to the other by reversing the orientation of 1–simplices.

24 Representations and Duality

A net of von Neumann algebras over a partially ordered set $\mathcal{K}$ is a mapping $\mathcal{O} \to \mathfrak{A}(\mathcal{O})$ from elements of $\mathcal{K}$ to von Neumann algebras such that when $\mathcal{O}_1 \subset \mathcal{O}_2$, $\mathfrak{A}(\mathcal{O}_1)$ is, in a coherent way, a von Neumann subalgebra of $\mathfrak{A}(\mathcal{O}_2)$.

By a representation π of a net $\mathfrak{A}$ of von Neumann algebras over $\mathcal{K}$ we mean a representation $\pi_\mathcal{O}$ of $\mathfrak{A}(\mathcal{O})$, for each $\mathcal{O} \in \mathcal{K}$, on a common Hilbert space $\mathcal{H}^\pi$ such that

$$\pi_{\mathcal{O}_1}(A) = \pi_{\mathcal{O}_2}(A), \quad \mathcal{O}_1 \subset \mathcal{O}_2, \quad A \in \mathfrak{A}(\mathcal{O}_1).$$

π is locally normal if $\pi_\mathcal{O}$ is normal for all $\mathcal{O}$. A bounded linear operator T from $\mathcal{H}^{\pi_\infty}$ to $\mathcal{H}^{\pi_\epsilon}$ intertwines π_1 and π_2 if

$$T\pi_{1,\mathcal{O}}(A) = \pi_{2,\mathcal{O}}(A)T, \quad A \in \mathfrak{A}(\mathcal{O}), \quad \mathcal{O} \in \mathcal{K}.$$

The set of intertwiners from π_1 to π_2 is denoted by (π_1, π_2).

The theory of superselection sectors singles out a particular representation π_0, the reference representation. To simplify the notation, we usually omit the symbol π_0 and consider $\mathfrak{A}$ as a concrete net on a Hilbert space $\mathcal{H}_I$. We denote the commutant of (π_0, π_0) by $\mathcal{R}$ and note that $\mathfrak{A}$ is a net of von Neumann subalgebras of $\mathcal{R}$. Up to now π_0 has always been supposed to be irreducible and the reader wishing to stick to this case need only ignore all mention of $\mathcal{R}$, cf. Sec. 12.

Two nets $\mathfrak{A}$ and $\mathfrak{B}$ of *-subalgebras of $\mathcal{R}$ over $\mathcal{K}$ are said to be *relatively local* if

$$\mathfrak{A}(\mathcal{O}_1) \subset \mathfrak{B}(\mathcal{O}_2)', \quad \text{whenever } \mathcal{O}_1 \perp \mathcal{O}_2.$$

There is a maximal net, the *dual net* $\mathfrak{A}^d$, in $\mathcal{R}$ which is relatively local to $\mathfrak{A}$. It is given by
$$\mathfrak{A}^d(\mathcal{O}) = \cap\{\mathfrak{A}(\mathcal{O}_1)' : \mathcal{O}_1 \perp \mathcal{O}\} \cap \mathcal{R}.$$

Since $\mathfrak{A}^{dd}$ is the largest net local relative to $\mathfrak{A}^d$, $\mathfrak{A} \subset \mathfrak{A}^{dd}$. However $\mathfrak{A} \subset \mathfrak{B}$ implies $\mathfrak{B}^d \subset \mathfrak{A}^d$, so that $\mathfrak{A}^d = \mathfrak{A}^{ddd}$. A net $\mathfrak{A}$ is said to be *local* if $\mathfrak{A} \subset \mathfrak{A}^d$ and then $\mathfrak{A}^{dd} \subset \mathfrak{A}^d = \mathfrak{A}^{ddd}$ so that $\mathfrak{A}^{dd}$ is local, too. The double dual is given by:

$$\mathfrak{A}^{dd}(\mathcal{O}) = \cap_{\mathcal{O}_1 \perp \mathcal{O}} \mathfrak{A}^d(\mathcal{O}_1)' \cap \mathcal{R} = \cap_{\mathcal{O} \perp \mathcal{O}_1}(\vee_{\hat{\mathcal{O}} \perp \mathcal{O}_1} \mathfrak{A}(\hat{\mathcal{O}}) \vee \mathcal{R}') \cap \mathcal{R}.$$

We recall the following definitions.

24.1 Definition A concrete net $\mathfrak{A}$ is said to satisfy *duality* if $\mathfrak{A} = \mathfrak{A}^d$ and *essential duality* if $\mathfrak{A}^{dd} = \mathfrak{A}^d$.

The assumption of duality or essential duality for the reference representation is the characteristic assumption of the theory of superselection sectors. The concepts themselves extend in an obvious manner to other representations.

Let me attempt a physical interpretation: if we consider just the normal states of the representation π then the self-adjoint elements of $(\pi, \pi)'$ are those operators which can be considered in an idealized sense as bounded observables. If such an element is in $\pi_{\mathcal{O}_1}(\mathfrak{A}(\mathcal{O}_1))'$ for all $\mathcal{O}_1 \perp \mathcal{O}$ then it commutes with all observables that can be measured in the causal complement of $\mathcal{O}$ and is therefore a candidate for an observable which can be measured in $\mathcal{O}$. Our equation asserts that every such element can, in an idealized sense, be measured in $\mathcal{O}$ because its expectation values in the normal states of π can be approximated arbitrarily well by observables localized in $\mathcal{O}$.

24.2 Lemma *The set of representations satisfying essential duality is closed under direct sums and subrepresentations.*

Proof. To be concise, if π is a representation, let us write $\pi(\mathcal{O})$ for the von Neumann algebra $\pi(\mathfrak{A}(\mathcal{O}))$ and $\pi^d(\mathcal{O})$ and $\pi^{dd}(\mathcal{O})$ for the von Neumann algebras associated with $\mathcal{O}$ in the dual and bidual nets. Let $W \in (\pi', \pi)$ be an isometry then the σ–continuous mapping $X \mapsto W^*XW$ has the following properties it maps $(\pi, \pi)'$ into $(\pi', \pi')'$, (π, π) into (π', π'), $\pi(\mathcal{O}_1)'$ into $\pi'(\mathcal{O}_1)'$ and $\pi(\hat{\mathcal{O}})$ into $\pi'(\hat{\mathcal{O}})$. Thus it induces a homomorphism

$$\pi^d(\mathcal{O}) := \cap_{\mathcal{O}_1 \perp \mathcal{O}} \pi(\mathcal{O}_1)' \cap (\pi, \pi)'$$

into $\pi'^d(\mathcal{O})$ and of

$$\pi^{dd}(\mathcal{O}) = \cap_{\mathcal{O}_1 \perp \mathcal{O}} (\vee_{\hat{\mathcal{O}} \perp \mathcal{O}_1} \pi(\hat{\mathcal{O}}) \vee (\pi, \pi)) \cap (\pi, \pi)'$$

into $\pi'^{dd}(\mathcal{O})$. Similarly, the σ–continuous homomorphism $Y \mapsto WYW^*$ maps $(\pi', \pi')'$ to $(\pi, \pi)'$, (π', π') to (π, π), $\pi'(\mathcal{O}_1)'$ to $\pi(\mathcal{O}_1)'$ and $\pi'(\hat{\mathcal{O}})$ into $\pi(\hat{\mathcal{O}}) \vee (\pi, \pi)$. It therefore maps $\pi'^d(\mathcal{O})$ into $\pi^d(\mathcal{O})$ and $\pi'^{dd}(\mathcal{O})$ into $\pi^{dd}(\mathcal{O})$. After these observations, the proof is simple. Let $W \in (\pi', \pi)$ be an isometry and suppose π satisfies essential duality and pick $Y \in \pi'^d(\mathcal{O})$, then $WYW^* \in \pi^d(\mathcal{O}) = \pi^{dd}(\mathcal{O})$ so that $Y = W^*WYW^*W \in \pi'^{dd}(\mathcal{O})$ and π' satisfies essential duality. Now let $W_i \in (\pi_i, \pi)$ be isometries with $\sum_i W_i W_i^* = 1_\pi$, where each π_i satisfies essential duality. Pick $X \in \pi^d(\mathcal{O})$, then $W_i^*XW_i \in \pi_i^d(\mathcal{O}) = \pi_i^{dd}(\mathcal{O})$ and noting that $W_i^*XW_j = 0$ for $i \neq j$,

$$X = \sum_{i,j} W_iW_i^*XW_jW_j^* = \sum_i W_iW_i^*XW_iW_i^* \in \pi^{dd}(\mathcal{O})$$

so that π satisfies essential duality.

As the above result would suggest, essential duality is a property of the quasiequivalence class of the representation. In fact, if we have an isomorphism $\varphi : \pi(\mathfrak{A})^- \to \pi'(\mathfrak{A})^-$ with $\varphi \circ \pi = \pi'$ then φ maps $\pi(\mathcal{O})$) onto $\pi'(\mathcal{O})$), hence $\pi^d(\mathcal{O})$ onto $\pi'^d(\mathcal{O})$ and $\pi^{dd}(\mathcal{O})$ onto $\pi'^{dd}(\mathcal{O})$. Thus π satisfies duality or essential duality if and only if π' satisfies duality or essential duality. As above, one can show that the set of representations satisfying duality is closed under subrepresentations but it is not closed under direct sums because (π, π) can acquire a non–trivial centre on taking direct sums and this centre is in $\pi^d(\mathcal{O})$ but, as a rule, not in $\pi(\mathcal{O})$. Thus a theory of sectors treating reducible reference representations should be based on essential duality rather than duality.

25 The Selection Criterion II

25.1 Definition A representation π of the net $\mathfrak{A}$ on $\mathcal{H}_0$ is said to be a representation in $\mathcal{R}$ if $\pi_{\mathcal{O}}(\mathfrak{A}(\mathcal{O})) \subset \mathcal{R}$ for all $\mathcal{O} \in \mathcal{K}$. An intertwining operator T is required to be an element of $\mathcal{R}$.

The reference representation π_0 will be treated as a representation in $\mathcal{R}$.

25.2 Definition A representation π of the net $\mathfrak{A}$ in $\mathcal{R}$ is said to satisfy the *selection criterion* if

$$\pi \upharpoonright \mathcal{O}^\perp \simeq \pi_0 \upharpoonright \mathcal{O}^\perp, \quad \mathcal{O} \in \mathcal{K}.$$

This means that for each $\mathcal{O}$ there is a unitary $V_{\mathcal{O}} \in \mathcal{R}$ such that

$$V_{\mathcal{O}} \pi_{\mathcal{O}_1}(A) = A V_{\mathcal{O}}, \quad A \in \mathfrak{A}(\mathcal{O}_1), \quad \mathcal{O}_1 \in \mathcal{O}^\perp,$$

where, to simplify notation in the sequel, we have omitted the symbol π_0 for the reference representation. We let $\mathrm{Rep}^\perp \mathfrak{A}$ denote the W^*–category whose objects are the representations of $\mathfrak{A}$ in $\mathcal{R}$ satisfying the selection criterion and whose arrows are the intertwiners between these representations.

To be able to work easily with the selection criterion, we need to make an assumption somewhat atypical of the theory of superselection sectors. It concerns a property first derived by Borchers in Minkowski space as a consequence of additivity, locality and the spectrum condition.

25.3 Definition. A net $\mathfrak{A}$ satisfies *Property B* if given $\mathcal{O} \in \mathcal{K}$, there is $\mathcal{O}_1 \in \mathcal{K}$, with $\mathcal{O}_1 \subset \mathcal{O}$ such that given any projection $E \neq 0$ in $\mathfrak{A}(\mathcal{O}_1)$, there is an isometry $W \in \mathfrak{A}(\mathcal{O})$ with $WW^* = E$.

Here we need to assume Property B for the dual net. It then means that, restricting the reference representation to $\mathcal{O}_1^\perp$, any of its non-trivial subrepresentations become equivalent to the vacuum representation in restriction to $\mathcal{O}^\perp$. It is used just to prove the following lemma.

25.4 Lemma *Suppose $\mathfrak{A}^d$ satisfies Property B. Then the set of representations satisfying the selection criterion is closed under direct sums and (non-trivial) subrepresentations. In other words, the W^*–category $\mathrm{Rep}^\perp \mathfrak{A}$ has direct sums and (non–zero) subobjects.*

Remark Any reasonable definition of physically relevant representations should satisfy the analogue of Lemma 25.4.

Proof of Lemma 25.4: Let π_E denote the subrepresentation of π corresponding to the projection $E \in (\pi, \pi)$, $E \neq 0$. Given $\mathcal{O} \in \mathcal{K}$, pick $\mathcal{O}_1 \in \mathcal{K}$ with $\mathcal{O}_1 \subset \mathcal{O}$ so that non–zero projections in $\mathfrak{A}^d(\mathcal{O}_1)$ are equivalent to I in $\mathfrak{A}^d(\mathcal{O})$. $\pi_E \upharpoonright \mathcal{O}_1^\perp \simeq \pi_{0,F} \upharpoonright \mathcal{O}_1^\perp$ for some $F \neq 0 \in \mathfrak{A}^d(\mathcal{O}_1)$ and by Property B, $\pi_{0,F} \upharpoonright \mathcal{O}^\perp \simeq \pi_0 \upharpoonright \mathcal{O}^\perp$ so π_E satisfies the selection criterion. Given π_1 and π_2 satisfying the selection criterion $\pi_1 \oplus \pi_2$ is unitarily equivalent to $\pi_0 \oplus \pi_0$ in restriction to any $\mathcal{O}^\perp$, $\mathcal{O} \in \mathcal{K}$. Hence it suffices to show that $\pi_0 \oplus \pi_0$ satisfies the selection criterion. If $\mathcal{O}_1$ is as above, given a projection $E \neq 0, I$ in $\mathfrak{A}^d(\mathcal{O}_1)$, we have isometries W_1 and W_2 in $\mathfrak{A}^d(\mathcal{O})$ with

$W_1 W_1^* + W_2 W_2^* = I$, by Property B. But this means that π_0 is unitarily equivalent to $\pi_0 \oplus \pi_0$ in restriction to $\mathcal{O}^\perp$, completing the proof.

We still know too little about the validity of Property B. As mentioned, its proof in Minkowski space involves positivity of the energy but an alternative argument would be needed in curved spacetime. A stronger and simpler hypothesis, that $\mathfrak{A}^d$ is a net of type III factors, might prove to be valid in practice [79].

I hope you will agree that the problem of classifying the representations satisfying the selection criterion is in the spirit of typical classification problems of non–commutative geometry but involves the net structure, too. An attack on the problem without assuming Property B might require us to delve further into the arsenal of non–commutative geometry,

26 The Cohomological Interpretation

We now provide a cohomological interpretation of superselection sectors leading to a proof of the Extension Theorem and to a description of tensor structure. To enter into the spirit of the cohomological interpretation, we regard $\mathcal{O}^\perp$, $\mathcal{O} \in \mathcal{K}$, as being a covering of $\mathcal{K}$, the *causal covering*. The selection criterion selects those representations that are trivial on the causal cover and these representations allow a cohomological description in analogy with locally trivial bundles. We shall suppose throughout that $\mathcal{K}$ is connected.

For each $a \in \Sigma_0$ we pick a unitary $V_a \in \mathcal{R}$ such that

$$V_a \pi_\mathcal{O}(A) = A V_a, \quad A \in \mathfrak{A}(\mathcal{O}), \quad \mathcal{O} \perp a$$

and set $z(b) := V_{\partial_0 b} V_{\partial_1 b}^*$, $b \in \Sigma_1$. Obviously if $\mathcal{O} \in |b|^\perp$, $z(b) \in \mathfrak{A}(\mathcal{O})' \cap \mathcal{R}$ thus $z(b) \in \mathfrak{A}^d(|b|)$. Furthermore,

$$z(\partial_0 c) z(\partial_2 c) = z(\partial_1 c), \quad c \in \Sigma_2$$

so that z is a unitary 1–cocycle with values in the dual net $\mathfrak{A}^d$. We consider such 1–cocycles as objects of a category $Z^1(\mathfrak{A}^d)$, where an arrow t in this category from z to z' is a $t_a \in \mathfrak{A}^d(a)$, $a \in \Sigma_0$, such that

$$t_{\partial_0 b} z(b) = z'(b) t_{\partial_1 b}, \quad b \in \Sigma_1.$$

This makes $Z^1(\mathfrak{A}^d)$ into a W^*–category. Note that $\|t_a\|$ is even independent of a since $\mathcal{K}$ is connected.

If we were to make a different choice V_a' of unitaries V_a, then setting $z'(b) := V_{\partial_0 b}' V_{\partial_1 b}'^*$ and $w_a := V_a' V_a^*$, we see that $w_a \in \mathfrak{A}^d(a)$ and $w_{\partial_0 b} z(b) = z'(b) w_{\partial_1 b}$. Thus $w \in (z, z')$ is a unitary and the 1–cocycle attached to π is defined up to unitary equivalence in $Z^1(\mathfrak{A}^d)$. More generally, if $T \in (\pi, \pi')$ and π and π' are trivial on the causal cover and z and z' are associated cocycles defined by unitaries V_a and V_a', as above, set

$$t_a := V_a' T V_a^*, \quad a \in \Sigma_0.$$

Then $t_a \in \mathfrak{A}^d(a)$ and

$$t_{\partial_0 b} z(b) = V_{\partial_0 b}' T V_{\partial_0 b}^* V_{\partial_0 b} V_{\partial_1 b}^* = V_{\partial_0 b}' T V_{\partial_1 b}^* =$$

$$= V_{\partial_0 b}' V_{\partial_1 b}'^* V_{\partial_1 b}' T V_{\partial_1 b}^* = z'(b) t_{\partial_1 b},$$

so that $t \in (z, z')$. Conversely, if $t \in (z, z')$ then $T := V_a'^* t_a V_a$ is independent of a so that

$$T \pi_{\mathcal{O}}(A) = \pi_{\mathcal{O}}'(A) T, \quad A \in \mathfrak{A}(\mathcal{O}), \quad \mathcal{O} \in \mathcal{K}.$$

and we clearly have a close relation between $Z^1(\mathfrak{A}^d)$ and the W^*–category $\mathrm{Rep}^\perp \mathfrak{A}$ of representations of $\mathfrak{A}$ trivial on the causal cover.

However, any cocycle z arising from such a representation has two special properties that may not be shared by a general 1–cocycle. First, z is trivial on $\mathcal{R}$, i.e. there are unitaries $V_a \in \mathcal{R}$, $a \in \Sigma_0$, on $\mathcal{H}_0$ such that $z(b) = V_{\partial_0 b} V_{\partial_1 b}^*$, $b \in \Sigma_1$.

If $\mathcal{K}$ is directed then, by Lemma 23.1, $\Sigma_*(\mathcal{K})$ admits a contracting homotopy. In this case every 1–cocycle of $\mathfrak{A}^d$ is trivial in $\mathcal{R}$. In general, if we consider the graph with vertices Σ_0 and arrows Σ_1 then the category generated by this graph has as arrows the paths in $\mathcal{K}$. Thus every 1–cocycle extends to a functor from this category. When z is trivial on $\mathcal{R}$ then $z(p)$ for a path p depends only on the endpoints $\partial_0 p$ and $\partial_1 p$ of the path. Conversely, if $z(p)$ just depends on the endpoints of p, then z is trivial on $\mathcal{R}$. To see this we pick a base point $a_0 \in \Sigma_0$, then $\mathcal{K}$ being connected, given $a \in \Sigma_0$, a path p_a with $\partial_0 p_a = a$ and $\partial_1 p_a = a_0$ and finally define $y(a) = z(p_a)$. Then $z(p) y(\partial_1 p) = y(\partial_0 p)$, so we have trivialized z in $\mathcal{R}$.

An elementary deformation of a path p consists in replacing a 1–simplex $\partial_1 c$ by a pair $\partial_0 c$, $\partial_2 c$ of 1–simplices, where $c \in \Sigma_2$, or, conversely, in replacing a consecutive pair $\partial_0 c$, $\partial_2 c$ of 1–simplices by a single 1–simplex, $\partial_1 c$. We say that two paths are equivalent if they can be obtained from one another by a set of elementary deformations. Equivalent paths have the same endpoints. If p and p' are equivalent, then $z(p) = z(p')$. We conclude that if any two paths with the same endpoints are equivalent then $z(p)$ depends only on the endpoints of p so that z can be trivialized in $\mathcal{R}$.

The second special property of cocycles z derived from representations satisfying the selection criterion is that, for any path p, $z(p) A z(p)^* = A$ whenever $A \in \mathfrak{A}(\mathcal{O})$ and $\partial_0 p, \partial_1 p \in \mathcal{O}^\perp$. The full subcategory of $Z^1(\mathfrak{A}^d)$ whose objects satisfy these two properties will be denoted by $Z_t^1(\mathfrak{A}^d)$.

The following simple result shows that the second condition is automatically satisfied in an important special case.

26.1 Lemma *If $\mathcal{O}^\perp$ is connected, then any object z of $Z^1(\mathfrak{A}^d)$, trivial on $\mathcal{R}$ satisfies*

$$z(p) A z(p)^* = A, \quad \partial_0 p, \partial_1 p \in \mathcal{O}^\perp, \ A \in \mathfrak{A}(\mathcal{O}).$$

Proof. Since $\mathcal{O}^\perp$ is connected, it suffices to prove the result when the path p is a 1–simplex b with $|b| \in \mathcal{O}^\perp$. But then, $z(b) \in \mathfrak{A}^d(|b|) \subset \mathfrak{A}(\mathcal{O})'$.

Having discussed these two conditions, we can give our cohomological characterization of the selection criterion.

26.2 Theorem *The W^*–categories $\mathrm{Rep}^\perp\mathfrak{A}$ and $Z^1_t(\mathfrak{A}^d)$ are equivalent.*

Proof. We pick unitaries V_a^π, $a \in \Sigma_0$, as above, for each object π of $\mathrm{Rep}^\perp\mathfrak{A}$. Given an arrow $T \in (\pi, \pi')$ in that category, we define for $b \in \Sigma_1$, $a \in \Sigma_0$

$$F(\pi)(b) = V^\pi_{\partial_0 b} V^{\pi*}_{\partial_1 b}; \quad F(T)_a := V_a^{\pi'} T V_a^{\pi*}.$$

Then F is a faithful *-functor and our computations above show that it is full. Hence, it remains to show that each object z of $Z^1_t(\mathfrak{A}^d)$, is equivalent to an object in the image of F. We show this by constructing a representation π^z. We pick unitaries $V_a \in \mathcal{R}$, $a \in \Sigma_0$, such that $z(b) = V_{\partial_0 b} V^*_{\partial_1 b}$, $b \in \Sigma_1$, and define

$$\pi^z_{\mathcal{O}}(A) = V_a^* A V_a, \quad a \in \mathcal{O}^\perp, \quad A \in \mathfrak{A}(\mathcal{O}).$$

This is well defined since $\mathcal{K}$ is connected and for any path p with $\partial_0 p$, $\partial_1 p \in \mathcal{O}^\perp$ we have $z(p) \in \mathfrak{A}(\mathcal{O})'$. Furthermore, the definition respects the net structure since

$$\pi^z_{\mathcal{O}_1}(A) = \pi^z_{\mathcal{O}_2}(A), \quad A \in \mathfrak{A}(\mathcal{O}_1), \quad \mathcal{O}_1 \subset \mathcal{O}_2.$$

Hence we get a representation of the net $\mathfrak{A}$, trivial on the covering by construction, and $V_{\partial_0 b} V^*_{\partial_1 b} = z(b)$ is an associated 1–cocycle. This completes the proof.

We now consider the problem of extending representations of a net $\mathfrak{A}$, trivial on the causal cover, to representations of the bidual net $\mathfrak{A}^{dd}$, again trivial on the causal cover.

26.3 Theorem *If each $\mathcal{O}^\perp$ is connected, every object π of $\mathrm{Rep}^\perp\mathfrak{A}$ admits a unique extension to an object of $\mathrm{Rep}^\perp\mathfrak{A}^{dd}$. Furthermore there is a canonical isomorphism of W^*–categories $\mathrm{Rep}^\perp\mathfrak{A}$ and $\mathrm{Rep}^\perp\mathfrak{A}^{dd}$.*

Proof. Let V_a, $a \in \Sigma_0$ be unitaries realizing the equivalence of π and π^0 on $a^\perp$. Then $z(b) := V_{\partial_0 b} V^*_{\partial_1 b}$, $b \in \Sigma_1$, is an associated object of $Z^1_t(\mathfrak{A}^d)$. Since each $\mathcal{O}^\perp$ is connected, z is at the same time an object of $Z^1(\mathfrak{A}^{ddd})$ since $\mathfrak{A}^d = \mathfrak{A}^{ddd}$. By Lemma 26.1, if we define

$$\tilde{\pi}_{\mathcal{O}}(A) := V_a^* A V_a, \quad A \in \mathfrak{A}^{dd}(\mathcal{O}), \quad a \in \mathcal{O}^\perp,$$

this gives a well defined element of $\mathrm{Rep}^\perp\mathfrak{A}^{dd}$ just as in the proof of Theorem 26.2. Furthermore, $\tilde{\pi}$ obviously extends π by the choice of the V_a. If we make another choice V_a' of the V_a then $V_a' V_a^* \in \mathfrak{A}^{dd}(a)$ so that $\tilde{\pi}$ remains unchanged and is consequently the unique extension of π to an object of $\mathrm{Rep}^\perp\mathfrak{A}^{dd}$. Passing to the extensions does not change the intertwiners by Theorem 26.2.

Some words on the situation when $\mathcal{O}^\perp$ is not connected are in order here. This happens in examples a) and f) of Sec. 23 when $s = 1$ when $\mathcal{O}^\perp$ splits up coherently into two components. A coherent choice of components here means that we can choose a component $\mathcal{O}_c^\perp$ of $\mathcal{O}^\perp$ for $\mathcal{O} \in \mathcal{K}$ in such a way that

$$\mathcal{O}_{2,c}^\perp \subset \mathcal{O}_{1,c}^\perp, \quad \text{for} \quad \mathcal{O}_1 \subset \mathcal{O}_2.$$

Given a 1–cocycle $z \in Z^1(\mathfrak{A}^d)$, we may construct a representation π^z as in Theorem 26.2 using $\mathcal{O}_c^\perp$ in place of $\mathcal{O}^\perp$. In general, this representation does not satisfy the selection criterion and depends on the choice of component. As was pointed out in [72], this is related to the phenomenon of solitons where the representations satisfy a modified selection criterion with reference representations depending on the component. Some information on work on solitonic sectors has been given in S18. Here it would be interesting to give a detailed account in the present context, but lack of time prevents me from giving more than a few structural remarks. The theory in a 2 dimensional Minkowski space takes the set $\mathcal{K}$ of double cones as its index sets and selects representations that are unitarily equivalent to one reference representation on the causal complement of right wedges and another on the causal complement of left wedges. Duality is assumed for wedges in the reference representations.

27 Tensor Structure

We now start looking for the missing algebraic structure. Our goal here is to show that sectors have a tensor structure. More precisely, we shall show that $Z_t^1(\mathfrak{A})$ has a canonical structure of a tensor W^*–category arising by adjoining endomorphisms.

If $\mathcal{A}$ is a C^*–algebra with unit then the category of its unital endomorphisms and their intertwiners, $\text{End}\mathcal{A}$, is a tensor C^*–category in a natural way. The tensor product of endomorphisms is just their composition, whilst if $R_1 \in (\rho_1, \rho_1')$ and $R_2 \in (\rho_2, \rho_2')$ then we write $R_1 \otimes R_2$ to denote $R_1\rho_1(R_2) = \rho_1'(R_2)R_1$ considered as an element of $(\rho_1\rho_2, \rho_1'\rho_2')$, cf. Sec. 13. Elaborations of this example will play a role in the sequel.

If $\mathfrak{A}$ is a net of von Neumann algebras, then there is an associated net $\mathcal{O} \mapsto \text{End}\mathfrak{A}(\mathcal{O})$ of tensor W^*–categories. $\text{End}\mathfrak{A}(\mathcal{O})$ has as objects the normal endomorphisms of the net $\mathcal{O}_1 \mapsto \mathfrak{A}(\mathcal{O}_1)$, $\mathcal{O}_1 \supset \mathcal{O}$ i.e. normal endomorphisms $\rho_{\mathcal{O}_1}$ of $\mathfrak{A}(\mathcal{O}_1)$ compatible with the net structure and defined over the partially ordered set

$$\mathcal{K} \upharpoonright \mathcal{O} := \{\mathcal{O}_1 \in \mathcal{K} : \mathcal{O}_1 \supset \mathcal{O}\}.$$

An arrow $T \in (\rho, \sigma)$ in $\text{End}\mathfrak{A}(\mathcal{O})$ is a $T \in \mathfrak{A}(\mathcal{O})$ such that

$$T\rho(A) = \sigma(A)T, \quad A \in \mathfrak{A}(\mathcal{O}_1), \quad \mathcal{O} \subset \mathcal{O}_1.$$

The tensor structure is defined as in the above example and the net structure is given by the obvious restriction mappings. Note the simplification when $\mathcal{K}$ is directed: the objects of each $\text{End}\mathfrak{A}(\mathcal{O})$ are then endomorphisms of the *–algebra $\cup_{\mathcal{O}}\mathfrak{A}(\mathcal{O})$.

We shall assume here that $\mathcal{K}$ is connected and we shall need to make use of a modified form of duality. A binary relation $\perp$ gives rise to two derived binary relations $\tilde{\perp}$ and $\hat{\perp}$ defined by supplementing $\mathcal{O}_1 \perp \mathcal{O}_2$ by requiring that there exists an $\mathcal{O}_3 \in \mathcal{K}$ such that

$$\mathcal{O}_1 \perp \mathcal{O}_3 \quad \mathcal{O}_2 \perp \mathcal{O}_3$$

or such that

$$\mathcal{O}_1, \ \mathcal{O}_2 \subset \mathcal{O}_3,$$

respectively. We have an inclusion $\mathcal{O}^{\hat{\perp}} \subset \mathcal{O}^{\tilde{\perp}} \subset \mathcal{O}^{\perp}$ of sieves. The operation of passing from $\perp$ to $\tilde{\perp}$ or $\hat{\perp}$ is idempotent. The derived relations coincide with $\perp$ when $\mathcal{K}$ is directed. On the circle, example c) of Sec. 23, $\perp \neq \tilde{\perp} = \hat{\perp}$. These derived relations are important in so far as they arise in proofs. Note that conditions a) and b) of Sec. 23 are automatically satisfied but c) is not and example b) of Sec. 23 provides a counter example. c) takes the explicit form

$\tilde{c}$) Given $\mathcal{O}_1 \in \mathcal{K}$, there are $\mathcal{O}_2, \mathcal{O}_3 \perp \mathcal{O}_1 \in \mathcal{K}$ with $\mathcal{O}_2 \perp \mathcal{O}_3$,

$\hat{c}$) Given $\mathcal{O}_1 \in \mathcal{K}$, there are $\mathcal{O}_2, \mathcal{O}_3 \in \mathcal{K}$ with $\mathcal{O}_1 \perp \mathcal{O}_2$ and $\mathcal{O}_1, \mathcal{O}_2 \subset \mathcal{O}_3$,

respectively and can be verified without difficulty in the interesting cases. Of course $\hat{c}$) implies $\tilde{c}$).

A net $\mathfrak{A}$ is said to satisfy $\tilde{\perp}$–duality if $\tilde{\perp}$ is a causal disjointness relation and

$$\mathfrak{A}(\mathcal{O}) = \cap_{\mathcal{O}_1 \tilde{\perp} \mathcal{O}} \mathfrak{A}(\mathcal{O}_1)' \cap \mathcal{R}, \quad \mathcal{O} \in \mathcal{K}$$

and $\hat{\perp}$–duality if $\hat{\perp}$ is a causal disjointness relation and

$$\mathfrak{A}(\mathcal{O}) = \cap_{\mathcal{O}_1 \hat{\perp} \mathcal{O}} \mathfrak{A}(\mathcal{O}_1)' \cap \mathcal{R}, \quad \mathcal{O} \in \mathcal{K}.$$

We shall first assume that our net $\mathfrak{A}$ satisfies $\tilde{\perp}$ duality but we shall later need the stronger assumption of $\hat{\perp}$–duality.

The construction of appropriate endomorphisms from a $z \in Z_t^1(\mathfrak{A})$ is just a variant on that already used to pass from a 1–cocycle $z \in Z_t^1(\mathfrak{A}^d)$ to a representation π^z. Given $a \in \Sigma_0$, and $A \in \mathfrak{A}(\mathcal{O})$, $a \subset \mathcal{O}$ pick a path p with $\partial_0 p = a$ and $\partial_1 p \in \mathcal{O}^{\perp}$ and set

$$y(a)(A) := z(p) A z(p)^*.$$

$y(a)(A)$ is independent of the choice of p since $z \in Z_t^1(\mathfrak{A}^d)$. Given $X \in \mathfrak{A}(\mathcal{O}_1)$ with $\mathcal{O}_1 \perp \mathcal{O}$, $\mathcal{O}_2$ with $\mathcal{O}_2 \perp \mathcal{O}$ and $\mathcal{O}_2 \perp \mathcal{O}_1$ and choosing $\partial_1 p = \mathcal{O}_2$, we see that $y(a)(A)$ and X commute so that $y(a)(A) \in \mathfrak{A}(\mathcal{O})$ by $\tilde{\perp}$–duality. Thus $y(a)$ is an object of $\text{End}\mathfrak{A}(a)$.

But $y(a)$ is not only localized in a in the sense of net endomorphisms but also in the sense of superselection theory in that $y(a)(A) = A$ whenever $A \in \mathfrak{A}(\mathcal{O}_1)$ where $\mathcal{O}_1 \in a^{\perp}$ and $\mathcal{O}_1, a \subset \mathcal{O}$, since the endpoints of p lie in $\mathcal{O}_1^{\perp}$. We write $\Delta(a)$ to denote the objects of $\text{End}\mathfrak{A}(a)$ satisfying this second localization condition and denote by $\mathcal{T}(a)$ the corresponding full tensor C^*–subcategory of $\text{End}\mathfrak{A}(a)$. Note that, in the presence of $\hat{\perp}$–duality, the condition $T \in \mathfrak{A}(a)$ in the definition of $T \in (\rho, \sigma)$ in $\mathcal{T}(a)$ may be omitted as it is a consequence of the intertwining relation.

27.1 Lemma *Let p be a path with $\partial_1 p$, $\partial_0 p \subset \mathcal{O}$ then*

$$z(p)y(\partial_1 p)(A) = y(\partial_0 p)(A)z(p), \quad A \in \mathfrak{A}(\mathcal{O}).$$

Proof. Given $A \in \mathfrak{A}(\mathcal{O})$ and a path p with $\partial_1 p, \partial_0 p \subset \mathcal{O}$, pick paths p', p'' with $\partial_0 p' = \partial_1 p, \partial_0 p'' = \partial_0 p$ and $\partial_1 p', \partial_1 p'' \in \mathcal{O}^\perp$, then

$$z(p)y(\partial_1 p)(A) = z(p)z(p')Az(p')^* = z(p'')Az(p'')^*z(p) =$$
$$= y(\partial_0 p)(A)z(p),$$

as required.

27.2 Corollary *Let p be a path in $\mathcal{K}|\mathcal{O} := \{\mathcal{O}_1 : \mathcal{O}_1 \subset \mathcal{O}\}$, then*

$$z(p)y(\partial_1 p)(A) = y(\partial_0 p)(A)z(p), \quad A \in \mathfrak{A}(\mathcal{O}).$$

Furthermore if $t \in (z, \hat{z})$, $A \in \mathfrak{A}(\mathcal{O})$ and p is a path with $\partial_0 p = a \subset \mathcal{O}$ and $\partial_1 p \subset \mathcal{O}^\perp$ then

$$t_a y(a)(A) = t_a z(p) Az(p)^* = \hat{z}(p) t_{\partial_1 p} Az(p)^* =$$
$$= \hat{z}(p) A\hat{z}(p)^* t_a = \hat{y}(a)(A) t_a.$$

In other words $t_a \in (y(a), \hat{y}(a))$. This same computation also gives the following result.

27.3 Lemma *Suppose we have a field $a \in \Sigma_0 \mapsto t_a \in \mathcal{R}$ and objects z, $\hat{z}$ of $Z_t^1(\mathfrak{A})$ such that*

$$\hat{z}(b) t_{\partial_1 b} = t_{\partial_0 b} z(b), \quad b \in \Sigma_1.$$

Suppose furthermore, that given $\mathcal{O} \supset a$, there is a path p with $\partial_0 p = a$, $\partial_1 p \perp \mathcal{O}$ and $t_{\partial_1 p} \in \mathfrak{A}(\mathcal{O})'$. Then

$$t_a y(a)(A) = \hat{y}(a)(A) t_a, \quad A \in \mathfrak{A}(\mathcal{O}), \mathcal{O} \supset a,$$

so that if $\hat{\perp}$–duality holds, $t_a \in (y(a), \hat{y}(a))$.

These results admit the following interpretation.

27.4 Theorem *Let $\mathfrak{A}$ be a net over $(\mathcal{K}, \perp)$ satisfying $\hat{\perp}$–duality. If z is a 1–cocycle of $\mathfrak{A}$ trivial in $\mathcal{R}$ then (y, z) is a 1–cocycle in the net $\mathcal{T}$ of tensor W^*–categories and the map $z \mapsto (y, z)$ together with the identity map on arrows is an isomorphism of $Z_t^1(\mathfrak{A})$ and $Z_t^1(\mathcal{T})$.*

Now, $\mathcal{T}$ being a net of tensor W^*–categories, $Z^1(\mathcal{T})$ is itself a tensor W^*–category. Given 1–cocycles (y_1, z_1) and (y_2, z_2), their tensor product $(y_1, z_1) \otimes (y_2, z_2)$ is the 1–cocycle (y, z), where

$$y(a) = y_1(a)y_2(a), \quad z(b) = z_1(b)y_1(\partial_1 b)(z_2(b)).$$

If both (y_1, z_1) and (y_2, z_2) are trivial in $\mathcal{R}$ then so is their tensor product. The tensor product on arrows is defined as follows: if t_i maps from (y_i, z_i) to (y_i', z_i') for $i = 1, 2$, then the tensor product $t_1 \otimes t_2$ is given by

$$(t_1 \otimes t_2)_a = t_{1,a} y_1(a)(t_{2,a}).$$

This achieves our goal of describing superselection structure in terms of a tensor W^*–category.

28 Localized Endomorphisms

We now define a class of endomorphisms including all the endomorphisms $y(a)$ derived from cocycles. An object ρ of $\mathcal{T}(\mathcal{O})$ is *transportable* if there is a map $a \mapsto \rho_a$, where ρ_a is an object of $\mathcal{T}(a)$ and $\rho_a = \rho$ when $a = \mathcal{O}$ and a map $\Sigma_1 \ni b \mapsto u(b)$, where $u(b)$ is a unitary from $\rho_{\partial_1 b}$ to $\rho_{\partial_0 b}$ in $\mathcal{T}(|b|)$. $b \mapsto u(b)$ need not be a 1-cocycle. We get in this way a net $\mathcal{O} \mapsto \mathcal{T}_t(\mathcal{O})$ of W^*-categories. $\Delta_t(\mathcal{O})$ denotes the set of objects of $\mathcal{T}_t(\mathcal{O})$. The tensor W^*-categories $Z_t^1(\mathcal{T})$ and $Z_t^1(\mathcal{T}_t)$ are canonically isomorphic.

When $\mathcal{K}$ is directed, we can study $\mathrm{Rep}^\perp \mathfrak{A}$ by studying $\mathcal{T}_t(a)$. In general, we can only assert that we have a faithful $*$-functor from $\mathrm{Rep}^\perp \mathfrak{A}$ to $\mathcal{T}_t(a)$ or a tensor $*$-functor from $Z_t^1(\mathfrak{A})$ to $\mathcal{T}_t(a)$ without knowing that these are equivalences. Thus $\mathcal{T}_t(a)$ may not yield a description of superselection sectors. Nevertheless, as we shall see, an analysis of localized endomorphisms still provides useful information.

If $\mathcal{O}_1 \subset \mathcal{O}_2$ then $\mathcal{K} \upharpoonright \mathcal{O}_2 \subset \mathcal{K} \upharpoonright \mathcal{O}_1$ and we have a faithful tensor $*$-functor of restriction from $\mathcal{T}_t(\mathcal{O}_1)$ to $\mathcal{T}_t(\mathcal{O}_2)$. We will use the same symbol for an object of $\mathcal{T}_t(\mathcal{O}_1)$ and its restriction to $\mathcal{T}_t(\mathcal{O}_2)$. How much information can be derived from localized endomorphisms depends on the properties of this restriction functor. We list here hypotheses to be referred to in the sequel.

a) $\mathcal{K} \upharpoonright \mathcal{O}_2$ is cofinal in $\mathcal{K} \upharpoonright \mathcal{O}_1$ for each $\mathcal{O}_1 \subset \mathcal{O}_2$ in $\mathcal{K}$.

b) (Local duality) Let $\mathcal{K}^\mathcal{O}$ denote the subset of $\mathcal{K}$ consisting of the $\mathcal{O}_1 \subset \mathcal{O}$ such that there exists an $\mathcal{O}_2 \subset \mathcal{O}$ with $\mathcal{O}_1 \perp \mathcal{O}_2$ with the induced order and binary relation $\perp$. Let $\mathfrak{A}^\mathcal{O}$ denote the restriction of $\mathfrak{A}$ to $\mathcal{K}^\mathcal{O}$ with reference representation obtained by restricting π_0 to $\mathcal{K}^\mathcal{O}$. Then for each $\mathcal{O} \in \mathcal{K}$, duality holds for $\mathfrak{A}^\mathcal{O}$ and this net generates $\mathfrak{A}(\mathcal{O})$ as a von Neumann algebra in the reference representation.

When $\mathcal{K}$ is directed, duality alone suffices to derive the results of this section. a) is equivalent to saying that each $\mathcal{K} \upharpoonright \mathcal{O}$ is directed, and $\hat{\perp}$-duality replaces duality. b) aims at avoiding the hypothesis that $\mathcal{K}$ is directed by working locally.

The next lemma derives properties of the embedding functor from hypotheses a) and b).

28.1 Lemma *If b) holds, then for each $\mathcal{O}_1 \subset \mathcal{O}_2$ in $\mathcal{K}$, all intertwiners in $\mathcal{T}_t(\mathcal{O}_2)$ between objects in the image of $\mathcal{T}_t(\mathcal{O}_1)$ are elements of $\mathfrak{A}(\mathcal{O}_1)$. If a) holds together with $\hat{\perp}$-duality, the restriction functor from $\mathcal{T}_t(\mathcal{O}_1)$ to $\mathcal{T}_t(\mathcal{O}_2)$ is an equivalence of tensor W^*-categories for each $\mathcal{O}_1 \subset \mathcal{O}_2$ in $\mathcal{K}$.*

Proof. Let ρ_1 and σ_1 be objects of $\mathcal{T}_t(\mathcal{O}_1)$ and T an intertwiner between their images in $\mathcal{T}_t(\mathcal{O}_2)$. If a) holds and $\mathcal{O}_3 \hat{\perp} \mathcal{O}_1$ then by cofinality there is an $\mathcal{O}_4 \in \mathcal{K}$ with $\mathcal{O}_2, \mathcal{O}_3 \subset \mathcal{O}_4$. Picking $A \in \mathfrak{A}(\mathcal{O}_3)$, we have

$$TA = T\rho_1(A) = \sigma_1(A)T = AT,$$

so that by $\hat{\perp}$-duality, $T \in \mathfrak{A}(\mathcal{O}_1)$. It now follows from cofinality that $T \in (\rho_1, \sigma_1)$ in $\mathcal{T}_t(\mathcal{O}_1)$ and our functor is full. The transportability condition guarantees that any object of $T_t(\mathcal{O}_2)$ is equivalent to an object in the image of $T_t(\mathcal{O}_1)$ so we have an equivalence of tensor W^*–categories. If b) holds, we note that $T \in \mathfrak{A}(\mathcal{O}_3)'$ whenever $\mathcal{O}_3 \perp \mathcal{O}_1$ with $\mathcal{O}_3 \subset \mathcal{O}_2$. Since $T \in \mathfrak{A}(\mathcal{O}_2)$, we may conclude from b) that $T \in \mathfrak{A}(\mathcal{O}_1)$ as required.

In the proof, all that was used of a) was given $\mathcal{O}_1 \subset \mathcal{O}_2$ and $\mathcal{O} \hat{\perp} \mathcal{O}_1$, there is a $\mathcal{O}_3 \supset \mathcal{O}_2, \mathcal{O}$. If the net $\mathfrak{A}$ is additive, a) could be replaced by requiring that the sieve

$$\{\mathcal{O} : \mathcal{O}, \mathcal{O}_2 \subset \hat{\mathcal{O}}, \text{ for some } \hat{\mathcal{O}} \in \mathcal{K}\},$$

generates $\mathcal{K}$.

The basic step in this analysis is to investigate the relation between causal disjointness and commutation of localized endomorphisms and their intertwiners. It is natural to say that an intertwiner $T \in T_t(\mathcal{O})$ is localized in $\mathcal{O}$, but we need a finer notion because we may have $T \in (\rho_1, \rho_0)$ in $T_t(\mathcal{O})$ where $\rho_i \in \Delta_t(\mathcal{O}_i)$ and $\mathcal{O}_i \subset \mathcal{O}$. In this case, we refer to $\mathcal{O}_1$ as an *initial support* and $\mathcal{O}_0$ as a *final support* of T.

28.2 Lemma *Suppose either $\hat{\perp}$ is a causal disjointness relation and b) holds or $\hat{\perp}$–duality together with a) and let $T_i \in (\rho_i, \rho_i')$ be arrows in $T_t(\mathcal{O})$, then*

$$T_0 \otimes T_1 = T_1 \otimes T_0$$

if there are $b, b' \in \Sigma_{1,c}^{\perp}$ so that $\partial_0 b$ and $\partial_1 b$ are initial supports of T_0 and T_1 and $\partial_0 b'$ and $\partial_1 b'$ are final supports of T_0 and T_1.

Proof. We first show that $T_0\rho_0(T_1) = T_1\rho_1(T_0)$ in the tensor W^*–category $T_t(\mathcal{O})$. By Lemma 28.1, this relation is trivial if T_0 and T_1 are causally disjoint in the sense that there is a $\hat{b} \in \Sigma_1^{\perp}(\mathcal{O})$ such that $\partial_0 \hat{b}$ contains an initial and final support of T_0 and $\partial_1 \hat{b}$ an initial and final support of T_1. The idea of the proof is to reduce to this trivial case. Replace T_0 and T_1 by $T_2 = T_0 \circ U_0$ and $T_3 = T_1 \circ U_1$, where $U_0 \in (\rho_2, \rho_0)$ and $U_1 \in (\rho_3, \rho_1)$ are unitary in $T_t(\mathcal{O}_1)$ for some $\mathcal{O}_1 \supset \mathcal{O}$. Then

$$T_2 \otimes T_3 = T_0 \otimes T_1 \circ U_0 \otimes U_1, \quad T_3 \otimes T_2 = T_1 \otimes T_0 \circ U_1 \otimes U_0$$

in $T_t(\mathcal{O})$. Thus if U_0 and U_1 are causally disjoint, the validity or not of our relation is unaffected by the passage from T_0, T_1 to T_2, T_3. But b and b' lie in a connected component $\Sigma_{1,c}^{\perp}$ by hypothesis, so that we can arrange that, after a finite number of steps, an initial and final support of each intertwiner coincides. This is again the trivial case so $T_0\rho_0(T_1) = T_1\rho_1(T_0)$, as required. It only remains to show that

$$\rho_0\rho_1 - \rho_1\rho_0 = 0.$$

The kernel of the left hand side does not change if we shift to ρ_2 and ρ_3 using U_0 and U_1 as above. We show that the kernel contains $\mathfrak{A}(\mathcal{O}_0)$ for any $\mathcal{O}_0 \supset \mathcal{O}$. By hypothesis ĉ), we may pick $\mathcal{O}_1, \mathcal{O}_2 \in \mathcal{K}$ with $\mathcal{O}_1 \perp \mathcal{O}_0$ and $\mathcal{O}_0, \mathcal{O}_1 \subset \mathcal{O}_2$. But now the deformation argument shows that the kernel does not change if we shift to $\rho_2 \in \Delta_t(\partial_0\hat{b})$ and $\rho_3 \in \Delta_t(\partial_1\hat{b})$ where $\hat{b} \in \Sigma_{1,c}^{\perp}$ has support in $\mathcal{O}_1$. But such ρ_2 and ρ_3 act trivially on $\mathfrak{A}(\mathcal{O}_0)$ since $\mathcal{O}_0, \mathcal{O}_1 \subset \mathcal{O}_2$, completing the proof.

28.3 Theorem *Suppose either* $\hat{\perp}$*–duality and a) or that* $\hat{\perp}$ *is a causal disjointness relation and b) holds and that we have a net* $\mathcal{O} \mapsto \Sigma_{1,c}^{\perp}(\mathcal{O})$*, where each* $\Sigma_{1,c}^{\perp}(\mathcal{O})$ *is a component of* $\Sigma_1^{\perp}(\mathcal{O})$ *then there is a unique intertwiner-valued function defined on pairs of objects of* $\mathcal{T}_t(\mathcal{O})$*,* $(\rho_0, \rho_1) \mapsto \varepsilon_{\mathcal{O}}^c(\rho_0, \rho_1) \in (\rho_0 \rho_1, \rho_1 \rho_0)$ *such that*

a) $\varepsilon_{\mathcal{O}}^c(\rho_0', \rho_1') \circ T_0 \otimes T_1 = T_1 \otimes T_0 \circ \varepsilon_{\mathcal{O}}^c(\rho_0, \rho_1)$, $T_i \in (\rho_i, \rho_i')$, $i = 0, 1$,
b) $\varepsilon_{\mathcal{O}}^c(\rho_0, \rho_1) = 1_{\rho_0 \rho_1}$, *if there is a* $b \in \Sigma_{1,c}^{\perp}(\mathcal{O})$ *such that* $\rho_i \in \Delta_t(\partial_i b)$, $i = 0, 1$.

Proof. The uniqueness claim tells us how to go about defining $\varepsilon_{\mathcal{O}}^c$: given ρ_0, ρ_1 pick $b \in \Sigma_{1,c}^{\perp}(\mathcal{O})$ and unitaries $U_i \in (\rho_i, \tau_i)$ where $\tau_i \in \Delta_t(\partial_i b)$ and we have no option but to set

$$\varepsilon_{\mathcal{O}}^c(\rho_0, \rho_1) = U_1^* \otimes U_0^* \circ U_0 \otimes U_1.$$

By Lemma 28.2, such a choice, however made, automatically satisfies b). We have $\varepsilon_{\mathcal{O}}^c(\rho_0', \rho_1') = U_1'^* \otimes U_0'^* \circ U_0' \otimes U_1'$, where $U_i' \in (\rho_i', \tau_i')$, $\tau_i' \in \Delta_t(\partial_i b')$ and $b' \in \Sigma_{1,c}^{\perp}(\mathcal{O})$. Set $S_i = U_i' \circ T_i \circ U_i^*$ then, by Lemma 28.2, $S_0 \otimes S_1 = S_1 \otimes S_0$ and rearranging this identity gives a) and completes the proof of the theorem.

28.4 Corollary *Under the hypotheses of Theorem 28.3,*

i) $\varepsilon_{\mathcal{O}}^c(\rho_1 \rho_2, \rho_3) = \varepsilon_{\mathcal{O}}^c(\rho_1, \rho_3) \otimes 1_{\rho_2} \circ 1_{\rho_1} \otimes \varepsilon_{\mathcal{O}}^c(\rho_2, \rho_3)$,
ii) $\varepsilon_{\mathcal{O}}^c(\rho_1, \rho_2 \rho_3) = 1_{\rho_2} \otimes \varepsilon_{\mathcal{O}}^c(\rho_1, \rho_3) \circ \varepsilon_{\mathcal{O}}^c(\rho_1, \rho_2) \otimes 1_{\rho_3}$,

If $b \in \Sigma_{1,c}^{\perp}(\mathcal{O})$ *implies* $\bar{b} \in \Sigma_{1,c}^{\perp}(\mathcal{O})$*, where* $|\bar{b}| = |b|$*,* $\partial_0 \bar{b} = \partial_1 b$ *and* $\partial_1 \bar{b} = \partial_0 b$*, then*

iii) $\varepsilon_{\mathcal{O}}^c(\rho_2, \rho_1) \circ \varepsilon_{\mathcal{O}}^c(\rho_1, \rho_2) = 1_{\rho_1 \rho_2}$.

Proof. These equalities follow easily from the formula

$$c_{\mathcal{O}}^c(\rho_1, \rho_2) = U_2^* \otimes U_1^* \circ U_1 \otimes U_2$$

used to define $\varepsilon_{\mathcal{O}}^c$ in the proof of Theorem 28.3.

As a consequence of i) and ii) or by direct computation, we also have

$$\varepsilon_{\mathcal{O}}^c(\rho, \iota) = \varepsilon_{\mathcal{O}}^c(\iota, \rho) = 1_\rho.$$

In virtue of i) and ii), the pair $(\mathcal{T}_t(\mathcal{O}), \varepsilon_{\mathcal{O}}^c)$ is a *braided* tensor W^*–category and when iii) holds, too, we get a *symmetric* tensor W^*–category. If $\Sigma_{1,c}^{\perp}(\mathcal{O}_1) \subset \Sigma_{1,c}^{\perp}(\mathcal{O}_2)$ when $\mathcal{O}_1 \subset \mathcal{O}_2$, then the braidings $\varepsilon_{\mathcal{O}_1}^c$ and $\varepsilon_{\mathcal{O}_2}^c$ are compatible with the inclusion $\mathcal{T}_t(\mathcal{O}_1) \subset \mathcal{T}_t(\mathcal{O}_2)$. Thus our hypotheses guarantee a net $(\mathcal{T}_t, \varepsilon^c)$ of braided tensor W^*–categories, giving the following result.

28.5 Theorem *Under the hypotheses of Theorem 28.3, the net* $\mathcal{T}_t$ *has a unique braiding* ε^c *such that* $\varepsilon_{\mathcal{O}}^c(\rho_0, \rho_1) = 1_{\rho_0 \rho_1}$ *whenever there is a* $b \in \Sigma_{1,c}^{\perp}(\mathcal{O})$ *with* $\rho_i \in \Delta_t(\partial_i b)$, $i = 0, 1$.

Under the hypotheses of Theorem 28.3 we have therefore derived a net of braided tensor W^*–categories $(\mathcal{T}_t, \varepsilon^c)$ and this in turn gives rise to a braided tensor W^*–category $(Z_t^1(\mathcal{T}_t), \varepsilon^c)$ by the following simple general result.

28.6 Lemma *Let* $(\mathcal{T}, \varepsilon)$ *be a net of braided tensor* C^**–categories then* $Z^1(\mathcal{T})$ *can be made into a braided tensor* C^**–category by setting for 1–cocycles* (y, z) *and* (y', z'),

$$\varepsilon((y,z),(y',z'))_a := \varepsilon(y(a),y'(a)).$$

Proof. The above expression obviously acts correctly on the arrows evaluated in a and the laws for a braiding hold for each a, the only point that has to be checked is that $\varepsilon((y,z),(y',z'))$ is an arrow from $(y,z) \otimes (y',z')$ to $(y',z') \otimes (y,z)$. However, if $b \in \Sigma_1$, $z(b) \in (y(\partial_1 b), y(\partial_0 b))$ in $\mathcal{T}(|b|)$ and similarly for $z'(b)$. Thus

$$z'(b) \otimes z(b) \circ \varepsilon(y(\partial_1 b), y'(\partial_1 b)) = \varepsilon(y(\partial_0 b), y'(\partial_0 b)) \circ z(b) \otimes z'(b),$$

as required.

29 Left Inverses

It might seem that we have all the basic structure to proceed with a purely mathematical analysis. Nevertheless, one item is missing. We need left inverses if we wish to proceed to the analysis of statistics. We examine, at this point, the notions involved and the relations between them. If π is a representation of $\mathfrak{A}$ in $\mathcal{R}$ then we define a left inverse φ of π to be given by unital positive linear mappings $\varphi_\mathcal{O}$ on $\mathcal{R}$ compatible with the net inclusions and satisfying

$$\varphi_\mathcal{O}(A\pi_\mathcal{O}(B)) = \varphi_\mathcal{O}(A)B, \quad A, B \in \mathfrak{A}(\mathcal{O}).$$

Note that if $\pi_\mathcal{O}(B) = B$ then $\varphi_\mathcal{O}(B) = B$. If π is localized in $\mathcal{O}$ in the sense that

$$\pi_{\mathcal{O}_1}(A) = A, \quad \mathcal{O} \perp \mathcal{O}_1, \quad A \in \mathfrak{A}(\mathcal{O}_1),$$

then φ is localized in $\mathcal{O}$ in the same sense. Furthermore, if $\mathcal{O}_1 \hat{\perp} \mathcal{O}_2$ and $\mathcal{O} \subset \mathcal{O}_2$, then $\varphi_{\mathcal{O}_2}(A)B = B\varphi_{\mathcal{O}_2}(A)$ for $A \in \mathfrak{A}(\mathcal{O}_2)$ and $B \in \mathfrak{A}(\mathcal{O}_1)$. In fact, picking $\mathcal{O}_3$ with $\mathcal{O}_1, \mathcal{O}_2 \subset \mathcal{O}_3$ we have

$$\varphi_{\mathcal{O}_2}(A)B = \varphi_{\mathcal{O}_3}(A)B = \varphi_{\mathcal{O}_3}(A\pi_{\mathcal{O}_3}(B))$$

$$= \varphi_{\mathcal{O}_3}(A\pi_{\mathcal{O}_1}(B)) = \varphi_{\mathcal{O}_3}(AB).$$

Since A and B commute, we interchange them and reverse the steps to conclude that $\varphi_{\mathcal{O}_2}(A)$ and B commute. This proves the following result.

29.1 Lemma *If φ is a left inverse for a representation π localized in $\mathcal{O}$ then φ is localized in $\mathcal{O}$ and if duality holds for the relation $\hat{\perp}$, $\varphi_{\mathcal{O}_1}\mathfrak{A}(\mathcal{O}_1) \subset \mathfrak{A}(\mathcal{O}_1)$ for $\mathcal{O} \subset \mathcal{O}_1$.*

The restriction of π to the net $\mathcal{O}_1 \mapsto \mathfrak{A}(\mathcal{O}_1)$, $\mathcal{O}_1 \supset \mathcal{O}$, is a localized endomorphism ρ and an object of the tensor W^*–category $\mathrm{End}\,\mathfrak{A}(\mathcal{O})$. The above notion of left inverse adapts easily to localized endomorphisms. If ρ is localized in $\mathcal{O}$, a *left inverse* of ρ is a family $\mathcal{O}_1 \supset \mathcal{O} \mapsto \varphi_{\mathcal{O}_1}$ of unital positive linear mappings on the $\mathfrak{A}(\mathcal{O}_1)$, compatible with the net inclusions and satisfying

$$\varphi_{\mathcal{O}_1}(A\rho_{\mathcal{O}_1}(B)) = \varphi_{\mathcal{O}_1}(A)B, \quad A, B \in \mathfrak{A}(\mathcal{O}_1).$$

Obviously, a left inverse for ρ considered as a representation yields a left inverse for the endomorphism ρ on restriction when $\hat{\perp}$–duality holds. If $\bar{\rho}$ is a conjugate(cf. [64]) for ρ then we get a left inverse φ for ρ by setting

$$\varphi_{\mathcal{O}_1}(A) := V^*\bar{\rho}_{\mathcal{O}_1}(A)V, \quad A \in \mathfrak{A}(\mathcal{O}_1), \ \mathcal{O}_1 \supset \mathcal{O},$$

where $V \in (\iota, \bar{\rho}\rho)$ is an isometry.

We now show that a left inverse φ for ρ induces a left inverse of ρ in the categorical sense [64]. In other words, we need a set

$$\varphi_{\sigma,\tau} : (\rho\sigma, \rho\tau) \to (\sigma, \tau),$$

of linear mappings where σ, τ are objects of the category. These have to be natural in σ and τ, i.e. given $S \in (\sigma, \sigma')$ and $T \in (\tau, \tau')$ we have

$$\varphi_{\sigma',\tau'}(1_\rho \otimes T \circ X \circ 1_\rho \otimes S^*) = T \circ \varphi_{\sigma,\tau}(X) \circ S^*, \ X \in (\rho\sigma, \rho\tau),$$

and furthermore to satisfy

$$\varphi_{\sigma\nu,\tau\nu}(X \otimes 1_\nu) = \varphi_{\sigma,\tau}(X) \otimes 1_\nu, \ X \in (\rho\sigma, \rho\tau)$$

for each object ν. We will require that φ is positive in the sense that $\varphi_{\sigma,\sigma}$ is positive for each σ and normalized in the sense that

$$\varphi_{\iota,\iota}(1_\rho) = 1_\iota.$$

We say that φ is faithful if $\varphi_{\sigma,\sigma}$ is faithful for each object σ.

Now, given $T \in (\rho\sigma, \rho\tau)$, we recall that $T \in \mathfrak{A}(\mathcal{O})$. Hence we set

$$\varphi_{\sigma,\tau}(T) = \varphi_{\mathcal{O}}(T)$$

and since $\varphi_{\mathcal{O}}(T) \in \mathfrak{A}(\mathcal{O})$ by Lemma 29.1, we conclude without difficulty that we get a left inverse for ρ in this way.

On the other hand, if we are dealing with a representation satisfying the selection criterion then we know that, by passing to an associated 1–cocycle, we get a field $a \mapsto y(a)$ of localized endomorphisms under the weaker assumption that duality holds for the relation $\tilde{\perp}$. In this case, we would actually like a left inverse for the 1–cocycle considered as an object of the tensor W^*–category $Z^1_t(\mathfrak{A})$. To this end, we pick, for each of the associated endomorphisms $y(a)$ a left inverse φ_a and ask whether $a \mapsto \varphi_a(t_a)$ is an arrow from z' to z'', whenever $a \mapsto t_a$ is an arrow from $z \times z'$ to $z \times z''$. Thus $t_a \in \mathfrak{A}(a)$ and

$$(z \times z'')(b)t_{\partial_1 b} = t_{\partial_0 b}(z \times z')(b).$$

It follows that

$$z''(b)\varphi_{\partial_1 b}(t_{\partial_1 b}) = \varphi_{\partial_1 b}(y(\partial_1 b)(z''(b))t_{\partial_1 b}) = \varphi_{\partial_1 b}(z(b)^* t_{\partial_0 b} z(b))z'(b)$$

and we deduce the following lemma.

29.2 Lemma *If $z \in Z_t^1(\mathfrak{A})$ and $a \mapsto y(a)$ is the associated field of endomorphisms. Then a field $a \mapsto \varphi_a$ of left inverses of the $y(a)$ defines a left inverse for z by the formula*

$$\varphi_{z', z''}(t)_a := \varphi_a(t_a)$$

provided $\varphi_{\partial_0 b} = \varphi_{\partial_1 b} \text{Ad} z(b)^*$ for $b \in \Sigma_1$.

There is no a priori reason to suppose that every left inverse for a 1-cocycle arises from such a field of left inverses. In particular a map $t \in (z, z') \mapsto t_a \in (y(a), y'(a))$ might not be surjective. We can also not just begin with a left inverse φ_a for $y(a)$ since it is not clear that we get a field of left inverses using the cocycle.

However, if we assume, that $\mathcal{K}$ has an asymptotically causally disjoint net $\mathcal{O}_n$, thus, given $\mathcal{O} \in \mathcal{K}$, there is an n_0 with $\mathcal{O}_n \perp \mathcal{O}$ for $n \geq n_0$, then we can construct left inverses for 1-cocycles. If z is an object of $Z_t^1(\mathfrak{A})$, we denote by $z(a, n)$, the evaluation of z on a path p with $\partial_0 p = a$ and $\partial_1 p = \mathcal{O}_n$. This is independent of the chosen path. We now define $\varphi_a(X)$ to be a Banach–limit over n of $z(a, n)^* X z(a, n)$. Then φ_a is a positive linear map satisfying

$$\varphi_a(X)A = \varphi_a(Xy(a)_{\mathcal{O}}(A)), \quad A \in \mathfrak{A}(\mathcal{O}).$$

Furthermore, from the cocycle identity we have

$$\varphi_{\partial_0 b} = \varphi_{\partial_1 b} \text{Ad} z(b)^*.$$

If $\hat{\perp}$–duality holds, then by Lemma 29.1, each φ_a defines a left inverse for $y(a)$ and we have constructed a left inverse for z by Lemma 29.2.

29.3 Lemma *If $\mathcal{K}$ has an asympotically causally disjoint net $\mathcal{O}_n$ and $\hat{\perp}$–duality holds, then every $z \in Z_t^1(\mathfrak{A})$ has a left inverse.*

$\mathcal{K}$ has an asymptotically causally disjoint net if it is directed or if it is the set of regular diamonds in a globally hyperbolic spacetime with non-compact Cauchy surfaces. However this method of constructing left inverses for cocycles or for transportable endomorphisms using an asymptotically causally disjoint net does not suffice, as it stands, when the Cauchy surface is compact and this is the first serious obstacle we have met in our attempt to develop the theory of sectors in curved spacetime. A left inverse is needed, at least locally, to be able to classify statistics and globally to show the existence of conjugate localized endomorphisms with finite statistics. I return to this problem in the next section.

Once we have solved the problem of proving the existence of a left inverse on a braided tensor C^*–category, we are in a position to apply existing results classifying the statistics. The following result is just a rewording of Theorem 5.5 of [35].

Theorem 29.4 *Let ρ be an object in a symmetric tensor C^*–category $(\mathcal{T}, \varepsilon)$ and φ a left inverse of ρ with $\varphi_{\rho, \rho} = \lambda 1_\rho$ for some scalar λ. Then $\lambda \in \{0\} \cup \{\pm d^{-1} : d \in \mathbb{N}\}$ and depends only on the equivalence class of ρ. The Young tableaux associated with the representations of $\mathbb{P}_n$ on (ρ^n, ρ^n), $n \geq 1$ are all Young tableaux:*

a) *whose columns have length $\leq d$, if $\lambda = d^{-1}$ (para-Bose statistics of order d);*
b) *whose rows have length $\leq d$ if $\lambda = -d^{-1}$ (para-Fermi statistics of order d);*
c) *without restriction, if $\lambda = 0$ (infinite statistics).*

Note that when ρ is irreducible, $\varphi_{\rho,\rho}(\varepsilon(\rho,\rho))$ is automatically a scalar, called the *statistics parameter* of ρ. d is referred to as the *statistics dimension* and the sign is the *statistics phase*, κ_ρ and corresponds to the Bose-Fermi alternative. In general, we say that ρ has *infinite statistics* if there is a left inverse φ with $\varphi_{\rho,\rho}(\varepsilon(\rho,\rho)) = 0$. Otherwise ρ is said to have finite statistics. Assuming our category $\mathcal{T}$ has subobjects, ρ has finite statistics if and only if ρ is a finite direct sum of irreducible objects with finite statistics. In the cases where we can have braid statistics there is, of course, no correspondingly complete classification, not even if we invoke the special setting of a two dimensional Minkowski space. However, many partial results are known in that case and the proofs presumably generalize without essential modification.

30 Change of Index Set

Comparing this discussion of superselection sectors based on nets over a partially ordered set $\mathcal{K}$ with a relation $\perp$ of causal disjointness with the original version where $\mathcal{K}$ is the set of double cones in Minkowski space, we see that the generalization has largely been successful: when $\mathcal{K}$ is directed, there are no problems. Otherwise, most of the desired results have been achieved at the cost of relatively few new hypotheses.

Let us review these hypotheses. $\tilde{\perp}$-duality was needed even to derive the tensor structure of superselection sectors whilst $\hat{\perp}$-duality or local duality is required before being able to talk about statistics or to show that left inverses are localized.

Furthermore, when $\mathcal{K}$ is the set of regular diamonds in a globally hyperbolic spacetime, the different assumptions of duality, other than local duality, are effectively equivalent. If $\mathcal{O}_1 \perp \mathcal{O}_2$ then $\mathcal{O}_1 \tilde{\perp} \mathcal{O}_3$ for a regular diamond $\mathcal{O}_3$ strictly smaller than $\mathcal{O}_2$ and based on the same Cauchy surface. Hence if the net is inner regular, duality and $\tilde{\perp}$–duality coincide. Similarly, see Proposition 5.6 of [50], if the net is additive all three versions of duality coincide. However, this latter result depends on a non–trivial geometric result, Lemma 2.1 of [50].

The real obstacle to progress when $\mathcal{K}$ is not directed is the absence of a left inverse without which we are not able to classify the statistics let alone prove the existence of conjugates in the case of finite statistical dimension. In what follows, we develop results designed to overcome this hurdle and circumvent local duality whose validity has not yet been verified in models.

The choice of the set of double cones in Minkowski space was a matter of convenience, never a matter of principle. The advantages were that they formed a base for the topology of Minkowski space, were diamonds based on some Cauchy hyperplane and easily visualized. In addition, duality held for double cones in models whereas it might not hold for disconnected or non-simply connected regions. Similar motives govern the choice of the set of regular diamonds in curved spacetime. Basically, we need $\mathcal{K}$ to be a base for the topology to ensure that we generate the correct theory and the elements of $\mathcal{K}$ should have topological and other properties sufficient to ensure duality. Other choices could be envisaged and, for consistency, one has to establish that such a change of base or index set does not change the superselection structure.

We begin by considering what happens if we make $\mathcal{K}$ smaller. We say that $\mathcal{J} \subset \mathcal{K}$ generates $\mathcal{K}$ if, given $\mathcal{O} \in \mathcal{K}$, we can find $\mathcal{O}_i \in \mathcal{J}$ with $\mathcal{O} = \vee_i \mathcal{O}_i$. We say that a net $\mathfrak{A}$ is *additive* if $\vee_i \mathcal{O}_i = \mathcal{O}$ implies $\vee_i \mathfrak{A}(\mathcal{O}_i) = \mathfrak{A}(\mathcal{O})$.

30.1 Theorem *Let $\mathcal{J} \subset \mathcal{K}$ be an inclusion of connected partially ordered sets with a causal disjointness relation $\perp$. Let $\mathfrak{A}$ be a net of von Neumann algebras over $\mathcal{K}$ Let $\mathfrak{A}_{\mathcal{J}}$ be the net over $\mathcal{J}$ with reference representation obtained by restricting $\mathfrak{A}$ and its reference representation, then if $\mathfrak{A}$ and $\mathfrak{A}_{\mathcal{J}}$ generate the same von Neumann algebra in the reference representation, restricting representations from $\mathcal{K}$ to $\mathcal{J}$ and acting identically on intertwiners defines a faithful functor from $Rep^\perp \mathfrak{A}$ to $Rep^\perp \mathfrak{A}_{\mathcal{J}}$. If $\mathfrak{A}$ is additive and $\mathcal{J}$ generates $\mathcal{K}$, then this functor is an equivalence.*

When $\mathcal{J}$ generates $\mathcal{K}$ and $\mathfrak{A}$ is additive, then $\perp$–duality or $\tilde{\perp}$–duality imply $\perp$–duality or $\tilde{\perp}$–duality for $\mathfrak{A}_{\mathcal{J}}$.

Looked at from the point of view of cohomology, the above functor corresponds to restricting cocycles. But in this context we have the following more interesting result.

30.2 Theorem *Let $\mathcal{J} \subset \mathcal{K}$ be an inclusion of connected partially ordered sets with a causal disjointness relation $\perp$ such that given $\mathcal{O}_1 \in \mathcal{K}$ there is a $\mathcal{O}_2 \in \mathcal{J}$ with $\mathcal{O}_1 \perp \mathcal{O}_2$. Let $\mathfrak{A}$ be a net of von Neumann algebras over $\mathcal{K}$ satisfying $\tilde{\perp}$–duality. Let $\mathfrak{A}_{\mathcal{J}}$ be the net over $\mathcal{J}$ with reference representation obtained by restricting $\mathfrak{A}$ and reference representation, then if $\mathfrak{A}$ and $\mathfrak{A}_{\mathcal{J}}$ generate the same von Neumann algebra in the reference representation and $\mathfrak{A}_{\mathcal{J}}$ satisfies $\tilde{\perp}$–duality, restricting from $\Sigma_*(\mathcal{K})$ to $\Sigma_*(\mathcal{J})$ defines a full and faithful tensor functor $Z_t^1(\mathfrak{A}) \to Z_t^1(\mathfrak{A}_{\mathcal{J}})$. If $\mathcal{J}$ generates $\mathcal{K}$ then we even get an equivalence.*

Proof. We obviously get a faithful functor in this way. It is full since $\mathcal{K}$ is connected. In fact, given objects z, z' of $Z_t^1(\mathfrak{A})$, and an arrow t in $Z^1(\mathfrak{A}_{\mathcal{J}})$ between their images, we pick a base point a_0 in $\Sigma_0(\mathcal{J})$ and for each $a \in \Sigma_0(\mathcal{K})$ a path p_a with $\partial_0 p_a = a$ and $\partial_1 p_a = a_0$. We then set

$$s_a := z'(p_a) t_{a_0} z(p_a)^*,$$

and verify that

$$s_{\partial_0 b} z(b) = z'(b) s_{\partial_1 b}, \quad b \in \Sigma_1(\mathcal{K})$$

and $s_a = t_a$ for $a \in \Sigma_0(\mathcal{J})$ using the cocycle property for z and z' and the connectivity of $\mathcal{J}$. The functor will therefore be full if we can show that $s_a \in \mathfrak{A}(a)$. However this follows from Lemma 27.3. We have a tensor functor since the field of localized endomorphisms associated with the image of z is just the restriction of y.

We now consider what happens when we extend $\mathcal{K}$. This is frequently done in Minkowski space in that in algebraic quantum field theory one considers not just double cones but other regions, such as wedges, light cones and spacelike cones, and their associated von Neumann algebras. These von Neumann algebras are defined to be generated by the smaller double cones, in accordance with the philosophy that the relevant physical information is already contained in the net of observables over $\mathcal{K}$.

In the context of the present formalism, where $\mathcal{K}$ is a partially ordered set, we use the notion of a sieve S, see Sec. 23, rather than a region, and consider the set $\tilde{\mathcal{K}}$ of sieves S of $\mathcal{K}$ such that neither S nor $S^{\perp}$ are the empty set, ordered under inclusion. The binary relation on $\tilde{\mathcal{K}}$ is defined by $S' \perp S$ if and only if $S' \subset S^{\perp}$. $\mathcal{K}$ is considered as a subset of $\tilde{\mathcal{K}}$ by identifying $\mathcal{O}$ with the sieve it generates. Note that when working with $\tilde{\mathcal{K}}$, $S^{\perp}$ denotes an element of $\tilde{\mathcal{K}}$ rather than a subset, the subset will be denoted

$$(S)^{\perp} := \{S' \in \tilde{\mathcal{K}} : S' \perp S\}.$$

On $\tilde{\mathcal{K}}$ we have $\tilde{\perp} = \hat{\perp}$. In fact if $S_1 \tilde{\perp} S_2$ then $[S_1 \cup S_2]^{\perp} = S_1^{\perp} \cap S_2^{\perp} \neq \emptyset$ so that $S_1 \hat{\perp} S_2$.

We get an extension of our net over $\mathcal{K}$ to a net over $\tilde{\mathcal{K}}$ by defining, for each sieve S, $\mathfrak{A}(S)$ to be the von Neumann algebra generated in the defining representation by the $\mathfrak{A}(\mathcal{O})$ with $\mathcal{O} \in S$.

We now show that a representation π of $\mathfrak{A}$ satisfying the selection criterion has a natural extension to a representation of the net $S \mapsto \mathfrak{A}(S)$. We pick for each $a \in \Sigma_0$ a unitary V_a such that

$$\pi_{\mathcal{O}}(A) = V_a^* A V_a, \quad A \in \mathfrak{A}(\mathcal{O}), \ \mathcal{O} \in a^{\perp},$$

and then define

$$\pi_S(A) := V_a^* A V_a, \quad A \in \mathfrak{A}(S), \ a \in S^{\perp}.$$

Note that this expression is well defined being independent of the choice of $a \in S^{\perp}$ since, if $a' \in S^{\perp}$, then

$$V_{a'} V_a^* \in \cap_{\mathcal{O} \in S} \mathfrak{A}(\mathcal{O})' = \mathfrak{A}(S)'.$$

In the same way, we see that π_S is independent of the choice of $a \mapsto V_a$. Note, too that we get a representation of the extended net in that if $S_1 \subset S_2$ then π_{S_1} is the restriction of π_{S_2} to $\mathfrak{A}(S_1)$. The extended representation obviously satisfies the selection criterion from the way it has been constructed. Obviously, an intertwiner $T \in (\pi, \pi')$ over $\mathcal{K}$ remains an intertwiner over $\tilde{\mathcal{K}}$ so that effectively $\text{Rep}^{\perp}\mathfrak{A}$ remains unchanged when we extend the net.

Nevertheless, the results obtained so far do not all apply to the extended net. At least $\mathcal{K}$ connected implies $\tilde{\mathcal{K}}$ connected, so that we have no problems on that score. However, even if the original net satisfies duality, we cannot expect the extended net to satisfy duality. Let us compute the dual of the extended net.

$$\mathfrak{A}^d(S) = \cap_{S_1 \perp S} \mathfrak{A}(S_1)' = \cap_{S_1 \perp S} \cap_{\mathcal{O} \in S_1} \mathfrak{A}(\mathcal{O})' = \cap_{\mathcal{O} \perp S} \mathfrak{A}(\mathcal{O})' = \mathfrak{A}(S^{\perp})'.$$

Thus duality for S takes the very simple form $\mathfrak{A}(S) = \mathfrak{A}(S^{\perp})'$ and implies duality for $S^{\perp}$ and $\mathfrak{A}(S) = \mathfrak{A}(S^{\perp\perp})$. Furthermore, restricting the extended dual net to $\mathcal{K}$ gives us the original dual net.

We get similar results for $\tilde{\perp}$–duality. In fact defining for $\mathcal{S} \in \tilde{\mathcal{K}}$,

$$\mathcal{S}^{\tilde{\perp}} := \{\mathcal{O} \in \mathcal{K} : \mathcal{O}\tilde{\perp}\mathcal{S}\},$$

the analogous computation shows that $\tilde{\perp}$–duality for $\mathcal{S}$ is expressed by $\mathfrak{A}(\mathcal{S}) = \mathfrak{A}(\mathcal{S}^{\tilde{\perp}})'$ and thus implies $\tilde{\perp}$–duality for $\mathcal{S}^{\tilde{\perp}}$. Furthermore, if $\tilde{\perp}$–duality holds for $\mathcal{O}$ in the original net, it holds for $\mathcal{O}$ in the extended net. As we need $\perp$–duality for superselection theory, we suppose in what follows that $\tilde{\perp}$–duality holds for each $\mathcal{O} \in \mathcal{K}$ and that given $\mathcal{O} \in \mathcal{K}$ there are $\mathcal{O}_1, \mathcal{O}_2 \subset \mathcal{O}$ with $\mathcal{O}_1 \perp \mathcal{O}_2$. Define $\mathcal{L}$ to be the partially ordered set of sieves in $\tilde{\mathcal{K}}$ for which $\tilde{\perp}$–duality holds, ordered under inclusion. Since $\mathcal{K} \subset \mathcal{L}$, $\mathcal{L}$ is obviously connected and coinitial in $\tilde{\mathcal{K}}$. However, it is also cofinal since if $\mathcal{S} \in \tilde{\mathcal{K}}$ there is an $\mathcal{O} \in \mathcal{K}$ with $\mathcal{O}\tilde{\perp}\mathcal{S}$ so $\mathcal{O}^{\tilde{\perp}} \supset \mathcal{S}$ and $\mathcal{O}^{\tilde{\perp}} \in \mathcal{L}$. Thus if we use the relation $\perp$ induced from $\tilde{\mathcal{K}}$, we shall continue to have $\tilde{\perp} = \hat{\perp}$ on $\mathcal{L}$. On the other hand, there will be less chance that all the restriction mappings $\mathcal{L} \upharpoonright \mathcal{S}_1 \to \mathcal{L} \upharpoonright \mathcal{S}_2$, $\mathcal{S}_2 \subset \mathcal{S}_1$, are cofinal. However, this condition is, presumably, of little practical importance. In particular, it does not hold if $\mathcal{K}$ is the set of regular diamonds in a globally hyperbolic spacetime with compact Cauchy surface.

We now consider the net $\mathfrak{A}^{\mathcal{L}}$ over $\mathcal{L}$ obtained by restricting the extended net. For clarity, the original net will be denoted by $\mathfrak{A}^{\mathcal{K}}$. The reference representation for $\mathfrak{A}^{\mathcal{L}}$ is just the restriction of that of the extended net.

Furthermore, $\tilde{\perp}$–duality holds for the net $\mathfrak{A}^{\mathcal{L}}$. In fact since each $\mathfrak{A}(\mathcal{S})$ is generated by the $\mathfrak{A}(\mathcal{O})$ with $\mathcal{O} \in \mathcal{S}$, and $\mathcal{O} \in \mathcal{L}$, the duality that holds for $\mathcal{S} \in \mathcal{L}$ in the extended net is not lost in passing to $\mathfrak{A}^{\mathcal{L}}$. Since $\tilde{\perp} = \hat{\perp}$ on $\mathcal{L}$, $\hat{\perp}$–duality holds for $\mathfrak{A}^{\mathcal{L}}$, too.

However, as regards the important hypothesis of local duality, the situation is less favourable. If local duality holds for $\mathfrak{A}^{\mathcal{K}}$, then a fortiori duality will continue to hold for nets of the form $(\mathfrak{A}^{\mathcal{L}})^{\mathcal{O}}$ with $\mathcal{O} \in \mathcal{K}$; it might well fail to hold for nets of the form $(\mathfrak{A}^{\mathcal{L}})^{\mathcal{S}}$ with $\mathcal{S} \in \mathcal{L}$.

Restricting from $\mathcal{L}$ to $\mathcal{K}$ takes us from $\mathfrak{A}^{\mathcal{L}}$ to $\mathfrak{A}^{\mathcal{K}}$ and induces an isomorphism of W^*–categories $\mathrm{Rep}^{\perp}\mathfrak{A}^{\mathcal{L}} \to \mathrm{Rep}^{\perp}\mathfrak{A}^{\mathcal{K}}$. It similarly induces an equivalence of tensor W^*–categories $Z_t^1(\mathfrak{A}^{\mathcal{L}}) \to Z_t^1(\mathfrak{A}^{\mathcal{K}})$ and indeed of tensor W^*–categories $Z_t^1(T^{\mathcal{L}}) \to Z_t^1(T^{\mathcal{K}})$. Thus, as one could anticipate, the superselection structure does not change when passing from $\mathfrak{A}^{\mathcal{K}}$ to $\mathfrak{A}^{\mathcal{L}}$. Recalling Theorem 30.2, we now have the following result.

30.3 Proposition *Let* $\mathcal{K} \subset \mathcal{M} \subset \mathcal{L}$ *be inclusions of partially ordered sets with a relation* $\perp$ *of causal disjointness,* $\mathcal{K}$ *and* $\mathcal{L}$ *as above, then restriction induces equivalences,*

$$Z_t^1(\mathfrak{A}^{\mathcal{L}}) \to Z_t^1(\mathfrak{A}^{\mathcal{M}}) \to Z_t^1(\mathfrak{A}^{\mathcal{K}}),$$

of tensor W^**–categories.*

Here is an application of this result. Let $\mathcal{K}$ denote the set of double cones in Minkowski space and $\mathcal{M}$ the set of regular diamonds, each ordered under inclusion and with the usual relation of causal disjointness. Then $\mathcal{K} \subset \mathcal{M}$ and $\mathcal{K}$ is both coinitial and cofinal. If we identify, in an obvious way, a regular diamond with the sieve in $\mathcal{K}$ that it generates, then we have an inclusion $\mathcal{M} \subset \tilde{\mathcal{K}}$. If the net $\mathfrak{A}^{\mathcal{M}}$

satisfies $\hat{\perp}$–duality and additivity then in fact $\mathcal{M} \subset \mathcal{L}$ and $\mathfrak{A}^{\mathcal{M}}$ is just the net induced from $\mathfrak{A}^{\mathcal{K}}$ by extending to sieves. We are now in the situation of Proposition 30.3 and deduce that we get the same sector structure whether we use double cones or regular diamonds.

Returning to the basic difficulty, the existence of a left inverse, I will explain how it should be possible to overcome it, although I have not even begun the task of verifying the hypotheses for a class of models in curved spacetime.

We first consider a spacetime with non–compact Cauchy surfaces. Let H be such a Cauchy surface and $\mathcal{K}_H \subset \mathcal{K}$ the set of regular diamonds based on H with the induced causal disjointness relation. Let $\mathfrak{A}_H$ be the net over $\mathcal{K}_H$ with reference representation obtained by restricting $\mathfrak{A}$ and its reference representation. Our formalism is sufficiently flexible to be applied without difficulty to the net $\mathfrak{A}_H$. Furthermore, since H is not compact, $\mathcal{K}_H$ has an asymptotically causally disjoint net and once $\mathfrak{A}_H$ satisfies duality in the reference representation all the standard results of the theory of superselection sectors apply. Proving duality for $\mathfrak{A}_H$ in a model or class of models should be no more difficult than proving duality for $\mathfrak{A}$. In fact, I know of no proof of duality which does not implicitly proceed by proving duality for $\mathfrak{A}_H$ for some class of Cauchy surfaces H.

We expect that, H being a Cauchy surface, all the relevant physical information pertinent to $\mathfrak{A}$ is already contained in $\mathfrak{A}_H$. The details that still have to be filled in concern the relationship between $\mathfrak{A}$ and $\mathfrak{A}_H$ with the aim of relating their sector structures as in Theorem 30.2. In view of causal propagation, we can expect this relationship to be close. First of all, $\mathfrak{A}$ and $\mathfrak{A}_H$ should generate the same von Neumann algebra $\mathcal{R}$ in the reference representation. This is true when $\mathcal{K}_H$ is cofinal in $\mathcal{K}$, but also if $\mathfrak{A}$ is additive and any sufficiently small regular diamond is contained in a regular diamond based on H. Note, too that if we foliate our spacetime into Cauchy surfaces H_t such that $\mathcal{K}_{H_t}$ covers $H_{t+\varepsilon}$ for $|\varepsilon|$ sufficiently small, then in the case of additive nets the von Neumann algebra generated by $\mathfrak{A}_{H_t}$ is independent of t.

Note that the use of Lemma 27.3 in Theorem 30.2 requires $\mathcal{O}^\perp \cap \mathcal{K}_H \neq \emptyset$. However the causal set of the closure of $\mathcal{O}$ intersects H in a compact subset so a regular diamond based on a set in the complement of that compact set will be in this intersection.

This is encouraging because it means that one can analyze the statistics of a object z of $Z_t^1(\mathfrak{A})$ by examining the statistics of its image in $Z_t^1(\mathfrak{A}_H)$. Admittedly, spacetimes with non–compact Cauchy surfaces present little problems because $\mathcal{K}$ has an asymptotically causally disjoint net, and is directed in practice.

Turning to the case of a globally hyperbolic spacetime with compact Cauchy surface, we pick such a surface H and a point $x \in H$ and let $\mathcal{K}_{H^x} \subset \mathcal{K}$ denote the set of regular diamonds based on H, the closure of whose bases does not contain x. We let $\mathfrak{A}_{H^x}$ denote the net over $\mathcal{K}_{H^x}$ with reference representation obtained by restricting $\mathfrak{A}$ and its representation. We expect that puncturing the Cauchy surface has no effect on the validity of duality since we expect cf. [83], [79] that

$$\cap_{x \in \mathcal{O}} \mathfrak{A}(\mathcal{O}) = \mathbb{C} \cdot I,$$

at least for additive nets. For the same reason, $\mathfrak{A}$ and $\mathfrak{A}_{H^x}$ should generate the same von Neumann algebra $\mathcal{R}$ in their reference representations and $\bar{\perp}$–duality should continue to hold for $\mathfrak{A}_{H^x}$. Having removed a point from H, $\mathcal{K}_{H^x}$ obviously has an asymptotically causally disjoint net that can be used to define left inverses.

We have therefore outlined how to obtain a left inverse that can be used to classify the statistics. However, there remains one other weak point in the original scheme. We have made use of the postulate of local duality in Sec. 28 to show the existence of a braiding or a symmetry ε. Whilst local duality is a plausible property with obvious affinities to duality for a theory over $\mathcal{O}$ considered as a spacetime with the induced metric, it is likely to prove hard, if not impossible, to verify. Passing to a Cauchy surface helps here too. Not only should it be possible to check local duality on a Cauchy surface but, as I indicate below, a substitute for Lemma 28.1 can be found, rendering local duality superfluous.

We rely on Lemma 2.1 of [50] or rather on a simpler situation also covered by the proof in [80].

30.4 Lemma *Let H be a Cauchy surface, $\mathcal{O} \in \mathcal{K}_H$ then*

a) *Given a neighbourhood of the base of $\mathcal{O}$ in H, there is a regular diamond with base in that neighbourhood containing the closure of $\mathcal{O}$.*

b) *Given $x \in H$ causally disjoint from $\mathcal{O}$, there is an $\mathcal{O}_1 \in \mathcal{K}_H$ with $\mathcal{O} \cup \{x\} \subset \mathcal{O}_1$.*

30.5 Lemma *Let H be a Cauchy surface and $\mathfrak{A}_H$ an additive net satisfying $\bar{\perp}$–duality, then for each $\mathcal{O}_1 \subset \mathcal{O}_2$ in $\mathcal{K}_H$, the restriction functor from $\mathcal{T}_t(\mathcal{O}_1)$ to $\mathcal{T}_t(\mathcal{O}_2)$ is an equivalence of tensor W^*–categories.*

Proof. As $\mathfrak{A}_H$ is additive, $\bar{\perp}$–duality and $\hat{\perp}$–duality coincide. Let ρ_1 and σ_1 be objects of $\mathcal{T}_t(\mathcal{O}_1)$ and $T \in (\rho_1, \sigma_1)$ in $\mathcal{T}_t(\mathcal{O}_2)$. Choose $\mathcal{O}_3 \in \mathcal{K}_H$ containing the closure of the base of $\mathcal{O}_2$ as in Lemma 30.4a. Then if $\mathcal{O} \perp \mathcal{O}_1$ and $\mathcal{O} \subset \mathcal{O}_3$ then

$$TA = T\rho_1(A) = \sigma_1(A)T = AT, \quad A \in \mathfrak{A}_H(\mathcal{O}).$$

The same computation is valid if $\mathcal{O} \hat{\perp} \mathcal{O}_2$. But such $\mathcal{O}$ cover the complement of the closure of the base of $\mathcal{O}_1$. Hence $T \in \mathfrak{A}_H(\mathcal{O}_1)$ as required. Since such $\mathcal{O}$ together with $\mathcal{O}_3$ cover the base of any $\mathcal{O}_4 \in \mathcal{K}_H \supset \mathcal{O}_1$, additivity shows that $T \in (\rho_1, \sigma_1)$ in $\mathcal{T}_t(\mathcal{O}_1)$ so our faithful functor is full. Transportability again ensures that it is an equivalence as in Lemma 28.1.

We have seen that the idea of changing $\mathcal{K}, \perp$ has been put to good use several times in the course of this discussion of sector theory on curved spacetime. As we have stressed, it is this structure that enters into algebraic quantum field theory. It suggests that the classical notion of a curved spacetime is not the right one for quantum field theory and should be replaced by a suitable class of partially ordered sets with a causal disjointness relation.

The interpretation of the AdS–CFT correspondence given by Rehren [70], [71] provides an example of a change of index set in that there is a bijection of partially ordered sets with a causal disjointness and which even preserves the action of some symmetry group. This bijection takes wedge regions in anti–de Sitter space into double cones on conformal Minkowski space. In a similar way, Buchholz, Mund and Summers [27] transplant nets of observables from de Sitter space to a large class of Robertson-Walker spacetimes to give examples where the condition of geometric modular action is satisfied. The resulting modular symmetry group can be larger than the isometry group of the spacetime.

References

1. H. Araki. Mathematical Theory of Quantum Fields, Oxford University Press, Oxford 1999.
2. H. Araki: "A Lattice of von Neumann Algebras associated with the Quantum Theory of a Free Bose Field," J. Math. Phys. **4**, 1343-1362 (1963)
3. H. Araki: "Von Neumann Algebras of Local Observables for the Free Scalar Field," J. Math. Phys. **5**, 1-13 (1964)
4. H. Araki, R. Haag: "Collision Cross Sections in Terms of Local Observables," Commun. Math. Phys. **4**, 77-91 (1967)
5. H. Baumgärtel. Operatoralgebraic Methods in Quantum Field Theory, Akademie Verlag, Berlin, 1995
6. H. Baumgärtel, M. Wollenberg. Causal Nets of Operator Algebras, Akademie Verlag, Berlin, 1992
7. J.J. Bisognano, E.H. Wichmann: "On the Duality Condition for an Hermitean Scalar Field," J. Math. Phys. **16**, 985-1007 (1975)
8. J.J. Bisognano, E.H. Wichmann: "On the Duality Condition for Quantum Fields," J. Math. Phys. **17**, 303-321 (1976)
9. N. Bohr, L. Rosenfeld: "Field and Charge Measurements in Quantum Electrodynamics," Phys. Rev. **78**, 794-798 (1950).
10. H.-J. Borchers: "Local Rings and the Connection of Spin with Statistics," Commun. Math. Phys. **1**, 281–307 (1965)
11. H.-J. Borchers: "A Remark on a Theorem of B. Misra," Commun. Math. Phys. **4**, 315-323 (1967)
12. H.-J. Borchers: Translation Group and Particle Representations in Quantum Field Theory, Springer–Verlag, Berlin Heidelberg New York 1996
13. R. Brunetti, K. Fredenhagen, "Interacting Quantum Fields in Curved Space: Renormalizability of φ^4," in the Proceedings of the Conference "Operator Algebras and Quantum Field Theory" held in Rome, July 1996, S. Doplicher, R. Longo, J.E. Roberts, L. Zsido eds, International Press, 1997
14. R. Brunetti, K. Fredenhagen, "Microlocal Analysis and Interacting quantum field theories: Renormalization on Physical Backgrounds," Commun. Math. Phys. **208**, 623-661 (2000)
15. R. Brunetti, K. Fredenhagen, M. Köhler, "The Microlocal Spectrum Condition and Wick Polynomials of Free Fields in Curved Spacetimes," Commun. Math. Phys. **180**, 633-652 (1996)
16. R. Brunetti, D. Guido, R. Longo, "Modular Localization and Wigner Particles," Rev. Math. Phys. **7** and **8**, 759-785 (2002)
17. D. Buchholz: "Product States for Local Algebras," Commun. Math. Phys. **36**, 287-304 (1974)

18. D. Buchholz: "Collision Theory for Massless Fermions," Commun. Math. Phys. **42**, 269-279 (1975)

19. D. Buchholz: "Collision Theory for Massless Bosons," Commun. Math. Phys. **52**, 147-173 (1977)

20. D. Buchholz: "The Physical State Space of Quantum Electrodynamics," Commun. Math. Phys. **85**, 49-71 (1982)

21. D. Buchholz: "Quarks, Gluons, Colour: Facts or Fiction?" Nucl. Phys. **B 469**, 333-353 (1996)

22. D. Buchholz, S. Doplicher, R. Longo: "On Noether's Theorem in Quantum Field Theory," Annals of Phys. **170**, 1-17 (1986)

23. D. Buchholz, S. Doplicher, G. Morchio, J.E. Roberts, F. Strocchi: "A Model for Charges of Electromagnetic Type," in the Proceedings of the Conference "Operator Algebras and Quantum Field Theory" held in Rome, July 1996, S. Doplicher, R. Longo, J.E. Roberts, L. Zsido eds, International Press, 1997

24. D. Buchholz, S. Doplicher, G. Morchio, J.E. Roberts, F. Strocchi: "Quantum Delocalization of the Electric Charge," Annals of Physics **290**, 53-66 (2001)

25. D. Buchholz, O. Dreyer, M. Florig, S.J. Summers: "Geometric Modular Action and Spacetime Symmetry Groups," Rev. Math. Phys. **12**, 475-560 (2000)

26. D. Buchholz, K. Fredenhagen: 'Locality and the Structure of Particle States," Commun. Math. Phys. **84**, 1-54 (1982)

27. D. Buchholz, J. Mund, S.J. Summers: "Transplantation of Local Nets and Geometric Modular Action on Robertson–Walker Space–times," Fields Institute Communications **30**, 65–81 (2001)

28. D. Buchholz, M. Porrmann and U. Stein: "Dirac versus Wigner: Towards a Universal Particle Concept in Local Quantum Field Theory," Phys. Lett. B **267**, 377-381 (1991).

29. A. Connes, J. Lott: "Particle Models and non commutative Geometry," Nucl. Phys. B. (Proc. Suppl.) **11B**, 19 (1990)

30. P. Deligne: "Catégories Tannakiennes." In: Grothendieck Festschrift, Vol. 2, 111-193 Birkhäuser, 1991

31. J. Dixmier, O. Maréchal: "Vecteurs totalisateurs d'une algèbre de von Neumann," Commun. Math. Phys. **22**, 44-50 (1971)

32. S. Doplicher: "Local Aspects of Superselection Rules," Commun. Math. Phys. **85**, 73-86 (1982)

33. S. Doplicher, K. Fredenhagen, J.E. Roberts: "Quantum Structure of Spacetime at the Planck Scale and Quantum Fields," Commun. Math. Phys. **172**, 187-224 (1995)

34. S. Doplicher, R. Haag, J.E. Roberts: "Fields, Observables and Gauge Transformations I," Commun. Math. Phys. **13**, 1-23 (1969), II, Commun. Math. Phys. **15**, 173-200 (1969)

35. S. Doplicher, R. Haag, J.E. Roberts: "Local Observables and Particle Statistics I," Commun. Math. Phys. **23**, 199-230 (1971) and II, Commun. Math. Phys. **35**, 49-85 (1974)

36. S. Doplicher, R. Longo: "Local Aspects of Superselection Rules II," Commun. Math. Phys. **88**, 399-409 (1983)

37. S. Doplicher, R. Longo: "Standard and Split Inclusions of von Neumann Algebras," Invent. Math. **75**, 493-536 (1984)

38. S. Doplicher, J.E. Roberts: "A New Duality Theory for Compact Groups," Invent. Math. **98**, 157-218 (1989)

39. S. Doplicher, J.E. Roberts: "Why there is a Field Algebra with a Compact Gauge Group describing the Superselection Structure in Particle Physics," Commun. Math. Phys. **131**, 51-107 (1990)

40. W. Driessler: "Duality and Absence of Locally Generated Superselection Sectors for CCR type Algebras." Commun. Math. Phys. **70**, 213-220 (1979)

41. K. Fredenhagen: "On the Existence of Antiparticles," Commun. Math. Phys. **79**, 141-151 (1981)

42. K. Fredenhagen: "Generalizations of the Theory of Superselection Theories." In: The Algebraic Theory of Superselection Sectors. Introduction and Recent Results, ed. D. Kastler, pp. 379-387. World Scientific, Singapore, New Jersey, London, Hong Kong 1990

43. K. Fredenhagen: "Superselection Sectors in Low Dimensional Quantum Field Theory," J. Geom. Phys. **11**, 337-348 (1993)

44. K. Fredenhagen, K.-H. Rehren, B. Schroer: "Superselection Sectors with Braid Group Statistics and Exchange Algebras," I, Commun. Math. Phys. **125**, 201-226 (1989); II, Geometric Aspects and Conformal Invariance, Rev. Math. Phys. **Special Issue**, 113-157 (1992)

45. J. Fröhlich: "New Super-selection Sectors ("Soliton States") in two dimensional Bose Quantum Field Theory Models," Commun. Math. Phys. **47**, 269-310 (1976)

46. J. Fröhlich: "The Charged Sectors of Quantum Electrodynamics in a Framework of Local Observables," Commun. Math. Phys. **66**, 223-265 (1979)

47. J. Fröhlich, F. Gabbiani, P.-A. Marchetti: "Braid Statistics in three-dimensional Local Quantum Theory." In: The Algebraic Theory of Superselection Sectors. Introduction and Recent Results, ed. D. Kastler, pp. 259-332. World Scientific, Singapore, New Jersey, London, Hong Kong 1990

48. D. Guido, R. Longo: "An Algebraic Spin and Statistics Theorem," Commun. Math. Phys. **172**, 517-533 (1995)

49. D. Guido, R. Longo: "The Conformal Spin and Statistics Theorem," Commun. Math. Phys. **181**, 11-35 (1996)

50. D. Guido, R. Longo, J.E. Roberts, R. Verch: "Charged Sectors, Spin and Statistics in Quantum Field Theory on Curved Spacetimes," Rev. Math. Phys. **13**, 125-198 (2001)

51. R. Haag: "Quantum Field Theories with Composite Particles and Asymptotic Conditions," Phys. Rev. **112**, 669-673 (1958)

52. R. Haag: Discussion des "axIomes" et des propriétés axiomatiques d'une théorie des champs locale avec particules composeés. Colloques Internationaux du CNRS, Lille 1957. CNRS Paris 1959

53. R. Haag: Local Quantum Physics, 2nd ed., Springer, Berlin, Heidelberg, New York, 1996

54. R. Haag, D. Kastler: "An Algebraic Approach to Quantum Field Theory," J. Math. Phys. **5**, 848-861 (1964)

55. R. Haag, H. Narnhofer, U. Stein: "On Quantum Field Theory in a Gravitational Background," Commun. Math. Phys. **94**, 219-238 (1984)

56. P.D. Hislop, R. Longo: "Modular Structure of Local Algebras associated with the Free Massless Scalar Field Theory," Commun. Math. Phys. **84**, 71-85 (1982)

57. A. Jaffe, A. Lesniewski, K. Osterwalder: "Quantum K-theory. I. The Chern character," Commun. Math. Phys. **118**, 1-14 (1988)

58. D. Kastler: "Cyclic Cocycles from Graded KMS Functionals," Commun. Math. Phys. **121**, 345-350 (1989)

59. Y. Kawahigashi, R. Longo: "Classification of Local Conformal Nets. Case $c < 1$."

60. Y. Kawahigashi, R. Longo, M. Müger: "Multi-interval Subfactors and Modularity of Representations in Conformal Field Theory," Commun. Math. Phys. **219**, 631-669 (2001)

61. W. Kunhardt: "On Infravacua and the Localization of Sectors," J. Math. Phys. 39, 6353-6361 (1998)

62. R. Longo: "Index of Subfactors and Statistics of Quantum Fields. I," Commun. Math. Phys. **126**, 217-247 (1989), II, "Correspondences, Braid Group Statistics and Jones Polynomial," Commun. Math. Phys. **130**, 285-309 (1990)

63. R. Longo: "An Analogue of the Kac–Wakimoto Formula and Black Hole Conditional Entropy," Commun. Math. Phys. **186**, 451-479 (1997)

64. R. Longo, J.E. Roberts: "A Theory of Dimension," K-Theory **11**, 103-159 (1997)

65. M. Müger: "Superselection Structure of Massive Quantum Field Theories in 1+1 Dimensions," Rev. Math. Phys. **10**, 1147-1170 (1998)

66. M. Müger: "On Soliton Automorphisms in Massive and Conformal Theories," Rev. Math. Phys. **11**, 337-359 (1999)

67. J. Mund: "The Bisognano–Wichmann Theorem for Massive Theories." Ann. Henri Poincaré **2**, 907-926 (2001)

68. M. Porrmann: "The Concept of Particle Weights in Local Quantum Field Theory," arXiv:hep-th/0005057.

69. K.-H. Rehren: Braid Group Statistics and their Superselection Rules. In: The Algebraic Theory of Superselection Sectors. Introduction and Recent Results, ed. D. Kastler, pp. 333-354. World Scientific, Singapore, New Jersey, London, Hong Kong 1990

70. K.-H. Rehren: "Algebraic Holography," Ann. Henri Poincaré **9**, 607-623 (2001)

71. K.-H. Rehren: "Local Quantum Observables in the AdS–CFT Correspondence," Phys. Lett. **B493**, 383-388 (2000)

72. J.E. Roberts: "Local Cohomology and Superselection Structure," Commun. Math. Phys. **51**, 107-119 (1976)

73. J.E. Roberts: Lectures on Algebraic Quantum Field Theory. In: The Algebraic Theory of Superselection Sectors. Introduction and Recent Results, ed. D. Kastler, pp. 1-112. World Scientific, Singapore, New Jersey, London, Hong Kong 1990

74. H. Roos: "Independence of Local Algebras in Quantum Field Theory," Commun. Math. Phys. **16**, 238-246 (1970)

75. S. Schlieder: "Einige Bemerkungen über Projektionsoperatoren," Commun. Math. Phys. **13**, 216-220 (1969)

76. D. Schlingemann: "On the Existence of Kink– (Soliton–) States in Quantum Field Theory," Rev. Math. Phys. **8**, 1187-1206 (1996)

77. S.J. Summers, R. Verch: "Modular Inclusion, the Hawking Temperature, and Quantum Field Theory in Curved Spacetime," Lett. Math Phys. **37**, 145 (1996)

78. V. Toledano Laredo: "Fusion of Positive Energy Representations of $LSpin_{2n}$", Thesis, University of Cambridge 1997

79. R. Verch: "Continuity of Symplectically Adjoint Maps and the Algebraic Structure of Hadamard Vacuum Representations for Quantum Fields in Curved Spacetime," Rev. Math. Phys. **9**, 635-674 (1997)

80. R. Verch: "Notes on Regular Diamonds," preprint, available as ps-file at http://www.lqp.uni-goettingen.de/lqp/papers/

81. A. Wassermann: "Operator Algebras and Conformal Field Theory III: Fusion of Positive Energy Representations of SU(N) using Bounded Operators," Invent. Math. **133**, 467-538 (1998)

82. G.C. Wick, A.S. Wightman, E.P. Wigner: "On the Intrinsic Parity of Elementary Particles," Phys. Rev. **88**, 101-105 (1952)

83. A.S. Wightman: Ann. Inst. de Henri Poincaré **1**, 403-420 (1964)

List of Participants

1. Almeida Paul
2. Bahns Dorothea
3. Benameur Maulay
4. Blaga Paul
5. Camassa Mario
6. Ciolli Fabio
7. Connes Alain (lecturer)
8. Conti Roberto
9. Cuntz Joachim (lecturer)
10. Dabrowski Ludwik
11. Deicke Klaus
12. Doplicher Sergio (editor)
13. Esteves Joao
14. Fischer Robert
15. Garcia-Bondia Jose'
16. Giorgi Giordano
17. Girelli Florian
18. Gualtieri Marco
19. Guentner Erik
20. Higson Nigel (lecturer)
21. Husemoeller Dale
22. Kaminker Jerry (lecturer)
23. Katsura Takeshi
24. Konderak Jerzy
25. Kopf Tomas
26. Krajewski Thomas
27. Lamourex Michael
28. Lauter Robert
29. Longo Roberto (editor)
30. Martinetti Pierre
31. Minervini Giulio
32. Morsella Gerardo
33. Nekliudova Valentina
34. Okajasu Rui
35. Onn Uri
36. Oyono-Oyono Herve
37. Paolucci Annamaria
38. Perrot Dennis
39. Piacitelli Gherardo
40. Pisante Adriano
41. Posthuma Hessel
42. Przybyszewska Agata Hanna
43. Puschnigg Michael
44. Roberts John Elias (lecturer)
45. Ruzzi Giuseppe
46. Scheck Florian
47. Steger Tim
48. Taylor Keith
49. Teleman Nicolae
50. Thom Andreas
51. Tomassini Luca
52. Valette Alain
53. Varisco Enzo
54. Vasselli Ezio
55. Vaz Ferreira Armeo
56. Voigt Christian
57. Wahl Charlotte
58. Wulkenhaar Raimar
59. Zanoni Alberto
60. Zito Pasquale

LIST OF C.I.M.E. SEMINARS

Fondazione C.I.M.E.

Centro Internazionale Matematico Estivo
International Mathematical Summer Center
http://www.math.unifi.it/~cime
cime@math.unifi.it

2004 COURSES LIST

Representation Theory and Complex Analysis

June 10–17, Venezia

Course Directors:

Prof. Enrico Casadio Tarabusi (Università di Roma "La Sapienza")
Prof. Andrea D'Agnolo (Università di Padova)
Prof. Massimo A. Picardello (Università di Roma "Tor Vergata")

Nonlinear and Optimal Control Theory

June 21–29, Cetraro (Cosenza)

Course Directors:

Prof. Paolo Nistri (Università di Siena)
Prof. Gianna Stefani (Università di Firenze)

Stochastic Geometry

September 13–18, Martina Franca (Taranto)

Course Director:

Prof. W. Weil (Univ. of Karlsruhe, Karlsruhe, Germany)

Druck und Bindung: Strauss Offsetdruck GmbH

Lecture Notes in Mathematics

For information about Vols. 1–1654
please contact your bookseller or Springer-Verlag

Vol. 1751: N. J. Cutland, Loeb Measures in Practice: Recent Advances. XI, 111 pages. 2001.

Vol. 1752: Y. V. Nesterenko, P. Philippon, Introduction to Algebraic Independence Theory. XIII, 256 pages. 2001.

Vol. 1753: A. I. Bobenko, U. Eitner, Painlevé Equations in the Differential Geometry of Surfaces. VI, 120 pages. 2001.

Vol. 1754: W. Bertram, The Geometry of Jordan and Lie Structures. XVI, 269 pages. 2001.

Vol. 1755: J. Azéma, M. Émery, M. Ledoux, M. Yor (Eds.), Séminaire de Probabilités XXXV. VI, 427 pages. 2001.

Vol. 1756: P. E. Zhidkov, Korteweg de Vries and Nonlinear Schrödinger Equations: Qualitative Theory. VII, 147 pages. 2001.

Vol. 1757: R. R. Phelps, Lectures on Choquet's Theorem. VII, 124 pages. 2001.

Vol. 1758: N. Monod, Continuous Bounded Cohomology of Locally Compact Groups. X, 214 pages. 2001.

Vol. 1759: Y. Abe, K. Kopfermann, Toroidal Groups. VIII, 133 pages. 2001.

Vol. 1760: D. Filipović, Consistency Problems for Heath-Jarrow-Morton Interest Rate Models. VIII, 134 pages. 2001.

Vol. 1761: C. Adelmann, The Decomposition of Primes in Torsion Point Fields. VI, 142 pages. 2001.

Vol. 1762: S. Cerrai, Second Order PDE's in Finite and Infinite Dimension. IX, 330 pages. 2001.

Vol. 1763: J.-L. Loday, A. Frabetti, F. Chapoton, F. Goichot, Dialgebras and Related Operads. IV, 132 pages. 2001.

Vol. 1764: A. Cannas da Silva, Lectures on Symplectic Geometry. XII, 217 pages. 2001.

Vol. 1765: T. Kerler, V. V. Lyubashenko, Non-Semisimple Topological Quantum Field Theories for 3-Manifolds with Corners. VI, 379 pages. 2001.

Vol. 1766: H. Hennion, L. Hervé, Limit Theorems for Markov Chains and Stochastic Properties of Dynamical Systems by Quasi-Compactness. VIII, 145 pages. 2001.

Vol. 1767: J. Xiao, Holomorphic Q Classes. VIII, 112 pages. 2001.

Vol. 1768: M.J. Pflaum, Analytic and Geometric Study of Stratified Spaces. VIII, 230 pages. 2001.

Vol. 1769: M. Alberich-Carramiñana, Geometry of the Plane Cremona Maps. XVI, 257 pages. 2002.

Vol. 1770: H. Gluesing-Luerssen, Linear Delay-Differential Systems with Commensurate Delays: An Algebraic Approach. VIII, 176 pages. 2002.

Vol. 1771: M. Émery, M. Yor (Eds.), Séminaire de Probabilités 1967-1980. A Selection in Martingale Theory. IX, 553 pages. 2002.

Vol. 1772: F. Burstall, D. Ferus, K. Leschke, F. Pedit, U. Pinkall, Conformal Geometry of Surfaces in S^4. VII, 89 pages. 2002.

Vol. 1773: Z. Arad, M. Muzychuk, Standard Integral Table Algebras Generated by a Non-real Element of Small Degree. X, 126 pages. 2002.

Vol. 1774: V. Runde, Lectures on Amenability. XIV, 296 pages. 2002.

Vol. 1775: W. H. Meeks, A. Ros, H. Rosenberg, The Global Theory of Minimal Surfaces in Flat Spaces. Martina Franca 1999. Editor: G. P. Pirola. X, 117 pages. 2002.

Vol. 1776: K. Behrend, C. Gomez, V. Tarasov, G. Tian, Quantum Comohology. Cetraro 1997. Editors: P. de Bartolomeis, B. Dubrovin, C. Reina. VIII, 319 pages. 2002.

Vol. 1777: E. García-Río, D. N. Kupeli, R. Vázquez-Lorenzo, Osserman Manifolds in Semi-Riemannian Geometry. XII, 166 pages. 2002.

Vol. 1778: H. Kiechle, Theory of K-Loops. X, 186 pages. 2002.

Vol. 1779: I. Chueshov, Monotone Random Systems. VIII, 234 pages. 2002.

Vol. 1780: J. H. Bruinier, Borcherds Products on O(2,1) and Chern Classes of Heegner Divisors. VIII, 152 pages. 2002.

Vol. 1781: E. Bolthausen, E. Perkins, A. van der Vaart, Lectures on Probability Theory and Statistics. Ecole d' Eté de Probabilités de Saint-Flour XXIX-1999. Editor: P. Bernard. VIII, 466 pages. 2002.

Vol. 1782: C.-H. Chu, A. T.-M. Lau, Harmonic Functions on Groups and Fourier Algebras. VII, 100 pages. 2002.

Vol. 1783: L. Grüne, Asymptotic Behavior of Dynamical and Control Systems under Perturbation and Discretization. IX, 231 pages. 2002.

Vol. 1784: L.H. Eliasson, S. B. Kuksin, S. Marmi, J.-C. Yoccoz, Dynamical Systems and Small Divisors. Cetraro, Italy 1998. Editors: S. Marmi, J.-C. Yoccoz. VIII, 199 pages. 2002.

Vol. 1785: J. Arias de Reyna, Pointwise Convergence of Fourier Series. XVIII, 175 pages. 2002.

Vol. 1786: S. D. Cutkosky, Monomialization of Morphisms from 3-Folds to Surfaces. V, 235 pages. 2002.

Vol. 1787: S. Caenepeel, G. Militaru, S. Zhu, Frobenius and Separable Functors for Generalized Module Categories and Nonlinear Equations. XIV, 354 pages. 2002.

Vol. 1788: A. Vasil'ev, Moduli of Families of Curves for Conformal and Quasiconformal Mappings. IX, 211 pages. 2002.

Vol. 1789: Y. Sommerhäuser, Yetter-Drinfel'd Hopf algebras over groups of prime order. V, 157 pages. 2002.

Vol. 1790: X. Zhan, Matrix Inequalities. VII, 116 pages. 2002.

Vol. 1791: M. Knebusch, D. Zhang, Manis Valuations and Prüfer Extensions I: A new Chapter in Commutative Algebra. VI, 267 pages. 2002.

Vol. 1792: D. D. Ang, R. Gorenflo, V. K. Le, D. D. Trong, Moment Theory and Some Inverse Problems in Potential Theory and Heat Conduction. VIII, 183 pages. 2002.

Vol. 1793: J. Cortés Monforte, Geometric, Control and Numerical Aspects of Nonholonomic Systems. XV, 219 pages. 2002.

Vol. 1794: N. Pytheas Fogg, Substitution in Dynamics, Arithmetics and Combinatorics. Editors: V. Berthé, S. Ferenczi, C. Mauduit, A. Siegel. XVII, 402 pages. 2002.

Vol. 1795: H. Li, Filtered-Graded Transfer in Using Noncommutative Gröbner Bases. IX, 197 pages. 2002.

Vol. 1796: J.M. Melenk, hp-Finite Element Methods for Singular Perturbations. XIV, 318 pages. 2002.

Vol. 1797: B. Schmidt, Characters and Cyclotomic Fields in Finite Geometry. VIII, 100 pages. 2002.

Vol. 1798: W.M. Oliva, Geometric Mechanics. XI, 270 pages. 2002.

Vol. 1799: H. Pajot, Analytic Capacity, Rectifiability, Menger Curvature and the Cauchy Integral. XII,119 pages. 2002.

Vol. 1800: O. Gabber, L. Ramero, Almost Ring Theory. VI, 307 pages. 2003.

Vol. 1801: J. Azéma, M. Émery, M. Ledoux, M. Yor (Eds.), Séminaire de Probabilités XXXVI. VIII, 499 pages. 2003.

Recent Reprints and New Editions

4. Manuscripts should in general be submitted in English. Final manuscripts should contain at least 100 pages of mathematical text and should always include
   - a general table of contents;
   - an informative introduction, with adequate motivation and perhaps some historical remarks: it should be accessible to a reader not intimately familiar with the topic treated;
   - a global subject index: as a rule this is genuinely helpful for the reader.

5. Lecture Notes volumes are, as a rule, printed digitally from the authors' files. We strongly recommend that all contributions in a volume be written in the same LaTeX version, preferably LaTeX2e. To ensure best results, authors are asked to use the LaTeX2e style files available from Springer's web-pages at

   www.springeronline.com

   [on this page, click on <Mathematics>, then on <For Authors> and look for <Macro Packages for books>]. Macros in LaTeX2.09 and TeX are available on request from: lnm@springer.de. Careful preparation of the manuscripts will help keep production time short besides ensuring satisfactory appearance of the finished book in print and online. After acceptance of the manuscript authors will be asked to prepare the final LaTeX source files (and also the corresponding dvi-, pdf- or zipped ps-file) together with the final printout made from these files. The LaTeX source files are essential for producing the full-text online version of the book. For the existing online volumes of LNM see: http://www.springerlink.com .
   The actual production of a Lecture Notes volume takes approximately 8 weeks.

6. Volume editors receive a total of 50 free copies of their volume to be shared with the authors, but no royalties. They and the authors are entitled to a discount of 33.3 % on the price of Springer books purchased for their personal use, if ordering directly from Springer.

7. Commitment to publish is made by letter of intent rather than by signing a formal contract. Springer-Verlag secures the copyright for each volume. Authors are free to reuse material contained in their LNM volumes in later publications: A brief written (or e-mail) request for formal permission is sufficient.

Addresses:

Professor J.-M. Morel, CMLA,
École Normale Supérieure de Cachan,
61 Avenue du Président Wilson, 94235 Cachan Cedex, France
E-mail: Jean-Michel.Morel@cmla.ens-cachan.fr

Professor F. Takens, Mathematisch Instituut,
Rijksuniversiteit Groningen, Postbus 800,
9700 AV Groningen, The Netherlands
E-mail: F.Takens@math.rug.nl

Professor B. Teissier, Université Paris 7
Institut Mathématique de Jussieu, UMR 7586 du CNRS
Équipe "Géométrie et Dynamique", 175 rue du Chevaleret
75013 Paris, France
E-mail: teissier@math.jussieu.fr

Springer-Verlag, Mathematics Editorial, Tiergartenstr. 17,
69121 Heidelberg, Germany,
Tel.: +49 (6221) 487-8410
Fax: +49 (6221) 487-8355
E-mail: lnm@springer.de